Symbian OS C++
for Mobile Phones

Symbian OS C++ for Mobile Phones

Richard Harrison

With

Alan Robinson, Arwel Hughes, Carol Holmes, Colin Anthony, Dan Daly, David Cunado, Dominic Pinkman, Elisabeth Mawe, Ian Bunning, Ian McDowall, John Crickett, John Davis, John McAleely, John Pagonis, Laura McLeod, Malcolm Box, Nao Kagabu, Stuart Fisher

Reviewed by

Andrew Thoelke, Colin Turfus, Dave Crookes, David Amos, Duncan Skelton, Jal Panvel, Jason Parker, Jelte Liebrand, Julia Segal, Kevin Dixon, Mark Shackman, Martin de Jode, Martin Tasker, Neil Hepworth, Nick Tait, Phil Spencer, Stephen Burden, Tim Ocock

Managing editor

Phil Northam

WILEY

Published by John Wiley & Sons Ltd,
        The Atrium, Southern Gate, Chichester,
        West Sussex PO19 8SQ, England

        National   01243 779777
        International (+44) 1243 779777

e-mail (for orders and customer service enquiries): cs-books@wiley.co.uk
Visit our Home Page on http://www.wileyeurope.com or http://www.wiley.com

British Library Cataloguing in Publication Data

A catalogue record for this book is available from the British Library

ISBN 0-470-85611-4

Typeset in 10/12pt Optima by Laserwords Private Limited, Chennai, India
Printed and bound in Great Britain by Biddles Ltd, Guildford and King's Lynn
This book is printed on acid-free paper responsibly manufactured from sustainable forestry, for which at least two trees are planted for each one used for paper production.

CONTENTS

About this book

Symbian OS C++ for Mobile Phones draws on the experience of Symbian's own engineers to provide a thorough grounding in writing C++ applications for Symbian OS phones. It won't teach you *everything* you need to know about developing Symbian OS applications – no single book could do that. However, it will take you a long way along the road to being an effective Symbian OS developer and give you a deep understanding of the fundamental principles upon which Symbian OS is based. The text is complemented by a specially developed suite of examples.

The book is organized into four sections, each of which starts with a chapter that describes a working example application, followed by chapters that expand on some of the issues raised by the example. The advantage of this approach is that, at all times, you can see where you are going and have a working example to refer to.

- **Section one** (Chapters 1 to 3) provides a general introduction to Symbian OS. In addition to explaining the main development tools, it introduces you to the overall system structure and the way that Symbian OS uses object orientation and C++.

- **Section two** (Chapters 4 to 8) explains the basic classes, resources, APIs, and programming idioms that you need to create a simple GUI (graphical user interface) application.

- **Section three** (Chapters 9 to 15) is about writing non-trivial stand-alone applications. It starts by taking a deep look at the effective use of the graphics and file APIs before moving on to the extremely practical topics of packaging applications for delivery to the end user and ensuring that your application code is as device-independent as possible.

- **Section four** (Chapters 16 to 20) treats the related topics of system programming, communications and event-handling that, together, allow you to develop sophisticated and responsive applications for Symbian OS phones.

Symbian OS is used in a variety of phones with widely differing screen sizes. Some have full alphanumeric keyboards, some have touch-sensitive screens, and some have neither. In order to enable this kind of variation, a range of user interface designs is required. As far as possible, the material in this book is independent of any particular user interface. However, real applications run on real phones so, where necessary, we have chosen to take the user interface known as UIQ and the Sony Ericsson P800 phone as concrete examples. Where we need to refer to a specific compilation tool, we use the Metrowerks C++ compiler and the Metrowerks CodeWarrior IDE.

Symbian OS C++ for Mobile Phones complements Symbian OS software development kits. When you've put this book down, the UIQ SDK will be your first resource for reference information on the central Symbian OS APIs that we cover here. For more specialized and up-to-date information relating to a specific mobile phone, you will probably need to refer to a phone-specific SDK, available from the relevant manufacturer.

The SDKs contain valuable guide material, examples, and source code, which together add up to an essential developer resource. We've pointed to these where they tie in with the book content. But as a general rule, look in the SDK anyway: you'll usually find additional information that explains things further than we could in this one book.

Who Is This Book For?

If you've programmed, at any level, in C++, it's for you. As a real and comprehensive system written in C++ from the ground up and targeted at the high-growth area where computers and mobile communications converge, Symbian OS gives you unparalleled opportunities in mass-market, enterprise, and system programming.

Besides C++ programmers, this book is interesting to other audiences:

- any other programmer or manager looking to exploit the potential of mobile solutions with Symbian OS technology

- consultants, trainers, and authors who think of basing their activity on Symbian OS technology

- anyone with an interest in system design, since Symbian OS is a full and interesting example in its own right

Conventions

To help you get the most from the text and keep track of what's happening, we've used a number of conventions throughout the book.

> **These boxes hold important, not-to-be forgotten information that is directly relevant to the surrounding text.**

This style is used for asides to the current discussion.

We use several different fonts in the text of this book:

When we refer to words you use in your code, such as variables, classes and functions, or refer to the name of a file, we use this style: `iEikonEnv`, `ConstructL()`, or `e32base.h`

URLs are written like this: ***www.symbian.com/developer***

And when we list code, or the contents of files, we use the following convention:

```
Lines that show concepts directly related to the surrounding
text are shown on a gray background
```

But lines that do not introduce anything new, or that we have seen before, are shown on a white background.

We show commands typed at the command line like this:

```
abld build winscw udeb
```

Foreword

David Wood, Executive Vice President, Partnering, Symbian

It has been my pleasure to have personally coached several hundreds of developers over the years in the art of Symbian OS C++. From this experience, I know that learning Symbian OS C++ can be something of a roller-coaster ride.

Symbian OS is a deep and subtle software system. Its designers have kept several demanding principles carefully in mind over many operating system iterations: the software should be long-lived, flexible, customizable, robust, high-performance, efficient, communications-centric, and future-proof. These goals have been admirably fulfilled. Witness the ever-increasing number of state-of-the-art mobile phones that take advantage of the rich benefits of Symbian OS. Witness also the rush of innovative applications, tools, and services that independent companies are bringing to the market, adding to the appeal of the Symbian OS phones that run them.

C++ is the native programming language for Symbian OS. It exposes the full power of the operating system to software developers. Unsurprisingly, this degree of depth and subtlety poses challenges to first-time Symbian OS developers, who often need to unlearn various programming idioms from other software contexts. The Symbian OS architecture is sophisticated and makes broad use of advanced features of C++ and object-oriented design. There's lots to savor and lots to ponder.

The book you now hold in your hand is packed with practical guidance to ease software developers up the Symbian OS learning curve. It contains insights and advice from many seasoned in-house Symbian engineers. Like Symbian OS itself, it's the outcome of a very substantial collaborative effort. Also like Symbian OS, it builds productively on the work of previous versions – specifically, the book *Professional Symbian Programming* published in February 2000. The benefits that readers can gain from *Symbian OS C++ for Mobile Phones* have their roots in the intense hard work of that pioneering writing team.

Simply better phones

The nature of the mobile phone market has changed. Sales are being driven by innovative new features combining voice, data, imaging and new wireless communication technologies such as wireless packet data and Bluetooth. Symbian OS provides both the technologies needed for this new market phase and the flexibility that enables mobile phone manufacturers to sustain innovation in their phone designs, which meet the needs of their many and varied customers.

Symbian OS C++ for Mobile Phones will help you understand the fundamental concepts behind programming in C++ for Symbian OS phones. Mobile phone manufacturers licensing Symbian OS have the advantage of being able to customize their products while maintaining interoperability between their own designs and those of their competitors, growing the entire wireless economy. Understanding the technologies that are common to Symbian OS phones will allow you to target phones from different manufacturers, regardless of the version of the operating system or the form factor of the phone. This introduction gives you a brief overview of the Symbian OS economy and explains how this book will benefit you.

Symbian, which formed in June 1998, is the result of an unprecedented level of collaboration in the wireless industry. It is owned by the industry, for the industry, and counts among its shareholders Ericsson, Nokia, Matsushita (Panasonic), Motorola, Psion, Siemens, and Sony Ericsson. All major mobile phone manufacturers now license Symbian OS for smartphone development.

Symbian develops Symbian OS – the open, industry standard operating system for mobile phones. This advanced operating system is the foundation of a generation of mobile phones that are evolving new ways to communicate, play, and work while on the move. With tightly integrated personal information management and rich communications capabilities, Symbian OS is an opportunity for software developers to deliver feature-rich applications and services to mobile phone users that number in the millions.

Symbian OS phones

Symbian OS C++ for Mobile Phones focuses on C++ programming and, specifically, on the core C++ APIs and programming patterns used on all Symbian OS phone designs and in Symbian OS itself. Its purpose is to equip you for programming in any Symbian OS environment – whatever the tool product and whatever the user interface, even in contexts such as developing low-level technology components, where there is no user interface at all. You can use what you learn here for application programming, although you'll also want to consult other works on user interface design and the specifics of the user interface you're targeting. You can practise what you learn here with whichever tool product you use, though you'll want to consult the tool's documentation for features unique to that product.

At the time of publication, Symbian OS phones on the market are based on three user interfaces open to C++ programmers – Nokia Series 80 Platform (Nokia 9200 Series Communicator), Nokia Series 60 Platform (Nokia 7650, Nokia 3650, N-Gage and Siemens SX-1), and UIQ (Sony Ericsson P800). All these designs and also Symbian OS phones from Japanese manufacturers are open to Java programming. Tooling available to C++ programmers includes CodeWarrior for Symbian OS from Metrowerks, C++Builder Mobile Set from Borland, and, on a legacy basis, Visual Studio from Microsoft. Borland and Metrowerks are adding unique value in terms of integrated development environments, ease-of-use, debugging, and support of Symbian OS features.

Mobile phones with a numeric keypad:

These phones are designed for one-handed use and require a flexible UI that is simple to navigate with a joystick, softkeys, jogdial, or any combination of these. The best current example of this form factor is the Series 60 Platform, which is the basis of the Nokia 7650, Nokia 3650, and Nokia N-Gage. Series 60 is also licensed to Panasonic, Samsung, Sendo, and Siemens.

Mobile phones with touch screens:

These mobile phones tend to have larger screens than the first category of phone and can dispense with a numeric keypad altogether. A larger screen is ideal for viewing content, working on the move and pen-based interaction and gives new opportunities to users and developers. The best current example of this form factor is UIQ, which is the platform for the Sony Ericsson P800. The P800 actually combines elements of full-screen access and more traditional mobile phone use by including a detachable numeric keypad.

Mobile phones with a full keyboard:

These mobile phones have the largest screens of all Symbian OS phones and can have a full keyboard and could also include a touch screen. With this type of mobile phone, developers may find enterprise applications particularly attractive. A current example of this form factor is the Series 80 Platform from Nokia. This UI is the basis of the Nokia 9200 Series, which has been used in the Nokia 9210 and Nokia 9210i.

To make this book as practical to use as possible, we've chosen to demonstrate the core Symbian OS APIs and programming patterns through a consistent development environment throughout the book. We chose the UIQ user interface, the Sony Ericsson P800 phone, and

Metrowerks CodeWarrior for Symbian OS tools. However, Symbian is equally committed to all its customers and all its tools partners, and this is not an application-programming book. So, on the one hand, we don't cover real-world UIQ application programming, we don't go into too many specifics of the CodeWarrior tooling, and if you don't have a P800 phone, you can still build and test almost all examples on the UIQ emulator. On the other hand, we mention specifics of Nokia Series 80 or Series 60, where these would motivate the use of core Symbian OS APIs or programming patterns that are less relevant for UIQ, and we cover all available tools in the appendix.

When you're ready to use the Symbian OS programming skills that you've learned in this book, you'll want an up-to-the-minute picture of available phones, user interfaces, and tools. For the latest information, access ***www.symbian.com/developer***, which gives to pointers to partner websites. If you're developing technology that could be used on any Symbian OS phone, you can find more information about partnering with Symbian at ***www.symbian.com/community***.

We wish you an enjoyable experience programming with Symbian OS and lots of commercial success.

This book includes a 30-day Evaluation Edition of Codewarrior Development Studio for Symbian OS Personal Edition, v2, featuring:

- **UIQ SDK for Symbian OS v7.0 Sony Ericsson P800 smartphone**
- **Windows x86 emulation debugging support**
- **Symbian descriptor presentation in debugger**
- **Updated Symbian OS build components, including AIF, resource compiler, bitmap compiler, and .sis file compiler.**

About the Authors

Richard Harrison, Lead Author

Richard has spent the majority of his time at Symbian in system integration, building up and leading the SI team. He joined Psion in 1983 after several years, teaching maths, physics and computer science. During that time he wrote a Forth language implementation for Acorn Computers and accompanying user manuals for the Acorn Atom and BBC Micro.

During his career he has produced user software documentation for the Sinclair QL and Psion's PC application software. Other assignments include coauthoring of the Organiser II spreadsheet, being the principal designer and author of the Psion Series 3 and 3a word processors, and the lead author of the Psion Sibo SDK team. He has also written system software for the Psion Organiser I, and developed the source code translator for the original version of OPL.

Educated at Balliol College, Oxford with an MA in Natural Science (Physics), Richard also gained an MSc in Astronomy from Sussex University, and spent a further two years of postgraduate research in the Astronomy Group at Imperial College.

Alan Robinson

Alan Robinson joined Symbian shortly after its formation in 1998 and has mostly worked on documentation and examples in messaging and communications. A key contributor to this book, Alan is an accomplished author, and has previously contributed to *Wireless Java for Symbian Devices* (Wiley, 2001).

A graduate of Cambridge University with a BA in literature and philosophy, he became interested in applying logical theory and took a Computing MSc at Middlesex University. He has worked on developer kits for a start-up company's messaging middleware platform, and for IBM's MQSeries.

Arwel Hughes

Arwel joined Psion in 1993, working on documentation for the Series 3a and also some software development. Since the formation of Symbian, he has contributed documentation and examples for Symbian OS. This is rather like painting the famous Forth Bridge: just when you think you can see the end. . .

Arwel previously worked on IBM mainframes in roles including programmer and systems programmer for a number of companies including GKN, Prudential Assurance, Shell and Chase Manhattan Bank (now renamed to J P Morgan Chase). He has a BSc in Applied Mathematics from Sheffield University.

Carol Holmes

Carol first joined Psion as a graduate software engineer in 1987 with a BSc in Maths. She was part of the original development team for the Series 3 product family, joining the company as it pioneered the use of object-oriented software development. She spent several years working for a management consultancy on large development projects, before happily returning to Psion, to work on the email software for the Series 3a.

Carol went on to lead the team that developed the messaging software for the Psion Series 5. Since then, Carol has led other large development groups in Symbian, but now works from home doing analysis and research on Symbian's development processes and their improvement. She has a reputation for being very organized, enjoys making things (from cards to quilts), and her favorite color is purple.

Colin Anthony

Since joining Symbian, Colin has worked with the systems integration team on various releases of connectivity software and most recently as a developer consultant assisting Symbian partners.

Colin began his career as an apprentice electronics engineer with an international medical company. He worked with the production engineering team (amongst others) on new production systems for medical equipment. During this time he became involved in the deployment of a new IT infrastructure in the company. This gave him new direction in software engineering and the idea of going back into education, where he graduated from Southampton University with a BSc in Computer Science.

Outside of work there's nothing he enjoys more than getting away from the distractions of London by doing a spot of rock climbing, scuba diving and snowboarding.

Dan Daly

In his time in developer relations at Symbian, Dan Daly has provided technical consultancy to the third party community on all areas of

Symbian OS. Recently, Dan joined the Partner Projects Group, which provides support and bespoke development to partners and advice and support for technical issues.

Dan worked for several years for GEC Marconi Defence Systems Division, now known as BAE Systems, developing Aircraft Flight analysis and replay software, before joining Symbian in July 2000. He has a BSc (Hons) degree in Software Engineering from the University of Westminster.

Dr David Cunado

David joined Symbian's browsing team in 1999. After working on the web browser he moved on to the WAP browser, both of which were released on the Nokia 9210. Since then he has been working on the Symbian OS transport framework. This includes enabling it to support WSP, receive HTTP requests and send pipelined HTTP requests.

David previously took part in biometrics research at Southampton University, investigating person recognition by gait, using a new feature extraction technique called the Velocity Hough Transform. The work produced several publications, including collaboration in the book Biometrics: *Personal identification in networked society* by A Jain. R Bolle and S Pankanti.

Dominic Pinkman

Dominic joined Psion in October 1995 as a technical author, remaining with the company as it evolved into Psion Software and then Symbian. He has worked on writing and maintaining the documentation for APIs throughout Symbian OS, in particular those in the application engines, base, application frameworks and graphics subsystems.

He has an MSc Computer Science from the University of Kent and a BA Modern Language studies from Leicester University. His interests include indoor hockey and playing the mandolin.

Elisabeth Måwe

Elisabeth joined the system documentation team in 2000 and has since been involved in designing and writing the Symbian Developer Library, specializing in operating system customization, kits and build tools.

Elisabeth has a BA (Hons) in Technical Communication/Information Design from Mälardalens Högskola and Coventry University, as well as an MA in Contemporary English Language and Linguistics from Reading University. After graduating in 1996 she worked as a technical author, information designer and web editor for various IT companies in the UK, producing documentation for both network management and market research software.

Ian Bunning

Ian joined Symbian after graduating from the University of Cambridge in 2001. He is now an engineer on the Personal Area Networking team, dealing primarily with infrared and OBEX, but also with USB and Bluetooth development.

While at University, Ian held the post of Director of Production at *The Cambridge Student*- the Student Union's weekly newspaper. He assisted two editorial teams by ensuring that their articles had pictures and reached the printers. He also acted as an interview photographer – his subjects included John Madden, Gail Porter and Big Brother's 'Nasty' Nick Bateman. Outside of Symbian, Ian is a keen photographer and occasional jeweler.

Ian McDowall

Ian joined Symbian in 2000 and is currently a technology architect responsible for connectivity. He has filled roles ranging from developer, through project manager to technical manager by way of quality manager and process consultant (including presentations at international conferences).

He has an MA in Computer Sciences from Cambridge University and an MBA from Warwick University. As a software engineer for over twenty years he has been with a number of software companies and has worked on more than fifteen operating systems, developing software ranging from enterprise systems to embedded software. He is married to Lorraine and they have two children, Ross and Kelly, and a number of pets.

John Crickett

John is an experienced independent software developer, specializing in object-oriented software engineering. He is a member of the British Standards Institute C++ Language Panel, and regularly writes articles on software development with C++ and Java for the Association of C and C++ Users magazines. He can be contacted through his website ***www.crickett.co.uk***.

John Davis

John has worked for Symbian for six months. He has written many in-house guides for leading edge C++-based applications and has eight years industrial experience, where he began as a C programmer writing communications packages.

Educated in Dublin and Aberystwyth, he enjoys travel and has published many travel articles on the web. You can find his archives at **www.heldencrow.com**.

John McAleely

John works in the partner projects group at Symbian, providing technical support to key partners as they prepare products for Symbian OS. Recent partner development projects include audio engine development, ports of Symbian OS to new hardware platforms and delivering technical training material to developers around the world.

John has worked for Symbian since 2000, and previously worked on DVD software for PCs while writing programs for Psion computers in his spare time. He has a Master's degree in Software Engineering from Imperial College.

John Pagonis

John joined Symbian in 1998, where he worked on the development of the Ericsson R380 smartphone before joining the personal area networking team, which implemented Bluetooth for Symbian OS.

These days, John works in the emerging technologies and telephony group. John is also on an academic quest as part-time PhD student, focusing on the issue of information overload. Other interests include computer and wireless personal area networks, software agents, software engineering, organizational process patterns, object technology, wearable and wireless information devices, resource-constrained systems and user interfaces.

John graduated from the University of Essex with a BEng in Computers and Networks and an MSc in Computers and Information Networks. A part-time martial artist, open source and freeware advocate, John believes it is better that certain things are kept unread.

Laura McLeod

Laura joined Symbian in 2001 as a technical author. She has written a range of reference documentation for the Symbian Developer Library, including for Versit (an application engine) and the file system. Before joining Symbian, she was a technical author in C++ software development, and also worked in customization of financial middleware software.

Malcolm Box

Malcolm joined the Symbian base team in 1998. After working on the kernel for the Ericsson R380 smartphone, he moved over into the personal area networks team. He was the technical architect and lead for the Symbian OS Bluetooth stack, which shipped in the Nokia 7650 and Sony Ericsson P800. Recently he has been working in Japan, leading a system architecture group. Now in the UK, Malcolm is involved with product development with Symbian's Japanese licensees. Malcolm has a MEng degree in Microelectronics and Software Engineering from Newcastle University. He started his career at Nortel, where he designed VLSI circuits for ATM and exchange equipment.

Outside work, Malcolm maintains the LXR code cross-referencer project and contributes to other freeware. An adventure sports enthusiast, Malcolm is learning to white-water kayak, which involves a lot of time swimming in cold rivers. Thanks are due to David Amos and Tim Ocock for their help in reviewing and Andrew Thoelke for the server startup code. He also thanks his family for putting up with him spending weekends writing rather than with them.

Nao Kagabu

Nao joined the research group at Symbian Japan in December 2000. After working on several research projects, he was transferred to Symbian UK. Since then he has been providing technical support to key partners, integrating their technologies into Symbian OS. Recently, he joined to the Partner Projects Group.

Nao previously worked on GPS car navigation for six years and has a Master's degree in Software Engineering from Toyohashi University of Technology, Japan.

Acknowledgements

Eagle-eyed readers will observe that *Symbian OS C++ for Mobile Phones* resembles one of its predecessors, *Professional Symbian Programming*, published in February 2000. Symbian OS has evolved a great deal since *Professional Symbian Programming* was written, but the previous publication served as an excellent framework for the new book. Therefore, the current authors owe a debt of gratitude to the original writers – lead author Martin Tasker and his team. Thanks also to the Laughing Gravy for providing vital fuel to us all. Cover design by Jonathan Tastard.

1

Getting Started

It seems to be traditional to start a book on computer programming with a 'Hello World' example and, although this is a book that is more about an operating system (OS) than a programming language, I'll follow that tradition. In the process I'll introduce you to the emulator, and to the tools for building C++ programs, so that by the end of the chapter you will have found out the basic information on how to build and run a Symbian OS application. I won't get *too* involved in describing Symbian OS programming conventions, application programming interface (API) functions, and so forth. Instead, I'll concentrate on the tools you need, and how to use them, leaving the more specific details until later chapters.

First, I'll briefly describe the **emulator**. Most Symbian OS software is developed first on the emulator and only then on real target hardware. The emulator also includes a number of Symbian OS applications, and so mimics a real Symbian OS phone very closely. You will need to get familiar with the emulator and, while doing so, we can use the opportunity to take a look at the applications and some of the distinctive features of UIQ, one of the various graphical user interfaces (GUIs) used by Symbian OS.

Then we'll create a program. The easiest things to build are text-mode console programs, so that's the form of the classic 'Hello World' application that we'll use. I'll demonstrate how to compile it for either the emulator or a target Symbian OS mobile phone, and how to launch and debug it using the Metrowerks CodeWarrior IDE.

1.1 Using the Emulator

The emulator is a fundamental tool for all the Symbian OS SDKs, so it's vital that you get to know it and learn how to use it.

Symbian OS C++ for Mobile Phones.   Edited by Richard Harrison
© 2003 John Wiley & Sons, Ltd   ISBN: 0-470-85611-4

If you've never seen Symbian OS before, the emulator offers an opportunity to get to know some Symbian OS basics from a user's perspective, so we'll look at these straight away. Later, you'll want to look at the Emulator Reference in the appendix to learn enough about how the emulator works so that you can begin to make effective use of it as a developer.

If you're used to Symbian OS, then you may want to skip the next section and go straight to *Hello World – Text Version* and start building an application.

1.1.1   Launching the Emulator

Once you've installed the Symbian OS UIQ C++ SDK, you can launch the emulator in either of the following ways:

- From Windows Explorer: find directory \epoc32\release\winscw\ udeb\ and launch epoc.exe
- From the command line: put \epoc32\release\winscw\udeb\ into your path, prefixed by the correct drive letter, launch a prompt, and just type epoc

Alternative initialization of the emulator is described in Appendix 4.

However, you choose to start it, the first thing you'll see in the UIQ emulator is the application launcher.

As its name suggests, the application launcher that you see in Figure 1.1 enables you to launch applications. Its menus allow you to view or change system settings, and it also has a control panel. It's very easy for end users to get to know the application launcher: you don't really need a manual. Just tap here and there, and you'll soon find out what it has to offer.

1.1.2   GUI Style

As you browse around the application launcher on the emulator, you'll begin to see how UIQ, as a particular example of the range of GUIs available for Symbian OS, is optimized for the pen-based mobile phone form factor.

UIQ is designed as a 'read mostly' user interface, to be used mainly for browsing and for making a selection from a range of options with a single tap of a pen. Other GUIs – such as the Series 60 interface used, for example, on the Nokia 7650 – are optimized for the different hardware resources of the devices on which they are intended to run. Although the different GUIs may have a superficially different appearance, they all rely on a common set of underlying features, some of which are briefly described below.

On the emulator, you click with a mouse, but on a real Symbian OS phone you would tap with a pen. For reasons we'll discuss later, the difference is important, so I'll always say, 'tap', just to remind you.

Figure 1.1

Screen layout

The UIQ screen layout, illustrated in Figure 1.2, includes the following areas (from top to bottom of the screen):

The **application picker** contains icons that allow you to switch applications. Tapping on any icon brings the application it represents to the foreground. The application launcher icon, at the far right, brings the application launcher to the foreground, allowing you to launch applications that are not displayed on the application picker. If you wish, you can customize the application picker to launch your own preferred set of applications.

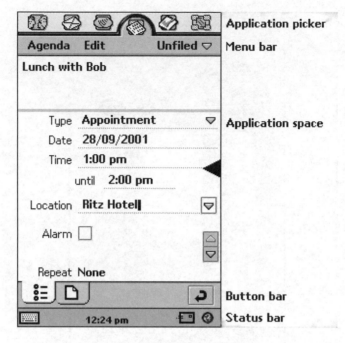

Figure 1.2

The **menu bar** contains one or more menus, whose names – and the selection of items that each menu contains – change from application to application, and also as you change view within a particular application. In UIQ, the menu bar usually contains two menus on the left and may optionally have a folder menu on the right.

The **application space** is the central area of the screen, where an application's view is displayed. Applications use this area in whichever way is appropriate to the information that they display.

Optionally, an application displays a **button bar** at the bottom of the application space. The most common use is to provide buttons to move between the application's various views. In UIQ, a detail view, such as the one above that shows the detail of a single Agenda entry, usually has a special button, in the lower right corner, to return you to the main view.

The **status bar** displays information such as battery charge, time of day, signal strength and notification of incoming messages. The P800's status bar includes a keyboard icon in the lower left corner that is used to display a virtual keyboard for text input if you do not wish to use handwriting recognition.

Most of these screen layout elements can be recognized in other GUIs used with Symbian OS, though they may differ significantly in appearance, or be located in different areas of the screen.

Menus

Have a look at the menu in the emulator's application launcher or, as illustrated in Figure 1.3, in the Agenda application. Again, the menu bar is different (but not too different) from menu bars in desktop GUIs.

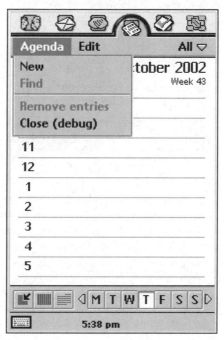

Figure 1.3

Cascaded menu items can be used both to hide less common options, and to reduce the vertical space required by menu panes. This feature is used sparingly in UIQ applications, in which menu content changes with context and each menu is designed to contain as small and as simple a set of menu commands as possible. Cascaded menu items are used slightly more frequently in the Series 60 interface, where they appear as illustrated in Figure 1.4.

In the UIQ emulator, use the menu with the pen, and you'll see some interesting visual cues to confirm the option you selected: the option will flash very briefly before the menu disappears. It took a long time to get that effect just right!

As with all other elements of a Symbian OS GUI and applications in which a keyboard is available (including the keyboard of a PC that is running the emulator) you can drive the menus with the keyboard as well as with the pen. You can use the arrow keys and *Enter* to select items. You can also use cursor keys and the *Confirm* button on real target hardware and emulators that support such features.

Figure 1.4

When writing an application you also have the option to assign a shortcut key to any menu item, which allows you to invoke the relevant function directly from a keyboard, without going through the menus at all. Although they can be defined in any Symbian OS application, shortcut keys are clearly not usable on mobile phones without keyboards (except when the application is running the emulator) so neither the UIQ nor the Series 60 user interfaces display shortcut key information in their menus.

Using a pen

From the UIQ application launcher, tap (once) on an icon that is not already selected. Your pen tap first selects the item, highlighting its name, and then opens the application that it represents. This behavior makes the user interface fast and easy to use, and is appropriate on a device where the most commonly performed activities do not create or modify data.

In other Symbian OS pen-based GUIs, particularly ones in which creating or modifying data are more common activities, this behavior may be modified so that a tap on an unselected item only selects it; an item is only opened if you tap it when it is already selected.

In this case, two taps are required to both select and open an item, but this is not equivalent to double-clicking with a mouse. When you double-click the mouse, the first click selects and the second click opens, but only *if it occurs within a certain time after the first click*, and *if it's in the same place*. This is easy to do with a mouse, but quite difficult with a pen.

Double-tapping can be made to work with a pen: Windows CE's GUI requires double-taps, and uses them just like the desktop versions of

Windows. However, since Symbian OS was *designed* for pen use, it uses more natural idioms.

There are a number of other ways in which a Symbian OS GUI deals with the difference between a mouse and a pen:

- You can't hover with a pen. That rules out tooltips, so Symbian OS buttons are big enough to include either text or a carefully designed image to indicate their action. Also, there's no pointer to change to a hand, an hourglass, or a crosshair, based on its position and the state of the program. That has a more subtle effect on the GUI and your application design.

- You can't right click with a pen. We could have invented some odd combination – such as *Alt* + tap – to simulate a right click, but we didn't. Symbian OS simply doesn't use the metaphor.

- A pen isn't as accurate as a mouse, so pen-selectable items have to be bigger than mouse-selectable items, and it's useful to provide some reassurance about what has been selected: check out the behavior of buttons in dialogs (or menu items) that briefly animate as you select them, confirming which item was selected.

So: a pen-based GUI is not a mouse-based GUI. A Symbian OS GUI is not a desktop GUI – it is different in ways that really matter. On the other hand, a Symbian OS GUI isn't different merely for the sake of it. Desktop GUIs are familiar to many people, and Symbian OS includes many familiar concepts.

1.2 Hello World – Text Version

Now that you've started to get to grips with the emulator, it's time to get your first Symbian OS C++ program running. Even though Symbian OS is primarily a system for developing GUI applications, the simplest kind of programs use a text interface, so for our first task we'll learn how to build a program that writes 'Hello world!' to a text console. That will introduce you to the tools required for building applications for both the emulator and a real machine, so that later on you'll be ready for a program with a GUI.

If you want to follow this chapter through at your desktop with the UIQ SDK, make sure that you've installed all the tools you need, and the example source for the book. See the appendices for more information.

1.2.1 The Program: `hellotext`

Here's the program we're going to build. It's your first example of Symbian OS C++ source code:

```
// hellotext.cpp

#include <e32base.h>
#include <e32cons.h>

LOCAL_D CConsoleBase* gConsole;

// Real main function
void MainL()
    {
    gConsole->Printf(_L("Hello world!\n"));
    }

// Console harness
void ConsoleMainL()
    {
    // Get a console
    gConsole = Console::NewL(_L("Hello Text"),
                        TSize(KConsFullScreen, KConsFullScreen));
    CleanupStack::PushL(gConsole);

    // Call function
    MainL();

    // Pause before terminating
    User::After(5000000);  // 5 second delay


    // Finished with console
    CleanupStack::PopAndDestroy(gConsole);
    }

// Cleanup stack harness
GLDEF_C TInt E32Main()
    {
    __UHEAP_MARK;
    CTrapCleanup* cleanupStack = CTrapCleanup::New();
    TRAPD(error, ConsoleMainL());
    __ASSERT_ALWAYS(!error, User::Panic(_L("SCMP"), error));
    delete cleanupStack;
    __UHEAP_MARKEND;
    return 0;
    }
```

Our main purpose here is to understand the Symbian OS tool chain, but while we have the opportunity, let's notice a few things in the source code above. There are three functions:

- The actual 'Hello world!' work is done in `MainL()`
- `ConsoleMainL()` allocates a console and calls `MainL()`
- `E32Main()` allocates a trap harness and then calls `Console-MainL()`

On first sight, this looks odd. Why have three functions to do what most programming systems can achieve in a single line? The answer is simple:

real programs, even small ones, aren't one-liners. So there's no point in optimizing the system design to deliver a short, subminimal program. Instead, Symbian OS is optimized to meet the concerns of real-world programs – namely, to handle and recover from memory allocation failures, with minimal programming overhead. There's a second reason why this example is longer than you might expect: real programs on a user-friendly machine use a GUI framework rather than a raw console environment. If we want a console, we're on our own, and have to construct it ourselves, along with the error-handling framework that the GUI would have included for us.

Error handling is of fundamental importance in a machine, with limited memory and disk resources, such as those for which Symbian OS was designed. Errors are going to happen, and you can't afford not to handle them properly. I'll explain the error-handling framework and its terminology, such as **trap harness**, **cleanup stack**, **leave**, **heap marking**, and so on in Chapter 6.

Every single program that appears in this book is fully error-checked. The Symbian OS error-checking framework is easy to use, so the overheads for the programmer are minimal. You might doubt that, judging by this example! After you've seen more realistic examples, you'll have better grounds for making a proper judgement.

Back to `hellotext`. The real work is done in `MainL()`:

```
// Real main function
void MainL()
    {
    console->Printf(_L("Hello world!\n"));
    }
```

The `printf()` that you would expect to find in a C 'Hello World' program has become `console->Printf()` here. That's because Symbian OS is object-oriented: `Printf()` is a member of the `CConsoleBase` class.

The `_L()` macro turns a C-style string into a Symbian OS-style **descriptor**. We'll find out more about descriptors – and a better alternative to the `_L()` macro – in Chapter 5.

Symbian OS always starts text programs with the `E32Main()` function. `E32Main()` and `ConsoleMainL()` build two pieces of infrastructure needed by `MainL()`: a **cleanup stack** and a **console**. Our code for `E32Main()` is:

```
// Cleanup stack harness
GLDEF_C TInt E32Main()
```

```
{
__UHEAP_MARK;
CTrapCleanup* cleanupStack = CTrapCleanup::New();
TRAPD(error, ConsoleMainL());
__ASSERT_ALWAYS(!error, User::Panic(_L("PEP"), error));
delete cleanupStack;
__UHEAP_MARKEND;
return 0;
}
```

The declaration of E32Main() indicates that it is a global function. The GLDEF_C macro, which in practice is only used in this context, is defined as an empty macro in E32def.h. By marking a function GLDEF_C, you show that you have thought about it being exported from the object module. E32Main() returns an integer – a TInt. You could have written int instead of TInt, but since C++ compilers don't guarantee that int is a 32-bit signed integer, Symbian OS uses typedefs for standard types to make sure they are guaranteed to be the same across all Symbian OS implementations and compilers.

E32Main() sets up an error-handling framework. It sets up a cleanup stack and then calls ConsoleMainL() under a trap harness. The trap harness catches errors – more precisely, it catches any functions that **leave**. If you're familiar with the exception handling in standard C++ or Java, TRAP() is like try and catch all in one, User::Leave() is like throw, and a function with L at the end of its name is like a function with throws in its prototype. It's very important to know what these functions are: the Symbian OS convention is to give them a name ending in L(). Here's ConsoleMainL():

```
// Console harness
void ConsoleMainL()
    {
    // Get a console
    gConsole = Console::NewL(_L("Hello Text"),
                        TSize(KConsFullScreen, KConsFullScreen));
    CleanupStack::PushL(gConsole);

    // Call function
    MainL();

    // Pause before terminating
    User::After(5000000);  // 5 second delay

    // Finished with console
    CleanupStack::PopAndDestroy(gConsole);
    }
```

This function allocates a console before calling MainL() to do the Printf() of the 'Hello world!' message. After that, it briefly pauses and then deletes the console again. If we were creating this example for a target machine that had a keyboard, we could have replaced the delay code:

```
// Pause before terminating
    User::After(5000000); // 5 second delay
```

with something like

```
// Wait for key
    console->Printf(_L("[ press any key ]"));
    console->Getch();    // Get and ignore character
```

so that the application would wait for a keypress before terminating.

There is no need to trap the call to MainL() because a leave would be handled by the TRAP() in E32Main(). The main purpose of the cleanup stack is to prevent memory leaks when a leave occurs. It does this by popping and destroying any object that has been pushed to it. So if MainL() leaves, the cleanup stack will ensure that the console is popped and destroyed. If MainL() doesn't leave, then the final statement in ConsoleMainL() will pop and destroy it anyway.

In fact, in this particular example, MainL() cannot leave, so the L isn't theoretically necessary. But this example is intended also as a starting point for other console mode programs, including programs that *do* leave. I've left the L there to remind you that it's acceptable for such programs to leave if necessary.

If you're curious, you can browse the headers: e32base.h contains some basic classes used in most Symbian OS programs, while e32cons.h is used for a console interface and therefore for text-mode programs – it wouldn't be necessary for GUI programs. You can find these headers (along with the headers for all Symbian OS APIs) in \epoc32\include\ on your SDK installation drive.

1.2.2 The Project Specification File

We are going to build the program for two environments:

- The emulator
- A target machine.

Like all C++ development under Symbian OS, we'll start by building the project to run under the emulator (that is, for an ×86 instruction set) using the Metrowerks CodeWarrior C++ compiler. We use a debug build, so that we can see the symbolic debug information, and to get access to some useful memory-leak checking tools.

Later, we'll use the GNU C++ Compiler (GCC) to build the project for a target Symbian OS phone, using an ARM instruction set. At that stage we'll use the release build since that is what you would eventually do to create your final, usable, application. You would need to make a debug build for the ARM target if you wanted to debug on the target machine, but we'll leave that topic for now and come back to it in Chapter 4.

So, we will need to build the same source code twice. In fact, for demonstration purposes, we're going to build it three times, because you can compile the code from the command line, or you can build it in the CodeWarrior IDE.

Each type of build requires a different project file. To simplify matters, you put all the required information into a single, generic, **project specification file**, and then use the supplied tools to translate that file into the makefiles or project files for one or more of the possible build environments. Project specification files have a .mmp extension (which stands for 'makmake project'). The contents of the one for the HelloText project is as follows:

```
// hellotext.mmp
TARGET          HelloText.exe
TARGETTYPE      exe
SOURCEPATH      .
UID             0
SOURCE          hellotext.cpp
USERINCLUDE     .
SYSTEMINCLUDE   \epoc32\include
LIBRARY         euser.lib
```

This is enough information to specify the entire project, enabling configuration files to be created for any platform or environment.

- The TARGET specifies the executable to be generated, and the TARGETTYPE confirms that it is a .exe
- The UID information is irrelevant for .exes, but I specify it explicitly as zero here, to suppress a build warning. I'll have more to say on UIDs in Chapter 4, since they are more important for GUI programs.
- SOURCEPATH specifies the location of the source files for this project.
- SOURCE specifies the single source file, hellotext.cpp (in later projects, we'll see that SOURCE can be used to specify multiple source files).
- USERINCLUDE and SYSTEMINCLUDE specify the directories to be searched for user (included with quotes; the SYSTEMINCLUDE path

is searched as well) and system include files (included with angle brackets; only the SYSTEMINCLUDE path is searched). All Symbian OS projects should specify \epoc32\include\ for their SYSTEM-INCLUDE path.

- LIBRARY specifies libraries to link to – these are the .lib files corresponding to the shared library DLLs whose functions you will be calling at runtime. In the case of this very simple program, all we need is the E32 user library, euser.lib.

1.2.3 The Component Definition File

The Symbian OS build tools require one further **component definition file** to be present. This file always has the name bld.inf and contains a list of all the project definition files (frequently, there is only one) that make up the component. In more complex cases it will usually contain further build-related information, but the one for HelloText is simply

```
// BLD.INF
PRJ_MMPFILES
hellotext.mmp
```

1.2.4 Building from the Command Line

Make sure you've installed the example project source from the Symbian website: see the appendices for how to do this.

To start the command-line build, open up a command prompt, change to your installation drive, and issue the following command to get into the source directory for this example:

```
cd \scmp\hellotext
```

Next, you need to invoke bldmake by typing

```
bldmake bldfiles
```

After a short pause, this command will return – by default, bldmake doesn't tell you anything. However, if you check the contents of the source directory, you'll notice a new file, abld.bat, that is used to drive the remainder of the build process. You will also find that the epoc32 directory tree contains a new \epoc32\build\scmp\hellotext directory that contains a number of generated files that relate to the various types of build that the build tools support.

Next, use `abld` to run the rest of the build by typing:

```
abld build winscw udeb
```

The `winscw` parameter specifies that we are building for the emulator, using the CodeWarrior compiler, and the `udeb` parameter means we are using a (unicode) debug build. The command generates the following output:

```
  make -r -f "\EPOC32\BUILD\SCMP\HELLOTEXT\EXPORT.make" EXPORT VERBOSE=-s
Nothing to do
  make -r -f "\EPOC32\BUILD\SCMP\HELLOTEXT\WINSCW.make" MAKEFILE VERBOSE=-s
perl -S makmake.pl -D \SCMP\HELLOTEXT\HELLOTEXT WINSCW
  make -r -f "\EPOC32\BUILD\SCMP\HELLOTEXT\WINSCW.make" LIBRARY VERBOSE=-s
make -s -r -f
"\EPOC32\BUILD\SCMP\HELLOTEXT\HELLOTEXT\WINSCW\HELLOTEXT.WINSCW" LIBRARY
  make -r -f "\EPOC32\BUILD\SCMP\HELLOTEXT\WINSCW.make" RESOURCE CFG=UDEB
VERBOSE=-s
make -s -r -f
"\EPOC32\BUILD\SCMP\HELLOTEXT\HELLOTEXT\WINSCW\HELLOTEXT.WINSCW"
  RESOURCEUDEB
  make -r -f "\EPOC32\BUILD\SCMP\HELLOTEXT\WINSCW.make" TARGET CFG=UDEB
VERBOSE=-s
make -s -r -f
"\EPOC32\BUILD\SCMP\HELLOTEXT\HELLOTEXT\WINSCW\HELLOTEXT.WINSCW" UDEB
  make -r -f "\EPOC32\BUILD\SCMP\HELLOTEXT\WINSCW.make" FINAL CFG=UDEB
VERBOSE=-s
```

The build is split into six phases:

- The export phase copies exported files to their destinations. This typically includes copying public header files into the `\epoc32\include` directory. For many applications, as in the current case, this stage will need to do nothing.

- The makefile phase creates the necessary makefiles or IDE workspaces.

- The library phase creates import libraries.

- The resource phase creates the application's resources files, bitmaps and application information files (aifs).

- The target phase essentially creates the application's main executables.

- The final phase is present to perform any final actions that need to be done after the main executables have been created. For most applications, this phase will do nothing.

These phases, and other possible options available with the `abld` tool, are fully described in the Build Tools Guide and Build Tools Reference sections of the Developer Library documentation supplied with

Symbian OS SDKs. Typing 'abld help' also gives a useful summary of the options available.

The result of running the abld tool is that the hellotext project is built into the emulator startup directory as `\epoc32\release\winscw\udeb\hellotext.exe`. To run the program, you can start it right from there, using either the command prompt or Windows Explorer. The emulator will boot, and you'll see Hello world! on the screen as is shown in Figure 1.5.

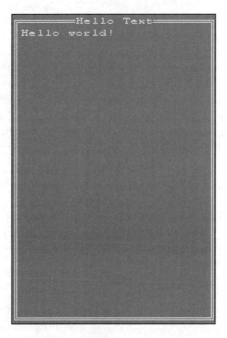

Figure 1.5

After a five second pause the emulator will close.

1.2.5 Building in the Metrowerks IDE

Now that we know our tool chain is working, let's build the project from the IDE and debug through the example. First, start up the Metrowerks CodeWarrior IDE and select `File | Import Project from .mmp File`.

Select the appropriate SDK and then browse to the `\scmp\hellotext` directory and select `hellotext.mmp`. Click on Finish and, after a short time, the `hellotext.mcp` CodeWarrior project will have been created.

In the project's Source folder you'll see two files: hellotext.cpp which is a real source file, and `hellotext.pref`, which was generated by CodeWarrior.

You can build the project straight away by selecting **Project | Make**, or simply F7 – the default target is **WINSCW UDEB**, as we used when previously building from the command line.

You can launch the emulator from the IDE using **Project | Run**, or Ctrl-F5. Alternatively, you can debug through the code by selecting **Project | Debug**, or simply F5. You can use any of the usual debug techniques – run to cursor, step over a whole line of code, step into each of the functions on a line, step out of the current function, run to breakpoint, and so on.

If you're curious, you might want to try debugging through line-by-line. You'll begin to get a feel for what's worth doing and what's not, and it will give you some more insight into the system structure. On the other hand, there's no need to jump in this deep right now: I'll explain what you really need to know through the next few chapters. The main point to note just now is that the CodeWarrior IDE provides an excellent debugger, and as a Symbian OS developer you can take full advantage of it.

You can debug through the most important function calls using the source code and debugging information provided with the Symbian OS UIQ C++ SDK. Here are some tips:

- Don't try to debug any executive calls (`Exec::Xxx()`).
- Be careful when debugging traps – it's usually best to set a breakpoint on the target of the trap, and run to that.
- If you find yourself in a function for which the SDK doesn't provide the source, jump out of that function, hopefully back into your own code or some SDK-provided source code.

1.2.6   Running on the Target Machine

The emulator is a good enough development environment on which to learn Symbian OS using only a PC and a Symbian OS C++ SDK. But if you're developing for a real Symbian OS phone, the emulator isn't your ultimate target; once you've built and debugged your application to a certain stage on the emulator, you'll want to start running it on the real machine also. This is a simple process: rebuild the same source code using the GCC compiler for ARM, and copy the resulting binaries to your Symbian OS phone.

Before we get too heavily involved in the building process, make sure that your machine is connected to your PC using the appropriate connectivity software.

*For simplicity, throughout this book I'll refer to the connectivity software that runs on your PC as **Symbian OS Connect.***

If you've set Symbian OS Connect to backup or synchronize on connection, let all that happen before you start using Symbian OS. Connect to transfer the programs you have built.

If you're going to be connected to the machine all day, you'll probably want to use mains power rather than batteries. In that case, you don't want your machine turning itself off just before you send it your new program build, so use the control panel's **Display** item to set the **Power save** option to **Off**. Now you're ready (but don't forget to reenable the power save option when you later disconnect your machine from the mains power lead).

You can perform target machine builds either from the command line or from within the CodeWarrior IDE. In the IDE, use **Project | Set Default Target** either to select the required target (e.g. ARMI **UREL**) or to choose **Build All**, and then build with **Project | Make**.

To perform a target build from the command line, first move to the directory containing the bld.inf file. If you have not done so previously, type

```
bldmake bldfiles
```

Then, perform the build using, for example,

```
abld build armi urel
```

This will invoke the GNU and Symbian OS toolchain to generate the following sequence of characters on your console.

```
  make -r -f "\EPOC32\BUILD\SCMP\HELLOTEXT\EXPORT.make" EXPORT VERBOSE=-s
Nothing to do
  make -r -f "\EPOC32\BUILD\SCMP\HELLOTEXT\ARMI.make" MAKEFILE VERBOSE=-s
perl -S makmake.pl -D \SCMP\HELLOTEXT\HELLOTEXT ARMI
  make -r -f "\EPOC32\BUILD\SCMP\HELLOTEXT\ARMI.make" LIBRARY VERBOSE=-s
make -s -r -f "\EPOC32\BUILD\SCMP\HELLOTEXT\HELLOTEXT\ARMI\HELLOTEXT.ARMI"
LIBRARY
  make -r -f "\EPOC32\BUILD\SCMP\HELLOTEXT\ARMI.make" RESOURCE CFG=UREL
VERBOSE=-s
make -s -r -f "\EPOC32\BUILD\SCMP\HELLOTEXT\HELLOTEXT\ARMI\HELLOTEXT.ARMI"
RESOURCEUREL
  make -r -f "\EPOC32\BUILD\SCMP\HELLOTEXT\ARMI.make" TARGET CFG=UREL
VERBOSE=-s
make -s -r -f "\EPOC32\BUILD\SCMP\HELLOTEXT\HELLOTEXT\ARMI\HELLOTEXT.ARMI"
UREL

PETRAN - PE file preprocessor V01.00 (Build 174)
Copyright (c) 1996-2001 Symbian Ltd.

  make -r -f "\EPOC32\BUILD\SCMP\HELLOTEXT\ARMI.make" FINAL CFG=UREL
VERBOSE=-s
```

This is similar to what we saw when compiling for the emulator, except that it additionally calls a Symbian OS-specific `petran` tool, which translates `hellotext.exe` into a compact form suitable for loading at runtime by the Symbian OS loader.

The net result is that `hellotext.exe` is built into `\epoc32\release\armi\urel\`.

To get the program to run, you have to copy it to your Symbian OS phone and start it from there. The convention for `.exe` programs like this one is to copy them to `c:\System\Programs\` and to run them from there (though another directory will do just as well).

If your combination of smartphone and connectivity software support this functionality, open Windows Explorer, check that your connection is functioning correctly, navigate to your Symbian OS phone's `c:\system\` folder and create a `Programs` subdirectory. Then navigate to `\epoc32\release\armi\urel\`, and copy `hellotext.exe` to the clipboard. Paste the file into the `c:\System\Programs\` directory. Symbian OS Connect will recognize that this file doesn't need to be converted, and copy it straight over.

If your connectivity software does not allow you to browse your smartphone's filing system, you will have to make an installable package and install it. From the command line, navigate to the `\scmp\hellotext` directory and type

```
makesys hellotext.pkg
```

This creates the file `hellotext.sis`, which you can install onto your smartphone.

There is more information about creating installation packages in Chapter 14.

In either case, it doesn't take long to copy the program: it's only about 990 bytes! If your Symbian OS machine has a file manager application, all you have to do is to use it to navigate to `c:\System\Programs\`, locate the `hellotext.exe` file and open it. However, machines like the Sony Ericsson P800 or the Nokia 7650 do not expose the filing system to the user and therefore do not have a file manager application. On such machines, after you've copied the file to your Symbian OS machine, you will find that there is a slight problem as there is no obvious way to start the program running.

One solution to this problem is to install and use a simple application that will locate and run the file for you. An installation file for such an application is supplied along with the example programs you will have downloaded from the Symbian website, as explained in Appendix 2. This program is supplied for use with the P800 as an installable package.

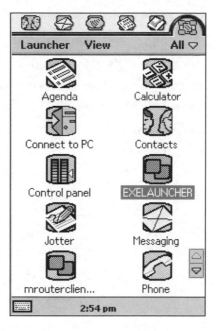

Figure 1.6

While your PC is still connected to your Symbian OS phone, use Windows Explorer to locate the installation file, \scmp\exelauncher\ group\exelauncher.sis in your SDK installation drive, double-click on it and follow the installation process to install it on drive C.

The ExeLauncher application, as shown in Figure 1.6 should now appear in the Application Launcher's view. When you run it, it lists all the .exe files that are in the c:\System\Programs directory – obviously on this occasion there will only be one, hellotext.exe, which will be highlighted. Run the highlighted file by selecting **Launch** from the **ExeLauncher** menu.

The hellotext program will run and you'll see, in tiny print, the famous hello 'Hello world!' message. *Congratulations!* You've got your first Symbian OS program working on both the Symbian OS emulator and a real Symbian OS mobile phone.

1.3 Summary

In this chapter, we've not gone very heavily into code, but have instead focused on the tools that come with the SDK and how to use them to build and test a simple project.

The topics we've looked at are the following:

- How to use the emulator
- A hint as to what support EPOC offers to help you code safely
- The basic structure of the project specification (.mmp) file
- Using and configuring the Metrowerks IDE and command line tools
- Building and running applications on both the emulator and the ARM platform.

2

System Structure

We've seen how to build programs for Symbian OS. Now we need some extra background information to understand how to write them.

In this chapter, we'll introduce as many issues as we need for the following chapters, without getting deeply involved in much C++ code. All these issues are fundamental for Symbian OS system design, and all will be essential background as we move to look at coding conventions, the user library, and other basic APIs in the next few chapters.

We'll start by reviewing the hardware of a typical Symbian OS machine from a developer's point of view. Hardware is more limited than that provided by a typical desktop computer, which, on the one hand creates many new market opportunities, and on the other hand imposes many constraints on software and software development.

We'll look at the four types of software that need to run on Symbian OS:

- applications
- servers
- engines
- kernel.

We'll cover the facilities provided by the Symbian OS base (the kernel plus lowest-level APIs) to support programming. Much of this is standard fare in system design, and will be familiar to anyone who has worked with an operating system like Windows NT or Unix.

Then I'll introduce something that is quite unique to Symbian OS: its optimized event-handling system using active objects and the client-server framework. Although we won't be tackling the details of this framework until much later in the book, everything we do will use it, and its existence is a key enabler for the performance, compactness and robustness of Symbian OS.

Symbian OS C++ for Mobile Phones. Edited by Richard Harrison
© 2003 John Wiley & Sons, Ltd   ISBN: 0-470-85611-4

I'll conclude the chapter with two overview sections on the big picture, to give you an idea of what's in the rest of the book. Firstly, I'll summarize the Symbian OS v7.0 APIs that we'll be covering. Then I'll do a quick tour through the application suite, so you can see how they use the APIs and get an idea on how you might use them yourself.

2.1 Hardware Resources

Symbian OS is intended to run on mobile phones. This profoundly affects the design of its software system. So let's begin this section by looking at the hardware facilities more closely.

Here are the main ingredients of a Symbian OS phone:

- a *CPU*: Symbian OS is designed for 32-bit CPUs, running at lower speeds compared with CPUs in desktops or workstations. Available Symbian OS systems are based on 190 MHz and 206 MHz Stron- gARM CPUs, with some on ARM 9. Future Symbian OS machines may use faster chips.

- a *ROM*: the system ROM contains the OS and all the built-in middleware and applications. Compare this to a PC in which only a small bootstrap loader and BIOS are built into ROM, with OS and applications loaded from the hard disk. The system ROM is mapped as the z: drive. Everything in the ROM is accessible both as a file on z: and directly by reading the data from ROM. So programs are executed in-place, rather than being loaded into RAM and then executed as PC programs are Symbian OS v7.0 machines use around 20 MB of ROM, though these days flash RAM is used more often.

- *system RAM*: the system RAM is used for two purposes: RAM for use by active programs and the system kernel, and RAM used as 'disk' space accessed as the c: drive, which as the cost of flash RAM has fallen has become more common. The system uses as much as is needed for these purposes: you don't have to preallocate some RAM for one purpose and some for another. Usually, there is also some free RAM. But since the total RAM on a typical machine is only around 8 MB or 16 MB, there is a real possibility that RAM may get exhausted, resulting in an out-of-memory error or (if the problem occurs when a file is being written) a disk full error.

- I/O devices, including a screen with 'digitizer' for pen input, a keyboard that may need to be even more compact than those found on laptops – in modern Symbian OS phones, these are increasingly mobile phone style keyboards, a memory-card slot for additional 'disks' accessed as d:, a serial port for RS232, and connection to a PC; an infrared port, and Bluetooth for wireless transfer of data between Symbian OS phones and others such as Palm PDAs or Nokia Communicators, or for convenient wireless access to data modems on a suitable mobile phone and other devices.

- Power sources, including main batteries and external power (from a mains adapter).

Figure 2.1

The Nokia 9210 (Figure 2.1) has 14 MB ROM, and 8 MB (SD-RAM), a 16 MB MMC card in the standard sales package, 4096 color screen, and full keyboard.

The Sony Ericsson P800 (Figures 2.2, 2.3) has 16 MB RAM and 16 MB of flash, with support for extra memory through the use of Sony's Memory Stick. It features a 640 × 480 digital camera, and a 24-million color display. User input is achieved through the digitizer pen and use of the mobile phone keyboard.

Symbian OS OEMs have freedom to vary this basic specification quite widely. The Psion Revo has a smaller screen (480 × 160), and an in-built rechargeable battery instead of main and backup batteries. However, while it uses infrared for communication with other Symbian OS machines and compatible mobile phones, and a docking cradle for communication with PCs, the Psion Revo has no memory-card slot. The Psion Series 7 uses a 190 MHz StrongARM CPU, and has a 640 × 480 color screen. Modern devices such as the Nokia 7650 include new devices such as Digital Cameras and support for Picture Messaging along with increased support for networking, in particular Personal Area Networks (PANs), through support for Bluetooth.

OEMs also have flexibility in the way they manage the ROM and deliver built-in applications. Applications can be delivered on a CD-ROM with the devices, to be installed into RAM if the user wants them, instead of

Figure 2.2

Figure 2.3

being built into ROM. There's room for ingenuity in managing the ROM itself: the devices have only a very small bootstrap loader in ROM. On cold boot, they load a ROM image from the memory card into the system RAM using as much RAM as needed. After this, the RAM used is marked as read-only and behaves exactly like a ROM. This arrangement allows the ROM to be configured by an enterprise IS department, replaced in the field by the user.

Symbian OS supports this variability by essentially the same techniques used in the PC industry: it uses a device driver architecture and offers an abstracted API for each device to programs that use that device.

But Symbian OS is otherwise very different from PC operating systems:

- Resources are constrained: the CPU is slower and there is less memory.
- There is no hard disk: we can't do disk-backed virtual memory, and we can't assume there is an infinite amount of room in which to place our program or data files.
- Power management is critical: modern devices are expected to function for several hours between recharges and there may be other uses of the power such as making a phone call.

To summarize, you have to make the software compact, and you have to tackle errors such as out-of-memory and others, because Symbian OS systems virtually never reboot. Symbian OS is designed from the ground up to help you do this, but you must learn the disciplines this involves and implement them in the software you write.

2.2 Software Basics

Symbian OS and its applications can be divided into various types of components, with different types of boundaries between them, in Figure 2.4:

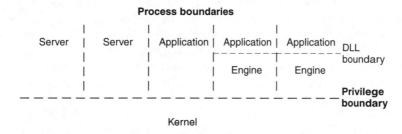

Figure 2.4

The **kernel** manages the machine's hardware resources such as system RAM and hardware devices. It provides and controls the way all other software components can access these resources. The kernel uses hardware-supported **privilege** to gain access to the resources. That is, the CPU will perform certain privileged instructions for only the kernel.

It runs other programs – so-called **user-mode** programs – without privilege, so that they can only access system resources through the kernel APIs. The boundary between the kernel and all other components is a privilege boundary.

An **application** is a program with a user interface. Each application runs in a separate **process** with its own virtual address space, so the boundary between one application and another is a process boundary. One application cannot accidentally overwrite another's data because their address spaces are entirely separate.

A **server** is a program without a user interface. A server manages one or more resources. It provides an API so that **clients** can gain access to its services. A server's clients may be applications, or other servers. Each server generally runs in its own process so that the boundary between a server and its clients is a process boundary. This provides a strong assurance of server integrity.

Actually, for performance reasons, certain closely related servers may run in the same process: we'll cover this in more detail below.

The isolation between a server and its clients is of the same order as the isolation between a kernel and user-mode programs – but servers are much easier to program and work with. Symbian OS uses servers to provide many services that, on other systems, are provided by the kernel or device drivers.

An **engine** is the part of an application that manipulates its data, rather than the part that interacts directly with the user. Often, you can easily divide an application into an engine part and a GUI part. Exactly where you draw this line is part of the art of software engineering. Most built-in Symbian OS applications, and the larger applications I'll be developing in this book, have engines of sufficient complexity that it's worth drawing the boundary explicitly. An application engine may be a separate source module, a separate DLL, or even a number of separate DLLs. The boundary between engine and application is a module or DLL boundary, whose main purpose is to promote good software design – in contrast to a process or privilege boundary, whose main purpose is to prevent unwanted interactions.

That gives us four component types and three boundary types. DLL or module boundaries are very cheap to cross: they promote system integrity by modularization and encapsulation. The privilege boundary is a little more expensive to cross: it promotes system integrity by hiding the kernel and devices from user-mode code. Process boundaries are most expensive of all to cross: they promote integrity by isolating programs' private RAM from each other.

If you're an application programmer, you'll spend most of your time writing applications and – if your application is big enough – engines.

I use the word 'system programming' to refer to the art of writing server or kernel software. I'll be covering servers in this book because they're important for communications programming. I'll give an overview of the kernel and what it does in this chapter, but I won't otherwise be covering how to program the kernel side of the privilege boundary.

2.3 Processes, Threads and Context Switching

The **process** is a fundamental unit of protection in Symbian OS. Each process has its own **address space**. The virtual addresses used by programs executing in that process are translated into physical addresses in the machine's ROM and RAM. The translation is managed by a **memory management unit** or MMU, so that read-only memory is shared, but the writable memory of one process is not accessible to the writable memory of another.

The **thread** is the fundamental unit of execution in Symbian OS. A process has one or more threads. Each thread executes independently of the others, but within the same address space. A thread can, therefore, change memory belonging to another thread in the same process – deliberately, or accidentally. Threads are not as well isolated from each other as processes.

Threads are **preemptively scheduled** by the Symbian OS kernel. The highest-priority thread that is eligible to run at a given time is run by the kernel. A thread that is ineligible is described as **suspended**. Threads may suspend to wait for events to happen and may **resume** when one of those events does happen. Whenever threads suspend or resume, the kernel checks for the thread that is now the highest one and **schedules** it for execution. This can result in one thread being scheduled even while another one is running. The consequent interruption of the running thread by the higher-priority thread is called **preemption**, and the possibility of preemption gives rise to the term **preemptive multitasking**.

The process of switching execution between one thread and another is **context switching**. Like any other system, Symbian OS's scheduler is written carefully to minimize the overheads involved in context switching.

Nevertheless, context switching is much more expensive than, say, a function call. The most expensive type of context switch is between a thread in one process and a thread in another because a process switch also involves many changes to the MMU settings, and various hardware caches must be flushed. It's much cheaper to context switch between two threads in the same process.

Typically, each Symbian OS application uses its own process that consists of just one main thread. Each server also uses its own process and just one thread. The decision to use a *separate* process for each application and server is one of security – a process cannot deliberately or accidentally change another process' memory. In cases where servers are designed to cooperate closely together, they are sometimes packaged into a single process so that context switching between them is cheaper. Thus, all the major communications-related servers in Symbian OS – serial, sockets, and telephony – run in the same process.

How can an application or server run effectively in a single thread? Don't sophisticated applications need to perform background tasks? And shouldn't a server have a single thread for each client? Symbian OS implements sophisticated applications and servers using only a single thread because it has a good event-handling system based on **active objects**. I'll return to that in the *Event Handling* section below.

2.4   Executable Programs

As far as the CPU is concerned, a C++ program is just a series of instructions. But if we want to manage software development and deployment effectively, we have to group code in more convenient packages. The packages Symbian OS uses are closely based on those used by Windows NT and similar systems. They are as follows:

- a .exe – a program with a single main entry point E32Main(). When the system launches a new .exe, it first creates a new process. The entry point is then called in the context of the main thread of that process.

- A **dynamic link library** or DLL – a library of program code with potentially many entry points. The system loads a DLL into the context of an existing thread (and therefore an existing process).

Both these are **executables**. I'll use 'executable' when I mean either a .exe or a DLL. I'll never use just 'executable' if I mean a .exe specifically – I'll use '.exe'.

There are two important types of DLL, which are as follows:

- A **shared library DLL** provides a fixed API that can be used by one or more programs. Most shared library DLLs have the extension .dll. Executables are marked with the shared libraries they require and, when the system loads the executable at runtime, the required shared libraries are loaded automatically. This happens recursively, so any shared libraries needed by the shared libraries are also loaded, until everything required by the executable is ready.

- A **polymorphic DLL** implements an abstract API such as a printer driver, sockets protocol, or an application. Such DLLs typically use an extension other than .dll – .prn, .prt, or .app, for instance. In Symbian OS, polymorphic DLLs usually have a single entry point, which allocates and constructs a derived class of some base class associated with the DLL. Polymorphic DLLs are usually loaded explicitly by the program that requires them.

2.4.1 The Place of Execution

To be executed, an executable has to be **loaded**. This means that its program and data areas must be prepared for use. There are two cases here:

- The first case is an executable in ROM (drive z:). ROM-based executables are executed in-place.

- Executables not in ROM must first be loaded into RAM. This applies to executables on memory cards (removable media drive d:), or in the system flash RAM disk (drive c:).

2.4.2 Loading and Sharing

Executables contain three types of binary data:

- program code
- read-only static data
- read/write static data.

Symbian OS handles .exes and DLLs differently.

.exes are not shared. If a .exe is loaded into RAM, it has its own areas for code, read-only data, and read/write data. If a second version of the same .exe is launched, new areas will be allocated for each of these. There is a small optimization: ROM-based .exes allocate a RAM area only for read/write data – the program code and read-only data are read directly from ROM.

DLLs are shared. When a DLL is first loaded into RAM, it is relocated to a particular address. When a second thread requires the same DLL, it doesn't have to load it – it merely **attaches** the copy already there. The DLL appears at the same address in all threads that use it. Symbian OS maintains reference counts so that the DLL is only unloaded when no more threads are attached to it. ROM-based DLLs, like ROM-based .exes, are not actually loaded at all – they are simply used in-place in ROM.

2.4.3 Cutting Down the Size

Symbian OS optimizes the formats used for DLLs in order to make them as compact as possible in ROM and RAM.

- Most systems supporting DLLs or analogous concepts offer two options for identifying the entry points in them. You can refer to the entry points either by name or by ordinal number. Names are potentially long and wasteful of ROM and RAM. So Symbian OS uses link-by-ordinal exclusively.

- Loading into RAM can involve locating the executable at an address that cannot be determined until load time: this means that relocation information has to be included in the executable format. Loading into ROM happens effectively at build time. So Symbian OS ROM-building tools perform the relocation and strip the DLLs of their relocation information to make them still smaller.

The Symbian OS link-by-ordinal scheme affects the disciplines used for binary compatibility (a future release of a DLL must use exactly the same ordinals as the previous release). The preloading scheme means, among other things, that you can't take an executable out of the ROM and deliver it in another package for RAM loading. These are largely matters for Symbian OS OEMs and I shan't be describing them further in this book.

2.4.4 Launching Applications and Servers

Most servers use their own .exe to generate their own process. For instance, **ewsrv.exe** is the window server and **efsrv.exe** is the file server.

As we saw earlier, some servers piggyback into the process of others to minimize context-switching overheads. The main server in such a group uses its own process – for instance, **c32exe.exe** launches the serial communications server. Other servers use a DLL and launch their own thread within the main server thread.

A console application, such as **hellotext.exe**, is built into its own .exe. A console application must create its own console, which it can then use to interact with the user.

Most GUI applications are like **hellogui.app**. They are actually poly-morphic DLLs whose main entry point, `NewApplication()`, creates and returns a `CEikApplication`-derived object. The application process is created by a small .exe, **apprun.exe**, to which the .app name is passed as a parameter. If the application wants to edit an embedded document, it can do so *without creating a new process* by loading the .app for the embedded document directly in the same thread.

2.5 Power Management

Power management is probably the single most difference between a desktop system and a portable system:

- Power has to be used efficiently. Battery life – even with rechargeable batteries – makes a difference to how the user thinks of the device. So also does battery weight. Symbian OS needs to work effectively on lower-speed, lower-power hardware, than that used by desktop PCs.

- Certain parts of the system should still be able to run while the system is apparently off. For example, when an alarm is due, the machine should be turned on so that the alarm can sound. For modern smartphones, the phone may need to react to an incoming call or SMS. This means that the system should switch from a low power mode to being in an active state.

- Even if all power is removed suddenly, the system should do what it can to save critical information, so that there is a possibility of a warm boot when power returns (rather than a cold boot).

As an application or server programmer, your task is easier than that of the kernel. But, you still get involved in power management. You have to write your programs efficiently to make the best use of a Symbian OS phone's scarce resources – this applies as much to power as to available RAM, CPU speed, and so on.

The deeper you delve into the system, the more complicated power management becomes. For instance, as a device driver programmer, you can see that power management is more complex than simply machine on/off. The user thinks the machine is off if the display is turned off. But each hardware component is responsible for its own power management. A communications link driver should turn the physical device off if it's not needed. The kernel scheduler even turns the CPU off if all threads are waiting for an event. Every possible step is taken to save power and ensure that user data is retained even in the most difficult power-loss situations.

2.6   The Kernel and E32

The most fundamental component of Symbian OS is E32. E32 consists of the kernel and user library. The kernel is entirely privileged. The user library, **euser.dll**, is the lowest-level user-mode code. It offers library functions to other user-mode code and controlled access to the kernel.

The kernel itself has two major components:

- The **kernel executive** runs privileged code in the context of a thread that usually executes in user mode. Executive code can therefore be preempted by higher priority user-mode threads, or by the kernel server.

- The **kernel server** is the main thread of its own process and always runs privileged. The kernel server is the highest-priority thread in the

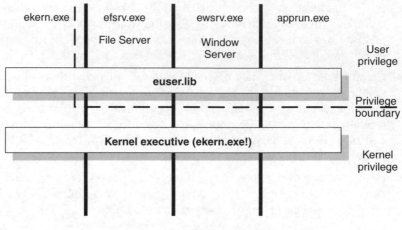

Figure 2.5

system. It allocates and deallocates kernel-side resources needed by the system and by user programs. It also performs functions on behalf of user-mode programs. The kernel server is a single thread: it handles user requests in sequence, non-preemptively.

We'll be describing euser.dll's most important facilities in detail over the next few chapters. For now, let's note that it offers three types of functions:

- Functions that execute entirely user-side, such as most functions in the array and descriptor classes (descriptors are the Symbian OS version of strings).

- Functions that require privilege, and so cross into the executive, such as checking the time or locale settings.

- Functions that require the services of the kernel server: these go through the user library, via the executive, to the server.

The functions that operate entirely on the user-side can also be used safely by any kernel-side code. Kernel-side code that needs access to kernel facilities can (and must) use these facilities directly rather than through the user library interface.

In this book, we'll be writing user-side code exclusively. So, although the distinction between the types of function in the user library helps you to understand the system design, it's not essential for you to know all the possible circumstances in which kernel-side code might be called.

2.7 Device Drivers

System devices such as screen, keyboard, digitizer (for the pen), sound codec, status LEDs, power sensors, serial port, CF-card, and so on are all driven by low-level device drivers. It's possible to add devices and write drivers for them. Symbian OS OEMs usually do this: Symbian OS phones are not typically user-expandable in the same way that PCs are.

A device driver is implemented in several parts as shown in Figure 2.6:

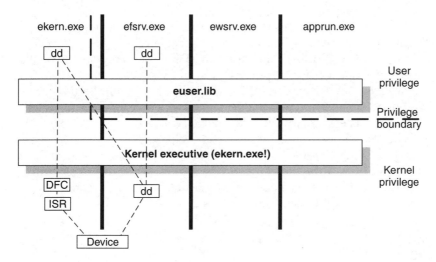

Figure 2.6

The kernel executive contains support for device drivers so that a user program can issue a request to device driver code running kernel-side, in either the kernel executive or kernel server. Such requests typically initiate a device operation, or tell the driver that the requesting program is waiting for something to happen on the device.

Drivers also process device interrupts and then tell the user (or kernel) program that an earlier request is complete. Interrupt handling works at two levels.

- First-level handling is done by an **interrupt service routine** (ISR). ISRs must be short, and can't do very much, because they could occur at any time, even in the middle of a kernel server operation. Usually, they simply acknowledge the device that raised the interrupt and then set a flag to request the kernel to run a **delayed function call** (DFC) for second-level processing.

- The kernel schedules the DFC when it is in a more convenient state – immediately, if user-mode code was executing when the interrupt occurred; otherwise, when the kernel would have otherwise crossed the privilege boundary back to user code. DFCs can use most kernel APIs. DFCs typically do a small amount of processing and then post to a user thread to indicate that an I/O request has completed.

2.8   Timers

The kernel supports a tick interrupt at 64 Hz on ARM, and 10 Hz on the PC-based emulator.

The tick interrupt is used to drive round robin scheduling of equal highest-priority threads. It can also be accessed (via `User::After()` and `RTimer::After()` function calls) by user programs. The tick interrupt suspends during power-off, so that if you request a timer to expire after 5 s, and then turn the machine off 2 s later, the timer event will occur 3 s after you turn the machine back on again – or even later, if you immediately turn the machine back off!

The kernel also supports a date/time clock, which you can access using `User::At()` and `RTimer:At()`. This timer expires at exactly the time requested. If the machine was turned off when the timer expires, it is turned on. This behavior makes it a suitable time for alarms, as the system will power up and can then alert the user.

2.9   Memory

System memory is managed by the memory management unit (MMU).

ROM handling is easy. The ROM consists entirely of files in a directory tree on drive z: and is mapped to a fixed address so that the data in every file can be accessed simply by reading it. Programs can be executed in-place, and bitmaps and fonts can be used in-place for on-screen blitting, without all the data going through the file server.

RAM management is more interesting. Physical RAM is divided into 4 k pages by the MMU. Each physical page can be allocated to the following:

- A user process' virtual address space: there may be many of these, as each process has it's own.

- The kernel server process' virtual address space.

- The RAM disk used as c:. Such RAM can only be accessed by the file server process.

- DLLs loaded from a non-ROM filing system: RAM for DLLs is marked read-only after the DLL has been loaded. Each DLL appears at exactly the same virtual address for all threads that use it.

- Translation tables for the MMU: the MMU is carefully optimized to keep these small. But there is no practical limit on the number of processes and threads allowed in Symbian OS.

- The free list of pages not yet allocated for any of the above purposes.

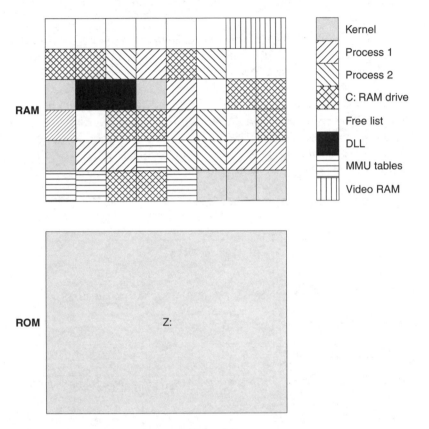

Figure 2.7

There is no virtual memory backed up by a swap file on a large hard disk. So any page needed for user processes, the kernel, or the RAM disk is taken from the free list. When the free list runs out, the next request for memory that requires a new page to be allocated will cause an out-of-memory error – or a disk full error, if the request came from a file write.

2.9.1 Process and Thread Memory

When a .exe is launched, it creates a new process with a single main thread. During the lifetime of a process, other threads may also be created.

The process' address space includes regions for

- system-wide memory, such as the system ROM and RAM-loaded shared DLLs,

- process-wide memory, such as the .exe image and its writable static data,

- memory for each thread, for a very small stack and a default heap (which can grow up to a limit set by the Symbian OS OEM: for instance, it's 2 MB on a Psion Series 5MX).

A thread's stack cannot grow after the thread has been launched. The thread will be panicked – terminated abruptly – if it overflows its stack. The usual initial stack size is 12 Kb. The stack is used for C++ automatic variables in each function. So you have to avoid using large automatics. Instead, put all large variables on the heap.

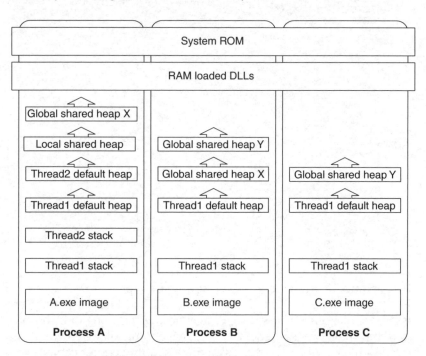

Figure 2.8

A thread's default heap is used for all allocations using C++ operator new, and user library functions such as `User::Alloc()`. If possible, memory is allocated from existing pages committed to the heap. If that's not possible, the heap manager requests additional pages from the system free list. If the system free list has insufficient pages, the allocation will fail, giving an out-of-memory error.

Because each thread makes allocations on its own nonshared heap, allocation and deallocation is very efficient. If an allocation can be satisfied without growing the heap, only a few instructions are required, no privilege boundaries need be crossed, and no synchronization with allocations by other threads is needed.

You can put small objects on the stack, such as integers or rectangles,

```
TInt x;
TRect region;
```

but most objects – especially larger ones – should go on the heap:

```
CEikDialog* dialog=new CGameSettingsDialog;
```

Objects whose class name begins with C can only go on the heap. Objects whose class name begins with T, however, can be either members of other classes, or automatics on the stack. Don't put them on the stack unless they're quite small. Beware especially of TFileName:

```
TFileName fileName;
```

A filename is 256 characters – 512 bytes in Symbian as it uses unicode. There isn't room for too many of them on a 12-Kb stack.

You can control the stack size in a .exe. This can apply to console programs, servers, or programs with no GUI – but not to UIQ programs, since they are launched with **apprun.exe**. You can also control the stack size when you launch a thread explicitly from within your program. If you have an application with an algorithm that requires a large stack such as a heavily recursive game-tree search, you may have to encapsulate the algorithm in a .exe of its own, or a separate thread.

Since each user heap eats into a scarce system resource – the free page list – and since applications and servers run for months or years without being restarted, it's vital that programs detect heap failure due to a lack of memory. It's also vital that programs release unneeded memory as soon as possible. This is the domain of the Symbian OS cleanup framework, which is covered in Chapter 6.

Threads have independent default heaps in the sense that each thread always allocates from its own heap. But since all heaps are in the same process' address space, each thread in a process can access objects on other heaps in that process – provided suitable synchronization methods are used.

In addition to the default heap, threads can have other heaps. But these introduce new complications, so you should use them only if you have to. For any nondefault heap, you must provide a specific C++ operator new() to allocate objects onto it. For local shared heaps–shared with other threads in the same process – you have to introduce synchronization using mutexes or the like. For global shared heaps – shared with threads in other processes – the heap is mapped to a different address in each process, so you have to introduce a smart reference system rather than straightforward pointers. All these things are possible if necessary – but they're rarely necessary. Usually, it's better to use a server to manage shared resources, rather than a shared heap. There's an overview of servers below, and they are covered more thoroughly, including performance optimization, in Chapters 18, 19 and 20.

A thread's nonshared heaps are allocated into a 256 MB region of a process' virtual address space. By limiting the maximum size to 2 MB, there is an implied maximum of 128 threads per process.

2.9.2   No Writable Static Data in DLLs

> **Symbian OS DLLs do not support writable static data.**

DLLs only support read-only data and program code. Writable static data is supported only by. exes.

This imposes some design disciplines on native Symbian OS code. It does make life more difficult when porting code, which often assumes the availability of writable static data.

The easiest workaround is to use a .exe to contain the ported code. The .exe is packaged as a Symbian OS server, which allows it to be shared between multiple programs. By using a separate process, we also gain the benefit of isolation.

Symbian OS doesn't support writable static data, as every DLL that supports writable static would require a separate chunk of RAM to be allocated in every process that uses the DLL. There are about 100 DLLs in Symbian OS v7.0: perhaps the typical application uses 60 of them. Say, I usually have about 20 applications running concurrently in my 12 MB RAM machine, and there are about 10 system servers working on behalf of those applications. The smallest unit of physical memory allocation in conventional MMUs is 4 Kb (and smaller wouldn't be at all sensible). If each DLL used even a single word of writable static data, it would require $4\,Kb \times (20\text{ app processes} + 10\text{ server processes}) \times 60$ DLLs each $= 7\,MB$ of RAM just for the writable static data!

I quite often have about 20 apps running on my PC too: 7 MB on a PC isn't unacceptable – most of it is paged out to disk anyway – but for a handheld system, this overhead or anything approaching it is out of the question.

You could argue that most DLLs wouldn't use writable static data, so these figures are exaggerated. But the Symbian OS architects' response was that if the facility was there, most people *would* use it, without even knowing that they were doing so. We would only find out at system integration time, and by then it would be too late to fix any problems. So, writable static data was not implemented by the Symbian OS loader.

Symbian OS v7.0 does provide a workaround for the writable static limitation, intended for system components only. Future versions of Symbian OS may further ease the restriction.

Even if the rules are loosened up, the underlying economics won't change: at least 4 Kb of RAM will be consumed by each process that loads each DLL that requires writable static. Using writable static data isn't environment friendly. Don't do it without being aware of the consequences.

In fact, Symbian OS associates a single machine word of writable static per thread with each DLL. This is **thread-local storage** or TLS. You can use the TLS word as an anchor for what would have been your writable static. There are no MMU granularities to worry about here – just a small performance implication, since getting the TLS pointer involves a system call which takes perhaps 20 or so instructions rather than the single instruction required to get a normal pointer. Not all DLLs use TLS, but the system allocates the word anyway. In my scenario above, TLS would account for only 1.8 Kb – which is perfectly acceptable.

2.10 Files

Let's summarize what we've already seen about files.

Symbian OS phones have no hard disk, as found on PCs. But Symbian OS always has two disks present, and may have more.

C:	Flash RAM disk – full read/write file system. Contents are initialized to empty on a cold boot. Data is maintained as long as there is power to refresh the RAM. Data is recovered in a warm boot, unless it has been corrupted beyond recovery. Files can be extended indefinitely so long as there are RAM pages to allocate to them from the system free list. RAM pages are subdivided into 512 byte sectors, so that small files are managed more efficiently.

Z: ROM – read-only file system. Contents are built by the
                      Symbian OS OEM when building the device. Some
                      machines, such as the Psion netBook or Series 5MX
                      Pro, load a ROM image on cold boot so that 'ROM' can
                      be updated and replaced by enterprise IS departments,
                      distributors and so on.

D: Memory card – removable read/write media, supported
                      by some Symbian OS phones.

                      Careful power management is used by the Symbian OS
                      memory-card file system to ensure that 512 byte sector
                      writes are **atomic** – they either complete fully, or don't
                      even start. File formats such as those used by the
                      persistent file store are written and extensively tested to
                      assume and support, atomic sector writing. These files
                      can be recovered if failure occurs on any sector write.
                      Memory cards are slower than the RAM disk, but their
                      nonvolatility and higher capacity – 20 to 200 MB or
                      so – makes them attractive.

                      Memory cards are an industry standard, slightly smaller
                      than PC cards used on laptops. You can buy PC-card to
                      memory-card adapters to insert a memory card into a
                      laptop and thus share data between your laptop and
                      Symbian OS phone. Memory cards are also used in
                      other devices such as digital cameras. Symbian OS can
                      share data with any other device that supports memory
                      cards, provided it uses standard DOS partitions and the
                      FAT (or VFAT) filing system.

The Symbian OS file server supports installable file systems that can
be loaded at runtime without any kind of reboot. Additional drive letters
and additional media types can also be supported, depending on system
and user requirements.

Data management is covered more extensively in Chapter 13.

2.11   Event Handling

> **Perhaps the most fundamental design decision in Symbian OS was
> to optimize the system for efficient event handling. Each native
> Symbian OS application or server is a single event-handling thread.**
> Active objects **are used to handle events non-preemptively.**

In the old days, programs were written such that every so often the program would deign to check for user input, and would then process it. With GUI systems, though, the user is in control – their input is the focus of the application's existence. This requires us to focus our programming on event-driven responses to users.

Symbian OS supports some fundamental building blocks for event-handling systems – active objects and the client-server architecture. You won't need to understand them in any more detail than this, until you need to write your own active objects and servers. These topics are covered in Chapters 17 and 18.

2.12 Perspectives on Event Handling

Say you are using the Symbian OS Word application. If you press a key, a small cascade of events will occur, which are handled by at least three Symbian OS threads, as in Figure 2.9:

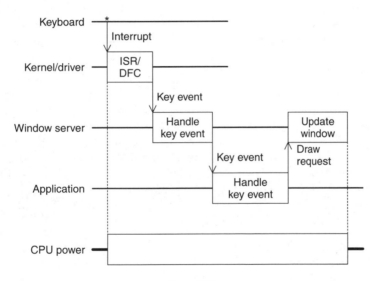

Figure 2.9

Let's look at this cascade from a couple of perspectives. Firstly, from the whole-system point of view:

- The I/O device responsible for looking after the keyboard generates an interrupt.

- The Interrupt Service Routine (ISR), translates this to an event. It interrogates the device, works out what ASCII key code to assign, and creates an event for whichever program is interested in raw key events and in any real Symbian OS system that's the window server.

- The window server then works out the application that is currently receiving keystrokes and sends the event to the application, in this case Symbian OS Word.

- Symbian OS Word then handles the key – perhaps by adding text to the document and then updating the display.

- The window server updates the display in response to the application's requests.

From the power-management point of view, power is needed for the CPU only while it's doing something. Power is turned on to handle an interrupt and turned off again when no more threads are eligible to run.

You can also look at each of the tasks in the diagram and ask, 'what other events might this task have to handle?'

- The keyboard driver handles an interrupt, does minimal processing, and notifies a user-mode thread – in this case, the window server. The keyboard driver must also handle requests from the window server for key events. So, the keyboard driver is an event-handling task that handles two types of an event: requests from a user-mode thread and hardware events from the keyboard.

- The window server handles the key, does enough processing to identify the application that is currently taking keys, and then notifies the application. The window server, like the keyboard driver, also handles requests from the application for key presses. And the window server also performs screen drawing on behalf of all applications. So, the window server is an event-handling task that handles these three event types (key events, requests to be notified about key events, and screen drawing) plus many more (for instance, pointer events, and requests to be notified about them).

- The application is an event-handling task that handles key events (and more, for instance, pointer events).

So each task is an event handler. In Symbian OS, events are handled using active objects.

2.13   Active Objects

All native Symbian OS threads are essentially event handlers, with a single **active scheduler** per thread cooperating with one or more **active objects** to handle events from devices and other programs.

Each active object has a virtual member function called `RunL()`. `RunL()` gets called when the event happens for which the particular active object is responsible, and must be implemented to handle the event. Usually, it starts with some preprocessing to analyze the event. It may complete the handling of the event without calling any other functions. But in a framework, `RunL()` will usually call one or more virtual functions that the programmer implements to provide specific behavior. The most important frameworks are for GUI applications and servers:

- An application, such as the one in our example above, uses the GUI framework. The framework analyzes input events, associates them with the correct control, and then calls virtual member functions such as `OfferKeyEventL()` to handle a key.

- A server, such as the window server above, uses the server framework to handle requests from client applications – including requests to draw on the screen, or to be notified about key events. Client requests are turned into messages that are sent to the server. The server framework analyzes these messages, associates them with the correct client, and calls `ServiceL()` on the server-side object representing the client, to handle the client's request.

A server also uses its own active objects to handle events other than client requests – for instance, key events from the kernel.

Active objects make life very easy for application programmers. All you have to do is to implement the correct framework function. Unless you need active objects for some other reason, you don't need to understand how they work. All you need to know is that your code must complete quickly (say, within a couple of milliseconds) so that your application is able to handle other events without undue delay.

You will eventually want to understand active objects and they are explained in Chapter 17. That's quite a lot later in the book, which proves my point: we don't need to get familiar with active objects until we get onto quite sophisticated programming.

2.14 Multitasking and Preemption

Symbian OS implements **preemptive multithreading** so that it can run multiple applications and servers simultaneously. Active objects are used to implement **non-preemptive multitasking** within the context of a single thread.

Active objects, like threads, have priorities that affect their scheduling. On completion of its execution (that's when the `RunL()` function returns), control returns to the `ActiveScheduler`, which then schedules the active objects according to the following rules:

- if there is just one object now eligible to run, then run it now;
- if there is more than one eligible object, then choose the one with the highest priority;
- if there are no eligible objects, then wait for the next event and then decide what to do based on these rules.

Some events are more important than others. It's much better to handle events in priority order than first in, first out (FIFO). Events that control the thread (key events to an application, for example,) can be handled with higher priority than others (for instance, some types of animation). But once a `RunL()` has started – even for a low-priority event – it runs to completion. No other `RunL()` can be called until the current one has finished. That's OK, provided that all your event handlers are short and efficient.

Non-preemptive multitasking is surprisingly powerful. Actually, there should be no surprise about this: it's the natural paradigm to use for event handling. For instance, the window server handles key and pointer events, screen drawing, requests from every GUI-based application in the system, and animations including a flashing text cursor, sprites, and self-updating clocks. It delivers all this sophistication using a single thread, with active-object-based multitasking.

And a sophisticated application such as Symbian OS Word uses active objects to handle status display update and text pagination at lower priority than more critical events – responding to editing events at and around the cursor position.

In many systems, the preferred way to multitask is to multithread. In Symbian OS, the preferred way to multitask is to use active objects.

In a truly event-handling context, using active objects is pure win–win over using threads:

- You lose no functionality over threads, as events occur sequentially, and can be handled in priority order.

You gain convenience over threads, because you know you can't be pre-empted: you don't need to use mutexes, semaphores, critical sections, or any kind of synchronization to protect against the activities of other active objects in your thread. Your `RunL` is guaranteed to be an atomic operation.

- It's more efficient. You do not occur the overheard of a **context switch** when switching between active objects.

Non-preemptive multitasking in Symbian OS is not the same as **cooperative multitasking**. The 'cooperation' in cooperative multitasking is that one task has to say, 'I am now prepared for another task to run', for instance, by using `Yield()` or a similar function. What this really means is, 'I am a long-running task, but I now wish to yield control to the system so it can get any outstanding events handled if it needs to'. Active objects don't work like that: during `RunL()`, you have the system to yourself until your `RunL()` has finished.

All multitasking systems require a degree of cooperation so that tasks can communicate with each other where necessary. Active objects require less cooperation than threads because they are not preemptively scheduled. They can be just as independent as threads: a thread's active scheduler manages active objects independently of one another, just as the kernel scheduler manages threads independently of one another.

2.15 Servers

Most multithreaded programming in Symbian OS uses the client-server framework, shown in Figure 2.10:

- A **server** thread is responsible for managing one or more related resources.
- One or more **client** threads may use the server to perform functions that use those resources.

The two most critical servers in Symbian OS are the **file server**, which handles all files, and the **window server**, which handles user input and drawing to screen. A wide range of other servers is used to manage communications, databases, schedule, contacts, and the like. The kernel also acts as a kind of server. A client program may be either another server, or an application.

Client-server programming involves two potentially difficult issues:

- it involves multithreaded programming disciplines, which are difficult to get right,
- it involves crossing process boundaries, which are a key guardian of system integrity.

In order to minimize any difficulty associated with these issues, Symbian OS constrains the client/server interface to something that is small enough to maintain confidence in the usefulness of the process boundary. It is built in such a way that you don't need to use thread synchronization as either the user or even the implementer of a server. The key elements of the interface are as follows:

- *The client interface*: each server provides an API to its clients – the client interface, which disguises all the client-server communications, so that clients can use the server easily without knowing the specifics of the client-server framework.

- *Kernel-supported message passing*: if you're implementing a server (along with its client interface), this is the main method by which you pass requests from the client to the server, and handle them. The message-passing framework is powerful enough for the job – but no more complex than it needs to be.

- *Kernel-supported interthread read and write*: messages can't convey much information from client to server, and even less from server to client. To pass more information, a server can read from, or write to, a client's address space.

Most client classes that access server-based resources have names beginning with R. Two examples are `RFile` (a file, with functions such as `Read()`, `Write()`, `Open()` etc.) and `RWindow` (an on-screen window, with functions such as `SetSize()`, `BeginRedraw()` etc.). These client interface classes are implemented (by the server designer) using message passing and interthread read and write.

Clearly servers are event handlers. The central classes in any server are a single `CServer`-derived class to implement the behavior of the whole server and a number of `CSession`-derived classes to handle requests on behalf of each active client. `CServer` is an active object whose `RunL()` interprets incoming messages, creates or destroys `CSessions` as needed, and calls their `ServiceL()` function to handle routine client requests. Most servers use more active objects to handle other events – such as key and pointer events in the case of the window server, or disk-door-opened events in the case of the file server.

The kernel server uses a similar framework. The `RTimer` class, and many other R classes in the user library, implement their APIs by message passing similar to that used by servers. The kernel server's framework is different from the standard server framework to take account of the privilege-mode environment and the fact that there's only one kernel. But the principles are the same.

Device drivers also use a message-passing system similar to that used by servers.

There's a lot more to say about servers. Chapter 18 explains the message-passing framework in more detail, and also provides many tips for getting the best performance from servers.

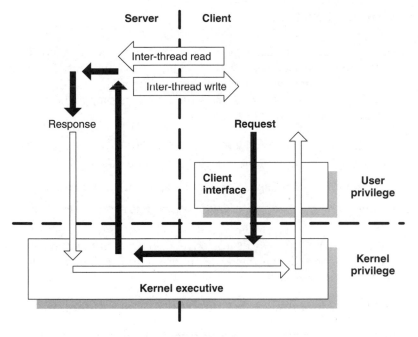

Figure 2.10

2.16 Where Threads Matter

Most tasks are event handlers. The design of Symbian OS is optimized for event handling, with good results for ease of programming, system efficiency, and robustness.

But some tasks really are long-running threads. Game engine calculations, spreadsheet recalculation, background printing, and the like can be particularly long running. Status display updates, animations, and the like are only slightly less demanding.

Symbian OS has broadly two approaches to handling tasks that really are long-running threads.

- Simulate them using active objects, and chains of pseudoevents. Split the task into short increments, generate a low-priority pseudoevent that will be handled if no real events (such as user input) need handling; handle an increment, and if that doesn't complete the task, generate another pseudoevent.

- Really use multithreading. Launch a background thread and work out some scheme of communication between the application's (or server's) main thread and the background thread.

If it's possible, the first approach is strongly preferred because it's more efficient.

2.17 APIs Covered in the Book

Now we've reviewed the type of hardware on which Symbian OS operates, the base facilities for constructing programs, and the event-handling system including the client-server framework.

Symbian OS APIs divide into categories corresponding to those we use for different types of program:

- the kernel exposes an API through the user library
- system servers expose APIs through their client interfaces
- application engines expose APIs to the applications that use them
- middleware components are APIs in perhaps the purest and simplest sense
- other API types such as device drivers, sockets protocol implementations, printer drivers, and so on are associated with particular system components.

These divide into several broad groupings:

Group	Description
Base	Provides the fundamental APIs for all of Symbian OS, which I've described in this chapter.
Middleware	Graphics, data, and other components to support the GUI, engines, and applications.
UI	The system GUI framework including the Shell (in UIQ, the Application Launcher) application.
Applications	Application software can be divided into GUI parts (which use UIQ) and engines (which don't deal with graphics). Some applications are simply thin layers over middleware components: others have substantial engines.
Communications	Industry-standard communications protocols for serial and sockets-based communication, dial-up networking, TCP/IP, and infrared.

Language systems	The Java runtime environment.
Symbian OS Connect	Communications protocols to connect to a PC, and services such as file format conversion, data synchronization for contacts, schedule entries and e-mail, clipboard synchronization, printing to a PC-based printer.

The table below shows the main C++/C APIs that we will be covering in the book. It introduces the issue of naming conventions for Symbian OS APIs I've used.

- A friendly title, which I'll normally use in the book, unless I need to be more precise.

- The DLL name: add .dll to this for the DLL name to use at runtime and add .lib for the import library that you must specify in your .mmp file at build time.

- The top-level project name in the source tree: this is the main system used internally by Symbian to refer to APIs and often corresponds to the DLL name – though not always, since some projects produce more than one DLL, while others produce none at all. In any case, only very few components are shipped with source in the developer SDKs.

Throughout the book I'll use sometimes one form, sometimes another. You can always tell which form I'm using because the presentation is different for each form. For good measure, I've included a category for each API (base, middleware etc.).

I've also included the header file naming convention. You'll find the corresponding header file(s) in \Symbian OS32\include\, with the names indicated. In the case of the Symbian OS C standard library, the header files are isolated into their own directory – \Symbian OS32\include\libc\.

Title	DLL	Source	Group	Headers	Description
User library	euser	E32	Base	e32def.h, e32std.h, e32base.h, e32*.h	Utility and kernel-object APIs. See Chapter 5 for strings and descriptors. Chapter 6 for resource cleanup. Chapter 17 for active objects.

Title	DLL	Source	Group	Headers	Description
					Chapter 18 for the principles behind client-server framework. Chapter 19 for an example server.
File server	efsrv	F32	Base	f32file.h	File and device management. See Chapter 13.
GDI	gdi	GDI	Middleware	gdi.h	Abstract graphical device interface. See Chapter 11 for intro to drawing. Chapter 15 for other facilities with the emphasis on device independence.
Window server	ws32	WSERV	Middleware	w32std.h	Shares screen, keyboard and pointer between all applications. See Chapter 11 and Chapter 12 for details, along with coverage of CONE.
CONE	cone	CONE	Middleware	coe*.h	Control environment: works with window server to enable applications to use controls.
Stream store	estor	STORE	Middleware	s32*.h	Stream and store framework and main implementations. See Chapter 13.
C standard library	estlib	STDLIB	Middleware	libc*.h	Provides functions found in POSIX-compliant C programming environments, mostly as thin layers over base and sockets server APIs. See Chapter 8.

Resource files	bafl	BAFL	Middleware	ba*.h	Once grandly titled 'basic application framework library', its most useful aspect is resource files, though it also contains other APIs. See Chapter 4 and Chapter 7.
Application architecture	apparc	APPARC	Middleware	apa*.h	Governs file formats and application launching. Briefly mentioned in Chapters 4 and 13.
Qikon + Uikon	qik*uik*	QIKON, UIKON and others	UI	qik*.h eik*.h	The system GUI. Qikon provides the UIQ-specific layer over Uikon. See Chapters 4, 9, 15 and 16 for main write-up, and information throughout the book.
Sockets server	esock	ESOCK	Comms	es_*.h	Sockets-based comms using protocols such as TCP/IP, infrared and others. See Chapter 20.
Telephony server	etel	ETEL	Comms	etel*.h	Voice, data, address book etc. on landline or mobile phones and modems. See Chapter 20.

Some of the components of Symbian OS v7.0 are parts of the corresponding API – for example, client-server and so on are provided by the user library, and zooming is provided by the GDI. Others aren't C++ or C APIs – for example, the emulator.

I'll give you a good head start on the main APIs in this book. The SDK contains much additional valuable information on the APIs I do cover and those I don't.

2.18 Summary

In this chapter, we've surveyed the component parts of the Symbian OS system. We've seen

- the impact of hardware on the design of Symbian OS and applications
- the four system component types – kernel, applications, servers, and engines
- privilege, process, and DLL boundaries
- the difference between a .exe, a .dll, and .prn, .prt and .app files
- using DLLs for Symbian OS applications allows multiple applications to run in one process for embedded documents
- DLL optimizations for Symbian OS – link-by-ordinal, address sharing, and lack of writable static
- launching applications and servers using .exes
- kernel and device driver overviews
- how Symbian OS handles memory
- event handling is at the core of Symbian OS – using a combination of active objects and a client/server framework to allow non-preemptive multitasking
- communications between client and server through messages and interthread read and write
- the main C/C++ APIs.

3

C++ and Object Orientation

In the previous two chapters, we've built a simple program, and looked at the architecture of Symbian OS. In the next chapter, we'll be taking our first look at writing GUI applications. Before we do that, it's a good idea to take a look at how Symbian OS approaches aspects of design and programming in C++. That's the purpose of this chapter.

The use of C++ in Symbian OS is not exactly the same as C++ in other environments:

- C++ does more than Symbian OS requires – private inheritance, for instance, and full-blown multiple inheritance.
- C++ does less than Symbian OS requires – it doesn't insist on the number of bits used to represent basic types, and it doesn't know anything about DLLs.
- Different C++ communities do things differently because their requirements are different. In Symbian OS, large-scale system design is combined with a focus on error handling and cleanup, and efficiency in terms of ROM and RAM budgets.

The fundamental design decisions of Symbian OS were taken in 1994–1995, and its toolchain for emulator and ARM builds was essentially stable by early 1996. Since then the compilers have changed, but the coding style still bears the hallmarks of the early GCC and Microsoft Visual C++ compilers.

3.1 Fundamental Types

Let's start with the basic types. `e32def.h` (in `\epoc32\include`) contains definitions for 8-, 16-, and 32-bit integers and some other basic types that map onto underlying C++ types such as `unsigned int`, and which are guaranteed to be the same regardless of C++ implementation:

Symbian OS C++ for Mobile Phones.   Edited by Richard Harrison
© 2003 John Wiley & Sons, Ltd   ISBN: 0-470-85611-4

	Related Types		**Description**
TInt8	TUint8		Signed and unsigned 8-bit integers
TInt16	TUint16		Signed and unsigned 16-bit integers
TInt32	TUint32		Signed and unsigned 32-bit integers
TInt	TUint		Signed and unsigned integers: in practice, this means a 32-bit integer
TReal32	TReal64	TReal	Single- and double-precision IEEE 754 floating-point numbers (equated to float and double). TReal is equated to TReal64
TText8	TText16		Narrow and wide characters (equated to unsigned char and unsigned short int)
TBool			Boolean – actually equated to int due to the early compilers used. Some code depends on this so it has not been changed with the new compilers
TAny			Equated to void, and usually used as TAny* (a 'pointer to anything')

For integers, use TInt unless you have a good reason not to. Use unsigned integer types only for flags, or if you know exactly what you're doing with unsigned types in C++. Use specific integer widths when exchanging with external formats, or when space optimization is paramount.

A TInt64 is also available. It's a class defined in e32std.h, rather than a typedef. There is no TUint64.

Symbian OS is designed for little-endian CPU architectures and will probably never be ported to an exclusively big-endian architecture.

Don't use floating point unless you have to. Machines running Symbian OS are very unlikely to include hardware floating-point units, so floating-point operations will be much slower than integer operations. Most routine calculations in Symbian OS GUI or communications programs can be done using integers. If you're using floating point, use TReal for routine scientific calculations: conventional wisdom has it that TReal32 isn't precise enough for serious use. Use TReal32 when speed is of the essence and when you know that it's sufficiently precise for your problem domain.

Use TBool to specify a Boolean return value from a function, rather than TInt. This conveys more information to anyone trying to read your code.

To represent Boolean values, don't use the TRUE and FALSE constants that are defined for historical reasons in e32def.h; rather, use ETrue and EFalse defined in e32std.h. Be aware, though, that ETrue is mapped to 1, but C++ interprets any integral value as 'true' if it is nonzero, so never compare a value with ETrue:

```
TBool b = something();
if(b == ETrue)              // Bad!
```

Instead, just rely on C++'s interpretation of Booleans:

```
if(b) { ... };
```

Always use the Symbian OS typedefs, rather than native C++ types, to preserve compiler independence. The one exception to this rule is related to C++ void, which can mean either 'nothing' (as in void Foo()) or 'anything at all' (as in void* p). We use void for the 'nothing' case:

```
void Foo();                 // Returns no result
```

And TAny* for the 'pointer to anything' case:

```
TAny* p;                    // A pointer to anything
```

Fundamental types also include characters and text. We'll cover them in Chapter 5, along with descriptors, which are the Symbian OS version of strings.

3.2   Naming Conventions

Like any system, Symbian OS uses naming conventions to indicate what is
important. The Software Development Kit (SDK) and Symbian OS source
code adhere to these conventions. Naming conventions are funny: people
tend either to love or hate them. Either way, I hope you'll find that the
established naming conventions make understanding Symbian OS code
much easier, and that, as these things go, they're not too burdensome.

The fundamental rule is, *use names to convey meaning*. Don't abbreviate too much (use real English), but don't make names too long and
unwieldy. Another basic rule is that *application programming interfaces
(APIs) use American English spelling*. American English is the international language of APIs, so expect to see `Color`, `Center`, `Gray` and
`Synchronize` rather than `Colour`, `Centre`, `Grey` and `Synchronise`.

3.2.1   Class Names

Classes use an initial letter to indicate the basic properties of the class.
The main ones are as follows:

Category	Examples	Description
T classes, types	`TDesC`, `TPoint`, `TFileName`	`T` classes don't have a destructor. They act like built-in types. That's why the `typedef`s for all built-in types begin with `T`. `T` classes can be allocated as automatics (if they're not too big), as members of other classes, or on the heap.
C classes	`CConsoleBase`, `CActive`, `CBase`	Any class derived from `CBase`. `C` classes are *always* allocated on the default heap. `CBase`'s operator `new()` initializes all member data to zero when an object is allocated. `CBase` also includes a virtual destructor, so that by calling `delete` on a `CBase*` pointer any `C` object it points to is properly destroyed.

R classes	`RFile, RTimer, RWriteStream, RWindow`	Any class that owns resources other than on the default heap. Usually allocated as member variables or automatics; in a few cases, can be allocated on the default heap. Most `R` classes use `Close()` to free their associated resources.
M classes, interfaces	`MGraphicsDevice-Map, MGameViewCmd-Handler, MEikMenuObserver`	An interface consisting of pure virtual functions and with no member data. A class implementing this interface should derive from it. `M` classes are the only approved use of multiple inheritance in Symbian OS; they act similarly to `interfaces` in Java. The old technical term was 'mixin', hence the use of `M`.
Static classes	`User, Math, Mem, ConeUtils`	A class consisting purely of `static` functions that can't be instantiated into an object. Such classes are useful containers of library functions.
Structs	`SEikControlInfo`	A C-style `struct`, without any member functions. There are only a few of these in Symbian OS; most later code uses `T` classes even for `structs`.

Some other prefixes are occasionally used for classes, in rare circumstances. The only one we'll encounter in this book is `HBufC`, for heap-based descriptors. Kernel-side programming uses `D` for kernel-side `CBase`-derived classes.

The distinction between `T`, `C`, and `R` is very important in relation to cleanup properties, which I'll cover in detail in Chapter 6.

Lastly, always ensure that class names are nouns: classes are for objects, not actions.

`CBase`-derived classes should not be allocated on the stack as automatics. The zero initialization will not work, so the class may not behave as expected. These classes are designed exclusively to be used on the heap; their behavior is undefined if used in a stack context. There are also situations in Symbian OS in which the destructor of an object may not be called. The methods used to resolve all this assume that the object is heap-based. To prevent this situation, most CBase classes have private constructors that are called from standard static functions (generally called `NewL` or `NewLC`).

3.2.2   Data Names

These also use an initial letter, excepting automatics.

Category	Examples	Description
Enumerated constant	EMonday, ESolidBrush	Constants in an enumeration. If it has a name at all, the enumeration itself should have a T prefix, so that EMonday is a member of TDayOfWeek. When we cover resource files, we'll also find some #defined constants use an E prefix in circumstances in which the constants belong to a logically distinct set.
Constant	KMaxFileName, KRgbWhite	Constants of the #define type or const TInt type. KMax-type constants tend to be associated with length or size limits: KMaxFileName, for instance, is 256 (characters).
Member variable	iDevice, iX, iOppFleetView	Any nonstatic member variable. The i prefix refers to an 'instance' of a class.

Arguments	`aDevice, aX,` `aOppFleetView`	Any variable declared as an argument. The `a` stands for 'argument', not the English indefinite article. Don't use `an` for words that begin with a vowel!
Automatics	`device, x,` `oppFleetView`	Any variable declared as an automatic.

Static members aren't used in native Symbian OS code. Global variables, such as `console`, are sometimes used in `.exes` (though not in DLLs). Globals have no prefix. Some authors use initial capitals for globals, to distinguish them from automatics. I haven't got very strong views on the right way to do things here, preferring to avoid the issue by not using globals.

The `i` convention is important for cleanup. The C++ destructor takes care of member variables, so you can spot overzealous cleanup code, such as `CleanupStack::PushL(iMember)` by using this naming convention.

As with class names, you should use nouns for value names, since they are objects, not functions.

3.2.3 Function Names

It's not the initial letter that matters so much here, as the final letter.

Category	Examples	Description
Nonleaving function	`Draw(),` `Intersects()`	Use initial capital. Since functions do things, use a verb rather than a noun.

Leaving function	`CreateL(),`
	`AllocL(),`
	`NewL(),`
	`RunL()`
		Use final `L`. A leaving function may need to allocate memory, open a file, and so on – generally, to do some operation that might fail because there are insufficient resources or for other environment-related conditions (not programmer errors). When you call a leaving function, you must

		always consider what happens both when it succeeds and when it leaves. You must ensure that both cases are handled. Symbian OS's cleanup framework is designed to allow you to do this. This is Symbian OS's most important naming convention.
LC functions	`AllocLC()`, `CreateLC()`, `OpenLC()` `NewLC()`	Allocate an object, and push it to the cleanup stack. If the function fails (which it might, since it involves allocation) then leave.
Simple getter	`Size()`, `Device()`, `Component-Control()`	Get some property or member data of an object. Often getters are used when the member is private. Use a noun, corresponding with the member name.
Complex getter	`GetTextL()`	Get some property that requires more work, and perhaps even resource allocation. Resource-allocating getters should certainly use `Get` as a prefix; other than that, the boundary between simple and complex getters is not hard-and-fast.
Setter	`SetSize()`, `SetDevice()`, `SetCommand-Handler()`, `SetChar-FormatL()`	Set some property. Some setters simply set a member. Some involve resource allocation, which may fail, and are therefore also `L` functions.

Leaving functions will be described in detail later, but they're basically a lighter-weight version of C++ exceptions. They provide a way to unwind a call stack to a known good state, and with the help of the cleanup stack, cleaning up allocations along the way.

3.2.4 Macro Names

Symbian OS uses the usual conventions for C preprocessor macro names:

- Use only upper case and split words with underscores, creating names such as `IMPORT_C` and `EXPORT_C`.

- For build-dependent symbols, use two leading and trailing under-scores (`__SYMBIAN32__`, `__WINS__`). The symbols `_DEBUG` and `_UNICODE` are notable exceptions to this rule. Double underscores are also used to indicate guards.

3.2.5 Layout

It's not a naming issue, but all Symbian OS code also uses a common layout convention. Rather than explain it, it's used throughout the book. Whatever layout convention you use for code in other environments, you'll find your Symbian OS code is easier to share if you use the Symbian OS convention.

3.2.6 Summary

Naming conventions are to some extent arbitrary: the only good thing you can say about most conventions is that life is better if everybody does the same thing.

But you'll probably have noticed that Symbian OS naming conventions do address one particular issue: cleanup. That's a key topic, which will be covered in detail in Chapter 6. The distinction between C and T is fundamental to cleanup. R classes combine aspects of both C and T. The i prefix for members makes a fundamental distinction that is also cleanup-related. And the suffix L on leaving functions indicates functions that may require cleanup.

In other areas, Symbian OS naming conventions are neither worse nor better than anyone else's. All Symbian OS system and example code uses these conventions (with perhaps a couple of exceptions in the very oldest code), so your work will be made easier if you use them too.

3.3 Functions

Function prototypes in C++ header files can convey a lot of information including

- whether it is imported from a DLL (indicated by `IMPORT_C`), `inline` and expanded from a header, or neither of these – that is, the function is private to a DLL;

- whether it is `public`, `protected`, or `private` in the C++ sense (you have to scan up the file to see this, but it's effectively part of the prototype even so);
- whether it is virtual (you have to scan down the base classes to be sure about this, but it's part of the signature) – and, if virtual, whether it's pure virtual;
- whether it is `static`;
- the return type (or `void`);
- the name – usually a good hint at what the function does;
- whether it can leave (`L` at the end of the name);
- the type and method of passing for all the arguments (with an optional name that hints at purpose, though the name is not formally part of the signature);
- whether there are any optional arguments;
- whether it is `const`.

If a function and its arguments (and class) have been named sensibly and if the right type of parameter passing has been used, you can often guess what a function does just by looking at its prototype. For example, the function `TInt RFile::Write(const TDesC8& aBuffer)` is the basic function for writing data to a file – you can even guess that `TDesC8` is a type suitable for data buffers by looking at this signature. The `TInt` return is an error code, while the `aBuffer` parameter is a descriptor containing the data and is not modified by the function.

Most of this is standard C++ fare. The exceptions are leaving functions, which we've already mentioned (and will explain fully in Chapter 6), and the naming conventions associated with DLLs. These are very important and aren't covered by C++ standards: I'll cover the significance of `IMPORT_C` later in the chapter.

3.3.1 Function Parameters

Each parameter's declaration gives valuable information about whether that parameter is to be used for input or output and a clue about whether the parameter is large or small. If a parameter is of basic type `X`, there are five possibilities for specifying it in a signature:

	By Value	By & Reference	By * Reference
Input	X	const X&	const X*
Output		X&	X*

For 'input' parameters, there is a fundamental distinction between passing by value and passing by reference. When you pass by value, C++ copies the object into a new stack location before calling the function. You should pass by value only if you know the object is small – a built-in type, say, or something that will be shorter than two machine words (64 bits). If you pass by reference, only a 32-bit pointer is passed, regardless of the size of the data.

If you pass by reference, you have to choose between * and &. Usually & is better. Use * if you have to, especially where a null value is possible, or you're transferring ownership of the object. It's more usual to pass C types with *, and R and T types directly or with &.

You have to use & for C++ copy constructors and assignment operators, but it's rare to need to code such things in Symbian OS. Some Symbian OS APIs use & for C types to indicate that a null value is not acceptable.

3.4   APIs

If you have a component X, then its API allows you to use X. In addition, X's API lets you allow X to use you. In the old days, components were simple libraries. X would specify **library functions** and you would call them to get X to do what you wanted.

Event-driven GUI systems are often associated with **frameworks**, which call your code to allow you to do things supported by the framework. For this, X specifies **framework functions** and you implement them.

For a while, framework functions were called **callbacks**: the basic theory was that your code was really in control, but the library needed to call you back occasionally so you could complete a function for it. But the truth these days is that the framework is essentially in control and it lets you do things. The framework functions are actually the main functions that allow you to do anything. 'Callback' is quite inappropriate for this. The word is not used for Symbian OS's major frameworks: only for a couple of situations in which the old callback scenario really applies or in relation to a couple of the oldest classes in Symbian OS.

We can loosely classify a class or even an entire API as either a **library API**, or a **framework API**. A library mainly contains functions that you call, while a framework consists mainly of functions that call you. Many APIs contain a good mixture of both: the GUI, for instance, calls you so that you can handle events, but provides functions that you call to draw graphics.

3.4.1   Types of Function

I've defined library functions and framework functions. But, throughout the book I use other terms to describe the role of different types of function.

Of course, there's the **C++ constructor**. I almost always use the full term, including 'C++', because, as we'll see in Chapter 6, there's also a **second-phase constructor**, usually called `ConstructL()`, in many classes. There's only one **destructor**, though, so in that context I don't usually feel the need to say 'C++ destructor'.

Convenience functions are trivial wrappers for things that could otherwise be done with a smaller API. If a class contains two functions `Foo()` and `Bar()`, which do all that's required by the class, but you often find that code using your API contains sequences such as

```
x.Foo();
x.Bar();
```

or this

```
x.Foo(x.Bar());
y = (x.Foo() + x.Bar()) / 2;
```

then you may wish to code some kind of convenience function `FooAnd-Bar()` that represents the sequence. This will reduce code size, reduce mistakes, and make code easier to read.

The cost of convenience is another function to design and document, and the risk of being tempted to produce many convenience functions that aren't really all that necessary or even convenient – and then being forced to maintain them for ever more, because people depend on them. This is a fine judgment call: sometimes we provide too few convenience functions and sometimes too many.

3.4.2 DLLs and Other API Elements

An object-oriented system delivers APIs mainly as C++ **classes**, together with all their member functions and data. Classes that form part of an API are declared in header files, implemented in C++ source files, and delivered in DLLs.

Library APIs (or the library parts of a framework API) are delivered in shared library DLLs with a `.dll` file extension. The DLL's exported functions are made available in a `.lib` file to the linker at program build time.

Framework APIs are usually defined in terms of C++ classes containing virtual functions and an interface specification for a polymorphic DLL with an extension other than `.dll` (for instance, `.app` for a GUI-based application).

It's important to make sure that only the *interface* – not the implementation – is made available to programs that use the API. Classes that are not part of the API should not be declared in API header files, and their functions should not be exported from the DLLs that implement them. Functions and data that belong to the API classes, but are not part of the API, should be marked `private`.

Besides classes, C++ APIs may contain enumerations, constants, template functions, and even nonmember functions.

3.4.3 Exported Functions

For a nonvirtual, noninline member function to be part of an API, it must be

- declared public in a C++ class that appears in a public header file
- exported from its DLL.

You will see exported functions marked in their header files with `IMPORT_C`, like this

```
class RTimer : public RHandleBase
    {
public:
    IMPORT_C TInt CreateLocal();
    IMPORT_C void Cancel();
    IMPORT_C void After(TRequestStatus& aStatus,
                        TTimeIntervalMicroSeconds32 anInterval);
    IMPORT_C void At(TRequestStatus& aStatus,
                     const TTime& aTime);
    IMPORT_C void Lock(TRequestStatus& aStatus,
                       TTimerLockSpec aLock);
    };
```

The `IMPORT_C` macro says that the function must be imported from a DLL by the user of that API. In the corresponding implementation, the function will be marked `EXPORT_C`, which means that it will be exported from the DLL. A function without `IMPORT_C` is not exported from its DLL and cannot, therefore, be part of the public API.

These macros are defined in `e32def.h`. Their implementations are compiler-dependent and differ between CodeWarrior and GCC.

Virtual and inline functions don't need to be exported – they form a part of the API, even without `IMPORT_C` in the header file.

If you're writing an API to be delivered in a DLL for use by other DLLs, you'll need to mark your `IMPORT_C`s and `EXPORT_C`s carefully.

If you're not writing APIs – or you're not encapsulating them in DLLs for export – then you needn't worry about how to use `IMPORT_C` and `EXPORT_C`. It's enough to understand what they mean in Symbian OS SDK headers.

3.4.4 Virtual Functions and APIs

C++ isn't well designed for API delivery. There is no way to prevent further override of a specific virtual function, and there is no way to guarantee that a function is not virtual without looking down all the base classes to check for the `virtual` keyword. You can simulate Java's `final` at the class level though, by making the constructor private.

C++'s access control specifiers aren't good for API delivery either. The meaning of `public` is clear enough, but `protected` makes a distinction between derived classes and other classes that doesn't put the boundary in the right place, since derivation is by no means the most important vehicle for code reuse in OO. `private` is not private when it comes to virtual functions: you can override private virtual functions whether or not this was intended by the designer of an intermediate class (derived from a base class).

C++ has no language support for packaging APIs except classes and header files. So Symbian had to invent its own rules for DLLs.

These design issues are most awkward when it comes to virtual functions. Best practice in Symbian OS C++ includes the following guidelines:

- Declare a function virtual in the base class and in any derived class from which it is intended to further derive and override (or implement) this function.
- When declaring a virtual function in a derived class, include a comment such as `// from CCoeControl`, to indicate where the function is defined.
- Use `private` in a base class to indicate that your base class (or its friends) calls this function – this is usually the case for framework functions. If you don't like friends or the framework function is designed to be called from another class, then make it `public` in the base class.
- Use `private` in a derived class for a framework function that is implementing something in a framework base class.

These guidelines are admittedly incomplete, and they're not always honored in Symbian OS code. But they're good for most cases.

Finally, there's another issue with virtual functions: if your class has virtual functions and you need to invoke the default C++ constructor from a DLL other than the one your class is delivered in, then you need to specify, and export, a default C++ constructor:

```
class CFoo : public CBase
    {
public:
    IMPORT_C CFoo();
    ...
    };
```

And then in the source code:

```
EXPORT_C CFoo::CFoo()
    {
    }
```

If you don't do this, a program that tries to create a default C++ constructor for your class won't be able to, because constructors need to create the virtual function table and the information required is all inside your DLL. You'll get a link error.

3.5 Templates

Symbian OS uses C++ templates extensively for collection classes, fixed-length buffers, and utility functions. Symbian OS use of templates is optimized to minimize the size in 'expanded' template code – basically, by ensuring that templates never get expanded at all. The **thin template** pattern is the key to this.

Symbian OS also uses numeric arguments in templates to indicate string and buffer sizes.

3.5.1 The Thin Template Pattern

The thin template pattern uses templates to provide a type-safe wrapper around type-unsafe code. It works like this: code a generic base class, such as `CArrayFixBase` that deals in 'unsafe' `TAny*` objects. This class is expanded into real code that goes in a DLL. Then, code a template class that derives from this one, and uses inline type-safe functions such as the following:

```
template <class T>
inline const T& CArrayFix<T>::operator[](TInt anIndex) const
    {
    return (*((const T*)CArrayFixBase::At(anIndex)));
    }
```

This returns the `anIndexth` item of type `const T&` in the `CArrayFix<T>` on which it is invoked. It acts as a type-safe wrapper around `At()` in the base class, which returns the `anIndexth` pointer of type `TAny*`.

This code looks pretty ugly, but the good news is that application programmers don't have to use it. They can simply use the template API:

```
CArrayFix<TFoo>* fooArray;
...
TFoo foo = (*fooArray)[4];
```

The template guarantees that this code is type-safe. The fact that the `operator[]()` is expanded inline means that no more code is generated when the template is used than if the type-unsafe base class had been used.

3.5.2   Numbers in Templates

Sometimes, the parameter to a template class is a number rather than a type. The declaration of `TBuf`, a buffer of variable length is as follows:

```
template <TInt S> class TBuf : public TDes
    {
    ...
    };
```

You can then create a five-character buffer with:

```
TBuf<5> hello;
```

This uses the thin template pattern too: The inline constructor is as follows:

```
template <TInt S>
inline TBuf<S>::TBuf() : TDes(0,S)
    {
    }
```

It calls the `TDes` base class constructor, passing the right parameters, and then completes the default construction of a `TBuf` (a couple of extra instructions).

3.6   Casting

Casting is a necessary evil. Old-style C provides casting syntax that enables you to cast anything to anything. Over time, different casting patterns have emerged, including

- cast away `const`-ness (but don't change anything else)
- cast to a related class (rather than an arbitrary cast)
- reinterpret the bit pattern (effectively, old-style C casting).

C++ provides individual casting keywords for each of these types of cast: `const_cast<>()`, `static_cast<>()` and `reinterpret-cast<>()`. These should be used in preference to old style C casting, as by doing this you get the benefit of C++ cast checking. Previous versions of the SDK defined macros that expanded to these keywords, for compatibility with older versions of GCC that did not support C++ casting. These are now deprecated – current compilers can use the standard C++ definitions.

3.7 Classes

As you'd expect, classes are used to represent objects, abstractions, and interfaces. Relationships between classes are used to represent relationships between objects or abstractions. The most important relationships between classes are the following:

- *uses-a*: if class A *uses-a* class B, then A has a member of type B, B&, const B&, B*, or const B*, or a function that can easily return a B in one of these guises. A can then use B's member functions and data.
- *has-a*: *has-a* is like *uses-a*, except that A takes responsibility for constructing and destroying the B as well as using it during its lifetime.
- *is-a*: if class A *is-a* class B, then B should be an abstraction of A. *is-a* relationships are usually represented in C++ using public derivation.
- *implements*: if class A implements an **interface** M, then it implements all M's pure virtual functions. Interface implementation is the only time multiple inheritance is used in Symbian OS.

Sometimes, abstract/concrete notions are blurred. CEikDialog is concrete as far as its implementation of CCoeControl is concerned, but in fact CEikDialog is an essentially abstract base class for user-specified dialogs or Uikon standard dialogs such as a CEikInfoDialog. Some classes (such as both CCoeControl and CEikDialog) contain no pure virtual functions, but are still intended to be derived from. They provide fall-back functionality for general cases, but it is likely that one or two functions will need to be overridden to provide a useful class.

3.7.1 Interfaces

Symbian OS makes quite extensive use of interface classes (originally called mixins). An interface is an abstract base class with no data and *only* pure virtual functions.

APIs that have both library and framework aspects often define their library aspect by means of a concrete class and their framework by means of an interface class. To use such an API, you need to use the concrete class and implement the interface.

The Symbian OS PRINT API provides an example. In addition to library classes to start the print job, there are framework classes for printing part of a page, and for notifying the progress of the job to an application. MPrintProcessObserver is the interface for notifying progress:

```
class MPrintProcessObserver
    {
public:
    virtual void NotifyPrintStarted(
                TPrintParameters aPrintParams) = 0;

    virtual void NotifyBandPrinted(
                TInt aPercentageOfPagePrinted,
                TInt aCurrentPageNum,
                TInt aCurrentCopyNum) = 0;
    virtual void NotifyPrintEnded(TInt aErrorCode) = 0;
    };
```

This interface definition includes functions for reporting the beginning and end of a print job and its progress at intervals throughout. The print preview image is designed to be owned by a control in a print preview dialog. It provides a standard print preview image and its definition starts as follows:

```
class CPrintPreviewImage : public CBase,
                    private MPrintProcessObserver,
                    private MPageRegionPrinter
    {
    ...
```

You can see that this uses C++ multiple inheritance, in this case inheriting from CBase and *two* interface classes.

Another example, which is more relevant for UIQ applications, appears in the view architecture, whose use is described in Chapter 9. The MCoeView class provides support for identifying the view and for view activation and deactivation:

```
class MCoeView
    {
public:
    virtual TVwsViewId ViewId() const=0;
private:
    virtual void ViewActivatedL(const TVwsViewId& aPrevViewId,
                Tuid aCustomMessageId,const TDesC8& aCustomMessage)=0;
    virtual void ViewDeactivated()=0;
    ...
```

It appears in the definition of an application's views, which starts as follows:

```
class CFleetView : public CCoeControl,
                public MCoeView
    {
    ...
```

You can read this as meaning 'CFleetView *is-a* control and it *implements* the view architecture interface'.

Incidentally, when using multiple inheritance in this manner, place the CBase (or CBase-derived) class first in the list or some subtle OS level details will cause problems.

> **The only encouraged use of multiple inheritance in Symbian OS is to implement interfaces. Any other use is unnecessary, and is strongly discouraged. Standard classes will not have been designed with multiple inheritance in mind.**

Java uses interfaces too: in Java, a class can explicitly have zero or one base class, and can implement zero, one, or more interfaces. If no base class is specified, an implicit relationship to the Object class is assumed. Java has `interface` and `implements` keywords, rather than a naming convention. Multiple inheritance isn't part of Java.

3.7.2   Bad Practices

Many C++ features that look attractive at first sight are not used in Symbian OS – or, at least, they're not encouraged in anything other than very specific situations:

- *Private inheritance*: Inheritance should only be used for *is-a* relationships. Private inheritance (the default in C++) is used to mean *has-a*, so that the private base class effectively becomes a private data member.

- *Multiple inheritance*: Except in the case of interfaces, full-blown C++ multiple inheritance is more confusing than useful.

- *Overriding nontrivial virtual functions*: Base classes with virtual functions in them should either specify trivial behavior (doing nothing, for example), or leave them purely virtual. This helps you to be clear about the purpose of the virtual function.

- *'Just-in-case' tactics*: Making functions virtual 'just in case' they should be overridden or protected 'just in case' a derived class wishes to use them is an excuse for unclear thinking.

There may be times when these practices can be used for good reason. The thin template pattern is really a C++ technical trick, so it's fair game to use C++ technical tricks such as private inheritance to help implement it. But if you want your C++ to be a straightforward implementation of good object-oriented system design, you should use the object-oriented features of C++ rather than murky technical tricks.

3.8   Design Patterns

Object orientation supports good design, using the *uses-a, has-a, is-a,* and *implements* relationships.

Through good design, object orientation also supports good code reuse. That's particularly attractive for Symbian OS, since minimizing the amount of code you require to implement a particular system is a very important design goal.

But code reuse isn't the only form of reuse. Often, you find yourself doing the same thing again and again, but somehow you can't abstract it into an API – even using the full power of object orientation and templates in C++. Or, you succeed in abstracting an API, but it's more difficult to use the API than it is to write the repeated code in the first place.

This is a good time to think in terms of reusing **design patterns**. Design patterns are ways of designing things rather than objects or APIs that you can reuse or glue together.

Symbian OS contains many frequently used design patterns. The most unique Symbian OS design patterns relate to cleanup (see Chapter 6) and active objects (see Chapter 17). However, most of the patterns used in Symbian OS are standard patterns used elsewhere in the software industry.

3.9   Class Diagrams and UML

Object orientation is more about relationships than anything else. When you are working on understanding a software system you don't know, the key questions to ask are the following:

- What are the main classes – objects, abstractions, and interfaces?
- What are the intended relationships between them?

Only then is it worth asking anything about the functions and data members in individual classes.

Header files are very good at telling you what the functions and data members are, but pretty hopeless at telling you the relationships between classes (though you can sometimes work them out with a little detective work). In other words, header files, even for well-designed APIs, aren't the place to get the broad picture.

UML is a visual notation that shows classes and their relationships very clearly. It's used to illustrate the structure of many applications and

Symbian OS APIs throughout the book. You can get good information on UML from the Rational website at ***http://www.rational.com***.

You can use UML to convey detailed design information for the system visually (instead of, say, through header files). But that's not the purpose in this book. The UML diagrams in this book are not intended to allow the construction of a header file, but rather to convey the main features of a system design, for the purposes of the current explanation. Irrelevant functions may be omitted, as may private and virtual specifiers if they can be guessed. The principle behind the diagrams in this book is that the relationships are the most important aspects. For complete details, see the header files.

Here's a lightning tour of the main features of UML.

3.9.1   Describing APIs

UML can describe basic API relationships:

- The existence, name, and major classes in an API
- Whether and how one API uses another.

For instance, you can read the Figure 3.1 as, 'CONE uses WSERV. The main classes in CONE are CCoeEnv, CCoeControl, and CCoeAp-pUi. The main classes in WSERV are RWsSession, RWindow, and TWsEvent.'

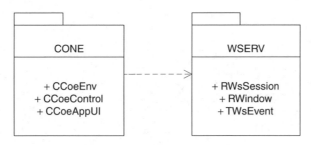

Figure 3.1

The dotted line between the packages shows that the CONE API uses the WSERV API.

3.9.2   Describing Classes

UML can describe classes and their content:

- The existence, name, major data members, and function members of a class
- Whether these are private or public (if that's interesting).

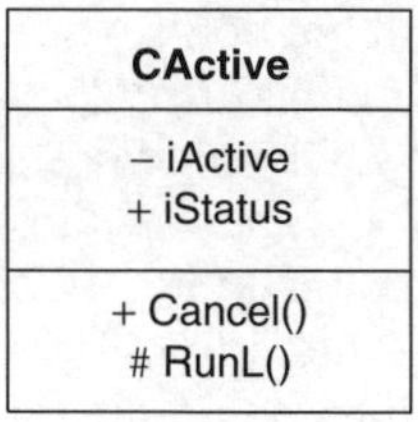

CActive
− iActive + iStatus
+ Cancel() # RunL()

Figure 3.2

For instance, you can read the Figure 3.2 as, 'CActive has public member iStatus and private member iActive, a public Cancel() function, and a protected RunL() function.'

3.9.3   Describing Relationships between Classes

UML can describe the relationships between classes as

- *has-a*: this uses a solid diamond next to the class
- *uses-a*: this uses an open diamond
- Whether the class on the other end of a *has-a* or *uses-a* relationship is intended to know anything about the class that has or uses it: if it's important to show that the class *doesn't* know anything, then an arrow is used to emphasize that the relationship is one-way.

Figure 3.3 shows a few of these ideas:

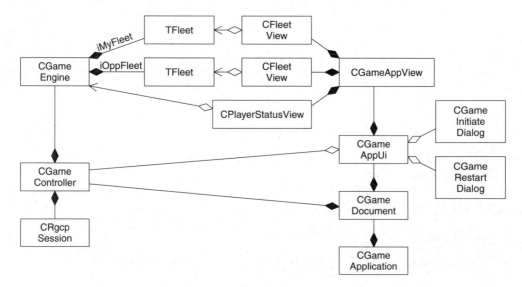

Figure 3.3

You can do as follows:

- Trace *has-a* relationships: the application *has-a* document, which *has-a* application user interface, which *has-a* application view. There are plenty of other *has-a* relationships in the rest of the diagram. Even if you don't know what the names mean yet, the relationships all sound very reasonable, and based on the names and relationships, you can make a guess at their purpose that probably won't be too wrong. I'll be explaining this diagram in detail in Chapter 9, by which time it'll be easier to understand.

- Trace major *uses-a* relationships, indicated by open diamonds. There may be other *uses-a* relationships besides those shown, but the ones we see highlighted are that the application UI uses the controller and a couple of dialogs and that the various view classes use the various engine classes.

- Note when relationships are strictly one-way: I've used one-way arrows for the view-engine relationships to emphasize the point that the engine doesn't know anything about the views.

3.9.4   Describing Derivation

UML can convey derivation relationships:

- *is-a*: this is indicated by an open arrowhead
- *implements*: UML provides a dotted line with an arrowhead for this purpose, You can also tell it's an *implements* relationship because the 'base' class name begins with M.

For instance, you can read Figure 3.4 as, 'The GSDP session *has-a* receive handler, which *is-a* active object. The receive handler *uses-a* GSDP packet handler interface. The client *has-a* GSDP session, and implements the GSDP handler interface.'

Incidentally, this is a standard version of the M class pattern I described earlier: the GSDP API provides a library class (`RGsdpSession`) and a framework class (`MGsdpReceiveHandler`). The client program uses the one and implements the other.

3.9.5   Cardinality

UML can convey **cardinality** – the number of objects expected at both ends of a relationship.

Unless otherwise stated, 1-to-1 cardinality is implied: this means that one of the using objects has exactly one of the used objects, perhaps as a member variable or a pointer.

If the cardinality is not 1-to-1, then you need to indicate its value with UML adornments. You usually write the number of objects intended on one end of the UML connector. Common values are as follows:

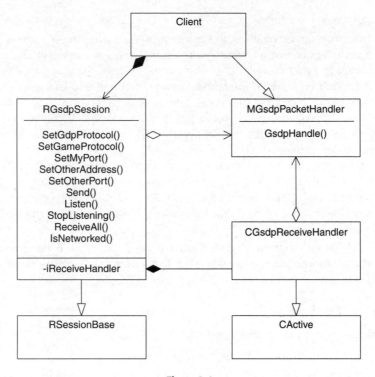

Figure 3.4

- 0..1: the object is optional – perhaps a pointer is used to refer to it

- 1..*n*, 0..*n* or simply *n*: an arbitrary number of objects is intended – a list, array, or other collection class may be used to contain them

- 10: exactly 10 objects are intended.

For instance, you can read Figure 3.5 as, 'The engine's `iMyFleet` fleet always has 10 ships. The engine's `iOppFleet` fleet may have anything from 0 to 10 ships.'

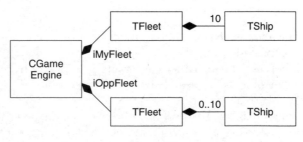

Figure 3.5

3.10 Summary

In this chapter, we've looked at the features of C++ that Symbian OS uses, those it avoids, and those it augments. We've also seen the coding standards that are used in Symbian OS. Specifically, we've seen the following:

- The use of Symbian OS fundamental types to guarantee consistent behavior between compilers and platforms.

- Naming conventions – the core ones help with cleanup. T, C, and R distinguish between different cleanup requirements for classes, i refers to member variables and L to leaving functions that may fail for reasons beyond the control of the programmer.

- M prefixes interfaces (consisting of pure virtual functions).

- Good design of function prototypes – how to code input and output parameters, and suggested use of references and pointers in Symbian OS.

- The difference between library and framework DLLs – the former provide functions that you call, the latter provide functions for you to implement that the framework will call.

- Exporting nonvirtual, noninline functions using the IMPORT_C and EXPORT_C macros.

- How to handle virtual functions.

- Symbian OS use of templates and the thin template pattern to keep code size to a minimum.

- The four relationships between classes and how to represent them in UML.

- The use of mixins to provide interfaces in the only use of multiple inheritance in Symbian OS.

4

A Simple Graphical Application

In Chapter 1, we built a text-mode version of the 'Hello World' program, and made the point that no one really wants to run text-mode programs on a user-friendly system like Symbian OS. Text-mode programming is useful for testing and for learning, but production programs use a GUI. We shall therefore now start looking at real graphics and GUI programming in Symbian OS.

After a brief introduction to the Symbian OS graphics architecture, I'll spend the body of this chapter talking about `hellogui`, the graphical 'Hello world!' program. I'll start by walking through the code, explaining how its design fits in with the Uikon framework. Then, I'll take you through a session with the debugger, which will reinforce the code design and also introduce some ideas about controls, pointer, and key event handling, which we'll return to in Chapter 12.

4.1 What's in a Name?

In Symbian OS v5, the graphics framework was known as Eikon. That version was originally intended to support both narrow (8-bit characters) and Unicode (16-bit characters) builds but, in the end, only the narrow build was supported. Version 5.1 contained the changes necessary to support Unicode builds and, from that version onwards, Unicode became the only supported build. To reflect this change, the name of the graphics framework was changed from Eikon to Uikon.

All later versions of Symbian OS support customization of the user interface (UI), depending on the features of the target machine, such as the size and aspect ratio of the screen, whether it supports keyboard and/or pen-based input and the relative significance of voice-centric or

Symbian OS C++ for Mobile Phones. Edited by Richard Harrison
© 2003 John Wiley & Sons, Ltd   ISBN: 0-470-85611-4

data-centric applications. These differences are largely implemented by the creation of additional UI layers above Uikon.

The Series 60 UI, based on Symbian OS v6.1, uses an additional layer known as Avkon to modify the behavior and/or appearance of the underlying Uikon framework. UIQ, based on Symbian OS v7.0, uses Qikon to perform a similar task. In this chapter we shall see only one or two consequences of these changes, mainly related to how a UIQ application is initialized and closed. Most of the code described in this chapter is applicable to any Symbian OS GUI application and I shall point out the cases in which it is specific to a particular UI.

4.2 Introduction to the Graphics Architecture

The most important of the Symbian OS graphics and GUI components, and their main relationships, are shown in Figure 4.1:

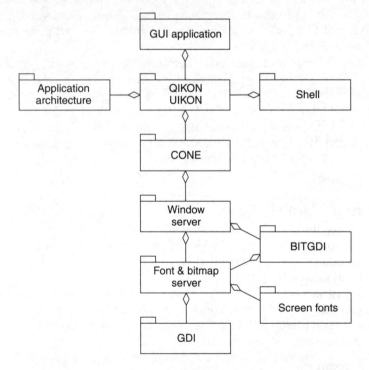

Figure 4.1

At the bottom is the GDI, which defines the drawing primitives and everything necessary to achieve device-independent drawing. The GDI is an entirely abstract component that has to be implemented in various contexts – for on-screen pixel graphics, for instance, or in a printer driver. I'll be covering the main drawing functions defined by the GDI in Chapter 11, and its support for device-independent drawing in Chapter 15.

The BITGDI handles optimized rasterizing and bit blitting for on-screen windows and off-screen bitmaps. The font and bitmap server (FBS) manages fonts and bitmaps – potentially large graphics entities – for optimal space efficiency.

Support for user interaction starts with the window server, which manages the screen, pointer or other navigation device, and any keypad or keyboard on behalf of *all* GUI programs within the system. It shares these devices according to windowing conventions that are easily understood by the average end user. A standard window is represented in the client application programming interface API by the RWindow class.

The window server is a single server process that provides a basic API for client applications to use. CONE, the control environment, runs in each application process and works with the window server's client-side API, to allow different parts of an application to share windows, key and pointer events. A fundamental abstract class delivered by CONE is CCoeControl, a **control**, which is a unit of user interaction that uses any combination of screen, keyboard, and pointer. Many controls can share a single window. Concrete control types are derived from CCoeControl.

CONE doesn't provide any concrete controls: that's the job of Uikon, the system GUI, and any UI-specific layer, such as Qikon. Together, they specify a standard look-and-feel, and provide reusable controls and other classes that implement that look-and-feel.

4.3 Application Structure

By studying the code and the execution of the GUI version of 'Hello world!', we'll begin to understand how Symbian OS GUI applications fit together, using the frameworks provided by Uikon and the application architecture (APPARC).

The source code in the three subdirectories of \scmp\hellogui\ consists of the definitions and implementations of four classes (and a resource file). You *have* to implement these four classes; anything less than this, and you don't have a Symbian OS GUI application. The classes are as follows:

- **An application**: the application class serves to define the *properties* of the application, and also to manufacture a new, blank, document. In the simplest case, as here, the only property that you have to define is the application's unique identifier, or UID.

- **A document**: a document represents the data model for the application. If the application is file-based, the document is responsible for storing and restoring the application's data. Even if the application is not file-based, it must have a document class, even though that class doesn't do much apart from creating the application user interface (app UI).

- *An app UI*: the app UI is entirely invisible. It creates an application view (app view) to handle drawing and screen-based interaction. In addition, it provides the means for processing commands that may be generated, for example, by menu items.

- *An app view*: this is, in fact, a concrete control, whose purpose is to display the application data on screen and allow you to interact with it. In the simplest case, an app view provides only the means of drawing to the screen, but most application views will also provide functions for handling input events.

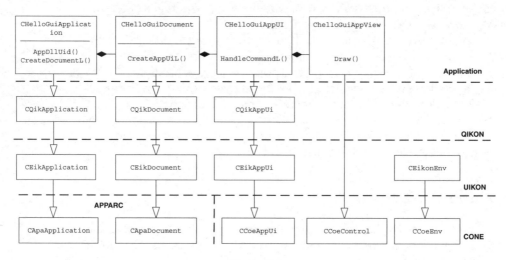

Figure 4.2

Three of the four classes in a UIQ application are derived from base classes in Qikon which themselves are derived from Uikon classes. In applications written for other target machines, these three classes may be derived directly from the Uikon classes or from an alternative layer that replaces Qikon.

In turn, Uikon is based on two important frameworks:

- CONE, the **control environment**, is the framework for graphical interaction. I'll be describing CONE in detail in Chapters 11 and 12.

- APPARC, the **application architecture**, is the framework for applications and application data, which is described in Chapter 13.

Uikon provides a library of useful functions in its environment class, CEikonEnv, which is in turn derived from CCoeEnv, the CONE environment, and I've shown these in the diagram as well. Fortunately, the application can use these functions without the need to write yet another derived class.

Remember that this is a diagram for a *minimal* application. The application is free to use any other class it needs, and it can derive and implement more classes of its own, but no GUI application can get smaller than this. More realistic UIQ applications that have more than one view, and want to take advantage of the **view architecture**, must additionally derive their view classes from CONE's MCoeView interface class. I haven't shown this relationship in the diagram because it isn't essential for a simple application with only one view. You will find more information on the view architecture in Chapter 9.

In addition to the four C++ classes, you have to define elements of your application's GUI – the menu, possibly some shortcut keys, and any string resources you might want to use when localizing the application – in a **resource file**. In this chapter we'll briefly describe the simple resource file used for hellogui: we'll take a closer look at resource files in Chapter 7.

> *The stated design goal of Symbian OS is to make real, potentially large, programs easier to write and understand. Compared with this aim, programming a minimal application really isn't very important. As a consequence, the document class in nonfile-based applications doesn't do much, and the idea that the app UI is there to 'edit the document' is a bit overengineered.*

Let's have a look at how the classes in hellogui implement the requirements of APPARC, Uikon, and CONE, and how the resource file is used to define the main elements of the GUI.

4.4 A Graphical Hello World

4.4.1 The Program

Our example program for this section is hellogui, a graphical version of 'Hello world!'. With the aid of this example, you'll learn how to build an application for the Uikon GUI. Since the application is intended to run on a P800 phone, I've used UIQ classes instead of Uikon ones, but I'll point out the differences as they arise.

It will take more time to learn how a GUI program works and in this chapter we'll only briefly comment on the C++ code itself. There's nothing particularly difficult about it, but it will be easier to cover how such programs use C++ when we have seen more of how Symbian OS C++ works. It's still worth getting to grips with the tools and techniques though, because we'll be using GUI programs to show off some of the basics of Symbian OS in the next few chapters.

Despite all the differences in the code, the build process is similar to that for `hellotext`: we start with a `.mmp` file, turn it into the relevant makefile or project file, open up the IDE, build the program for the emulator and check that it works. Then we rebuild for an ARM target, copy to the target machine and run it there instead. There are, however, some important differences:

- GUI programs must be built in such a way that Symbian OS can recognize them and launch them. To ensure this, a program called `hellogui` *must* be built into a path such as `\system\apps\hellogui\hellogui.app`.

- GUI programs use the Symbian OS unique identifier (UID) scheme to verify that they *are* Symbian OS GUI applications, and also to associate them, if necessary, with file types identified by the same UID.

- GUI programs consist not only of the executable `.app` file, but also GUI data in a resource file.

In this example, we'll start by looking at the project file for `hellogui.app`. We'll do a command-line build for the emulator, to generate `hellogui.app` together with its resource file, `hellogui.rsc`. Then we'll show how you can build the project from the IDE. Finally, we'll build for an ARM target and transfer both files to a target machine using Symbian OS Connect.

4.4.2 The Project Specification File

The project specification file for `hellogui` is as follows:

```
// HelloGui.mmp
//
// Copyright (c) 2002 Symbian Ltd. All rights reserved.
//


TARGET          HelloGui.app
TARGETTYPE      app
UID             0x100039ce 0x101f74a8
TARGETPATH      \system\apps\HelloGui

SOURCEPATH      ..\group
USERINCLUDE     ..\inc
RESOURCE        HelloGui.rss
```

```
SOURCEPATH          ..\src
SYSTEMINCLUDE       \epoc32\include

SOURCE              HelloGui.cpp
SOURCE              HelloGui_Application.cpp
SOURCE              HelloGui_AppUi.cpp
SOURCE              HelloGui_AppView.cpp
SOURCE              HelloGui_Document.cpp

LIBRARY             euser.lib
LIBRARY             apparc.lib
LIBRARY             cone.lib
LIBRARY             eikcore.lib
LIBRARY             qikctl.lib
```

Compared with the .mmp file for hellotext, whose listing appears in Chapter 1, there are some interesting differences here:

- The TARGET is HelloGui.app, and TARGETTYPE is app – that is, an application. The build tools know what to do to make the right kind of executable in the proper target directory.

- This time, the UIDs are nonzero. The first one you have to enter is 0x100039ce, and in fact this is the same for all GUI applications. The second should be obtained by you from Symbian- I'll show how to do that shortly.

- The TARGETPATH specifies where HelloGui.app will be generated. On the emulator, the emulated z: drive's path will be used as a prefix, so \epoc32\release\winscw\udeb\z\system\apps\ hellogui\hellogui.app will be the path used for the emulator debug build.

- As well as a number of source files, a resource file- hellogui.rss- is included in the project.

- Many more .lib files are involved this time.

In addition, as indicated by the SOURCEPATH and USERINCLUDE statements, the source files are distributed in different subdirectories of \scmp\hellogui. The C++ source files are in a\scmp\hellogui\ src directory, whereas the header files are in \scmp\hellogui\inc. A third subdirectory, \scmp\hellogui\group, contains the bld.inf component definition file, the .mmp file and the resource file, hel- logui.rss. This is probably an overkill for such a simple project but it becomes increasingly helpful for larger and more complex projects.

4.4.3 Getting a UID

Every GUI application should have its own UID. This allows Symbian OS to distinguish files associated with that application from files associated with other applications. A UID is a 32-bit number, which you get as you need from Symbian.

Microsoft uses 128-bit 'globally unique IDs', GUIDs. Programmers allocate their own, using a tool incorporating a random-number generator and distinguishing numbers (such as network card ID and current date and time) to ensure uniqueness. Symbian OS uses 32-bit UIDs for compactness, but this rules out the random-number generation approach. That's why you have to apply for UIDs from Symbian.

Getting a UID is simple enough. Just send an e-mail to uid@symbiandevnet.com, titled 'UID request', and requesting clearly how many UIDs you want – ten is a reasonable first request. Assuming your e-mail includes your name and return e-mail address, that's all the information Symbian needs. Within 24 hours, you'll have your UIDs by return e-mail.

If you're impatient, or you want to do some experimentation before using real UIDs, you can allocate your own UIDs from a range that Symbian has reserved for this purpose: 0x01000000-0x0fffffff. However, you should never release any programs with UIDs in this range.

Don't build different Symbian OS applications with the same UID – even the same test UID – on your emulator or Symbian OS machine. If you do, the system will only recognize one of them, and you won't be able to launch any of the others.

4.4.4 Building the Application

To build a debug application for the emulator from the command line, we follow the same steps as with `hellotext`; from the directory containing the `bld.inf` file, type

```
bldmake bldfiles
abld build winscw udeb
```

The output from the command-line build is broadly similar to that for `hellotext`, showing the same six basic stages:

```
make -r -f "\EPOC32\BUILD\SCMP\HELLOGUI\GROUP\EXPORT.make" EXPORT
VERBOSE=-s
```

```
Nothing to do
  make -r -f "\EPOC32\BUILD\SCMP\HELLOGUI\GROUP\WINSCW.make" MAKEFILE
VERBOSE=-s
perl -S makmake.pl -D \SCMP\HELLOGUI\GROUP\HELLOGUI WINSCW
  make -r -f "\EPOC32\BUILD\SCMP\HELLOGUI\GROUP\WINSCW.make" LIBRARY
VERBOSE=-s
make -s -r -f
"\EPOC32\BUILD\SCMP\HELLOGUI\GROUP\HELLOGUI\WINSCW\HELLOGUI.WINSCW" LIBRARY
  make -r -f "\EPOC32\BUILD\SCMP\HELLOGUI\GROUP\WINSCW.make" RESOURCE
CFG=UDEB VERBOSE=-s
make -s -r -f
"\EPOC32\BUILD\SCMP\HELLOGUI\GROUP\HELLOGUI\WINSCW\HELLOGUI.WINSCW"
RESOURCEUDEB
  make -r -f "\EPOC32\BUILD\SCMP\HELLOGUI\GROUP\WINSCW.make" TARGET
CFG=UDEB VERBOSE=-s
make -s -r -f
"\EPOC32\BUILD\SCMP\HELLOGUI\GROUP\HELLOGUI\WINSCW\HELLOGUI.WINSCW" UDEB
  make -r -f "\EPOC32\BUILD\SCMP\HELLOGUI\GROUP\WINSCW.make" FINAL
CFG=UDEB
VERBOSE=-s
```

When building has finished, the application will be in `\epoc32\release\winscw\udeb\z\system\apps\hellogui\`. In Windows Explorer, you'll see `hellogui.app` and `hellogui.rsc`, which are the real targets of the build.

This time, unlike for hellotext, you can't launch the application directly by (say) double-clicking on `hellogui.app` from Windows Explorer. Instead, you have to launch the emulator using `\epoc32\release\winscw\udeb\epoc.exe`. You'll find that `hellogui` is displayed as an application, and you can launch it by clicking on its icon. This is what it looks like.

You can see the world-famous text in the center of the screen, together with a menu that happened to be popped up when I did the screenshot.

4.4.5 Building in the CodeWarrior IDE

The main phase of real application development is writing, building, and debugging from within the Metrowerks CodeWarrior IDE – not the command line. To see how to build a GUI program from the IDE, let's start with a clean sheet. If you tried out the command-line build above, from the command line type:

```
abld reallyclean
```

which will get rid of all files from the build. (It may also report errors and issue not-found messages as it tries to get rid of files that might have been produced, but weren't.)

You can then launch the IDE and select **File | Import Project From .mmp File** to build the CodeWarrior project file. Select the appropriate

Figure 4.3

SDK and then browse to the \scmp\hellogui directory and select hel-
logui.mmp. Click on **Finish** and, after a short time, the hellogui.mcp
CodeWarrior project will have been created.

Select **Project | Set Default Target** and make sure that the **HELLOGUI
WINSCW UDEB** target is selected, then select **Project | Make** (or press
F7) to build the project.

You should then be able to start up the winscw udeb emulator and run
the application in exactly the same way as you did after building from
the command line.

4.4.6   The Source Code

From within the CodeWarrior IDE project window, check the source
files for the project, and you'll see there are seven files. The first two
are hellogui.pref, the source file generated by the IDE from the
.mmp file, containing the UID information, and hellogui.resources,
another generated file that references the source text file for the application's resources. There are also five C++ source files. The C++ source
file hellogui.cpp simply contains two nonclass functions and the

remaining four C++ files contain the source for the four classes that make up the application.

For each of the five C++ source files, there is a corresponding `.h` header file (in the `\scmp\hellogui\inc` directory). Again, for a project of this size, it is somewhat excessive to have a separate header file for each class but it makes it clearer, for demonstration purposes, which source is related to which class. Such a division becomes increasingly helpful, if not essential, for more complex projects – although you need to make a compromise between this type of clarity and the confusion that arises from having too many source files and too many header file inclusions. In a real application, you may want to use source files that contain collections of closely related classes, but that is a decision you'll have to make for yourself.

At this stage, I don't want to get too involved in the details of the code, so I'll just concentrate on the essentials of what is needed to start up an application. You'll see that the code obeys the class naming conventions described in Chapter 3 but, for the moment, you can ignore things like the precise way in which the classes are instantiated and constructed. For the sake of completeness, the descriptions include references to first-phase and second-phase construction; these terms are explained in Chapter 6 and you might want to review the descriptions given here after reading that chapter.

With that in mind, I'll run through the content of each of the source files and explain what it's doing. Then I'll sum up some of APPARC's features, and give some pointers to more information in the book.

DLL startup code

When you launch an application, the file that is launched is not your application, but one with the name `apprun.exe`. The application name and the application's file name are passed as command line parameters to `apprun.exe`, which then uses the APPARC to load the correct application DLL.

> **Every Symbian OS application is a DLL.**

After loading the DLL, the application architecture:

- checks that its second UID is `0x100039ce`,
- runs the DLL's first ordinal exported function to create an application object – that is, an object of a class derived from `CApaApplication`.

As we've already seen, the build tools support the generation of application DLLs. In order to understand those last two bullet points, let's take a closer look at the important features of a `.mmp` file. For our application, `hellogui.mmp` starts:

```
TARGET        HelloGui.app
TARGETTYPE    app
UID           0x100039ce 0x101f74a8
```

The `app` target type causes the build tools to generate instructions that force `NewApplication()` to be exported as the first ordinal function from the `hellogui.app` DLL.

All GUI applications specify a UID `0x100039ce`, which identifies them as applications, while UID `0x101f74a8` identifies my particular application, `hellogui.app`. I allocated this UID from a block of 10 that were issued to me by Symbian DevNet, as I described earlier in this chapter.

I said above that 0x100039ce was the second UID for this application. The build tools helpfully generate the first UID for you, a UID that indicates that the application is in fact a DLL.

The main purpose of the DLL-related code in `hellogui.cpp` is to implement `NewApplication()`:

```
// DLL interface code
EXPORT_C CApaApplication* NewApplication()
    {
    return new CHelloGuiApplication;
    }

GLDEF_C TInt E32Dll(TDllReason)
    {
    return KErrNone;
    }
```

> **You *must* code the prototype of `NewApplication()` exactly as shown, otherwise the C++ name mangling will be different from that expected by the build tools' first ordinal export specification, and your application won't link correctly.**

My `NewApplication()` function returns my own application class, after first-phase construction. In the rather unlikely event that there isn't enough memory to allocate and C++-construct a CHelloGuiApplication object, NewApplication() will return a null pointer and the application architecture loader will handle cleanup.

Note in passing that all Symbian OS DLLs *must* also code an E32Dll() function that *ought* to do nothing, as mine does above.

If it exists to do nothing, why is the function there and why does it take a parameter? In Windows terminology, this function is the DLL entry point, and it's called when the DLL is attached to or detached from a process or thread – at least, that's the theory, and we thought that would be useful for allocating DLL-specific data in the early days of Symbian OS development.

In practice, however, it isn't called symmetrically in Windows, so it doesn't work under the emulator, and therefore, is pretty useless for Symbian OS development. Furthermore, different versions of Symbian OS call E32Dll() a differing number of times and in slightly different circumstances, which are further reasons not to use this function to do any real work in your application. Calls to this function are likely to be withdrawn in future versions of Symbian OS.

If you need to allocate DLL-specific memory, use the thread-local storage functions Dll::Tls() and Dll::SetTls(): call them from your main class' C++ constructor and destructor, and point to your class rather than some other ad hoc global variable structure. See Chapter 8 and \release\generic\app-framework\cone\src\coemain.cpp (along with \release\generic\app-framework\cone\src\coetls.h) in the SDK for an example.

The application

Assuming that NewApplication() returned an application, the architecture then calls AppDllUid() to provide another check on the identity of the application DLL. That's all it needs to know about the application. It then calls CreateDocumentL(), which the application uses to create a default document. Both of these are virtual functions that I implement in my derived application class.

The definition of CHelloGuiApplication in HelloGui_Application.h is as follows:

```
class CHelloGuiApplication : public CQikApplication
    {
```

```
private: // Inherited from class CApaApplication
  CApaDocument* CreateDocumentL();
  TUid AppDllUid() const;
  };
```

Here is the first appearance of something that is specific to UIQ. Because this application is being written to run on a P800, I've derived this class from UIQ's `CQikApplication` rather than Uikon's `CEikApplication` class. In this case, however, it makes no difference to the implementation, which you'll find in `HelloGui_Application.cpp`:

```
TUid CHelloGuiApplication::AppDllUid() const
  {
  return KUidHelloGui;
  }

CApaDocument* CHelloGuiApplication::CreateDocumentL()
  {
  return new(ELeave) CHelloGuiDocument(*this);
  }
```

Given the explanation above, there's nothing too surprising here. I defined the UID in my `HelloGui_Application.h` file:

```
const TUid KUidHelloGui = { 0x101f74a8 };
```

I use the same UID here as in my `.mmp` file: the APPARC verifies that this is the same as my DLL UID, as a final check on the integrity of my application DLL.

In the code above, then, you can see that the application class has two purposes:

- it conveys some information about the *capabilities* of the application, including its UID,

- it acts as a factory for a default document.

These are little things, but important. Pretty much every Symbian OS GUI application contains these two functions coded just as I've implemented them here, changing only the class names and the UID from one application to the next.

The document

The really interesting code starts with the document class. In a file-based application (that is, one that handles and modifies persistent data), the document essentially represents the data in the file. You can do several things with that data, each of which makes different demands: printing it, for example, doesn't require an application – but editing it does. We'll see more of this in Chapter 15.

If your application handles modifiable data, the APPARC requires your document to create an application user interface (app UI), which is then used to 'edit' the document.

In a nonfile-based application, you still have to code a function in the document class to create an app UI, since it's the app UI that does the real work. Apart from creating the app UI, then, the document class for such an application is trivial.

The declaration of `CHelloGuiDocument` is as follows:

```
class CHelloGuiDocument : public CQikDocument
    {
public:
  // construct/destruct
  CHelloGuiDocument(CEikApplication& aApp);
private: // from CEikDocument
  CEikAppUi* CreateAppUiL();
    };
```

Again, because the application is intended to run on a P800, I'm deriving from a UIQ class. The implementation is as follows:

```
CHelloGuiDocument::CHelloGuiDocument(CEikApplication& aApp)
  : CQikDocument(aApp)
    {
    }
CEikAppUi* CHelloGuiDocument::CreateAppUiL()
    {
    return new(ELeave) CHelloGuiAppUi;
    }
```

We start with a C++ constructor that simply calls `CQikDocument`'s constructor, passing in the application as an argument. Next, there's the virtual `CreateAppUiL()` function which, to nobody's surprise, returns a new app UI. An application that is not UI-specific would simply replace the references to `CQikDocument` with ones to `CEikDocument`.

Note carefully that `CreateAppUiL()` is responsible only for *first-phase* construction (which excludes any initialization that might fail). We'll see the second phase, to perform the possibly failure-prone initialization, shortly.

The application UI

The GUI action proper starts with the app UI, which has two main roles:

- to get **commands** to the application,
- to distribute keystrokes to controls, including the application's main view, which is owned by the app UI.

A command is simply an instruction without any parameters, or any information about where it came from, which the program must execute. In any practical GUI program, you implement `HandleCommandL(TInt aCommandId)` in your derived app UI class. The command ID is simply a 32-bit integer.

The definition is deliberately vague, since commands can originate from a variety of places. In UIQ applications, commands usually originate from an application's menu or – at least, in the emulator – from a shortcut key. (Applications may assign a shortcut key to any command ID – regardless of whether it is shown on the menus. The conventional shortcut key for the Close command, for instance, is *Ctrl+E*.)

In other customized UIs, commands may originate from other sources. In the Series 80 (Crystal) UI application, for example, which shows commonly used commands on a toolbar, commands may come from a toolbar button.

A command may be available from any or all of these sources. It doesn't matter where it comes from, the app UI receives it as a 32-bit command ID in `HandleCommandL()` and simply executes the appropriate code to handle the command.

Some commands can be executed immediately, given only their ID. An Exit command exits the application. (If the application has unsaved data, we don't ask the user: we simply save the data. If the user didn't want to exit, all that is needed is to start the application again.)

Other commands need further UI processing. A Find command, for example, needs some text to find, and possibly some indication of where to look. Handling such commands involves **dialogs**, which are the subject of Chapter 10.

Not all input to applications comes from commands. If you type the letter X into a word processor while you're editing, it doesn't generate a command, but a key event. If you tap on an entry in an Agenda, it doesn't generate a command, but a pointer event. Handling pointer and key events is the business of the app view and other controls in the application. After a chapter on drawing to controls (Chapter 11), I cover key and pointer interaction in Chapter 12.

Some key and pointer events are captured by controls in the Uikon framework and turned into commands. Later in this chapter, we'll get a glimpse into how the menu and shortcut key controls convert these basic UI events into commands.

The declaration of CHelloGuiAppUi in HelloGui_AppUi.h is as follows:

```
class CHelloGuiAppUi : public CQikAppUi
  {
public:
  void ConstructL();
  ~CHelloGuiAppUi();
private: // from CEikAppUi
  void HandleCommandL(TInt aCommand);
private:
  CHelloGuiAppView* iAppView;
  };
```

As is usually the case, the second-phase constructor and the destructor reveal who owns what:

```
void CHelloGuiAppUi::ConstructL()
  {
  CQikAppUi::ConstructL();
  iAppView = CHelloGuiAppView::NewL(ClientRect());
  }

CHelloGuiAppUi::~CHelloGuiAppUi()
  {
  delete iAppView;
  }
```

Again, this class is derived from the UIQ CQikAppUi class, rather than the Uikon CEikAppUi equivalent, but as with the application class, there is no significant difference in the implementation.

ConstructL() performs second-phase construction of the base class, CQikAppUi, using its own ConstructL() function. This, in turn calls CEikAppUi's BaseConstructL(). It's this function that, among other things, reads the application's resource file and constructs the menu and shortcut keys for the application.

ConstructL() then constructs the main app view using a two-phase constructor. The ClientRect() passed as a parameter to the app view is the amount of screen left over after the menu bar and any other adornments set by CQikAppUi have been taken into account. Predictably, the destructor destroys the application view.

The `HandleCommandL()` function is as follows:

```
void CHelloGuiAppUi::HandleCommandL(TInt aCommand)
  {
  switch (aCommand)
    {
          // Just issue simple info messages to show that
          // the menu items have been selected
    case EHelloGuiCmd0:
      iEikonEnv->InfoMsg(R_HELLOGUI_TEXT_ITEM0);
      break;

    case EHelloGuiCmd1:
      iEikonEnv->InfoMsg(R_HELLOGUI_TEXT_ITEM1);
      break;

    case EHelloGuiCmd2:
      iEikonEnv->InfoMsg(R_HELLOGUI_TEXT_ITEM2);
      break;
          // Exit the application. The call is
            // implemented by the UI framework.
    case EEikCmdExit:
      Exit();
      break;
    }
  }
```

Again, there is no difference between the Uikon and UIQ implementations.

This function handles four commands. The first three are identified by application-specific values that are defined in `hellogui.hrh`. The fourth is defined by Uikon and identified by `EEikCmdExit`. Uikon's command constants are defined in `eikcmds.hrh`, which you can find in the `\epoc32\include\` directory.

Files with the `.hrh` extension are designed to be included in both C++ programs (which need them, as above, for identifying commands to be handled) and resource scripts (which need them, as we'll see below, to indicate commands to be issued). Uikon's standard command definitions include many of the commonly used menu commands. All Symbian OS-defined command IDs are in the range 0x0100 to 0x01ff.

My command constants are defined in `hellogui.hrh`, as follows:

```
enum THelloGuiMenuCommands
    {
    EHelloGuiCmd0 = 0x1000,
    EHelloGuiCmd1,
    EHelloGuiCmd2
    };
```

It's clearly important that the constants I choose should be unique with respect to Uikon's, so I started numbering them at 0x1000 – you're safe if you do this too.

To handle these commands, I call on Uikon:

- Ideal with the `EEikCmdExit` command by calling `Exit()`. That resolves to `CEikAppUi::Exit()`, which terminates the Uikon environment.
- Ideal with my own commands by calling `CEikonEnv::InfoMsg()` to display an **info-message** on the screen. `InfoMsg()` is just one of many useful functions in the Uikon environment; its argument is a resource ID that identifies a string to be displayed, for about three seconds, in the top right corner of the screen.

> *It is important that a UIQ application handles EEikCmdExit in its AppUi::HandleCommandL( ) by calling CEikAppUi::Exit( ), even if the application does not have a 'Close' item in its menu. This is the way applications are closed if there is an imminent out-of-memory problem, or the application is about to be uninstalled.*

A more substantial application will have to handle *many* commands in its app UI's `HandleCommandL()` function. Normally, instead of handling them inline as I have done here, you would code each case as a function call followed by `break`. Most Symbian OS applications (and the examples in this book) use a function named `CmdFoo()` or `CmdFooL()` to handle a command identified as `EAppNameCmdFoo`.

The app view

Anything that can draw to the screen is a **control**. Controls can also (optionally) handle key and pointer events.

> *Note that controls don't have to be able to draw. They can be permanently invisible. But that's quite unusual: a permanently invisible control clearly can't handle pointer events, but it could handle keys, as we'll see in Chapter 12.*

The app UI is *not* a control. It owns one or more controls, including such obviously visible controls as a UIQ application's button bar; we'll see a few others throughout the next few chapters.

In a typical GUI application, you write one control yourself. You size it to the size of the **client rectangle**, the area of the screen remaining after the toolbar and so on have been taken into account. You then use that control to display your application data, and to handle key and pointer events (which are not commands).

hellogui's application view is a control whose sole purpose is to draw the text 'Hello world!' on the screen in a reasonably pleasing manner. It doesn't handle key or pointer events.

Like all controls, CHelloGuiAppView is derived from CCoeControl, which has virtual functions that you override to implement a particular control's functionality. In this case, the only function of interest is Draw(). The definition, from HelloGui_AppView.h is as follows:

```
class CHelloGuiAppView : public CCoeControl
    {
public:
    static CHelloGuiAppView* NewL(const TRect& aRect);
    ~CHelloGuiAppView();
    void ConstructL(const TRect& aRect);
private:
    void Draw(const TRect& /* aRect */) const;
private:
    HBufC* iHelloText;
    };
```

Let's start with the implementation of the second-phase constructor:

```
CHelloGuiAppView* CHelloGuiAppView::NewL(const TRect& aRect)
  {
  CHelloGuiAppView * self = new(ELeave) CHelloGuiAppView;
  CleanupStack::PushL(self);
  self->ConstructL(aRect);
  CleanupStack::Pop();
  return self;
  }

void CHelloGuiAppView::ConstructL(const TRect& aRect)
  {
  CreateWindowL();
  SetRect(aRect);
  ActivateL();
        // Fetch the text from the resource file.
  iHelloText = iEikonEnv->AllocReadResourceL(R_HELLOGUI_TEXT_HELLO);
  }
```

ConstructL() uses CCoeControl() base-class library functions to create a window, set it to the rectangle offered, and activate it. It then reads the 'Hello World!' text from a resource file into an HBufC, which is allocated the appropriate length. This is a common pattern in application programming. The memory for the HBufC is, of course deleted, in the destructor:

```
CHelloGuiAppView::~CHelloGuiAppView()
    {
    delete iHelloText;
    }
```

The `Draw()` code is as follows:

```
void CHelloGuiAppView::Draw(const TRect& /*aRect*/) const
    {
        // Window graphics context
    CWindowGc& gc = SystemGc();

        // Start with a clear screen
    gc.Clear();
    TRect rect = Rect();
    rect.Shrink(10,10);
    gc.DrawRect(rect);
    rect.Shrink(1,1);
    const CFont* font = iEikonEnv->TitleFont();
    gc.UseFont(font);
    TInt baseline = rect.Height()/2 - font->AscentInPixels()/2;
    gc.DrawText(*iHelloText, rect, baseline, CGraphicsContext::ECenter);
    gc.DiscardFont();
    }
```

You can probably guess well enough what most of this code is doing. Note especially that the penultimate line is `DrawText()`, with the `iHelloText` string as an argument. It's *this* line that actually achieves our objective of saying hello to the watching world.

Although straightforward and sufficient for the needs of such a simple application, this code is inefficient. The purpose of this chapter, however, isn't to explain drawing or controls: I return to that in Chapter 11, in which I'll cover the `Draw()` function and the functions called by `ConstructL()` thoroughly.

4.5   The Resource File

As with the other source files, we'll take just a brief look at the resource file and I'll show an outline how the resources defined in it build up the app UI's menu and so on. This will give you a pretty good idea of how resource files work – good enough that you'll be able to read resource files and guess what they mean. But to use them seriously, you need more than that, so I've included a minireference in Chapter 7.

In the code we've examined so far, we've already seen references to things that must be implemented in the application's resource file:

- The strings R_HELLOGUI_TEXT_HELLO, R_HELLOGUI_TEXT_ITEM1, and so on.
- The enumerated constants EHelloGuiCmd1, and so on.

The other things we *haven't* seen in the code are the definitions necessary to construct some aspects of the application's GUI, such as the menu and any shortcut keys. They're defined in the resource file too.

4.5.1 The Header

Let's look through the resource file, hellogui.rss, to see exactly what gets defined, and how. First, there's some boilerplate material:

```
NAME HELO

#include <eikon.rh>
#include <eikcore.rsg>
#include "HelloGui.hrh"
```

The name supplied in the NAME statement is used to generate part of the ID for each of the resources within the file. If an application uses more than one resource file, you should make sure that each contains a different name, so that there is no ID conflict between resources from the different files. To ensure there is no danger of conflict with system resources (which are available for use by all applications) the name has to be distinct from the resource names used by Uikon and one or two other system components. There is no need for the names to be unique between different applications, except in the case in which two applications share resources.

The #include statements load definitions of structures and constants that are used within the resource file. The final #include refers to my own .hrh file. As we've already seen, this contains the enumerated constants for my application's commands. I need those commands in the C++ file so I can tell which command had been issued; I also need them here so that I can associate the commands with the right menus and shortcut keys.

After the NAME and #include lines, every GUI application resource file begins with three unnamed resources as follows:

```
RESOURCE RSS_SIGNATURE { }

RESOURCE TBUF { buf=""; }
```

```
RESOURCE EIK_APP_INFO
    {
    menubar = r_hellogui_menubar;
    hotkeys = r_hellogui_hotkeys;
    }
```

The RSS_SIGNATURE allows me to specify version information, but I don't want to use this facility here. The TBUF allows me to specify a friendly name for my default file, but I don't want to use that either, since this isn't a file-based application.

Of more interest is the EIK_APP_INFO resource, which, in this case, identifies the symbolic resource IDs of my menu and shortcut keys.

> *'Shortcut keys' is the style-guide-approved language for what most programmers call 'hotkeys'. We decided that 'hotkeys' was too ambiguous or frightening for end users, so we chose a friendlier term to be used in the user interface, help text, and so on.*

Normally, a UIQ application would omit the hotkeys keyword and so not define any shortcut keys, since UIQ devices do not have a keyboard. The shortcut codes are not displayed in UIQ application menus, and are only usable while running the application in the emulator (which is the only reason for including one in this example).

A UIQ application can specify a button bar by means of a toolbar keyword in the EIK_APP_INFO resource. I haven't included one here, since hellogui is a minimal application, but you will see examples of how this is done in later chapters.

4.5.2 Defining the Shortcut Keys and the Menu

The application's shortcut keys are defined in a HOTKEYS resource, identified by the symbolic ID r_hellogui_hotkeys, which ties in with the symbolic ID given above in the EIK_APP_INFO resource:

```
RESOURCE HOTKEYS r_hellogui_hotkeys
    {
    control =
        {
        HOTKEY
            {
            command = EEikCmdExit;
            key = "e";
            }
        };
    }
```

Those upper and lower case letters can make you seasick. The reason for them is lost in the mists of time: the resource compiler predates Symbian OS by several years.

The syntax is a bit bizarre (`control` = identifies all shortcut keys identified by a *Ctrl+* key combination, so that `EEikCmdExit` is on *Ctrl+E*).

As I pointed out earlier, you would not normally be defining shortcut keys for a UIQ or Series 60 application, given that these devices do not have keyboards.

Symbian OS menus don't support the kind of shortcut keys found in Windows and other desktop systems that allow you to select **File | Close** using *Alt+F, C*. In Symbian OS applications for UIs that support their use, you either have to use the shortcut key (e.g. *Ctrl+E* – and displayed as such alongside the corresponding menu item) or navigate manually to the item.

We considered the Windows way seriously but rejected it on the grounds that it makes the menus look ugly, and most average Windows users don't understand what the underscores mean anyway. [Displaying shortcuts in the form described above advertises the facility in a way that anyone can understand, without a manual or training].

The final resource that's promised by `EIK_APP_INFO` is as follows: the menu specification:

```
RESOURCE MENU_BAR r_hellogui_menubar
    {
    titles =
        {
        MENU_TITLE
            {
            menu_pane = r_hellogui_hello_menu;
            txt = "HelloGui";
            },
        MENU_TITLE
            {
            menu_pane = r_hellogui_edit_menu;
            txt = "Edit";
            }
        };
    }
```

This menu bar has two named menu panes, which are defined in the following resources:

```
RESOURCE MENU_PANE r_hellogui_hello_menu
    {
    items =
        {
        MENU_ITEM
            {
            command = EHelloGuiCmd0;
            txt = "Item 0";
            },
        MENU_ITEM
            {
            command = EEikCmdExit;
            txt = "Close (debug)";
            }
        };
    }

RESOURCE MENU_PANE r_hellogui_edit_menu
    {
    items=
        {
        MENU_ITEM
            {
            command = EHelloGuiCmd1;
            txt = "Item 1";
            },
        MENU_ITEM
            {
            command = EHelloGuiCmd2;
            txt = "Item 2";
            }
        };
    }
```

Each menu item is associated with some text, and a command. More advanced menu trees might have more options (cascading menus, check marks against options that are active, etc.), but I don't need them here. While such features are accepted – and even necessary – in other customized Symbian OS UIs, they should be avoided, or used sparingly, in UIQ applications, in which menu content should be kept as short and simple as possible.

Note that the text for the EEikCmdExit command is 'Close (debug)'. A UIQ application, by convention, does not have a close option in its menus, but – as mentioned earlier – relies on the APPARC to close down the application (by issuing an EEikCmdExit command) when necessary. However, it is useful to include a Close option for debugging purposes. In more realistic applications you might want to add this option to a menu dynamically, so that it only appears in debug builds.

4.5.3 String Resources

Finally, there are the string resources, which are very simple indeed:

```
RESOURCE TBUF r_hellogui_text_item0
    {
    buf = "Item 0";
    }

RESOURCE TBUF r_hellogui_text_item1
    {
    buf = "Item 1";
    }

RESOURCE TBUF r_hellogui_text_item2
    {
    buf = "Item 2";
    }

RESOURCE TBUF r_hellogui_text_hello
    {
    buf = "Hello world!";
    }
```

A properly constructed Symbian OS application should have *all* its translatable string resources in resource files, so that they can be translated without having to change the C++ code. This recommendation applies to string text within GUI elements (e.g. menu items) as well as those explicitly listed in TBUFs.

If you intend to translate your application into other languages, you should go further than this and define your strings in a separate file with a .rls (resource localizable string) extension, as described in Chapter 7.

4.6 Bringing it to Life

Now that you've seen how the different parts of the program fit together, let's bring it to life by running through the code with the Metrowerks CodeWarrior debugger. Doing so will also show us what debugging in the GUI environment is like – which, as a developer, is something you'll need to get used to anyway.

Start up CodeWarrior, and build the program according to the instructions earlier in this chapter. Then you can begin debugging, for example, by pressing *F5*. At the end of the launch sequence, the application launcher will be running on the emulator.

4.7 Launching the Application

Don't launch hellogui *yet.* When you *do* launch the program, its functions will be called by the Uikon framework. For now, though, let's pretend that we don't know what will happen, and put a breakpoint on every function in hellogui.cpp, hellogui_application.cpp, hellogui_appui.cpp, hellogui_appview.cpp and hellogui_document.cpp, using the *F9* key. When you've done that, launch hellogui from the application launcher.

The first breakpoint to be hit is the one in E32Dll(), which gets called twice during the startup process. As pointed out earlier, this function is called a different number of times and in slightly different circumstances in different versions of Symbian OS. You are strongly recommended not to use the E32Dll() function to do anything important.

Next, you'll see that NewApplication() is called, and the context in which this happens is interesting. If you take a look at the call stack in Figure 4.4, you can confirm that NewApplication() is called by the application architecture, working on behalf of the program loader. The UIQ C++ SDK includes source (in the \Release\Generic directory tree) and debug information for these frameworks, so you can hunt around in the functions that call NewApplication() if you want.

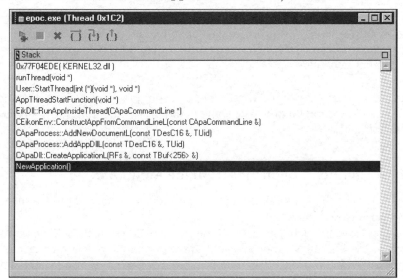

Figure 4.4

After NewApplication(), you'll see all the other initialization functions in the application, document, app UI, and app view classes called in the correct sequence. Eventually, you'll see CHelloGuiAppView::Draw() being called, and after that the emulator window is displayed.

4.8   Command and Event Handling

When you look at the code for `HandleCommandL()`, you might not think it's worth debugging – after all, it's just a simple `switch` statement, and all the `case` handlers are one-liners. In fact, it's worthwhile for a number of reasons:

- There are several ways of getting commands to `HandleCommandL()`, and it's informative to look at the call stack in the debugger to see how commands that come from different starting points arrive in the same place.

- Most of the case handlers display an info-message, and when that disappears your application view has to redraw itself. Because of this, your `Draw()` function gets called, with some surprises that we'll return to in Chapter 11.

- Most calls to `HandleCommandL()` are generated as a result of some user-initiated event (an exception for UIQ is a system-generated Exit command). As we saw in Chapter 2, this means active objects are involved in handling them. We can easily observe this by debugging through `HandleCommandL()`.

- You begin to get a feel for many of the relationships between the window server, controls, and your application. These relationships are the stuff of life for GUI programming, and I'll be explaining them in later chapters.

In the next few sections, we'll invoke commands from the menu and shortcut keys, and I'll point out some of the interesting things revealed by the debugger.

4.8.1   Pointer-generated Commands from the Menu Bar

Starting with the first of these, use the pointer to select and press a menu option, say Edit | Item 2.

You'll see the debugger stop at the `HandleCommandL()` breakpoint, where the Variables window reveals the value of the `aCommand` parameter to be 4098, or 0x1002, which is `EHelloGuiCmd2`. The Call Stack window reveals what is happening:

```
CCoeEnv::RunL()
CQikAppUi::HandleWsEventL(const TWsEvent &, CCoeControl *)
CEikAppUi::HandleWsEventL(const TWsEvent &, CCoeControl *)
CCoeAppUi::HandleWsEventL(const TWsEvent &, CCoeControl *)
CCoeControl::ProcessPointerEventL(const TPointerEvent &)
```

```
CEikMenuPane::HandlePointerEventL(const TPointerEvent &)
CEikMenuPane::ReportSelectionMadeL()
CEikMenuPane::ProcessCommandToAllObserversL(int)
CQikAppUi::ProcessCommandL(int)
CQHelloGuiAppUi::HandleCommandL(int)
```

In Chapter 2, we saw that Symbian OS is fundamentally an event-handling system, and that events are handled by active objects in their `RunL()` member function. That's exactly what we've got here: a `RunL()` for the event from the window server.

The first thing to establish, in `CCoeAppUi::HandleWsEventL()`, is that this event is a pointer event. CONE's app UI then hands the event to the appropriate control, which is one of the application's menu panes. The menu pane determines which menu item has been selected and converts that to the corresponding command ID, which in this case is 4098.

The menu pane informs its observer(s) – in this case, just the AppUi – which, in turn, calls `CQikAppUi::ProcessCommandL()` with the appropriate command ID value.

Thankfully, as an application programmer, you don't have to worry about any of that: it's all looked after for you. The point is that you get to field a call to `HandleCommandL()` with command ID 4098 – 0x1002, or `EHelloGuiCmd2`. If you debug through `HandleCommandL()`, you'll see how it's processed.

After `HandleCommandL()` completes processing, you'll see an info-message displayed on the emulator. When that has finished, your `CHelloGuiAppView::Draw()` function will be called to redraw the part of the window that had been covered by the info-message.

Actually, the code you see in `Draw()` draws the whole screen. But the window server clips drawing to the invalid region that actually needed to be redrawn. For some controls, it's worth optimizing to avoid drawing outside the invalid region. We'll see more on this in Chapter 11.

Also, you might be wondering why you didn't see a call to `Draw()` when the menu pane disappeared. I explain that at the end of the following section.

4.8.2 Keyboard-generated Commands from the Menu Bar

Next, use *F1* on your PC keyboard to pop up the `HelloGui` menu pane. Use the cursor keys to switch to the `Edit` menu pane and to highlight Item 2. Then press *Enter* to select that item. As before, you hit the breakpoint in `HandleCommandL()`, but this time the call stack is different:

```
CCoeEnv::RunL()
CQikAppUi::HandleWsEventL(const TWsEvent &, CCoeControl *)
CEikAppUi::HandleWsEventL(const TWsEvent &, CCoeControl *)
CCoeAppUi::HandleWsEventL(const TWsEvent &, CCoeControl *)
CCoeControlStack::OfferKeyL(const TKeyEvent &, TEventCode)
CEikMenuBar::OfferKeyEventL(const TKeyEvent &, TEventCode)
CEikMenuBar::DoOfferKeyEventL(const TKeyEvent &, TEventCode)
CEikMenuPane::OfferKeyEventL(const TKeyEvent &, TEventCode, int)
CEikMenuPane::DoOfferKeyEventL(const TKeyEvent &, TEventCode, int)
CEikMenuPane::ReportSelectionMadeL()
CEikMenuPane::ProcessCommandToAllObserversL(int)
CQikAppUi::ProcessCommandL(int)
CQHelloGuiAppUi::HandleCommandL(int)
```

We have an event being handled by an active object's `RunL()` function, but this time the raw window server event is identified as a *key* event. The app UI delegates keystroke distribution to the control stack, which offers it to the menu bar, which in turn offers it to the menu pane. The menu pane reports that a selection has been made, which results in the menu being dismissed (in `CQikAppUi::ProcessCommandL()`), and the command being handled by `HandleCommandL()`.

We'll see the control stack, and its role in key handling, in Chapter 12.

Why didn't the app view redraw when you switched from the **HelloGui** *to the* **Edit** *menu pane, or when the menu pane disappeared? The answer is that these panes are handled using windows that maintain a backup copy of whatever is underneath them – so-called 'backed-up behind' windows. When the window moves or is dismissed, the window server replaces whatever was underneath from the backup copy, without asking for the application to redraw. Before this feature was implemented (prior to the first release of Symbian OS), flipping between menu panes was very slow in all but the simplest applications.*

4.8.3   Commands from Shortcut Keys

The final way into `HandleCommandL()` is via a shortcut key. You know that you can close the application with *Ctrl+E*; try this, and you'll get:

```
CCoeEnv::RunL()
CQikAppUi::HandleWsEventL(const TWsEvent &, CCoeControl *)
CEikAppUi::HandleWsEventL(const TWsEvent &, CCoeControl *)
CCoeAppUi::HandleWsEventL(const TWsEvent &, CCoeControl *)
CCoeControlStack::OfferKeyL(const TKeyEvent &, TEventCode)
```

```
CEikMenuBar::OfferKeyEventL(const TKeyEvent &, TEventCode)
CEikMenuBar::DoOfferKeyEventL(const TKeyEvent &, TEventCode)
CQikAppUi::ProcessCommandL(int)
CQHelloGuiAppUi::HandleCommandL(int)
```

As before, RunL() handles the event, while HandleWsEventL() decides it's a key and gets it handled by the control stack. The control stack offers it to the menu bar, which recognizes shortcut keys and calls the application to process them.

4.9 Terminating the Application

By pressing *Ctrl+E*, we sealed the fate of our debugging session. As the application unwinds, you'll see the app UI's destructor, which explicitly calls the app view's destructor. You'll also see that E32Dll() is called for the final time before the application exits. You might as well end the debugging session by closing the emulator.

4.10 On-target Debugging

Having successfully debugged the application on the emulator, you might like to try repeating the process to debug the application on an actual phone.

Symbian OS v7.0, on which UIQ is based, supports two on-target debuggers – The GNU debugger (GDB) and the Codewarrior Debugger (MetroTRK). GDB can be used from Window's Command Line or from the graphical Insight UI. As you would expect, MetroTRK is driven from the CodeWarrior IDE.

Both of these debuggers require 'debug agent' software running on the target to communicate with the IDE on the host.

4.11 Setting Up MetroTRK

The MetroTRK (Metrowerk's Target Resident Kernel) software is the debug agent that handles the communication between the target and the Codewarrior Debugger on the host PC.

First, you must install the MetroTRK software on your target device. MetroTRK is composed of 5 executables, and 1 .ini file. These are available on your SDK or from the Metrowerks web site (***www.metrowerks. com***).

The MetroTRK install package will place the following files on your device.

```
C:\System\Libs\TRKENGINE.DLL
C:\System\Libs\TRKKERNELDRIVER.LDD
C:\System\Programs\MetroTRK.EXE
C:\System\Apps\MetroTRK\METROTRK.APP
C:\System\Apps\MetroTRK\METROTRK.RSS
C:\METROTRK.INI
```

4.11.1   Configuration

The METROTRK.INI file contains a number of configuration settings that you should modify to match your target hardware. If the installation was from a product SDK, then this will already have been done for you, and you can skip to the 'Launching MetroTRK' section.

The METROTRK.INI file has the following content:

```
[SERIALCOMM]
PDD EUART1
LDD ECOMM
CSY ECUART
PORT 2
RATE 115200
```

You should alter the PORT number to match the one on your platform. Other configuration options can be left alone.

4.12   Launching MetroTRK

Next, connect a serial cable from your PC to a free serial port on the target device. If your target device only has one serial port, you should disable the PC connectivity link.

Now the target has all the required software and physical connection, and you can launch the MetroTRK application from the shell.

You should see the following message:

```
Welcome to MetroTrk for Symbian OS
Version 1.7
Implementing MetroTrk API version 1.7
Press "Q" to quit.
```

4.13   Setting up the CodeWarrior IDE

Now we need to configure the Remote Connection Settings in CodeWarrior:

- Select the **Edit | Preferences** menu item to open the IDE preferences window.
- Select Remote Connections in the IDE Preference Panel window.
- Select Symbian MetroTRK in the Remote Connections list.

If the Symbian MetroTRK option is not available, then select the `Import Panel ...` button and use the Open dialog to locate the `remote_connections.xml` file. In my installation, this is located in a `Metrowerks \Codewarrior for Symbian Pro v2.0\Bin` directory.

- Select the `Change ...` button.
- From the Symbian MetroTRK dialog you can set the appropriate protocol settings. Change the connection type to Serial. The settings will typically be 115 200 baud, 8 data bits, No parity, 1 stop bit. You should also change the serial port number to match the physical serial connection to your device.

This completes configuration settings for the CodeWarrior Debugger. You can now debug your application on the target device.

4.14 Debugging Your Application

Select the ARMI UDEB build target from the project window. You can now press the Debug button, or select the **Project | Debug** menu item to launch your application under the control of the debugger.

The IDE builds your application, connects to MetroTRK on the target, and downloads your newly compiled binary. Extra files can be selected for automatic download to the target from the Remote Download panel in the project settings window. This is useful for automatically downloading your application's resource and data files.

MetroTRK will now launch your application and will halt execution in the `NewApplication()` function defined in your code. From here you can perform all the debug stepping, variable watching, and other debug tasks you are familiar with from debugging your application under the emulator.

4.15 Summary

In this chapter, we've seen

- how a Symbian OS GUI application is put together and how it interacts with the frameworks provided by Uikon, CONE, and the application architecture;

- that the app UI provides the framework for handling commands issued by the menu, toolbar, and shortcut keys;
- where commands come from, how to identify and handle them;
- the basic contents of a resource file;
- the sequence of calls in a GUI application during startup, when handling key and pointer events, when redrawing, and when closing down.

The application architecture's role in application launch is to

- load the DLL,
- call its first exported function to create a new 'application' – a `CApaApplication`-derived object,
- check the UID returned by the application object,
- ask the application to create a new default document,
- ask the document to edit itself, which Uikon implements by asking you to create an app UI.

By doing a live demonstration using the debugger, we've seen

- how the application launches, handles commands, and exits;
- an example of how all events in a Symbian OS application program are handled by an active object `RunL()` function;
- some insight into redrawing, which we'll cover in much more detail in Chapter 11;
- some insight into key and pointer handling, which we'll cover in much more detail in Chapter 12.

5

Strings and Descriptors

I once looked up the word 'computer' in a big dictionary in my high school library. It said, 'one who computes.' This view is somewhat old-fashioned: the truth is that most programmers expend far more effort on processing strings than they do computing with numbers, so it's a good idea to start getting to know Symbian OS APIs by looking at its string handling facilities.

In C, string processing is inconvenient. You have an awkward choice of `char*`, `char[]`, and `malloc()` with which to contend, just to allocate your strings. You get some help from such functions as `strlen()`, `strcpy()`, and `strcat()`, but little else. You have to pass around awkward maximum-length parameters to functions such as `strncpy()` and `strncat()` that modify strings with an explicit length limit. You have to add one and subtract one for the trailing NUL at the end of every string. If you get your arithmetic slightly wrong, you overwrite memory and produce bugs that are hard to track down. It's not much fun.

In Java, life is much easier. There is a String class with nice syntax such as a + operator for concatenating strings. Memory for strings looks after itself: new memory is allocated for new strings and memory for old string values or intermediate results is garbage collected when no longer needed.

In standard C++, a similarly useful string class is also available, though it came along quite a while after the C++ language itself.

In Symbian OS, strings are implemented by descriptors. Descriptors provide a safe and consistent mechanism for dealing with both strings and general binary data regardless of the type of memory in which they reside.

Like the string classes in Java and standard C++, they're much more comfortable to work with than C strings. However, Symbian OS doesn't take the same approach as either Java or standard C++, because memory

Symbian OS C++ for Mobile Phones.   Edited by Richard Harrison
© 2003 John Wiley & Sons, Ltd   ISBN: 0-470-85611-4

management is so important in Symbian OS. You have to be fully aware of the memory management issues when you're using descriptors.

I'll start this chapter off with a discussion of descriptors and memory management. Because C string handling gives you control of memory management, I'll compare descriptors and their memory management with C strings and their memories.

Then I'll move on to what you can do with descriptors. You need to know about both the concrete implementation classes and the two key abstract base classes, TDesC and TDes, which include a large number of convenience functions. TDesC is a one-word class (just the length), and its convenience functions are all const – that's what the C in TDesC stands for. TDes derives from TDesC, and adds an extra word (maximum length), and nonconst convenience functions.

Symbian OS is built to use 'wide' characters – 16 bits, using Unicode. I'll explain some of the implications of this.

Finally, I'll look at descriptors' role in describing data – which is where they got their name. Descriptors are fundamental to many data-related Application Programming Interfaces (APIs), such as the interthread reading and writing used by the client-server framework and reading and writing data to files.

5.1 Strings and Memory

To understand strings in any C or C++-based system, you have to understand memory management as it relates to strings. Essentially, there are three types of memory:

- *Program binaries*: In ROM, DLLs, and .exes (for the most part), program binaries are constant and don't change. Literal strings that we build into our program go into program binaries.

- *The stack (automatic objects)*: This is suitable for fixed-size objects whose lifetimes coincide with the function that creates them, and which aren't too big. Stack objects in Symbian OS shouldn't be too big, so they should only be used for *small* strings – tens of characters, say. A good rule of thumb is to put anything larger than a file name on the heap. It's quite acceptable to put pointers (and references) on the stack – even pointers to very large strings in program text or on the heap.

- *The heap (dynamic objects)*: Memory is allocated from the default heap as and when required. It is used for objects (including strings) that are built or manipulated at runtime, and which can't go on the stack because they're too big, or because their lifetimes don't coincide with the function that created them.

5.1.1 Strings in C

So, in C, there are three ways to allocate a string corresponding to whether they are held in program binaries, on the stack, or in heap memory.

A string in a program binary is represented thus:

```
static char hellorom[] = "hello";
```

You can get a *pointer* to this string, on the stack, simply by assigning the address of the string data into an automatic:

```
const char* helloptr = hellorom;
```

You can put the string itself onto the stack by declaring a character array of sufficient size on the stack and then copying the string data into this array:

```
char hellostack[sizeof(hellorom)];
strcpy(hellostack, hellorom);
```

And you can put the string onto the heap by allocating a heap cell of sufficient size and then copying the string data:

```
char* helloheap = (char*)malloc(sizeof(hellorom));
strcpy(helloheap, hellorom);
```

When the statements above have executed, the situation in memory – program binaries, heap, and stack – is shown in Figure 5.1.

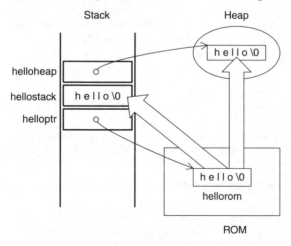

Figure 5.1

5.1.2 Strings in Symbian OS

Here's how Symbian OS does the same kind of thing. I'll go through the following program text more slowly. It's in \scmp\strings\, a Symbian OS project with a text-mode program based on hellotext. It's more interesting to run it from the debugger than to launch it from a console. To remind you, you can do this by opening the project in Metrowerks CodeWarrior using the File | Import Project from .mmp File command and then building and stepping through the code.

To get a string into program binaries, use the _LIT macro (short for 'literal'):

```
_LIT(KHelloRom, "hello");
```

This puts a literal descriptor into your program binaries. The symbol for the descriptor is KHelloRom, and its value is 'hello'. You can get a pointer descriptor to this string on the stack using

```
TPtrC helloPtr (KHelloRom);
```

TPtrC is a two-word object that includes both a pointer *and a length*. The statement above sets both of these in helloPtr. With a pointer and a length, you can perform any const function on a string – anything that doesn't modify its data. That's the significance of the C in TPtrC.

You can get the string data itself into the stack, if you first create a buffer for it. Here's how:

```
TBufC<5> helloStack(KHelloRom);
```

TBufC<5> is a 5-character buffer descriptor. This object contains a single header word saying how long it is (in this case, 5 characters), followed by 10 bytes (because Unicode needs 2 bytes per character) containing the data. As before, the C indicates that only const functions are allowed on a TBufC after its construction.

You can get the string data into a heap cell if you allocate a heap-based buffer and copy in the data:

```
HBufC* helloHeap = KHelloRom().AllocLC();
```

This statement is doing a lot of things. Let's take them in order:

- HBufC* is a pointer to a heap-based buffer descriptor. This is the only class in Symbian OS whose name begins with H. It's reasonable to have a unique name because, as we'll see, HBufC's properties are unique.

- By putting function brackets after `KHelloRom()`, I invoke an operator that turns it into the base class for all descriptors, `TDesC`. I need this because a literal descriptor is not derived from `TDesC`, for reasons I'll explain later.

- `AllocLC()`, on any descriptor class, allocates an `HBufC` of the required size on the default heap and copies the (old) descriptor contents into the (new) `HBufC`. `AllocLC()` also pushes the `HBufC*` pointer to the cleanup stack so that I can later delete the object using `CleanupStack::PopAndDestroy()`.

In short, we create a new heap cell and copy the string text into it. Unlike my C program, this code is also fully error-checked and memory leak proof:

- If allocation fails, `AllocLC()` leaves. Everything is trapped and cleaned up by the cleanup mechanisms built into the strings example's startup code.

- If a later function leaves, then the cleanup stack will cause the `helloHeap` object to be popped and destroyed.

- If I forget to deallocate this `HBufC*` and the one I allocate later in the example, the program panics on exit because of heap marking built into it.

In the next chapter, I'll go into these issues more thoroughly. For now, it's enough to note that we didn't have to do much, given the framework that I just copied from **hellotext**, to make our program's cleanup safe. After this code has run, our program memory looks like Figure 5.2:

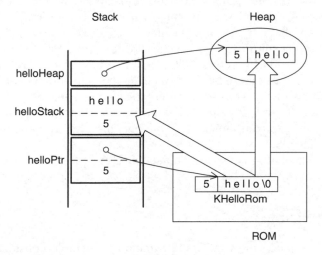

Figure 5.2

There are four descriptor types here, each corresponding to a different type of memory:

- A pointer descriptor, TPtrC, consisting of a length and a pointer to the data. This can be used where a const char* would be used in C.
- A buffer descriptor, TBufC, which contains the data itself and also its length. This can be used where a char[] would be used in C.
- A heap descriptor, always referred to by an HBufC* pointer, which is a heap cell containing the length and data (similar to a buffer). This is used where a malloc()'d cell would be used in C.
- A literal descriptor of type TlitC, which is hidden in the _LIT macro. Although not a true descriptor class, it can masquerade as a TBufC because it has the same layout. This is used where a static char[] would be used in C.

Except for TLitC, these descriptor classes are derived from TDesC, which contains a Length() function to get the current length and a Ptr() function to find the address of the data. The current data length is always the first machine word in a concrete descriptor class, so TDesC::Length() is implemented identically for all descriptor classes. In the case of a TPtrC, the address of the data is contained in the word after the length. For HBufC and TBufC, the address is simply the address of the object itself, plus 4 (for the length word).

With an address and a length, you have all you need for any const function on a string. TDesC provides those functions, and thus, so do the derived descriptor types.

On the face of it, Ptr() should be a virtual function, because it's an abstract interface that depends for its implementation on the concrete class. However, there is no need to use virtual functions because there are only five descriptor classes and there will never be any more. Avoiding them means that we save one machine word – 4 bytes – from the size of every descriptor. Instead, the length word reserves 4 bits to indicate the concrete version of descriptor class and TDesC::Ptr() uses a switch statement to check these bits and calculate the data address correctly.

Of course, using 4 bits for a descriptor identifier leaves 'only' 28 bits for the length and descriptor data is therefore constrained to around 250 million characters rather than 4 billion. This is not a serious restriction for Symbian OS.

Space efficiency matters in Symbian OS. Descriptors are space efficient and they allow you to be.

5.2 Modifying Strings

Having compared the way C and Symbian OS handle constant strings, let's look at the support they provide for manipulating strings.

5.2.1 Modifying C Strings

When you add something to the end of a string, you have to have enough room for the new text.

Doing this in program binaries isn't an option: they can't be modified if they're in ROM or even in a RAM-loaded Symbian OS DLL.

You can allocate enough space on the stack by declaring an array big enough for both strings: you can save a byte because you won't need two trailing NULs:

```
static char worldrom[] = "world!";
char helloworldstack[sizeof(hellorom) + sizeof(worldrom) - 1];
strcpy(helloworldstack, hellorom);
strcat(helloworldstack, worldrom);
```

You can do a similar thing on the heap:

```
char* helloworldheap = (char*)malloc(strlen(hellorom) +
    strlen(worldrom) + 1);
strcpy(helloworldheap, hellorom);
strcat(helloworldheap, worldrom);
```

This time, I've used `strlen()` rather than `sizeof()`, to emphasize that heap-based allocation can evaluate lengths at runtime: I can't use `strlen()` to evaluate the size of my stack buffer. As a consequence, I have to add a byte rather than subtracting, because `strlen()` doesn't include the trailing NUL. In memory, the result of these operations looks like Figure 5.3:

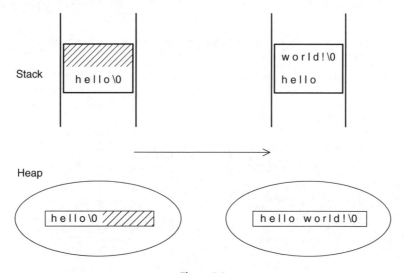

Figure 5.3

The issue here is that had I got my array size wrong (for `helloworld-stack`) or my heap cell size wrong (for `helloworldheap`), I might have overwritten the end of the string. It's not simply the irritation of having to add or subtract 1 for the trailing NUL – the real problem is that the string data might not fit into the memory allocated for it.

5.2.2 Modifying Symbian OS Strings

The **strings** example shows how to modify strings using descriptors. You can get a buffer suitable for appending one string to another by placing a `TBuf` on the stack:

```
_LIT(KWorldRom, "world!");
TBuf<12> helloWorldStack(KHelloRom);
helloWorldStack.Append(KWorldRom);
```

`TBuf<12>` is a modifiable buffer with a maximum length of 12 characters. After the constructor has completed, the data is initialized to 'hello' and the current length is set to 5.

The code is somewhat unsatisfactory in that I have used a magic number, 12, for the size of the buffer rather than calculating it. This is because you can't take the size of a `_LIT` constant. I could have avoided magic numbers, but it didn't seem worth it for this example.

`Append()` starts by checking the maximum length to ensure that there will be enough room for the final string. Then, assuming there is room, it appends the 'world!' string, and adjusts the current length.

You can see how it looks in memory in Figure 5.4.

The processing of `Append()` illustrates two fundamental aspects of descriptors:

- The descriptor APIs do not perform memory allocation: you have to allocate a descriptor, which is big enough.
- If you use the descriptor APIs, you can never overflow a buffer: if you try to do so, the system will panic your program. (That is, it will abort it with an error code. We'll look at panics in detail in the next chapter.)

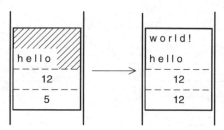

Figure 5.4

C's string APIs are awkward because they don't perform memory allocation, unreliable because they allowed you to write beyond the end of memory allocated for strings, and doubly unreliable because, with all those trailing NUL calculations, you are quite likely to get it wrong occasionally anyway.

Java's String and standard C++ string classes solve these problems and also manage the memory for you. The cost of this functionality is more bytes for string objects, which doesn't matter as much in Java's and standard C++'s intended application areas as it does for Symbian OS.

Symbian OS is a kind of halfway house: you have to do your own memory management, but you can't overwrite memory beyond the end of a string and, if you try to, you'll find out about it very early in your debugging cycle. So descriptors contribute significantly to the compactness and robustness of Symbian OS.

5.2.3 Modifying `HBufCs`

You might have thought there would be an `HBuf` class to make it easy for you to modify descriptors on the heap. But there isn't: if you want an `HBuf`, the best thing to do is to allocate a `TBuf` of the right size or use a `CBufBase`-derived class (see Chapter 8 for more on `CBufBase`).

You can modify an `HBufC` by using the pointer descriptor `TPtr` to address the memory it contains:

```
HBufC* helloWorldHeap = HBufC::NewLC(KHelloRom().Length() +
        KWorldRom().Length());
TPtr helloWorldAppend(helloWorldHeap->Des());
helloWorldAppend = KHelloRom;
helloWorldAppend.Append(KWorldRom);
```

`HBufC::NewLC()` allocates a heap cell of sufficient size to hold the requested number of characters (12 in this case) plus a descriptor header. It sets the current length to 0.

The `Des()` function returns a `TPtr` consisting of the address, current length, and heap cell length of the `HBufC` minus the length of its header. That means I can use the `TPtr` to change the `HBufC`'s content safely – this form of `TPtr` will also update the `HBufC`'s current length in step with the `TPtr`'s current length.

Then I do the data copying: I set the `TPtr` content to contain 'hello', which sets its current length to 5, and then append 'world!', which sets its current length to 12.

In memory, it looks like this:

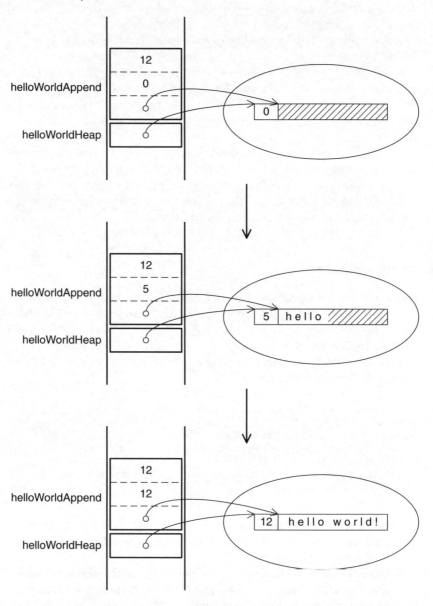

Figure 5.5

Admittedly, there is more code here than in the C case. But because the length and the maximum length are carried in the descriptors (rather than in an object that must be passed around separately), the code is just as safe as the TBuf code in Figure 5.5.

5.2.4   Descriptor Type Summary

We have now seen five concrete descriptor types:

- `TPtrC`, `TBufC`, and `HBufC`, all derived from `TDesC`. All have a pointer and a current length, and all inherit const convenience functions from `TDesC`. The essential difference between these classes is the way they contain or refer to the string data. Also, they differ in how you initialize them.

- `TPtr` and `TBuf`, both derived from `TDes`, which is in turn derived from `TDesC`. `TDes` adds a maximum length to `TDesC`, and a number of nonconst convenience functions.

We can picture the layout of each concrete descriptor as in Figure 5.6.

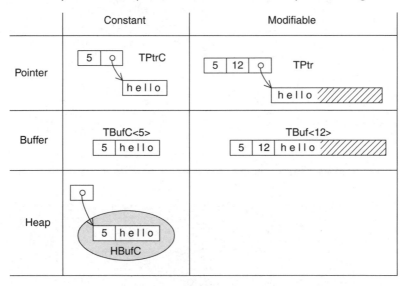

Figure 5.6

The other way to look at these classes is Figure 5.7.

5.2.5   Using the Abstract Classes in Interfaces

`TDesC` and `TDes` are abstract classes, so you can never instantiate them. But in functions designed to manipulate strings, you should always use these base classes as the arguments. Use

- const `TDesC&` to pass a descriptor for string data which you will read from, but not attempt to change,

- `TDes&` to pass a descriptor for string data, which you want to change.

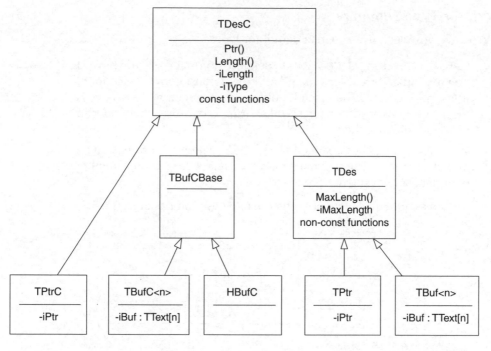

Figure 5.7

An example in **strings** is the `greetEntity()` global function:

```
void greetEntity(TDes& aGreeting, const TDesC& aEntity)
    {
    aGreeting.Append(aEntity);
    }
```

You pass one modifiable descriptor that contains the greeting text ('`hello`', perhaps), and you pass a nonmodifiable descriptor that contains the entity to be greeted (the world, maybe). The function simply appends the entity onto the greeting. You can call this function using code like this:

```
TBuf<12> helloWorld(_L("hello"));
greetEntity(helloWorld, _L("world!"));
```

I've used an alternative format for the literals used to initialize descriptors here: `_L` produces a `TPtrC`, which describes the text in the string. So `greetEntity()` gets called with `TBuf<12>` and `TPtrC` parameters, which suitably match its prototype requirements.

We saw when discussing `TDesC` how the current data length is always the first machine word of a concrete descriptor class, allowing the same

Length() implementation in all classes. In TDes and its derived classes, the maximum allowed data length is always the second machine word, so TDes::MaxLength() is implemented identically for all modifiable descriptor classes.

The address of the first byte of data therefore varies depending on the concrete type of the descriptor class: sometimes it's the first byte after the length; sometimes it's the first byte after the maximum length; sometimes it's contained in a pointer after the length. This is why TDesC::Ptr() is implemented differently for each descriptor class.

Every other TDesC or TDes function depends only on Ptr(), Length(), and MaxLength(), so they can be implemented without any dependence on the concrete descriptor class.

5.2.6 Literals Again

We've now seen two kinds of literal descriptor:

- _LIT, which associates a symbol with a literal value and produces a TLitC, which is not derived from TDesC
- _L, which produces a TPtrC from a literal value and can be used without a name.

At first sight, _L is more attractive, for precisely the two reasons I mentioned above. It was the only type of literal supported until Symbian OS v5.0, when _LIT was added. This is now preferred to using _L in most circumstances: let's see why.

Here's how _L works. It is defined as

```
#define_L(string) TPtrC((const TText*) string)
```

So that

```
const TDesC& helloRef = _L("hello");
```

It does three things:

- When the program is built, the string, including the trailing NUL, is built into the program code. We can't avoid the trailing NUL because we're using the C compiler's usual facilities to build the string.
- When the code is executed, a temporary TPtrC is constructed as an invisible automatic variable with its pointer set to the address of the first byte of the string and its length set to five.
- The address of this TPtrC temporary is then assigned to the const TDesC& reference, helloRef, a fully-fledged automatic variable.

This code works provided either that the reference is only used during the lifetime of the temporary – that is, during the lifetime of the function – or that the TPtrC pointer and length are copied if the descriptor is required outside this lifetime. This is usually OK, but you get the feeling that you're walking on eggshells.

Another issue we need to consider is that the second step above – constructing a TPtrC including the pointer, the length, and the 4-bit descriptor class identifier – is always required. Code has to be built and run, wasting code bytes and execution time.

The _LIT macro tackles these issues as follows:

```
#define _LIT(name, s) const static TLitC<sizeof(s)> name =
    { sizeof(s) - 1, s }
```

The macro builds into the program binary an object of the type TLitC, which stores the required string. The binary layout of TLitC is designed to be identical to TBufC: this allows a TLitC to be treated (by using the proper casts) as a TDesC.

So, if you code,

```
_LIT(KHelloRom, "hello");
```

and then use it like this,

```
const TDesC& helloDesC = KHelloRom;
```

then the TDesC& reference looks exactly like a reference to a TBufC. But no runtime temporary and no inline constructor code are generated, so ROM budget is significantly reduced for components that contain many string literals.

Figure 5.8 shows the difference between _LIT and _L:

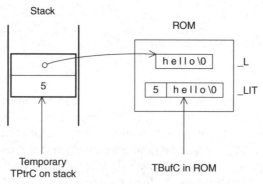

Figure 5.8

Most programs use resource files to contain their language-dependent literal strings: we'll cover resource files in Chapter 7. But many system components require literals for system purposes, which don't get translated for different locales.

For those programs that do use literals, _L is now deprecated, although it's still perfectly acceptable to use it in test programs, or other programs where you know memory use is not critical.

5.3 Standard Descriptor Functions

I mentioned that descriptors contain many convenience functions – const functions in TDesC and both const and nonconst functions in TDes. Here's a lightning tour to give you a taster. For more information, see the SDK. I also use many descriptor functions throughout the book.

If you wish, you can write your own descriptor functions: just put them into another class (or, exceptionally, no class at all) and pass descriptors to your functions as const TDesC& or TDes& parameters. Symbian OS itself provides utility classes that work this way; perhaps the most fundamental is TLex, which provides string-scanning functions and includes the code to convert a string into a number.

5.3.1 Basics

You can get at descriptor data using Ptr() to find its address, Length() to find out how many characters it is, and Size() to find out how many bytes of data it contains. Be careful not to use Size() where you really mean Length(); as for Unicode strings these values are different.

TDes provides MaxLength(), which says how many characters are allocated to the descriptor in total. Any manipulation function that would cause this to be exceeded will panic.

If you write your own string handling functions, you should usually construct them using the descriptor library functions as described here. In some circumstances, you may wish to access descriptor contents directly and manipulate them using C pointers – Ptr() allows you to do that – but be careful. In particular, make sure you honor MaxLength() if you're modifying descriptor data and make sure that any function you write panics if asked to do something that would overflow the allocated MaxLength() – the standard way of ensuring this is to call SetLength() before modifying the content, which will panic if the length exceeds the MaxLength().

5.3.2 Comparison

You can compare descriptors using

- Compare(): this compares the bytes without any locale sensitivity
- CompareC(): this is sensitive to locale-specific collation position of each character in the strings

- `CompareF()`: this folds lowercase into uppercase, ignores accents, and takes account of locale-specific collation.

The convenience operator`<()` and so on are provided, which call `Compare()` and return a `TBool`.

Collating is the process of removing any differences between characters so that they can be put in a sequence that allows for straightforward comparisons. *Folding* sets everything to uppercase and removes accents.

5.3.3 Searching

`TDesC`'s `Locate()` and `Find()` functions allow forward or reverse searching with optional case insensitivity and accent insensitivity.

5.3.4 Extracting

`TDesC`'s `Left()`, `Right()`, and `Mid()` functions can extract any part of a string.

5.3.5 Clearing and Setting

`SetLength()` allows the length to be specified to anything within range 0 to `MaxLength()`, and `SetMax()` sets the length to the maximum length.

Various `Fill()` functions allow a descriptor to be filled with data quickly.

5.3.6 Manipulating Data

`TDes`'s `Copy()` function copies data to the descriptor, starting at the beginning, while `Append()` can copy additional data to the descriptor, starting where the existing data stops. There are variants of the `Copy()` function that perform case or accent folding.

`Insert()` inserts data into any position, pushing up subsequent characters toward the end of the string. `Delete()` deletes any sequence of characters, moving down subsequent characters to close the gap. `Replace()` overwrites characters in the middle of a string.

5.3.7 Letter Manipulation

`TDes`'s `Fold()`, `Collate()`, `LowerCase()`, `UpperCase()`, and `Capitalize()` functions all manipulate characters in place.

Collating an individual string isn't very useful, and `TDes::`
`Collate()` is therefore deprecated.

5.3.8 Trimming and Justification

`TDes`'s `Trim()` functions shave whitespace from either or both ends
of a string. `Justify()` performs left, right, center, or no justification and
allows the fill character to be overridden. Various `AppendJustify()`
functions append and justify simultaneously.

`Justify()` is not really very useful, since real text justification is
done at the GUI level, not the individual string level.

5.3.9 Formatting

`TDes::AppendFormat()` is a bit like C `sprintf()`: it takes a format
string and a variable argument list and appends the result to an existing
descriptor. Functions such as `Format()` are implemented in terms of this
one as was `CConsoleBase::Printf()`, which we saw in hellotext
and strings.

Many lower-level functions exist to support `AppendFormat()`: various `Num()` functions convert numbers into text, and corresponding
`AppendNum()` functions append converted numbers onto an existing
descriptor. For simple conversions, the `AppendNum()` functions are
much more efficient than using `AppendFormat()` with a suitable format string.

In C, scanning functions are provided by `sscanf()` and packaged
variants such as `scanf()`, `fscanf()` and so on. Similar functions are
available in Symbian OS, through `TLex` and associated classes, which
scan data held in `TDesCs`. These functions are relatively specialized and
it was not thought appropriate to implement them directly in `TDesC`.

5.4 More Text APIs

Symbian OS uses descriptors for basic string handling. But a real GUI-
based system needs more than strings; it needs classes for text that can be
formatted conveniently into any size display area, edited conveniently,
displayed using a full range of character and paragraph formatting, and
include pictures. Symbian OS uses two sophisticated components, ETEXT
(text content) and FORM (text views) to provide these functions, and
convenient, reusable, text editors in the GUI libraries. Rich text also uses
a more powerful memory management scheme than any of the descriptor
classes: dynamic buffers, which are described in Chapter 8.

At the other end of the spectrum, an individual character has many attributes, such as uppercase, alphabetic, whitespace and so on. The `TChar` class represents a single character and includes functions to interrogate attributes. So you can write

```
TBool u = TChar(_L("Hello")[4]). IsUpper();
```

This sets u to `EFalse`: change the 4 to a 0, and it will set u to `ETrue`.

`TChar` is defined in `e32std.h`. `TLex` functions use `TChar` to find character attributes. `TChar` character attributes involve kernel executive calls (albeit quite efficient ones), so don't use `TChar` or `TLex` where fast, locale-independent parsing is required: hard-code your own character and attribute tables instead.

Symbian OS v7 introduced two new text-related APIs, which were primarily inspired by communications programming needs. The Internet Protocol Utility library (`InetProtUtil.lib`) has specialist classes for processing Internet-related strings such as URLs and dates. The String Pool library (`stringpool.h`, `bafl.lib`) provides a way of storing strings that makes comparison almost instantaneous at the expense of string creation. It is particularly efficient at handling string constants that are known at compile time. These properties make the API very useful for parsing text-based communication protocols such as HTTP.

5.5   Unicode and Character Conversion

Symbian OS was designed from the beginning to support worldwide locales. Psion had been unable to address considerable interest from the Far East for its SIBO models, because characters were represented within SIBO by every conceivable means – C strings, arrays, integers, assembler symbols – all of which could be used to contain other data. It was not commercially feasible to modify SIBO to support 16-bit characters. Symbian OS would not repeat the same mistake, so, although Symbian OS up to and including v5 used narrow 8-bit characters, its architects planned for 16-bit characters from the beginning.

With 16-bit characters, character values in the range 0–65 535 are allowed. These are mapped onto the Unicode code page whose 256 lowest characters match those of the ISO Latin 1 code page, which in turn is (with a few exceptions) the same as Windows Latin 1. Unicode is big enough to provide code points for most of the characters used by practically all the world's living languages – and many of its dead languages too. In addition to providing code points for character glyphs in languages such as Chinese, Japanese,

Korean, Thai, Hebrew, and Arabic, Unicode provides standards to support these languages' typesetting conventions, including cues for left-to-right and right-to-left changeovers.

The strategy for the technical foundation of Symbian OS was to define classes to represent text and to use them *everywhere* text was required, and *nowhere* else. By simply changing a compiler flag, it would then be possible to rebuild Symbian OS with 16-bit or 8-bit characters. Compiled code would be incompatible and so would application data files, but there would be virtually no source code changes.

So symbols were assigned for all the critical text classes such as `TText` and `TDesC`. In the narrow build, these were equated to `TText8` and `TDesC8`. In the wide build, they were equated to `TText16` and `TDesC16`. The setting of the `_UNICODE` macro would be used to control which was which.

Other classes and macros included in this scheme are as follows:

Symbol	Narrow	Wide	Meaning
TText	TText8	TText16	Character
_L	_L8	_L16	Literal (old-style)
_LIT	_LIT8	_LIT16	Literal (new-style)
TDesC	TDesC8	TDesC16	Nonmodifiable descriptor
TDes	TDes8	TDes16	Modifiable descriptor
TPtrC	TPtrC8	TPtrC16	Nonmodifiable pointer descriptor
TPtr	TPtr8	TPtr16	Modifiable pointer descriptor
TBufC	TBufC8	TBufC16	Nonmodifiable buffer descriptor
TBuf	TBuf8	TBuf16	Modifiable buffer descriptor
HBufC	HBufC8	HBufC16	Heap descriptor
TLex	TLex8	TLex16	Lexer

You can find these definitions throughout `e32def.h`, `e32std.h`, `e32des8.h`, and `e32des16.h`. The narrow classes are present even on wide builds.

Unless you are writing for the Psion PDAs that run Symbian OS v5, you don't have to worry about narrow builds any more. But the idioms that were created to help source compatibility between narrow and wide builds are still worth following, if only to make your code more readable:

- Code all your general-purpose text objects to use the neutral variants in the first column of the table above, for example, `TText` or `TDesC`.

- Where you are using descriptors to refer not to text but to binary data, code specifically 8-bit classes, for example, `TDesC8` – and use `TInt8` or `TUint8` rather than `TText8`, for an individual byte.

There are a few tricky cases. Some types of data are awkward, especially data in communications protocols that looks like a string, but is actually binary data that should always use 8-bit characters. Examples include the HELLO used to log on to a POP post office or the AT commands used for modems.

Another big issue with communications is that much of the text sent between a Symbian OS phone and the outside world, for example, in e-mails, will be encoded in non-Unicode character sets. For this situation, a dedicated library (charconv.h, charconv.lib) is provided to enable conversion, in both directions, between Unicode and other character sets. It also provides functionality for converting, again in both directions, between ordinary 2-byte Unicode and its two transformation formats UTF-7 and UTF-8 (these are ASCII-compatible encodings of Unicode that use sequences of multiple bytes to encode non-ASCII characters). The library can be extended with plug-ins to support whatever character sets are appropriate to the device.

5.6 Binary Data

You can use descriptors just as well for data as you can for strings. The main reason you can do this is because descriptors include an explicit length, rather than using NUL as a terminator.

The data API and string API offered by descriptors are almost identical. The main things to watch out for are the following:

- Use `TDesC8`, `TDes8`, `TBuf8`, `TPtr8`, and so on rather than their equivalent `TDesC`, `TDes`, `TBuf`, and `TPtr` character classes.

- Use `TUint8` or `TInt8` for bytes rather than `TText`.

- Locale-sensitive character-related functions such as `FindF()` and `CompareF()` aren't useful for binary data.

- You can use `Size()` to get the number of bytes in a descriptor, whether it's the 8-bit or 16-bit type. For `TDesC8` these are identical, and I prefer to always use `Length()`, and only use `Size()` to find the number of bytes in a text descriptor.

APIs for writing and reading or sending and receiving are often specified in terms of binary data. Here, for example, is part of `RFile`'s API:

```
class RFile : public RFsBase
  {
public:
  IMPORT_C TInt Open(RFs& aFs, const TDesC& aName, TUint aFileMode);
  IMPORT_C TInt Create(RFs& aFs, const TDesC& aName, TUint aFileMode);
  IMPORT_C TInt Replace(RFs& aFs, const TDesC& aName, TUint aFileMode);
  IMPORT_C TInt Temp(RFs& aFs, const TDesC& aPath,
         TFileName& aName, TUint aFileMode);
  IMPORT_C TInt Write(const TDesC8& aDes);
  IMPORT_C TInt Write(const TDesC8& aDes, TInt aLength);
  IMPORT_C TInt Read(TDes8& aDes) const;
  IMPORT_C TInt Read(TDes8& aDes,TInt aLength) const;

  ...
  };
```

The `Open()`, `Create()`, `Replace()`, and `Temp()` functions take a filename, which is text, and is specified as a `const TDesC&`.

The `Write()` functions take a descriptor containing binary data to be written to the file. For binary data, these classes are specified as `const TDesC8&` – the '8' indicating that 8-bit bytes are intended, regardless of the size of a character. The version of `Write()` that takes only a descriptor writes its entire contents (up to `Length()` bytes) to the file. Many APIs use descriptors exactly like this for sending binary data to another object.

There is another version of `Write()` here, which takes a length that is used to override the length in the descriptor. This is a simple convenience function. If it had not been provided in the API, and you wanted it, then instead of using `file.Write(buffer,length)`, you could use `file.Write(buffer.Left(length))`. `Left()` returns a `TPtrC` with the same address as its argument, but with its length shortened to the number of characters (or bytes) required.

Likewise, the `Read()` functions take a descriptor in which the data to be read will be put. These parameters are specified as `TDes&` types. `Read()`-like functions coded like this use one of three conventions:

- They fill the destination buffer to its `MaxLength()`.

- They fill the destination buffer with a fixed amount of data, but truncate if that would exceed `MaxLength()`.

- They fill the destination buffer with a fixed amount of data, but panic if the buffer isn't big enough.

`RFile` usually fills the entire buffer – except when reading has reached the end of the file. Many servers and device drivers follow the third convention and expect the buffer to be big enough.

Communications protocols at the stream level behave like an `RFile`. Communications protocols at the packet level behave like a server or an I/O device.

In Chapter 16 and those that follow, we'll learn a lot more about binary descriptors, as we get to grips with packet-based communications and servers for the Battleships game.

5.7   Summary

In this chapter, I've introduced you to the way Symbian OS handles strings and other data, using descriptors. The main advantages of descriptors are

- descriptors can handle both string and binary data
- they provide a uniform API for dealing with data, whether it is part of a program binary, on the stack, or on the default heap
- they prevent buffer overflow errors, but otherwise don't hide memory management issues – preserving the balance between safety and efficiency that underlies Symbian OS
- descriptors encapsulate the address and length of data
- 8-bit versions of descriptor classes allow binary data to be handled.

In terms of the classes we've seen

- for immutable strings and data, classes ending with C provide `const` functions, derived from `TDesC`
- when modifying strings or data, use classes derived from `TDes`, with its nonconst functions
- the `_LIT` and `_L` macros, the difference between them and why `_LIT` is preferred.

6

Error Handling and Cleanup

So far, we've seen three examples of code written for Symbian OS. Two of them, `hellotext`, in Chapter 1, and `strings`, in Chapter 5, use the same copy-and-paste framework code that, as I said back in Chapter 1, is there for handling errors and cleaning up. I didn't make much use of the framework in `hellotext`, but in `strings` I used it to handle out-of-memory errors when allocating `HBufCs`: I allocated them using `AllocLC()` and deleted them using `Cleanup-Stack::PopAndDestroy()`.

I also hinted in Chapter 3 that many of the naming conventions are related to cleanup – `C` for heap-based classes, `i` for member variables, and `L` for functions that can leave. The GUI example in Chapter 4 used these conventions and introduced terms such as two-phase construction, without going into the details of what is happening and why it is necessary.

In this chapter, I'm going to explain the error handling and cleanup framework. It's a vital part of Symbian OS and you'll need to become familiar with it over the course of the book. Every line of code you write – or read – will be influenced by thinking about cleanup. No other Symbian OS framework has that much impact: cleanup is a fundamental aspect of Symbian OS programming.

Because of this, we've made sure that error handling and cleanup are very effective, and very easy to do. My `strings` example was no more complicated than an equivalent example written in standard C, and yet it was fully error checked. You'll see that repeatedly, throughout the book.

Having given the subject a big build up, much of it won't become really useful until you've actually started programming with Symbian OS. Read this chapter at whatever pace that suits you, and don't try to memorize everything! You'll see the patterns I describe here again and again, and you can always come back for reference if you need to.

Symbian OS C++ for Mobile Phones.   Edited by Richard Harrison
© 2003 John Wiley & Sons, Ltd   ISBN: 0-470-85611-4

Error handling is really about producing reliable programs, and that's very important for Symbian OS. Besides the cleanup framework (which deals with environment-related errors), this chapter includes a brief section on program errors, and also on testing in relation to finding and handling environment-related errors.

6.1 What Kinds of Error?

It's easiest to start this chapter by focusing on out-of-memory (OOM) errors.

These days, desktop PCs come with at least 256 MB RAM, virtual memory swapping out onto 20 GB or more of hard disk, and users, who expect to perform frequent reboots. In this environment, running out of memory is rare, so you can be quite cavalier about memory and resource management. You try *fairly* hard to release all the resources you can, but if you forget then it doesn't matter too much – things will get cleaned up when you close the application, or when you reboot. That's life in the desktop world.

By contrast, Symbian OS phones have as little as 4 MB RAM, and usually no more than 16 MB. The RAM contains the equivalent of a PC's RAM *and* hard disk – there is no disk-backed virtual memory. Your users consider their devices to be more like mobile phones or paper organizers than desktop PCs: they are *not* used to having to reboot frequently.

You have to face some key issues here – issues that don't trouble modern desktop software developers:

- You have to program efficiently so that your programs don't use RAM unnecessarily. We've already begun to see how descriptors help with that.

- You have to release resources as soon as possible because you can't afford a running program to gobble up more and more RAM without ever releasing it.

- You have to cope with out-of-memory errors. In fact, you have to cope with potential out of memory in *every single operation* that can allocate memory because an out-of-memory condition can arise in any such operation.

- When an out-of-memory situation arises that stops some operation from happening, you must not lose any user data, but must roll back to an acceptable and consistent state.

- When an out-of-memory situation occurs partway through an operation that involves the allocation of several resources, you must clean up all those resources as part of the process of rolling back.

> **These considerations are fundamental to a successful handheld system. An operating system for mobile phones that doesn't provide the programmer with good support in these areas is doomed.**

Here are some examples of where you might run out of memory when using a real mobile phone.

When you launch a new application, resources are created first by the kernel (memory to hold information about the new process and thread used by the application), then by the application itself (the memory it requires), and then by various servers throughout the system. (The file server is constantly at the ready to load application resources, and the window server displays the application's graphics on-screen, and queues keyboard and pointer events to the application when it has focus.) At *any time* during this sequence, the attempt to allocate one of these resources may fail because of an out-of-memory condition. The whole application launch has to be called off, and the application, the kernel, and any servers that allocated resources during the aborted launch must then release them.

Application launch requires a lot of resources, so it's a good guess that the whole operation might fail if the system is running out of them. However, each of the resources required by the application is allocated *individually*, and potential failure must be detected in *each possible case*. Proper cleanup must occur wherever the failure occurred.

It's easy enough to clean up on application launch – you just kill everything and the application isn't launched. Nobody's data gets lost and the user understands what has happened.

Other examples are more demanding. Imagine that you're typing into a word processor document. Each key you press potentially expands the buffers used to store and format the rich text object that's at the heart of the word processor application. If you press a key that requires the buffers to expand, but there is insufficient memory available, the operation will fail. In this case, it would clearly be quite wrong for the application to terminate – all your typing would be lost. Instead, the document must roll back to the state it was in before the key was processed, and any memory that was allocated successfully during the partially performed operation must be freed.

In fact, the Symbian OS error handling and cleanup framework is good for more than OOM errors. Many operations can fail because of other environment conditions – reading and writing to files, opening files, sending and receiving over communications sessions. The error handling and cleanup framework can make it easier to deal with those kinds of errors too.

Even user input errors can be handled using the cleanup framework: as an example, code that processes the OK button on a dialog can allocate many resources before finding that an error has occurred. Dialog code can use the cleanup framework to flag an error and free the resources with a single function call.

There's just one kind of error that the cleanup framework can't deal with – programming errors. If you write a program with an error, you have to fix it. The best service Symbian OS can do you (and your users) is to kill your program as soon as possible when the error is detected, with enough diagnostics to give you a chance to identify the error and fix it – hopefully, before you release the program. In Symbian OS, this is a **panic**. We've heard about panics already – in this chapter, I'll explain them.

6.2   Handling Out-of-memory Errors

For application programmers, the resources you use mostly consist of the memory in your own application. With that in mind, we'll introduce the cleanup tools in the context of dealing with out-of-memory conditions in your own application (or library) code. We'll get confident with OOM handling, and then we'll look at handling cleanup with other kinds of resources.

The toolkit provided by Symbian OS for handling out of memory and for testing your OOM handling includes the following:

- The UIQ GUI framework's debug keys.

- Heap-checking tools, which check that all resources that were allocated by a function are also freed.

- Proper use of the C++ destructor to destroy any owned objects.

- Heap-failure tools, which produce deliberate out-of-memory errors for testing purposes.

- The leave mechanism, which is used to indicate an error. The fundamental function here is `User::Leave()`, which is at the heart of any `leaving` function. It does a job similar to C++'s and Java's `throw`.

- *The cleanup stack*: objects on the cleanup stack are deleted when a leave occurs.

- *The trap harness*: leave processing is caught by a trap. A leave aborts its function and any functions that called it, up to and excluding the first function that contains a trap harness. The trap harness does a similar job to C++'s and Java's `try-catch` mechanism.

- The `CBase` class, the ultimate base class of all `C` classes, which is recognized by the cleanup stack, and includes a virtual C++ destructor so that any `C` class can be cleaned up using the cleanup stack.

- The **two-phase construction** pattern necessary to ensure that C++ constructors never leave. C++ constructors must, therefore, perform only basic initialization. `CBase`-derived classes should include a `ConstructL()` function that will complete the initialization of the object.

- Naming conventions that indicate class resource-allocation patterns, whether a function has the potential to leave, and other behavior important for cleanup. These conventions go together with some rules that make it easy to address cleanup requirements in the vast majority of cases.

The toolkit may look strange at first, even if you have used another error-handling framework. Actually, it's quite easy to work with. It makes it possible to build very robust applications with only a few disciplines, and almost no programming overhead.

6.2.1 Uikon Debug Keys

Some of the memory management tools are quite conveniently accessible through debug builds of the Uikon GUI. Boot up the debug-mode emulator (`\epoc32\release\wins\udeb\epoc.exe`), start the Jotter application, and create a new note. Then type *Ctrl+Alt+Shift+C*. Uikon responds with 60 Window Server resources used.

> These *Ctrl+Alt+Shift+Key* combinations are called the Uikon debug keys. They are very useful for GUI application development, especially while you're getting used to Symbian OS.

Three Uikon debug keys give resource counts like this one:

Ctrl+Alt+Shift+A	Heap cells in use
Ctrl+Alt+Shift+B	File server resources
Ctrl+Alt+Shift+C	Window server resources

Next, try *Ctrl+Alt+Shift+P*. You should see a dialog that looks something like Figure 6.1.

In the true spirit of debugging tools developed by engineers, this dialog isn't too pretty, but it's effective. To begin with, we'll use only the top part of the dialog, which controls the **heap-failure tool**. There are three options here:

- **Off:** Allocations succeed unless there genuinely is insufficient memory.
- **Random:** Attempted allocations usually succeed (unless there really is insufficient memory), but randomly fail.

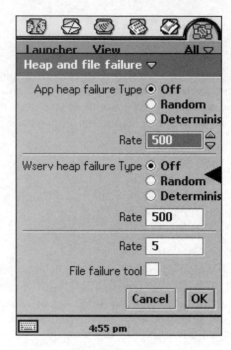

Figure 6.1

- **Deterministic:** You type in a number on the line below (say, 20). Then, allocations are guaranteed to fail on attempt number 20, 40, and so on. (They may also fail on other attempts because of genuine out of memory.)

Try setting deterministic failure every 50 attempts, and then use the Jotter application to do some typing. Every two or three letters, you'll get an out-of-memory error, but you'll lose no data. This demonstrates that Symbian OS applications are written to handle out-of-memory conditions properly.

It also demonstrates that the rich text object that you're editing in the Jotter is doing a lot of work for you, given that it's making around twenty allocations per key. Most of these are unallocated again before the key press has been completely handled. In addition, some of the allocations are connected with undo functionality.

If you like, you can try a few more experiments by setting the window server and file server failure modes as well. If you get stuck, use Ctrl+Alt+Shift+Q to reset all the failure modes to normal, so that operations fail only if there are genuine resource problems.

The debug keys are useful to have around. We'll see a few more, later in the book. Note that they only affect the operation of the main thread of the UIQ application in which you're currently working. So, you can't use debug keys to help with application launch, or with other threads. For that, you'll have to write your own code into the application you want to debug. It's not such a great restriction: native Symbian OS applications implement multitasking using active objects rather than multithreading, so usually, there won't be more than a single thread around.

6.2.2 The Memorymagic Application

We're going to use UIQ's debug build facilities – including the debug key technology we introduced above – to demonstrate how the cleanup framework works at low levels. To show these off effectively, I've written an application called memorymagic. Although it's a UIQ application, all you need to understand about UIQ for the moment is how to build a UIQ application. As a reminder,

```
cd \scmp\memorymagic\group
bldmake bldfiles
abld build winscw udeb
```

Alternatively you can use the CodeWarrior IDE to import the project from its **.mmp** file and build it.

Here's the initial screen display in Figure 6.2

Figure 6.2

The behavior described in this chapter assumes you are debugging the application from the CodeWarrior IDE. memorymagic is a UIQ application but you don't need to be familiar with UIQ to understand this example. The important material is in the declarations of classes CX, CY, and CZ in memorymagic.h, and in the implementation of these classes in memorymagic.cpp. The main action in the program happens in the CExampleAppUi class: HandleCommandL() contains small functions that are invoked when you press toolbar buttons, and the destructor ~CExampleAppUi() is also important.

It's fun to play with memorymagic, but if you're in a hurry you can get what you need from this chapter without browsing the source code or running the application, since relevant screenshots and extracts are all included.

The **Alloc Info** menu item (equivalent to *Ctrl+Alt+Shift+A*) tells you how many heap cells have been allocated – 455 on my system, totaling 19 248 bytes – immediately after the application has been launched. It should be pretty close to that on your system too.

6.2.3   Allocating, Destroying, and Heap Balance

Let's use memorymagic's **New 1** and **Delete 1** menu items to show some basic principles. If you check through the source code in the .cpp file, you'll see that **New 1** is handled by the following code:

```
case EMagicCmdNew1:
    iObject1 = new CX;
    iEikonEnv->InfoMsg(_L("New 1"));
    break;
case EMagicCmdDelete1:
    delete iObject1;
    iEikonEnv->InfoMsg(_L("Delete 1"));
    break;
```

So, the **New 1** menu item allocates and constructs a CX object (no need to worry, for now, what a CX is), and stores a pointer to it in CExampleAppUi::iObject1. **Delete 1** destroys the object.

Both commands use EIKON's environment class, which includes a large function library to display an info-message, so you know you've successfully invoked the function. The info-message flashes briefly on the top right corner of the screen.

Select **New 1** from the menu, and then check the number of heap cells allocated: it's up by one, as you'd expect. Select **Delete 1**, and then check the number of heap cells allocated again: it's down by one. Now close the application. Assuming you did nothing else, all will be well.

Heap-balance checking

Launch `memorymagic` again, select **New 1**, and then close the application. This time, it's different. You get:

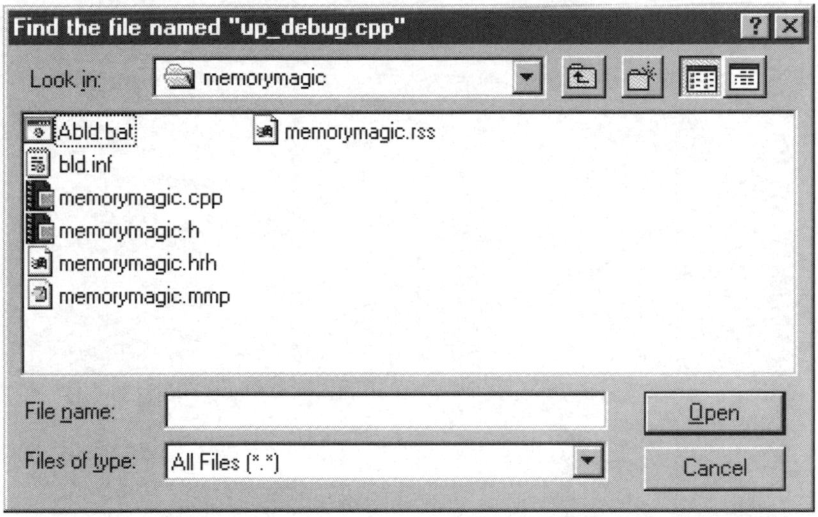

Figure 6.3

When you're debugging with the emulator, this is your first indication that a thread has panicked. CodeWarrior is asking you for the source code that contains the Symbian OS panic function; unfortunately, it isn't shipped with the SDK, so all you can do is press *Escape*.

You'll then get a whole pile of disassembly in the main debugger window (because you didn't have the source). That's not too useful, so select the `CCoeEnv` destructor from the call stack, as shown in Figure 6.4.

Remember that in the framework for `hellotext`, I included a `__UHEAP_MARK` macro at the beginning, and `__UHEAP_MARKEND` at the end. The GUI framework includes those macros for any GUI program, and the `__UHEAP_MARKEND` is called from its destructor, as you can see in the source code in the Figure 6.3.

`__UHEAP_MARKEND` is there to check whether the heap is properly balanced. If the heap doesn't have the same number of cells allocated as when `__UHEAP_MARK` was called, your program gets panicked. You can use `__UHEAP_MARK` and `__UHEAP_MARKEND` in your own code, and you can nest them. In this case, we know why the heap isn't balanced; we allocated a cell and didn't delete it again.

All GUI programs benefit from this check. Your program may work perfectly satisfactorily until it exits, but if you forgot to delete any objects you allocated, the framework will panic your program on exit.

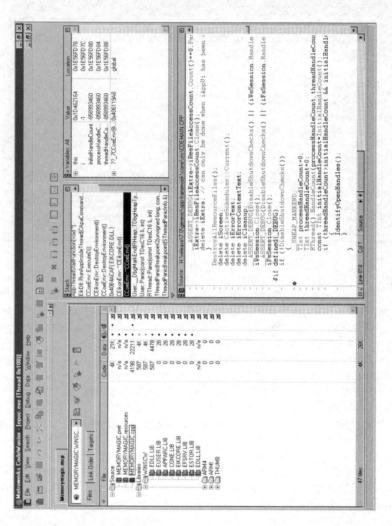

Figure 6.4

> This ensures that heap imbalance gets picked up early in the development cycle, a great service to all Symbian OS programs. You don't even have to be careful about testing in order to benefit from this check: you get this panic whether you asked for it or not.

This feature is built into all debug builds of Symbian OS – including the emulator, which is the normal development environment. It isn't built into release builds because by then it isn't needed. Also, heap marking requires an otherwise unnecessary counter to be updated, which slows things down.

Destroying objects properly

The solution to this problem is really pretty basic for C++ programmers:

> **Use a destructor to destroy objects you own.**

The right thing to correct the above problem would be to amend the destructor, `CExampleAppUi::~CExampleAppUi()`, to include

```
delete iObject1;
```

We haven't included this line in `memorymagic` (although we do have `delete` statements for other objects that get created in other circumstances, as we'll see). You might wish to add this line yourself, and then rerun the scenario above. Select **New 1**, and then exit. The destructor should ensure heap balance, and the program will exit cleanly.

To summarize, you should only delete objects that you own – that is, objects with which you have a *has-a* relationship.

Don't allocate twice

Once you've added `delete iObject1` to `CExampleAppUi`'s destructor, rebuild `memorymagic` and start it again. Select **New 1** *twice*, and then exit. You'll get another panic.

The reason for this should be obvious. The second time through, we simply created a second `CX`, stored its address in `iObject1`, and forgot the address of the `CX` object we allocated previously. There was no way for the destructor – or any other part of the C++ system – to find this object, so it couldn't be deleted.

We could have solved this by coding:

```
case EMagicCmdNew1:
    delete iObject1;
    iObject1 = new CX;
    iEikonEnv->InfoMsg(_L("New 1"));
    break;
```

Here, we're taking advantage of the zero-checking service that C++ provides as part of `delete`. If the object has already been allocated, we delete it and allocate it again. If it hasn't yet been allocated, the `delete` statement will do nothing.

If you build this new line into `memorymagic`, you should find that it exits cleanly no matter how many times you select **New 1**.

There is one other technique that can help us to avoid allocating twice. We can ensure that this does not happen by adding an ASSERT before allocating member variables, like this:

```
ASSERT(iObject1 == NULL);
iObject1 = new CX;
```

Don't delete twice

If double allocation is a serious crime, then deleting twice is a capital offense. If you *allocate* something twice, the result is a heap cell that doesn't get destroyed, which gets picked up eventually by a `__UHEAP_MARKEND`. If you *delete* something twice, the effects can be more subtle.

`memorymagic` can demonstrate this effect too. Select **New 1** from the menu, and then select **Delete 1** twice. In this case, I found that the application panics immediately, but in other situations I haven't always been so lucky: double deletion doesn't always cause an immediate crash, and sometimes it leaves side effects that only surface a long time after the real problem – the double delete – occurred. As a result, double deletes are very hard to debug.

On the other hand, double deletes are easy to avoid – just follow this little discipline:

> **C++ `delete` does not set the pointer to zero. If you delete any member object from outside its class's destructor, *you* must set the member pointer to NULL.**

The Symbian OS `i` naming convention for class members makes it even easier to see when this is required. In our case, we need to amend the handler for the `EMagicCmdDelete1` command:

```
case EMagicCmdDelete1:
    delete iObject1; iObject1 = NULL;
```

```
iEikonEnv->InfoMsg(_L("New 1"));
break;
```

*We do not have to do this in the destructor because you know there
that iObject1 is never going to be used again.*

Be clear about ownership. As I said above, your destructor should only
delete objects that you own. This is fairly easy to control in 99 percent
of cases in which ownership is not transferred. The majority of double-
delete bugs probably arise from misunderstanding the consequences of
ownership transfer – you just have to be careful.

Rely on zeros

Just as C++ provides a zero-checking service as part of `delete`, Symbian
OS provides a companion service as part of `new`, for any C class.

> **Symbian OS uses an overloaded `CBase::operator new()` to
> ensure that any object derived from `CBase` is zero initialized when
> first allocated with `new`. This means in particular that all pointers in
> any C class are zero initialized.**

This is a huge boon to Symbian OS C++ programmers. It means that
you don't have to set pointers to zero yourself from the C++ constructor,
just in case the objects pointed to are subsequently `deleted`. If you're
used to initializing every member of every C++ object you create, it may
be hard at first to trust Symbian OS to zero initialize C objects, but in a
context in which every attempted allocation may fail, this feature is here
to help you, as we'll see.

Finally, a word of warning: don't assume that this will happen to
other objects that you allocate on the heap! R and T objects need proper
construction. The only exception is if you can guarantee that all instances
of an R or T class will be members of a C class.

Note that zero initialization won't happen to C objects if they are
allocated on the stack. C objects are fundamentally not designed to go on
the stack – they should always be allocated on the heap.

6.2.4 Heap Failure

I've labored the points about using C++ destructors properly and relying
on zero initialization in C classes. Now we can begin to tackle out-of-
memory errors.

Leave if you can't allocate memory

memorymagic has two menu items, **New 2** and **Delete 2**, whose handlers are coded with all the lessons we learned earlier. In addition, we have an out-of-memory check. The command handlers look like this:

```
case EMagicCmdNew2:
    delete iObject2;
    iObject2 = new(ELeave) CX;
    iEikonEnv->InfoMsg(_L("New 2"));
    break;
case EMagicCmdDelete2:
    delete iObject2; iObject2 = NULL;
    iEikonEnv->InfoMsg(_L("Delete 2"));
    break;
```

Also, as you saw earlier, the CExampleAppUi destructor includes delete iObject2.

The following line needs expanding:

```
iObject2 = new(ELeave) CX;
```

It's equivalent to

```
{
CX* temp = ::new CX;
if(!temp)
    User::Leave(KErrNoMemory);
Mem::FillZ(temp, sizeof(CX));
iObject2 = temp;
}
```

In other words, if the allocation fails, the function leaves with an out-of-memory error. If it doesn't leave, it fills the CX with zeroes using Mem::FillZ().

new(ELeave) is an overloaded version of the new() operator. The C++ language allows the operator new() to be overridden to take a parameter that allows programmers to write their own memory allocator – something that Symbian OS exploits. There is an enumerated type called TLeave with a single constant, ELeave. CBase::operator new(TLeave) contains the code above. You invoke it by calling new(ELeave) classname; the value ELeave is thrown away, but its type is used to invoke the correct operator new(). The global ::operator new(TLeave) is overridden in the same way – but it doesn't zero fill.

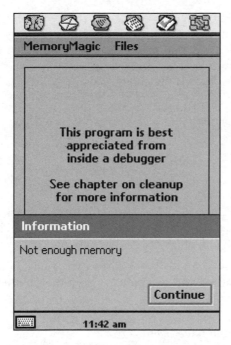

Figure 6.5

Try this in `memorymagic`. Start the program from the CodeWarrior IDE, press *Ctrl+Alt+Shift+P*, and set the **App heap** failure mode to **Deterministic, 1**. Then press 3 (New 2). This should fail, and you'll see

Reset the heap-failure mode with *Ctrl+Alt+Shift+Q*, and then select any combination of **New 2** and **Delete 2** again. Exit the program. Everything should be clean.

The **Not enough memory** message was produced by the Symbian OS framework, which contains a trap for all functions that leave – in this case, the `new` operator function. This means that, without any programmer support, you can be sure that leaves are trapped, and an error message is displayed. Later, we'll see what a trap looks like.

In most cases, you should use `new(ELeave)` rather than plain `new`. Symbian considered making the default behavior for `new` to leave, but rejected this on the grounds that it would break compatibility for code being ported. This was perhaps an unfortunate choice: as in Standard C++ the default is now to throw an exception.

Reallocate properly

Now try another experiment with `memorymagic`. Start the program, press *Ctrl+Alt+Shift+P*, and set the **App heap-failure** mode to **Deterministic, 2**.

Press 3 (**New 2**). Then press 3 again. This will fail, and you'll see the **Not enough memory** message. Then, close the application using Ctrl-e.

You'll get the panic that means that you have heap imbalance. Why is this? The answer is that the code for reallocating `iObject2` didn't anticipate failure. Here is the code again:

```
case EMagicCmdNew2:
    delete iObject2;
    iObject2 = new(ELeave) CX;
    iEikonEnv->InfoMsg(_L("New 2"));
    break;
```

If the `new(ELeave)` fails, then `iObject2` is still pointing to the previously allocated object, even though that object has been deleted. The result is going to be a double-deletion error, one way or another.

> **When reallocating, you must always zero the pointer after `delete` and before `new`.**

The code should have been

```
case EMagicCmdNew2:
    delete iObject2;
    iObject2 = NULL;
    iObject2 = new(ELeave) CX;
    iEikonEnv->InfoMsg(_L("New 2"));
    break;
```

Once again, there is no need to set pointers to zero after a `delete` in the destructor because there is no chance that the pointer will ever be used again. But *at all other times*, if you delete something, you should zero its pointer immediately.

6.2.5 How Does Leave Work?

The `User::Leave()` function causes execution of the active function to terminate, and on through all calling functions, until the first function is found that contains a `TRAP()` or `TRAPD()` macro.

`User::Leave()` is defined in `e32std.h`; you pass a single 32-bit integer error code.

```
class User
    {
```

```
public:
    IMPORT_C static Leave(TInt aErrorCode);
    ...
    };
```

The `TRAP()` and `TRAPD()` macros are also defined in `e32std.h`; edited slightly, they look like this:

```
#define TRAP(_r,_s)   { \
                        TTrap __t; \
                        if(__t.Trap(_r) == 0) \
                            { \
                            _s; \
                            TTrap::UnTrap(); \
                            } \
                        }
#define TRAPD(_r,_s) TInt _r; \
                        { \
                        TTrap __t; \
                        if(__t.Trap(_r) == 0) \
                            { \
                            _s; \
                            TTrap::UnTrap(); \
                            } \
                        }
```

Macros like this aren't really meant to be understood, and the `TTrap` class isn't something you need to know about. The main point here is that `TRAP()` calls a function (its second parameter) and returns its leave code in a 32-bit integer (its first parameter). If the function returns normally, without leaving, then the leave code will be `KErrNone` (which is defined as zero). `TRAPD()` defines the leave code variable first, saving you a line of source code, and then essentially calls `TRAP()`.

As we saw in Chapter 3, GUI applications are fundamentally event-driven. All code in a Symbian OS application runs under a `RunL()` function. If this leaves, the error is trapped by the active scheduler, using the following code:

```
TRAPD(r, pR->RunL());
if(r != KErrNone)
    pS->Error(r);
```

The active scheduler calls its `Error()` function, which is implemented to run the one interesting piece of code that runs outside a `RunL()` – its error dialog. The error dialog interprets all the standard error codes and displays an error message, for instance, `KErrNoMemory`, which is −4, is displayed as **Out of memory**.

Error codes

Symbian OS approach to error codes is very simple. Standard error values are listed in `e32std.h`. Here are some of them:

```
const TInt KErrNone = 0;
const TInt KErrNotFound = (-1); // Must remain set to -1
const TInt KErrGeneral = (-2);
const TInt KErrCancel = (-3);
const TInt KErrNoMemory = (-4);
const TInt KErrNotSupported = (-5);
const TInt KErrArgument = (-6);
...
const TInt KErrBadPower = (-42);
const TInt KErrDirFull = (-43);
```

L functions

> **It is very important to know whether a function might leave. Any function that could leave should have a name ending with `L`, so we get `RunL()`, `HandleCommandL()`, etc.**

If you're writing a function, and that function calls another one that might leave, then you should put an `L` in its name. This will remind you, your colleagues, or anyone else looking at your code, that this function might leave. Here are a few cases to think about:

- If you're writing a function that calls another, and the other is an `L` function, then your function must be an `L` function too (unless you trap the leave).

- If your function calls `new(ELeave)`, your function must also be an `L` function.

- If you're implementing a function provided by a framework, then it's important to know whether that framework function is an `L` function. If it is, you can allocate resources that can potentially leave. If not, your code must not leave – or, if it does, you must handle it privately. For example, `CEikAppUi::HandleCommandL()` allows you to leave, so user command handlers can allocate resources and potentially leave. `CCoeControl::Draw()` doesn't allow you to leave, so any drawing code you write must work with preallocated resources and not leave or, if for its own reasons it can't preallocate resources, it must trap potential leaves and handle them privately.

- If you're specifying a framework function for others to implement, think very carefully about whether you will allow an implementer to leave. This is an essential aspect of your function. Don't just code it as an L function to allow the implementer to do what they want; code L if it's needed, and don't code it if it's not.

The L naming convention gives the same kind of message to the programmer as the `throws` clause in Java or standard C++. However, L is not checked by the compiler, so if you forget to include it in the name, the compiler will not complain. However, if someone else then uses your function and doesn't realize that it might leave – they may not take appropriate precautions.

In the C++ Knowledge Base on Symbian's website, you'll find an L-correctness checking script that can be quite useful.

Nested traps

You don't often need to code your own traps, partly because the Symbian OS framework provides one for you, and partly because (as we'll see shortly) routine cleanup is handled by other means.

Sometimes, though, you'll need to perform a recovery action, in addition to some nonroutine cleanup. Or you'll need to code a nonleaving function, such as `Draw()` that allocates resources and draws successfully if possible, but otherwise traps any leaves and handles them internally. In these cases, you must code your own trap.

In a word processor, for example, processing a key press causes many things to happen – to name but two, there's the allocation of undo buffers, and the expansion of the document to take a new character. If anything like that goes wrong, you need to undo the operation completely. You could use code such as

```
TRAPD(error, HandleKeyL());
if(error)
    {
    RevertUndoBuffer();

    // Any other special cleanup
    User::Leave(error);
    }
```

This performs some specialized cleanup, and then leaves anyway, so that the Symbian OS framework can post the error message.

While they will not normally need to allocate any resources, some `Draw()` functions are rather complicated and it may be appropriate to code them in such a way that they do make allocations. In this case, you have to hide the fact from the Symbian OS framework by trapping any failures yourself.

```
virtual void Draw(const TRect& aRect) const
    {
    TRAPD(error, MyPrivateDrawL(aRect));
    }
```

In this case, we choose to keep quiet if `MyPrivateDrawL()` failed. The failing draw code should take some graceful action such as drawing in less detail, or blanking the entire rectangle, or not updating the previous display.

Don't use traps when you don't have to. Here's a particularly useless example:

```
TRAPD(error, FooL());
if(error)
    User::Leave(error);
```

The net effect of this code is precisely equivalent to

```
FooL();
```

But it's more source code, more object code, and more processing time whether or not `FooL()` actually does leave.

6.2.6 The Cleanup Stack

Now we're ready to look at the cleanup stack. The cleanup stack addresses the problem of cleaning up objects that have been allocated on the heap, but to which the *only* pointer is an automatic variable. If the function that has allocated the objects leaves, the objects need to be cleaned up. Since Symbian OS doesn't use C++ exceptions, it needs to use its own mechanism to ensure that this happens.

In `memorymagic`, the **Use 3** menu item is handled by the following code:

```
case EMagicCmdUse3:
    {
    CX* x = new(ELeave) CX;
    x->UseL();
    delete x;
    }
```

The `UseL()` function might leave. It's coded as

```
void CX::UseL()
    {
    TInt* pi = new(ELeave) TInt;
    delete pi;
    }
```

You can invoke these functions from `memorymagic` by selecting **Use 3**, so go ahead and try it. When you exit the application, everything should be OK.

The code above could go wrong in two places. Firstly, the allocation of `CX` might fail – if so, the code leaves immediately with no harm done. You can try this in `memorymagic` too; set the heap to fail on the first allocation, press 5 (**Use 3**), and you'll see the **Not enough memory message**. Exit the application (by pressing Ctrl-e), and all will be well.

But what if the allocation of the `TInt` fails in `UseL()`? In this case, the `CX`, which is pointed to only by an automatic variable `x` in `HandleCommandL()`, can never be deleted. Try it – start `memorymagic`, set the heap to fail on the *second* allocation, press 5, see the **Not enough memory** message, and then exit the application. The heap check will find a memory leak, and panic the program.

Use the cleanup stack if you need it

What's actually happening here is that after this line has been executed:

```
CX* x = new(ELeave) CX;
```

The automatic `x` points to a cell on the heap. But after the leave, the stack frame containing `x` is abandoned without deleting `x`. That means

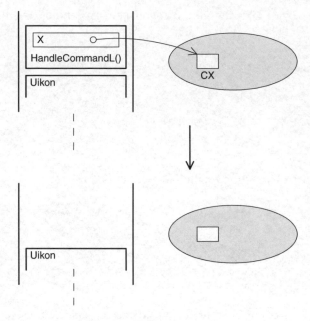

Figure 6.6

the CX object is on the heap but no pointer can reach it and it will never get destroyed. That's why the heap check fails.

The solution to this problem is to make use of the cleanup stack provided by the UIQ framework.

The cleanup stack can be used to hold a pointer to an object like this. Objects on the cleanup stack are destroyed when a leave occurs.

Here's how we should have coded the handler for **Use 3**:

```
case EMagicCmdUse3:
    {
    CX* x = new(ELeave) CX;
    CleanupStack::PushL(x);
    x->UseL();
    CleanupStack::PopAndDestroy(x);
    }
```

The cleanup stack class, CleanupStack, is defined in e32base.h. With these changes in place, here's what happens:

- Immediately after we have allocated the CX and stored its pointer in x, we also push a copy of this pointer to the cleanup stack.
- We then call UseL().

- If this doesn't fail, our code pops the pointer from the cleanup stack and deletes the object. We could have used two lines of code for this (`CleanupStack::Pop()`, followed by `delete x`), but this is such a common pattern that the cleanup stack provides a single function to do both.
- If `UseL()` *does* fail, then *as part of leave processing*, all objects on the cleanup stack are popped and destroyed anyway.

So our code works whether `UseL()` leaves or not. Change `memorymagic` to include this code, rebuild, set the heap to fail on the second allocation, invoke **Use 3**, and then exit the application. Once again, all will be well.

Of course, we could have done this without the aid of the cleanup stack by using code like this:

```
case EMagicCmdUse3:
    {
    CX* x = new(ELeave) CX;
    TRAPD(error, x->UseL()):
    if(error)
        {
        delete x:
        User::Leave(error);
        }
    delete x;
    }
```

However, this is much less elegant. The cleanup stack works particularly well for a long sequence of operations, such as

```
case EMagicCmdUse3:
    {
    CX* x = new(ELeave) CX;
    CleanupStack::PushL(x);
    x->UseL():
    x->UseL();
    x->UseL();
    CleanupStack::PopAndDestroy(x);
    }
```

Any one of the calls to `UseL()` may fail, and it would begin to look very messy if we had to surround *every* L function with a trap harness just to address cleanup.

Notice that we passed x to the `CleanupStack::PopAnd` *`Destroy()` function to indicate that we believe x is the variable that is being cleaned up. It isn't strictly necessary, but it can prove reassuring in practice.*

C++'s native exception handling addresses the problem of automatics on the stack by calling their destructors explicitly so that a separate cleanup stack isn't needed. C++ exception handling was not available at all on GCC, or reliably on Microsoft Visual C++, when Symbian OS was designed, so it wasn't an option to use it.

Don't use the cleanup stack when you don't need It

You only need to use the cleanup stack to prevent an object's destructor from being bypassed. If the object's destructor is going to be called anyway, then you must not use the cleanup stack.

> **If an object is a member variable of another class (rather than an automatic like x), then it will be destroyed by the class's destructor, so you should never push a member variable to the cleanup stack.**

Member variables are indicated by an i prefix, so a code like this is *always* wrong:

```
CleanupStack::PushL(iMember);
```

This is likely to produce a double deletion (once from the cleanup stack, once from the class's destructor).

What if `CleanupStack::PushL()` fails?

Pushing to the cleanup stack may potentially allocate memory, and therefore may itself fail! You don't have to worry about this because such a failure will be handled properly. But for reassurance, here's what happens under the covers.

Symbian OS addresses this possibility by always keeping at least one spare slot on the cleanup stack. When you do a PushL(), the object you are pushing is first placed on the cleanup stack (which is guaranteed to work because there was a spare slot). Then, a new slot is allocated. If *that* fails, then the object you just pushed is popped and destroyed.

The cleanup stack actually allocates more than one slot at once, and doesn't throw away slots that have been allocated when they are popped. So pushing and popping from the cleanup stack are very efficient operations.

Since the cleanup stack is used to hold fairly temporary objects, or objects whose pointers haven't been stored as member pointers in their parent object during the parent object's construction, the number of cleanup stack slots ever needed by a practical program is not too high. More than ten would be very rare. So the cleanup stack itself is very unlikely to be a contributor to out-of-memory errors.

CBase and the cleanup stack

When we pushed x to the cleanup stack, we actually invoked the function `CleanupStack::PushL(CBase* aPtr)` because CX is derived from CBase.

When a subsequent `PopAndDestroy()` happens, this function can only call the destructor of a CBase-derived object. We have already noted that `CBase::operator new()` zero initializes all member variables; now we meet CBase's second important property:

CBase's destructor is a virtual function.

This means that any object derived from CBase can be pushed to the cleanup stack and, when it is popped and destroyed, its destructor is called (as you would expect).

The cleanup stack and C++ destructors make it very easy for a programmer to handle cleanup. Use the cleanup stack for objects pointed to only by C++ automatics. Use the destructor for objects pointed to by member variables. It just works. You very rarely need to use TRAP(). The resulting code is easy to write, compact, and efficient.

6.2.7 Two-phase Construction

The cleanup stack is used to hold pointers to heap-based objects so that they can be cleaned up if a leave occurs. This means that you must have the opportunity to push objects to the cleanup stack. One key situation in which this would not be possible when using normal C++ conventions is in between the allocation performed by new, and the invocation of a C++ constructor that follows the allocation.

This problem requires us to invent a new rule that C++ constructors cannot leave. We also need a work-around: two-phase construction.

C++ constructors must not leave

Let's see what happens if we allow C++ constructors to leave.

memorymagic has a class called CY that contains a member variable that points to a CX. Using conventional C++ techniques, we allocate the CX from the constructor

```
class CY : public CBase
    {
public:
    CY();
    ~CY();
public:
    CX* iX;
    };
CY::CY()
    {
    iX = new(ELeave) CX;
    }
CY::~CY()
    {
    delete iX;
    }
```

memorymagic's **Use 4** menu item calls cleanup-friendly code, as follows:

```
case EMagicCmdUse4:
    {
    CY* y = new(ELeave) CY;
    CleanupStack::PushL(y);
    y->iX->UseL();
    CleanupStack::PopAndDestroy(y);
    }
```

Looks good, doesn't it? We have used C++ constructors in the usual way, and we've used the cleanup stack properly too. Even so, this code *isn't* cleanup safe. It makes *three* allocations as it runs through:

- The command handler allocates the CY: if this fails, everything leaves and there's no problem.

- The CY constructor allocates the CX: if this fails, *the code leaves, but there is no CY on the cleanup stack*!

- CX::UseL() allocates a TInt: by this time, the CY *is* on the cleanup stack, and the CX will be looked after by CY's destructor, so if this allocation fails, everything gets cleaned up nicely.

Try running it through (press 6 to invoke Use 4) and setting the heap to fail after one, two, and then three allocations. You can verify that, if the second allocation fails, you get a memory leak. Needless to say, this is bad news. The trouble is that the C++ constructor is called at a time when no pointer to the object is accessible to the program. This code

```
CY* y = new(ELeave) CY;
```

is effectively expanded by C++ to

```
CY* y;
CY* temp = User::AllocL(sizeof(CY));    // Allocate memory
temp->CY::CY();                         // C++ constructor
y = temp;
```

The problem is that we get no opportunity to push the CY to the cleanup stack between the User::AllocL() function, which allocates the memory for the CY, and the C++ constructor, which might leave. And there's nothing we can do about this.

> **It's a fundamental rule of Symbian OS programming that no C++ constructor should contain any functions that can leave.**

Use ConstructL() for construction that might leave

We have to work around this restriction by providing a separate function to do any initialization that might leave. We call this function the **second-phase constructor** and, we usually name it ConstructL().

In memorymagic, we have at last coded a class correctly: CZ. This class is *like* CY, but it uses a second-phase constructor. Here's the class definition:

```
class CZ : public CBase
    {
public:
    static CZ* NewL();
    static CZ* NewLC();
    void ConstructL();
    ~CZ();
public:
    CX* iX;
    };
```

Here are the ConstructL() and destructor functions (we'll return to NewL() and NewLC() later):

```
void CZ::ConstructL()
    {
    iX = new(ELeave) CX;
```

```
    }
CZ::~CZ()
    {
    delete iX;
    }
```

So `CZ::ConstructL()` performs the same task as `CY::CY()`, but the leaving function `new(ELeave)` is now in a second-phase constructor, rather than the C++ constructor. Here's how we can use it:

```
case EMagicCmdUse5:
    {
    CZ* z = new(ELeave) CZ;
    CleanupStack::PushL(z);
    z->ConstructL();
    z->iX->UseL();
    CleanupStack::PopAndDestroy(z);
    }
```

It's quite clear here that `ConstructL()` is cleanup safe. You can try it out by running `memorymagic` and setting the heap to fail on the first, second, and third allocations.

First- and second-phase constructors

A class's initialization often involves

- copying some constructor arguments to class member variables
- calling base-class constructors
- invoking some functions that cannot leave
- invoking some functions that can leave.

In C++, the rules are clear and basically simple (provided that you take a practical approach to software engineering – that is, you use only public single inheritance, and keep names distinct). With a second-phase constructor, you need to consider the following:

- Functions that can leave must *only* be called from the second-phase constructor.
- Constructor arguments and member initialization can happen from either constructor, or even from both.
- The base-class C++ constructors are automatically called by C++ in the usual way.

- If base classes have `ConstructL()` functions, you must call them explicitly from your `ConstructL()` – preferably before you do anything else. You will have to use explicit scoping (`CMyBase::ConstructL()`, for example).

- If you're a base class with a second-phase constructor, you might want to give your second-phase constructor a distinctive name, so that it can be called by derived classes without the explicit scoping. `CEikAppUi`, for example, has a `BaseConstructL()` function that's called by derived classes.

It's good to do all the initialization you can from the C++ constructor, but you *must* initialize anything that could leave from the second-phase constructor.

When you use someone else's class, be sure to check whether you are required to perform single-phase or two-phase construction. There is no naming convention for classes as such; you have to check for `ConstructL()` or similarly named functions manually.

Throughout the book, and indeed in the UIQ programs we've seen already, we'll see examples of all these things happening.

Wrapping up `ConstructL()`: `NewL()` and `NewLC()`

Working with the two-phase constructor pattern can be a bit inconvenient because the user of the class has to remember to call the second-phase constructor explicitly. This is less than the full support we are accustomed to from C++ constructors. A pattern that has emerged to address this problem is the use of functions called `NewL()`. `CZ` has a static `NewL()` function that's coded as follows:

```
CZ* CZ::NewL()
    {
    CZ* self = new(ELeave) CZ;
    CleanupStack::PushL(self);
    self->ConstructL();
    CleanupStack::Pop(self);
    return self;
    }
```

Because `CZ::NewL()` is static, you can call it without any existing instance of `CZ`. The function allocates a `CZ` with `new(ELeave)` and then pushes it to the cleanup stack so that the second-phase constructor can safely be called. If the second-phase constructor fails, then the object is popped and destroyed by the rules of the cleanup stack. Otherwise, the object is simply popped, and the pointer is returned. `CZ::NewL()`

operates as a **factory function** – a static function that acts as a kind of constructor.

You can see how this is used in `memorymagic`'s **Use 6** handler:

```
case EMagicCmdUse6:
    {
    iObject6 = CZ::NewL();
    iObject6->iX->UseL();
    delete iObject6;
    }
```

We didn't use the cleanup stack here because `iObject6` is a member variable. The important thing about the code above is that nothing can leave between the `CleanupStack::Pop()`, and the assignment of the `CZ*` to `iObject6`.

If we want to refer to the new `CZ` from an automatic variable, then we need the cleanup stack throughout the lifetime of the `CZ`. We could use the `NewL()` above and then push the object to the cleanup stack on return, but that would be wasteful. Instead, we use a static `NewLC()` function that operates like `NewL()` but *doesn't* pop the object from the cleanup stack before returning:

```
CZ* CZ::NewLC()
    {
    CZ* self = new(ELeave) CZ;
    CleanupStack::PushL(self);
    self->ConstructL();
    return self;
    }
```

This allocates the new `CZ` and, if all goes well, leaves it on the cleanup stack – that's what the `C` in `NewLC()` stands for. If anything fails, the function leaves and everything is cleaned up.

This is used from **Use 7**, whose handler is

```
case EMagicCmdUse7:
    {
    CZ* z = CZ::NewLC();
    z->iX->UseL();
    CleanupStack::PopAndDestroy(z);
    }
```

This is very convenient, but it's not necessary to use `NewL()` and `NewLC()` in *every* class. If your class is a one-off class, designed for use only in your application, then it's a waste to code `NewL()`-type functions, and it's easy enough to code `new` followed by `ConstructL()`

in the one place where your application constructs an instance of this class. For classes designed for frequent reuse, the case for encapsulating construction in `NewL()`-type functions is more compelling.

For classes with more than one C++ constructor such as application documents that may construct either a blank document, or a document loaded in from file, there is a very strong case to wrap up the various construction sequences in distinct `NewL()`-type functions.

During development, you sometimes change your mind about things like this and equip a class with `NewL()`-type functions, or occasionally even remove them. The presence or absence of these convenience functions is not a fundamental property of a class, which is probably why it's not reflected in a naming convention.

`CZ::NewL()` was actually coded wastefully. If you're coding both a `NewLC()` and a `NewL()`, write the `NewLC()` first and then code `NewL()` as

```
CZ* CZ::NewL()
    {
    CZ* self = CZ::NewLC();
    CleanupStack::Pop();
    return self;
    }
```

If you can push yourself to the cleanup stack from within `NewL()`, why not do it from the C++ constructor and allow leaves from within the constructor? Symbian considered this seriously, but for heavily derived classes there is no way to avoid repeatedly pushing and popping the class from the cleanup stack from all its base-class constructors.

6.3 Summary of Cleanup Rules

We have used the `memorymagic` application and a few pages of commentary to explain the principles of memory management and cleanup, but we could summarize all of it in a single rule:

- Whenever you see something that might leave, think about what happens (1) when it doesn't leave, and (2) when it does.

In practice, it's better to operate with a few safe patterns, so here's a more detailed summary of the rules we've seen:

- Always delete objects your class owns, from the class destructor.
- Don't delete objects that you don't own (that is, those that you merely use).
- Don't allocate twice (this will cause a memory leak).

- Don't delete twice (this will corrupt the heap).
- When you delete outside the destructor, immediately set the pointer to zero.
- When you are reallocating, you must use the sequence 'delete, set pointer to zero, allocate', just in case the allocate fails.
- Use `new(ELeave)` rather than plain `new`.
- Use `L` on the end of the name of any function that might leave.
- Use traps where you need to – for instance, when a function can't leave and must handle errors privately – and not where you don't.
- Push an object to the cleanup stack if (1) that object is otherwise only referred to by an automatic pointer, and (2) you are going to call a function that might leave during that object's lifetime.
- Never push a member variable to the cleanup stack – use the `iMember` naming convention.
- Give all heap-based classes a name beginning with `C`, and derive them from `CBase`. Trust `CBase` to zero initialize all data, including all pointers. Exploit `CBase`'s virtual destructor for cleanup purposes.
- Never leave from a C++ constructor.
- Put construction functions that might leave into a second-phase constructor such as `ConstructL()`.
- Optionally, use `NewL()` and `NewLC()` to wrap up allocation and construction.

It doesn't take long to get familiar with these rules and to work with them effectively. It's easy to trap many forms of misbehavior by using the debug keys provided by UIQ and the heap checking provided by CONE.

6.4   C and T Classes

We've already seen that the naming convention for classes has been chosen to indicate their main cleanup properties. So far, I've described the cleanup-related properties of C classes. I'll briefly review them, and then reintroduce T classes, which are quite similar.

C classes are derived from CBase and allocated on the heap. They must, therefore, be cleaned up when they are no longer needed. Most C classes have a destructor.

C classes are referred to by a pointer – a member variable of some class that owns it, a member variable of a class that uses it, or an automatic variable.

If a C class is referred to only by a single automatic, in a function that might leave, then the pointer should be pushed to the cleanup stack.

CBase offers just two things to any C class:

- Zero initialization, so that all member pointers and handles are initially zero, which is cleanup safe.
- A virtual destructor, so that CBase-derived objects can be properly destroyed from the cleanup stack.

By contrast, T types are *defined* as classes or built-in types that don't need a destructor. They don't need one because they own no data. Examples of T types are as follows:

- *Any built-in type*: these are given typedefs such as TInt for an unsigned integer.
- Any enumerated type, such as TAmPm, which indicates whether a formatted time of day is am or pm. All enumerations are Ts, though enumerated constants such as EAm or EPm begin with E.
- Class types that do not need a destructor, such as TBuf<40> (a buffer for a maximum of 40 characters) or TPtrC (a pointer to a string of any number of characters). TPtrC contains a pointer, but it only *uses* (rather than *has*) the characters it points to, and so it does not need a destructor.

T classes do not own any data, so they don't need a destructor. However, they *may* have pointers, provided that these are *uses-a* pointers rather than *has-a* pointers, like TPtrC's string data pointer.

T types are normally allocated as automatics, or as member variables of any other kind of class.

It's possible (but rare) to allocate T class objects explicitly on the heap. If so, you need to ensure that the heap cell is freed. You can push a T to the cleanup stack using code like this:

```
TDes* name = new(ELeave) TBuf<40>;   // TDes is a base of TBuf
CleanupStack::PushL(name);
DoSomethingL();
CleanupStack::PopAndDestroy(name)
```

This invokes CleanupStack::PushL(TAny*). When something pushed as a TAny* is popped and destroyed, its memory is deallocated, but no destructor is called.

If you forget to derive a C class from CBase, it will be pushed as a TAny*. The cleanup code will deallocate the class's data but won't call its destructor.

A `T` class object can usually be assigned using a bit-wise copy. Therefore, `T` types do not need copy constructors or assignment operators (except in specialized cases such as copying one `TBuf` to another, where the `TBuf`s' maximum lengths may differ).

Since `C` types reside on the heap and are referred to by pointers, `C` types are passed by reference – that is, by copying the pointer. Thus, `C` types do not need a copy constructor or an assignment operator.

As a result, C++ copy constructors and assignment operators are extremely rare in Symbian OS.

6.5   R Classes

Many class names begin with `R`, which stands for 'Resource'. Usually, `R` objects contain a handle to a resource that's maintained elsewhere. Examples include `RFile` (maintained by the file server), `RWindow` (maintained by the window server), and `RTimer` (maintained by the kernel).

The `R` object itself is typically small (at a minimum, it contains only a handle). A function in an `R` class doesn't usually change the member data of the `R` class itself; rather, it sends a message to the real resource owner, which identifies the real object using the handle, performs the function, and sends back a result. Functions such as `Open()`, `Create()`, and so on, allocate the resource and set the handle value. Typically, a `Close()` function frees the resource and sets the handle value to zero. A C++ constructor ensures that the handle is zero to begin with.

A few `R` classes do not obey the conventions described here. We will point out such classes as we encounter them.

`R` classes are like `T` classes in some ways, and like `C` classes in other ways:

- Like `T` classes, they can be automatics or class members. Also like `T` classes, they can be copied (which just copies the handle), and the copy can be used like the original.

- As with `T` classes, it is very rare to refer to `R`s using pointers. They are usually passed by value or by reference.

- Like `C` classes, they own resources. Although `R` classes don't usually have a destructor, they do have a `Close()` function that has a similar effect (including setting the handle to zero, which is rather like zeroing the pointer to a `C` object). It's safe to call `Close()` twice – on an already closed `R` object it has no effect.

- Like C classes, they zero initialize their handle value so that functions can't be used until the handle is initialized. But the zero initialization must be explicit in a C++ constructor. There is no RBase corresponding to CBase.

memorymagic includes some functions that demonstrate various approaches to cleanup for R classes.

- **Write file** opens a file (c:\test.txt) and writes the text **Hello world!** into it
- **Read file** reads the text and prints it as an information message
- **Delete file** deletes the file.

Each of these functions can be made to fail – you'll get a **Not found** error, for instance, if you delete the file when it hasn't been written, or it has already been deleted. You'll get **Access denied** if you write the file, set its properties to read only (using the file manager), and then try to write to it again or delete it.

6.5.1 R Classes as Member Variables

R class objects are often members of a C class object. Assuming that iFs is a member variable of CExampleAppUi, the memorymagic delete code could have been written as follows:

```
case EMagicCmdDeleteFile:
    {
    User::LeaveIfError(iFs.Connect());
    User::LeaveIfError(iFs.Delete(KTextFileName));
    iFs.Close();
    }
```

It's important that CExampleAppUi's C++ destructor includes a call to iFs.Close() so that the RFs is closed even if the delete operation fails.

Opening and closing server sessions is relatively expensive, so it's better to keep server sessions open for a whole program, if you know you're going to need them a lot.

We could do this by including the User::LeaveIfError(iFs.Connect()) in the CExampleAppUi::ConstructL(). Then, the **Delete file** handler would become very simple indeed:

```
case EMagicCmdDeleteFile:
    {
    User::LeaveIfError(iFs.Delete(KTextFileName));
    }
```

This is a common pattern for using R objects. At the cost of a small amount of memory needed to maintain an open session throughout the lifetime of the application, we save having to open and close a session for every operation that uses one.

In fact, it's so common to need RFs that the CONE environment provides one for you. I could have used

```
case EMagicCmdDeleteFile:
    {
    User::LeaveIfError(iCoeEnv->FsSession().Delete(KTextFileName));
    }
```

Then I wouldn't have needed to allocate my own RFs at all. It's a very good idea to reuse CONE's RFs, since an RFs object uses up significant amounts of memory.

6.5.2   Error Code Returns versus L Functions

RFs does not provide functions such as ConnectL() or DeleteL() that would leave with a TInt error code if an error was encountered. Instead, it provides Connect() and Delete(), which *return* a TInt error code (including KErrNone if the function returned successfully). This means you have to check errors explicitly using User::LeaveIfError(), which does nothing if its argument is KErrNone or a positive number, but leaves with the value of its argument if the argument is negative.

A few low-level Symbian OS APIs operate this way; their areas of application are a few important circumstances in which it would be inappropriate to leave.

- Many file or communications functions are called speculatively to test whether a file is there or a link is up. It is information, not an error, if these functions return with 'not found' or a similar result.
- Symbian's implementation of the C Standard Library provides a thin layer over the native RFile and communications APIs that returns standard C-type error codes. It's much easier to handle errors directly than by trapping them. In any case, standard library programs don't have a cleanup stack.

Granted, leaves could have been trapped. But that was judged undesirably expensive. Instead, when you want a leave, you have to call User::LeaveIfError(). That's a little bit costly too, but not as expensive as trapping leaves.

Some truly ancient code in Symbian OS was written assuming that there might not be a cleanup stack at all. But this isn't a sensible assumption these days, and is certainly no justification for designing APIs that don't use L functions. All native Symbian OS code should ensure it has a cleanup stack, and design its APIs accordingly. The GUI application framework, and the server framework, provide you with a cleanup stack, so you only have to construct your own when you're writing a text console program.

If you are using components with `TInt` error codes, don't forget to use `User::LeaveIfError()`, where you need it.

6.5.3   R Classes on the Cleanup Stack

Sometimes, you need to create and use an R class object as an automatic variable rather than as a member of a C class. In this case, you need to be able to push to the cleanup stack. There are two options available to you:

- Use `CleanupClosePushL()`

- *Do it directly*: make a `TCleanupItem` consisting of a pointer to the R object, and a static function that will close it.

If you have an item with a `Close()` function, then `CleanupClose-PushL()` (a global nonmember function) will ensure that `Close()` is called when the item is popped and destroyed by the cleanup stack. C++ templates are used for this, so `Close()` does not have to be virtual.

The code below, taken from `memorymagic`, demonstrates how to use `CleanupClosePushL()`:

```
case EMagicCmdDeleteFile:
    {
    RFs fs;
    CleanupClosePushL(fs);
    User::LeaveIfError(fs.Connect());
    User::LeaveIfError(fs.Delete(KTextFileName));
    CleanupStack::PopAndDestroy(&fs);
    }
```

You just call `CleanupClosePushL()` after you've declared your object, and before you do anything that could leave. You can then use `CleanupStack::PopAndDestroy()` to close the object when you've finished with it.

You can look up the `TCleanupItem` class in `e32base.h`: it contains a pointer and a cleanup function. Anything pushed to the cleanup stack is actually a cleanup item. `CleanupStack::PushL(CBase*)`,

`CleanupStack::PushL(TAny*)`, and `CleanupClosePushL()` simply create appropriate cleanup items and push them.

There are two other related functions:

- `CleanupReleasePushL()` works like `CleanupClosePushL()`, except that it calls `Release()` instead of `Close()`.

- `CleanupDeletePushL()` is the same as `CleanupStack::PushL(TAny*)`, except that in this case the class destructor is called. This is often used when we have a pointer to an M class object that needs to be placed on the cleanup stack.

You can also create your own `TCleanupItems` and push them if the cleanup functions offered by these facilities are insufficient.

6.6 User Errors

The cleanup framework is so good that you can use it to handle other types of errors besides resource shortages.

One common case is handling errors in user input. The function in a dialog that processes the **OK** button (an overridden `CEikDialog::OkToExitL()`) must

- get each value from the dialog's controls;
- validate the values (the controls will have done basic validation, but you may need to do some more at this stage, taking the values of the whole dialog into account);
- Pass the values to a function that performs some action.

A typical programming pattern for `OkToExitL()` is to use automatics to contain the T-type value, or to point to the C-type values, in each control.

If you find that something is invalid at any stage in the `OkToExitL()` processing, you will need to

- display a message to the user indicating what the problem is;
- clean up all the values you have extracted from the dialog controls – that is, anything you have allocated on the heap;
- return.

A great way to do this is to push all your control values, and any other temporary variables you need, to the cleanup stack, and then use `CEikonEnv`'s `LeaveWithInfoMsg()` function. This displays an info-message, and then leaves with `KErrLeaveNoAlert`. As part of standard leave processing, all the variables you have allocated will be cleaned up.

The active scheduler in Symbian OS traps the leave, as usual, but for this particular error code, instead of displaying a dialog containing error code it doesn't display anything.

Some people have realized independently that the framework is good for this, and tried to achieve the same effect by coding `User::Leave(KErrNone)`. *This appears to work because you don't get the error message. But in fact the error handler isn't called at all, so you don't get some other useful things either.*

So use `iEikonEnv->LeaveWithInfoMsg()` *or, if you don't need an info-message, use the same function but specify a resource ID of zero.*

In Chapter 13, the `streams` program includes an example of this pattern.

6.7 More on Panics

So far, I've dealt with how you can respond to errors that are generated by the environment your program runs in – whether out-of-memory errors, files not being there, or bad user input.

One type of error that can't be handled this way is programming errors. These have to be fixed by rewriting the offending program. During development, that's usually your program (though, like the rest of us, you probably start by blaming the compiler!). The best thing Symbian OS can do for you here is to panic your program – to stop it from running as soon as the error has been detected, and to provide diagnostic information meaningful enough for you to use.

The basic function here is `User::Panic()`. Here's a panic function I use in my Battleships game, from `\scmp\battleships\control-ler.cpp`:

```
static void Panic(TInt aPanic)
    {
    _LIT(KPanicCategory, "BSHIPS-CTRL");
    User::Panic(KPanicCategory, aPanic);
    }
```

`User::Panic()` takes a panic category string, which must be 16 characters or less (otherwise, the panic function gets panicked!), and a 32-bit error code.

On the emulator debug build, we've seen what this does – the kernel's panic function includes a `DEBUGGER()` macro that allows the debugger to be launched with the full context from the function that called panic. That gives you a reasonable chance of finding the bug.

On a release build, or on real hardware, a panic simply displays a dialog titled **Program closed**, citing the process name, and the panic category and the number you identified. Typically, it's real users who see this dialog, though you might be lucky enough to see it during development, before you release the program. To find bugs raised this way, you essentially have to guess the context from what the user was doing at the time, and the content of the **Program closed** dialog. You'll need inspiration and luck.

You can shorten the odds by being specific about the panic category and number and by good design and testing before you release.

Although technically it's a thread that gets panicked, in fact Symbian OS will close the entire process. On a real machine, that means your application will get closed. On the emulator, there is only one Windows process, so the whole emulator is closed.

The standard practice for issuing panics is to use **assert macros**, of which there are two: `__ASSERT_DEBUG` and `__ASSERT_ALWAYS`. There are various schools of thought about which one to use when – as a general rule, put as many as you can into your debug code and as few as you can into your release code. Do your own debugging – don't let your users do it for you.

Here's one of the many places where I might potentially call my `Panic()` function:

```
void CGameController::HandleRestartRequest()
    {
    __ASSERT_ALWAYS(IsFinished(), Panic(EHandleRestartReqNotFinished));

    // Transition to restarting
    SetState(ERestarting);
    }
```

The pattern here is `__ASSERT_ALWAYS`(*condition, expression*), where the *expression* is evaluated if the *condition* is not true. When the controller is asked to handle a restart request, I assert that the controller is in a finished state. If not, I panic with panic code `EHandleRestartReqNotFinished`. This gets handled by the `Panic()` function above so that if this code was taken on a production machine, it would show **Program closed** with a category of BSHIPS-CTRL and a code of 12. The latter comes from an enumeration containing all my panic codes:

```
enum TPanic {
    EInitiateNotBlank,
    EListenNotBlank,
    ERestartNotFinished,
    ESetGdpNotBlank,
    ESetPrefBadState,
    EHitFleetNotMyTurn,
    EAbandonNotMyTurn,
    EResendBadState,
    EBindBadState,
    ESendStartNoPrefs,
    EHandleRequestBadOpcode,
    EHandleResponseBadOpcode,
    EHandleRestartReqNotFinished,
    EHandleStartReqNotAccepting,
    EHandleAbandondReqNotOppTurn,
    EHandleHitReqNotOppTurn,
    EHandleStartRespNotStarting,
    EHandleHitRespNotOppTurn,
    };
```

Incidentally, I considered it right to assert the IsFinished() condition, even in production code. The Battleships controller is a complex state machine, responding to events from systems outside my control, and responding to software that's difficult to debug even though it's within my control. I might not catch all the errors in it before I release, even if I test quite thoroughly. In this case, I want to be able to catch errors after release, so I use __ASSERT_ALWAYS instead of __ASSERT_DEBUG.

6.8 Testing Engines and Libraries

The Symbian OS framework includes a cleanup stack, a trap harness, and (in debug mode) heap-balance checking and keys to control the heap-failure mode. This makes it practically impossible for a developer to create a program with built-in heap imbalance under nonfailure conditions, very easy to handle failures, and easy to test for correct operation, including heap balance under failure conditions.

Lower- and intermediate-level APIs don't use the Symbian OS framework and therefore don't get these tools as part of their environment. To test these APIs, they must be driven either from an application, or from a test harness that constructs and manipulates the heap failure, heap-balance checking, and other tools. Happily, this is quite easy, and test harnesses can be used to test engines very aggressively for proper resource management, even under failure conditions.

As an example of the kind of component for which you might wish to do this, take the engine for the spreadsheet application. The engine manipulates a grid of cells whose storage is highly optimized for compactness. Each cell contains an internal format representing the contents of that cell, perhaps including a formula that might be evaluated. The engine supports user operations on the sheet, such as entering new cells, deleting cells, or copying (including adjusting relative cell references). All of this is pure data manipulation – exactly the kind of thing that should be done with an engine module, rather than being tied to a GUI application. In this situation, you would want to test the engine, firstly to verify the accuracy of its calculations, and secondly for its memory management, both under success conditions and failure due to memory shortage.

Firstly, you'd develop a test suite for testing the accuracy of calculations. The test suite would contain one or more programs of the following form:

- A command-line test program that loads the engine DLL but has no GUI. The test program will need to create its own cleanup stack and trap harness such as `hellotexts` because there's no GUI environment to give us these things for free.

- A test function that performs a potentially complex sequence of operations, and checks that their results are as expected. Think carefully, and aggressively, about the kinds of things you need to test.

- Heap-check macros to ensure that the test function (and the engine it's driving) releases all memory that it allocates.

Secondly, you'd use that test suite during the entire lifetime of the development project – in early, prerelease testing and postrelease maintenance phases.

- Every time you release your engine for other colleagues to use, run it through all tests. Diagnose *every* problem you find before making a release. The earlier in the development cycle you pick up and solve problems, the better.

- If you add new functionality to the engine, but you have already released your engine for general use, then test the updated engine with the old test code. This will ensure that your enhancement (or fix) doesn't break any established, working function. This is **regression testing**.

Finally, you can combine a test function with the heap-failure tools to perform high-stress, out-of-memory testing. A test harness might look like this:

```
for(TInt i = 1; ; i++)
    {
    __UHEAP_MARK;
    __UHEAP_SETFAIL(RHeap::EFailNext, i);
    TRAPD(error, TestFunctionL());
    __UHEAP_MARKEND;

    TAny* testAlloc = User::Alloc(1);
    TBool heapTestingComplete = (testAlloc == NULL);
    User::Free(testAlloc);

    if ((error != KErrNoMemory) && heapTestingComplete)
        break;
    }
```

This loop runs the test function in such a way that each time through, one more successful heap allocation is allowed. It is guaranteed that there will be a KErrNoMemory error (if nothing else) that will cause a leave and cleanup. Any failure to clean up properly will be caught by the heap-balance checks. The loop terminates when the test function has completed without generating a memory-allocation failure.

6.9   Summary

Memory and other resources are scarce in typical Symbian OS environments. Your programs *will* encounter resource shortages, and must be able to deal with them. You must avoid memory leaks, both under normal circumstances and when dealing with errors. Symbian OS provides an industrial-strength framework to support you, with very low programmer overhead, and very compact code.

In this chapter, we've seen the following:

- How the debug keys, and their simulated failures, and the heap-checking tools built into CONE, mean that testing your code is easy.

- How naming conventions help with error handling and cleanup.

- Allocating and destroying objects on the heap – how to preserve heap balance, what to do if allocation fails, how to reallocate safely.

- How leaves work, the TRAP() and TRAPD() macros, and Symbian OS error codes.

- When to use your own traps.

- When and how to use the cleanup stack.

- Two-phase construction, using ConstructL() and the NewL() and NewLC() factory functions.

- What CBase provides for any C class to help with cleanup.
- Cleanup for R and T classes.
- How to panic a program, and test engines and libraries.

I've covered a lot of material in this chapter, and you could be forgiven for not taking it all in at once. Don't worry – if you only remember one thing from this chapter, remember that cleanup is vital. You'll see cleanup-related disciplines in all the code throughout the rest of the book, and you'll be able to come back here for reference when you need to.

7

Resource Files

We've seen enough of resource files to understand how they're used to define the main elements required by a Symbian OS application UI. In later chapters, we'll also be using resource files to specify dialogs.

In this chapter, we review resource files and the resource compiler more closely to better understand their role in the development of Symbian OS. This chapter provides a quick tour – for a fuller reference, see the SDK.

7.1 Why a Symbian-specific Resource Compiler?

The Symbian OS resource compiler starts with a text source file and produces a binary data file that's delivered in parallel with the application's executable. Windows, on the other hand, uses a resource compiler that supports icons and graphics as well as text-based resources, and which builds the resources right into the application executable so that an application can be built as a single package. Furthermore, many Windows programmers never see the text resource script nowadays because their development environment includes powerful and convenient GUI-based editors.

So, why does Symbian OS have its own resource compiler, and how can an ordinary programmer survive without the graphical resource editing supported by modern Windows development environments? Unlike Windows developers, Symbian OS developers target a wide range of hardware platforms, each of which may require a different executable format. Keeping the resources separate introduces a layer of abstraction that simplifies the development of Symbian OS, and the efforts required by independent developers when moving applications between different hardware platforms. Furthermore, and perhaps more importantly,

Symbian OS C++ for Mobile Phones. Edited by Richard Harrison
© 2003 John Wiley & Sons, Ltd ISBN: 0-470-85611-4

it provides good support for localization. In addition to facilitating the process of translation (made even simpler by confining the items to be translated to `.rls` files), a multilingual application is supplied as a single executable, together with a number of language-specific resource files.

An application such as `hellogui.app` uses a resource file to contain GUI element definitions (menus, dialogs, and the like) and strings that are needed by the program at runtime. The runtime resource file for `hellogui.app` is `hellogui.rsc`, and it resides in the same directory as `hellogui.app`.

7.1.1 Source-file Syntax

Because processing starts with the C preprocessor, a resource file has the same lexical conventions as a C program, including source-file comments and C preprocessor directives.

The built-in data types are as follows:

Data Type	Description
BYTE	A single byte that may be interpreted as a signed or unsigned integer value.
WORD	Two bytes that may be interpreted as a signed or unsigned integer value.
LONG	Four bytes that may be interpreted as a signed or unsigned integer value.
DOUBLE	Eight byte real, for double precision floating point numbers.
TEXT	A string, terminated by a null. This is deprecated: use LTEXT instead.
LTEXT	A Unicode string with a leading length byte and no terminating null.
BUF	A Unicode string with no terminating null and no leading byte.
BUF8	A string of 8-bit characters, with no terminating null and no leading byte. Used for putting 8-bit data into a resource.
BUF	A Unicode string containing a maximum of n characters, with no terminating null and no leading byte.

LINK	The ID of another resource (16 bits).
LLINK	The ID of another resource (32 bits).
SRLINK	A 32-bit self-referencing link that contains the resource ID of the resource in which it is defined. It may not be supplied with a default initializer; its value is assigned automatically by the resource compiler.

The resource scripting language uses these built-in types for the data members of a resource. It also uses STRUCT statements to define aggregate types. A STRUCT is a sequence of named members that may be of built-in types, other STRUCTs, or arrays. STRUCT definitions are packaged into .rh files in \epoc32\include\ (where 'rh' stands for 'resource header'). The STRUCT statement is of the form

```
STRUCT struct-name [ BYTE | WORD ] { struct-member-list }
```

where

Element	Description
struct_name	Specifies a name for the STRUCT in uppercase characters. It must start with a letter and should not contain spaces (use underscores).
BYTE | WORD	Optional keywords that are intended to be used with STRUCTs that have a variable length. They have no effect unless the STRUCT is used as a member of another STRUCT, when they cause the data of the STRUCT to be preceded by a length BYTE, or length WORD, respectively.
struct_member_list	A list of member initializers, separated by semicolons, and enclosed in braces { }. Members may be any of the built-in types or any previously defined STRUCT. It is normal to supply default values (usually a numeric zero or an empty string).

Uikon provides a number of common STRUCT definitions that you can use in your applications' resource files by including the relevant headers. The SDK contains full documentation on resource files, although the examples in this book aren't too hard to follow without it. Besides reading the SDK documentation, you can learn plenty by looking inside the .rh files in \epoc32\include, together with the .rss files supplied with the examples in this book, and in various projects in the SDK.

To ensure that your resource script and C++ program use the same values for symbolic constants such as EHelloGuiCmd0, the resource compiler supports enum and #define definitions of constants with a syntax similar to that used by C++. These constants map a symbolic name to the resource ID. By convention, these definitions are contained in .hrh include files, which can be included in both C++ programs and resource scripts. Incidentally, 'hrh' stands for 'h' and 'rh' together.

Legal statements in a resource file are as follows:

Statement	Description
NAME	Defines the leading 20 bits of any resource ID. Must be specified prior to any RESOURCE statement.
STRUCT	As we have already seen, this defines a named structure for use in building aggregate resources.
RESOURCE	Defines a resource, mapped using a certain STRUCT, and optionally given a name.
enum/ENUM	Defines an enumeration and supports a syntax similar to C's.
CHARACTER_SET	Defines the character set for strings in the generated resource file. If not specified, cp1252 (the same character set used by Windows 9x and non-Unicode Windows NT) is the default.

The most important statement in an application resource file is RESOURCE. The statement must take the form

```
RESOURCE struct_name [ id ] { member_list }
```

where

Element	Description
struct_name	Refers to a previously encountered STRUCT definition. In most application resource files, this means a STRUCT defined in a #included .rh file.
id	The symbolic resource ID. If specified, this must be in lower case, with optional underscores to separate words (for instance r_hellogui_text_hello). Its uppercase equivalent will then be generated as a symbol in the .rsg file using a #define statement, as in #define R_HELLOGUI_TEXT_HELLO 0x276a800b.
member_list	A list of member initializers, separated by semicolons, and enclosed in braces { }. It is normal to supply default values (usually a numeric zero or an empty string) in the .rh file that defines the struct. If you don't specify a value in the resource file, then the value is set to the default.

Punctuation rules

With a definition like this,

```
RESOURCE MENU_PANE r_example_other_menu
    {
    items =
        {
        MENU_ITEM { command = EExampleCmd1; txt = "Cmd 1"; },
        MENU_ITEM { command = EExampleCmd2; txt = "Cmd 2"; },
        MENU_ITEM { command = EExampleCmd3; txt = "Cmd 3"; },
        MENU_ITEM { command = EExampleCmd4; txt = "Cmd 4"; }
        };
    }
```

it is clear that there are specific rules for the placement and type of punctuation character to use after a closing brace – should it be a comma, a semicolon, or nothing at all? The rules are quite simple:

- If a closing brace is at the end of a member definition that started with something like items = { ... }, then put a semicolon after it.

- If a closing brace separates two items in a list such as the menu items in the `items` = member above, then put a comma after it.

- Otherwise (that is, for the last item in a list, or at the end of a resource), don't put anything after a closing brace.

7.1.2 Localizable Strings

If you intend to translate your application into other languages, then you should put all text strings that need to be localized into one or more separate files, with a `.rls` (resource localizable string) extension. A `.rls` file defines symbolic identifiers for strings, to which the resource file refers when it needs the associated string.

As an example, consider the following two resources taken from `HelloGui.rss`:

```
RESOURCE MENU_PANE r_hellogui_edit_menu
    {
    items=
        {
        MENU_ITEM
            {
            command = EHelloGuiCmd1;
            txt = "Item 1";
            },
        MENU_ITEM
            {
            command = EHelloGuiCmd2;
            txt = "Item 2";
            }
        };
    }

...

RESOURCE TBUF r_hellogui_text_hello
    {
    buf = "Hello world!";
    }
```

The three text strings in these resources would appear in, say, a `HelloGui.rls` file as follows:

```
rls_string STRING_r_hellogui_edit_menu_first_item "Item 1"
rls_string STRING_r_hellogui_edit_menu_second_item "Item 2"
rls_string STRING_r_hellogui_message_hello "Hello world!"
```

The keyword `rls_string` appears before each string definition, followed by a symbolic identifier, and then the string itself in quotes.

The resource file itself is then modified to include the `.rls` file and refer to the strings via their symbolic names:

```
#include "HelloGui.rls"

...


RESOURCE MENU_PANE r_hellogui_edit_menu
    {
    items=
        {
        MENU_ITEM
            {
            command = EHelloGuiCmd1;
            txt = STRING_r_hellogui_edit_menu_first_item;
            },
        MENU_ITEM
            {
            command = EHelloGuiCmd2;
            txt = STRING_r_hellogui_edit_menu_second_item;
            }
        };
    }

...


RESOURCE TBUF r_hellogui_text_hello
    {
    buf = STRING_r_hellogui_message_hello;
    }
```

The advantage in doing this is that it allows the translator to concentrate on the task of translation without having to be concerned with maintaining the structure and syntax of the full resource file. You can include C- or C++-style comments in the `.rls` file to inform the translator of the context in which each text item appears, or to give information about constraints such as the maximum length permitted for a string.

7.1.3 Multiple Resource Files

A single resource file supports 4095 resources, but a Symbian OS application may use multiple resource files, each containing this number of resources. The application identifies each resource by a symbolic ID comprising two parts:

- a leading 20 bits (five hex digits) that identifies the resource file,
- a trailing 12 bits (three hex digits) that identifies the resource (hence the 4095 resource limit).

```
#define R_HELLOGUI_HOTKEYS              0x276a8004
#define R_HELLOGUI_MENUBAR              0x276a8005
#define R_HELLOGUI_HELLO_MENU           0x276a8006
#define R_HELLOGUI_EDIT_MENU            0x276a8007
#define R_HELLOGUI_TEXT_ITEM0           0x276a8008
#define R_HELLOGUI_TEXT_ITEM1           0x276a8009
#define R_HELLOGUI_TEXT_ITEM2           0x276a800a
#define R_HELLOGUI_TEXT_HELLO           0x276a800b
```

The leading 20 bits are generated from the four-character name that was specified in the NAME statement in the resource file. You can tell what the 20 bits are by looking at the .rsg file. Here, for example, is hellogui.rsg, which has the NAME HELO:

Uikon's resource file NAME is EIK, and its resource IDs begin with 0x00f3b.

You don't have to choose a NAME that's distinct from all other Symbian OS applications on your system – with only four letters, that could be tricky. The resource files available to your application are likely to be only those from Uikon, UIQ (or other customized UI), and your application, so you simply have to avoid using EIK – or, for UIQ, anything beginning with Q – as a name. Avoid CONE, BAFL, and other Symbian component names as well, and you should be safe.

7.1.4 Compiling a Resource File

The resource compiler is invoked as part of the application build process, either from within the CodeWarrior IDE, or from the command line with, for example:

```
abld build winscw udeb
```

As we saw in Chapter 1, this command runs the build in six stages, one of which is resource compilation.

From the command line, you can invoke the resource compiler alone with a command of the form:

```
abld resource winscw udeb
```

but you must first have run the abld makefile command to ensure that the appropriate makefiles exist (type abld on its own to get help on the available options).

Building for the winscw target by any of these methods causes the .rsc file to be generated in the \epoc32\release\winscw\<variant>\z\ system\apps\<appname> directory (where <variant> is either udeb or urel). Building for any ARM target causes the .rsc file to be generated in the \epoc32\data\z\system\apps\<appname> directory.

To use a resource at runtime, a program must specify the resource ID. The resource compiler, therefore, generates not only the resource file but also a header file, with a **.rsg** extension, containing symbolic IDs for every resource contained in the file. This header file is written to the \epoc32\include\ directory and is included in the application's source code. That's why the resource compiler is run *before* you run the C++ compiler when building a Symbian OS program. The .rsg file is always generated to \epoc32\include\, but if the generated file is identical to the one already in \epoc32\include\, the existing one isn't updated.

Uikon has a vast treasury of specific resources (especially string resources) that are accessible via the resource IDs listed in eikon.rsg.

Conservative .rsg update

When you compile hellogui.rss, either from the CodeWarrior IDE, or from the command line, it generates both the binary resource file hellogui.rsc and the generated header file hellogui.rsg.

Now, the project correctly includes a dependency of, for example, hellogui.obj on hellogui.rsg (because hellogui.cpp includes hellogui.rsg via hellogui.h). So, if hellogui.rsg is updated, the hellogui.app project needs rebuilding. Scaling this up to a large application, and making a change in its resource file, and hence in the generated headers, could cause lots of rebuilding.

The resource compiler avoids this potential problem by updating the .rsg file only when necessary – and it isn't necessary unless resource IDs have changed. Merely changing the text of a string, or the placement of a GUI element won't cause the application's executable to go out of date.

This is why, when you run the resource compiler, you are notified if the .rsg file was changed.

Summary of processing and file types

In summary, the resource-related files for a typical application are as follows:

Filename	Description
Appname.rss	The application's resource file script.
Appname.rls	The application's localizable strings.
Appname.rsc	The generated resource file.

`Appname.rsg`	Generated header containing symbolic resource IDs that are included in the C++ program at build time.
`Appname.hrh`	An application-specific header containing symbolic constants, for example, the command IDs that are embedded into resources such as the menus, button bars and, if relevant, shortcut keys. Such header files are used by both resource scripts and C++ source files, in places such as `HandleCommandL()`.
`Eikon.rh (qikon.rh)`	Header files that define Uikon's (and UIQ's) standard STRUCTs for resources.
`Eikon.hrh (qikon.hrh)`	Header files that define Uikon's (and UIQ's) standard symbolic IDs, such as the command ID for `EeikCmdExit`, and the flags used in various resource STRUCTs.
`Eikon.rsg (qik*.rsg)`	Resource IDs for Uikon's (and UIQ's) own resource files, which contain many useful resources. Many of these resources are for internal use by Symbian OS, although some are also available for application programs to use.

These different file types are involved in the overall build process for a GUI application as illustrated in Figure 7.1.

7.1.5 The Content of a Compiled Resource File

The resource compiler builds the runtime resource file sequentially, starting with header information that identifies the file as being a resource file. It then appends successive resources to the end of the file, in the order of definition. Because it works this way, the resource compiler will not have built a complete index until the end of the source file is reached; hence, the index is the last thing to be built and appears at the end of the file. Each index entry is a word that contains the offset, from the

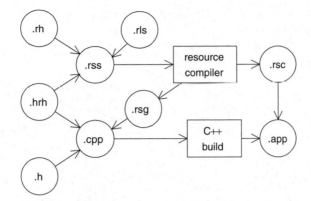

Figure 7.1

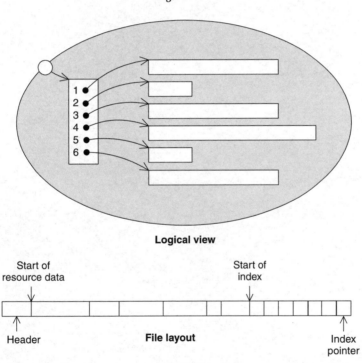

Logical view

Figure 7.2

beginning of the file to the start of the appropriate resource. The index contains, at its end, an additional entry that contains the offset to the byte immediately following the last resource, which is also the start of the index.

Each resource is simply binary data, whose length is found from its own index entry and that of the following resource. To see what a compiled

resource looks like, let's take a look at a resource we have seen before,
hellogui's Hello menu:

The menu pane and menu item resource STRUCTs are declared in
uikon.rh as

```
STRUCT MENU_PANE
    {
    STRUCT items[]; // MENU_ITEMs
    LLINK extension=0;
    }

STRUCT MENU_ITEM
    {
    LONG command=0;
    LLINK cascade=0;
    LONG flags=0;
    LTEXT txt;
    LTEXT extratxt="";
    LTEXT bmpfile="";
    WORD bmpid=0xffff;
    WORD bmpmask=0xffff;
    LLINK extension=0;
    }
```

and the menu pane resource itself is

```
RESOURCE MENU_PANE r_hellogui_hello_menu
    {
    items =
        {
        MENU_ITEM
            {
            command = EHelloGuiCmd0;
            txt = "Item 0";
            },
        MENU_ITEM
            {
            command = EEikCmdExit;
            txt = "Close (debug)";
            }
        };
    }
```

As you can see, each of the two menu items only specifies two of
the nine elements – the remaining seven take the default values that are
specified in the declaration.

A hex dump of the resource file gives the following result:

```
0000  6B 4A 1F 10 00 00 00 00   A8 76 02 00 8C 49 B7 4A   kJ.......v...I.J
0010  01 5C 00 F0 07 04 00 00   00 01 80 6A 27 04 80 6A   .\.........j'..j
0020  27 05 80 6A 27 00 00 00   00 00 00 00 00 00 00 00   '..j'...........
0030  00 00 00 00 00 00 00 01   00 00 01 00 00 65 00 00   .............e..
0040  00 00 00 00 07 02 00 06   80 6A 27 08 08 48 65 6C   .........j'..Hel
0050  6C 6F 47 75 69 12 00 00   00 00 00 FF FF FF FF 00   loGui...........
0060  00 00 00 07 80 6A 27 04   04 45 64 69 74 11 00 00   .....j'..Edit...
0070  00 00 00 FF FF FF FF 00   00 00 00 00 00 00 00 00   ................
0080  0F 02 00 00 10 00 00 00   00 00 00 00 00 00 00 06   ................
0090  06 49 74 65 6D 20 30 17   00 00 FF FF FF FF 00 00   .Item.0.........
00A0  00 00 00 01 00 00 00 00   00 00 00 00 00 00 0D 0D   ................
00B0  43 6C 6F 73 65 20 28 64   65 62 75 67 29 0E 00 00   Close.(debug)...
00C0  FF FF FF FF 00 00 00 00   00 00 00 00 00 0F 02 00   ................
00D0  01 10 00 00 00 00 00 00   00 00 00 00 06 06 49 74   ..............It
00E0  65 6D 20 31 17 00 00 FF   FF FF FF 00 00 00 00 02   em.1............
00F0  10 00 00 00 00 00 00 00   00 00 00 06 06 49 74 65   .............Ite
0100  6D 20 32 0E 00 00 FF FF   FF FF 00 00 00 00 00 00   m.2............
0110  00 00 06 49 74 65 6D 20   30 06 49 74 65 6D 20 31   ...Item.0.Item.1
0120  06 49 74 65 6D 20 32 0C   48 65 6C 6C 6F 20 77 6F   .Item.2.Hello.wo
0130  72 6C 64 21 15 00 1D 00   1D 00 35 00 43 00 7F 00   rld!......5.C...
0140  CC 00 12 01 19 01 20 01   27 01 34 01               
```

The highlighted bytes represent the data for the menu pane resource.

The most obviously recognizable parts of this resource are the strings, `"Item 0"` and `"Close (Debug)"`. However, on reflection, you might be slightly puzzled: I have said that Symbian OS applications use a Unicode build, whereas these strings appear to be plain ASCII text. Furthermore, if you count through the resource, identifying items on the way, you will find that there is an occasional extra byte here and there – for example, each of the strings appears to be preceded by two identical count bytes, instead of the expected single byte.

The answer to these puzzles is that the compiled resource is compressed in order to save space in Unicode string data. The content of the resource is divided into a sequence of runs, alternating between compressed data and uncompressed data. Each run is preceded by a count of the characters in that run, with the count being held in a single byte, provided that the run is no more than 255 bytes in length. There is more information about resource file compression in the UIQ SDK.

Taking this explanation into account, you can see from the following table how the data is distributed between the elements of which the resource is composed.

What you see is that, apart from the effects of data compression, the compiled resource just contains the individual elements, listed sequentially. A WORD, for example, occupies two bytes and an LTEXT is represented by a byte specifying the length, followed by the text. Where there are embedded STRUCTs, the effect is to flatten the structure. As we shall see later, the runtime interpretation of the data is the responsibility of the class or function that uses it.

Start of MENU_PANE

00	[compressed data run length]
0F	[uncompressed data run length]
02 00	array item count

First MENU_ITEM

00 10 00 00	command = EHelloGuiCmd0
00 00 00 00	cascade=0 (default)
00 00 00 00	flags=0 (default)
06	[string count byte]
06	[compressed data run length]
49 74 65 6D 20 30	txt = "Item 0"
17	[uncompressed data run length]
00	extratxt="" (default)
00	bmpfile="" (default)
FF FF	bmpid=0xffff (default)
FF FF	bmpmask=0xffff (default)
00 00 00 00	extension=0 (default)

Second MENU_ITEM

00 01 00 00	command = EEikCmdExit
00 00 00 00	cascade=0 (default)
00 00 00 00	flags=0 (default)
0D	[string count byte]
0D	[compressed data run length]
43 6C 6F 73 65 20	txt = "Close (Debug)"
28 64 65 62	
75 67 29	
0E	[uncompressed data run length]
00	extratxt="" (default)
00	bmpfile="" (default)
FF FF	bmpid=0xffff (default)
FF FF	bmpmask=0xffff (default)
00 00 00 00	extension=0 (default)

Completion of MENU_PANE

00 00 00 00	extension=0 (default)

APIs for reading resources

BAFL provides the basic APIs for reading resources:

- The oddly named RResourceFile class, in barsc.h, is used for opening resource files, finding a numbered resource, and reading its data. (RResourceFile behaves more like a C class than an R class.)

- `TResourceReader` in `barsread.h` is a kind of stream-oriented reader for data in an individual resource. `TResourceReader` functions are provided that correspond to each of the resource compiler's built-in data types.

As an example of the use of `TResourceReader` functions, here's the code to read in a `MENU_ITEM` resource, extracted from `eikmenup.cpp` (see `\Release\Generic\app-framework\uikon\coctlsrc \eikmenup.cpp` in the source supplied with the UIQ SDK):

```
EXPORT_C void CEikMenuPaneItem::ConstructFromResourceL
    (TResourceReader& aReader)
      {
    iData.iCommandId=aReader.ReadInt32();
    iData.iCascadeId=aReader.ReadInt32();
    iData.iFlags=aReader.ReadInt32();
    iData.iText=aReader.ReadTPtrC();
    iData.iExtraText=aReader.ReadTPtrC();
    TPtrC bitmapFile=aReader.ReadTPtrC();
    TInt bitmapId=aReader.ReadInt16();
    TInt maskId=aReader.ReadInt16();
    aReader.ReadInt32(); // extension link
    if (bitmapId != -1)
          {
        SetIcon(CEikonEnv::Static()->CreateIconL(
                bitmapFile, bitmapId, maskId));
          }
      }
```

In this code, a `TResourceReader` pointing to the correct (and uncompressed) resource has already been created by the framework, so all this code has to do is actually read the resource.

Notice how the code exactly mirrors the resource `STRUCT` definition: the `MENU_ITEM` definition starts with `LONG` command, so the resource reading code starts with `iData.iCommandId=aReader. ReadInt32()`. The next defined item is `LLINK` cascade, which is read by `iData.iCascadeId=aReader.ReadInt32()` and so on.

CONE builds on the services provided by BAFL to provide other functions, such as

- `CreateResourceReaderLC()`, which creates and initializes an object of type `TResourceReader` to read a single specified resource;

- `AllocReadResourceL()` and `AllocReadResourceLC()`, which allocate an `HBufC` big enough for the resource and read it in;

- `ReadResource()`, which simply reads a resource into an existing descriptor (and panics if it can't).

7.2   Summary

In this chapter, we've seen

- why a Symbian OS-specific resource compiler is needed,
- a brief explanation of the syntax,
- how to organize your resource text to ease the task of localization,
- how to use multiple resource files,
- the effect of compiling a resource file, and how to read the resulting data.

8

Basic APIs

We've now seen basic Symbian OS development tools, the overall structure of the system, and the three most fundamental programming frameworks – for descriptors, cleanup, and data management.

Before we move on to graphics and the GUI, here's some other useful information for developers: a few basic APIs, a guide to collection classes, and information about the Symbian OS implementation of the C standard library.

You'll find more reference information on most of these facilities in the SDK, and you'll find plenty of useful hints and tips on the Symbian Developer Network website.

8.1 A Few Good APIs

We've seen the descriptor and cleanup APIs from the E32 user library. In later chapters (Chapter 18 and Chapter 19), I'll cover the user library's other two important frameworks – those for active objects (AOs) and client-server programming.

Some other basic APIs, mostly from the user library, are also worth a mention – though I won't describe them in detail here.

8.1.1 `User` Class

`User` is a static class with more than 70 functions in various categories. We've already seen `User::Leave()`, `User::LeaveIfError()`, and `User::Panic()`.

To support memory handling, use `User::Alloc()`, which behaves like `malloc()` in C. `User::Free()` behaves like `free()`, while

---

Symbian OS C++ for Mobile Phones.   Edited by Richard Harrison
© 2003 John Wiley & Sons, Ltd   ISBN: 0-470-85611-4

`User::Realloc()` behaves like `realloc()`. Leaving variants (`User::AllocL()`, `User::AllocLC()` etc.) are also provided. Two major functions suspend a thread until a timer expires:

- `User::After()` suspends until *after* a given number of microseconds has elapsed. `User::After()` uses the hardware tick interrupt and is designed for short-term timing, GUI time-outs, and so on. The tick interrupt is turned off when the machine is turned off, so a `User::After()`'s completion is delayed until the machine is turned back on and the clock starts ticking again.

- `User::At()` suspends until a particular date and time. `At()` uses the date/time clock, which is always running, even when the machine is turned off. When an `At()` timer completes, it will turn the machine on if necessary. `At()` timers are for alarms and other events for which an accurate date and time are essential.

These functions are defined in `e32std.h`:

```
class User : public UserHeap
    {
public:
    // Execution control
    IMPORT_C static void Exit(TInt aReason);
    IMPORT_C static void Panic(const TDesC& aCategory,
        TInt aReason);

    // Cleanup support
    IMPORT_C static void Leave(TInt aReason);
    IMPORT_C static void LeaveNoMemory();
    IMPORT_C static TInt LeaveIfError(TInt aReason);
    IMPORT_C static TAny* LeaveIfNull(TAny* aPtr);
    IMPORT_C static TAny* Alloc(TInt aSize);
    IMPORT_C static TAny* AllocL(TInt aSize);
    IMPORT_C static TAny* AllocLC(TInt aSize);
    IMPORT_C static void Free(TAny* aCell);
    IMPORT_C static TAny* ReAlloc(TAny* aCell,TInt aSize);
    IMPORT_C static TAny* ReAllocL(TAny* aCell,TInt aSize);

    // Synchronous timer services
    IMPORT_C static void After(TTimeIntervalMicroSeconds32
        anInterval);
    IMPORT_C static TInt At(const TTime& aTime);
    ...
    };
```

Many other useful functions are provided. As usual, the SDK has the details.

The derivation of `User` from `UserHeap`, another static class, betrays `User`'s heritage as one of the oldest classes in Symbian OS.

8.1.2 Dynamic Buffers

CBufBase is an abstract base class for dynamic memory buffers, which store any number of bytes from zero upward, and which can be expanded and contracted at will. You can read or write bytes from the buffer, insert bytes into the buffer, or delete them from it.

Here's the declaration of CBufBase, from e32base.h:

```
class CBufBase : public CBase
    {
public:
    IMPORT_C ~CBufBase();
    inline TInt Size() const;
    IMPORT_C void Reset();
    IMPORT_C void Read(TInt aPos, TDes8& aDes) const;
    IMPORT_C void Read(TInt aPos, TDes8& aDes, TInt aLength) const;
    IMPORT_C void Read(TInt aPos, TAny* aPtr, TInt aLength) const;
    IMPORT_C void Write(TInt aPos, const TDesC8& aDes);
    IMPORT_C void Write(TInt aPos, const TDesC8& aDes, TInt aLength);
    IMPORT_C void Write(TInt aPos, const TAny* aPtr, TInt aLength);
    IMPORT_C void InsertL(TInt aPos, const TDesC8& aDes);
    IMPORT_C void InsertL(TInt aPos, const TDesC8& aDes, TInt aLength);
    IMPORT_C void InsertL(TInt aPos, const TAny* aPtr, TInt aLength);
    IMPORT_C void ExpandL(TInt aPos, TInt aLength);
    IMPORT_C void ResizeL(TInt aSize);

    // Pure virtual
    virtual void Compress() = 0;
    virtual void Delete(TInt aPos, TInt aLength) = 0;
    virtual TPtr8 Ptr(TInt aPos) = 0;
    virtual TPtr8 BackPtr(TInt aPos) = 0;
private:
    virtual void DoInsertL(TInt aPos, const TAny* aPtr, TInt aLength) = 0;
protected:
    IMPORT_C CBufBase(TInt anExpandSize);
protected:
    TInt iSize;
    TInt iExpandSize;
    };
```

You can find out how many bytes are in a CBufBase using Size(). Bytes are indexed from zero. You insert using InsertL(), specifying a byte position from which to start inserting, and a pointer descriptor containing data (in fact, all InsertL() functions are convenience functions for the private DoInsertL()). You can delete data using Delete(); you can write data to the buffer using Write() – this overwrites without inserting – and you can read from the buffer using Read().

Two types of buffer are provided, as shown in Figure 8.1, both derived from CBufBase:

CBufFlat, which puts all the bytes in a single heap cell. This means that access to any byte is quick (it just adds the byte index to the beginning of the buffer). However, memory allocation can be inefficient, so that it

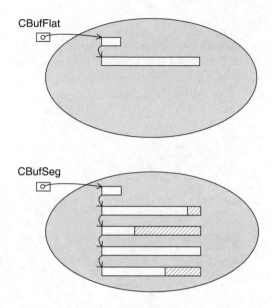

Figure 8.1

might not be possible to expand the buffer when desired, even though there may be more than enough bytes of unused heap available.

CBufSeg, which puts the bytes in multiple heap cells, each of which is a segment of the buffer. For large buffers that are constantly changing in size, and where insertions and deletions part-way through the buffer are frequent, segmented buffers are much more efficient than flat buffers. Finding a particular byte theoretically requires a walk of the entire segment structure: CBufSeg caches a reference to the last-used byte, which speeds up most operations.

The buffers example illustrates the functions available. The mainL() function is as follows:

```
void mainL()
    {
    CBufBase* buf=CBufSeg::NewL(4);
    CleanupStack::PushL(buf);
    //
    _LIT8(KTxt1,"hello!");
    buf->InsertL(0,KTxt1);
    printBuffer(buf);
    _LIT8(KTxt2," world");
    buf->InsertL(5,KTxt2); // "hello world!"
    printBuffer(buf);
    buf->Delete(2, 7); // "held!"
```

```
printBuffer(buf);
buf->Compress();
printBuffer(buf);
buf->ExpandL(2, 7); // "he.......ld!"
printBuffer(buf);
_LIT8(KTxt3,"llo wor");
buf->Write(2,KTxt3); // "hello world!"
printBuffer(buf);
//
CleanupStack::PopAndDestroy(); // buf
}
```

This example creates 8-bit descriptors (using the _LIT8 macro) and inserts them into the buffer. It is not a recommended practice to use buffers to hold text, because buffers can only store 8-bit values, but Unicode uses 16 bits per character. The example works only because the characters used can all be represented in 8 bits. However, the purpose of the example is to show the basics of insert, delete, compress, expand, and write with buffers.

The example also shows how a CBufSeg is allocated. NewL() takes a granularity, which is the maximum size of a segment. Oddly, there is no NewLC(), so I have to push the buffer explicitly onto the cleanup stack.

To scan all the data in a buffer, you need to know where the segment boundaries are. You can use the Ptr() function to get a TPtr descriptor for all the bytes from a given position, to the end of the segment that position is in. For a CBufFlat, that means all the data in the buffer. But for a CBufSeg, only the rest of the current segment is given.

printBuffer() in the example code shows how to do this:

```
void printBuffer(CBufBase* aBuffer)
    {
    _LIT(KMsgBufEquals,"buffer=");
    console->Printf(KMsgBufEquals);
    TInt i = 0;
    TPtrC8 p=aBuffer->Ptr(i);
    while(p.Length() > 0)
        {
        TBuf<12> temp;
        temp.Copy(p);
        _LIT(KBufFormatString,"[%S]");
        console->Printf(KBufFormatString, &temp);
        i += p.Length();
        p.Set(aBuffer->Ptr(i));
        };
```

```
    _LIT(KNewLine,"\n");
    console->Printf(KNewLine);
    }
```

The result is

```
buffer=[hell][o!]
buffer=[hell][o wo][rld!]
buffer=[he][ld!]
buffer=[held][!]
buffer=[held][ÌÌÌÌ][!ld!]
buffer=[hell][o wo][rld!]
```

Because I specified a granularity of 4, at first, the segments are all 4 bytes long.

When I delete the seven characters from the middle of the string, the segments are optimized for minimum data shuffling. After random operations on the buffer over a long period – say, operations controlled by a user typing and editing text in a word processor – the buffer can become very fragmented, with many segments containing less than the maximum amount of data. Compress() moves the data so that as few segments as possible are used.

ExpandL() and ResizeL() (which puts extra bytes at the end, if the new size is greater than the current size) are useful for making a sequence of InsertL() operations atomic and for improving their performance. Say you needed to insert six items: if you used six InsertL()s within your function InsertSixItemsL(), you would need to trap each one and, if it failed because of an out-of-memory error, you would need to delete all the items you had so far inserted. In addition, the use of repeated allocation for each InsertL() would impact performance and fragment the heap – especially for flat buffers. You can avoid all these problems by using ExpandL() or ResizeL(), and then a series of Write()s – which cannot leave.

If you're at your PC, replace the allocation of CBufSeg with a CBufFlat and run buffers again. You should have little difficulty predicting or explaining the results.

Dynamic buffers are used to store array elements in expandable arrays. Both flat and segmented arrays are provided, corresponding to the buffer types used to implement them. CArrayFixFlat<T> uses a flat buffer to store an array of T objects, while CArrayFixSeg<T> uses a segmented buffer. Ts are stored so that the nth T occupies sizeof(T) bytes from position $n*$sizeof(T) in the buffer. A CBufSeg granularity of a multiple of sizeof(T) is specified, so that a T never spans a segment boundary. These and other array types are described in Section 8.1.3.

Dynamic buffers are also used to store global and rich text, derived from `CEditableText`, part of the Text and Text Attributes API defined in `txt*.h` headers. Segmented buffers are particularly efficient at handling the kind of operations required by intensive word processing. The editable text APIs include `InsertL()`, `Read()`, and `Delete()` with specifications similar to those of `CBufBase`. However, the position argument in all these functions is a *character* position, not a byte position: Unicode uses two bytes per character. The SDK includes full documentation on rich text, and a pretty example project in `\Examples\AppFramework\Text`.

8.1.3   Collections

A collection class is one that holds a number of objects. In Symbian OS, collections are provided by a wide range of arrays and lists. The definitions for arrays and lists and for their supporting classes can be found either in `e32std.h` or in `e32base.h`, (descriptor arrays are defined in `badesca.h`). All are documented in the SDK, many with example code fragments showing how they are used. This section summarizes the types of collection classes available, describes the key properties of the concrete collection classes, and provides a quick selection guide.

What types of collection are available?

`RArray` classes:

- use flat storage only,
- support sorting and searching using a comparator function and can ensure uniqueness,
- provide specializations for common types, for instance, integers.

`CArray` classes

- provide a choice of either flat or segmented storage,
- support sorting and searching using a key specification and can ensure uniqueness,
- provide several variants, for instance, fixed or variable size elements, packed data
- are generally slower and can be less easy to use than the `RArray` classes.

Descriptor arrays

- provide a choice of either flat or segmented storage
- can contain 8-bit or 16-bit descriptors

- support sorting and searching and can ensure uniqueness
- provide variants that can store any type of descriptor or can hold pointers to the data.

Linked lists

- support iterators for scanning through the list
- are available as singly and doubly linked lists; the link object must be a member of the linked class.

TFixedArray

- used when the number of elements is fixed and known at compile time
- should be used instead of traditional C++ arrays because they provide bounds checking.

TArray

- used for representing any array data in a generic way.

What are the main concrete containers?

Linked lists

Class name	Description
TSglQue<T>	Singly linked list – elements of type T can be added at the start or end of the list, any element can be removed. The list can be iterated through in a single direction. *A link object (TSglQueLink), that connects each element to the next one, must be a member of the template class.
TDblQue<T>	Doubly linked list – elements of type T can be added to and removed from the list, at any position. The list can be iterated through in both directions. **A link object (TDblQueLink), that connects each element to the next and previous elements, must be a member of the template class.

*Using a TSglQueIter
**Using a TDblQueIter

Fixed size arrays

Class name	Description
TFixedArray<T,TInt S>	A fixed-size array where S specifies the array size and T specifies the type of elements that it can hold. This class is a thin wrapper over a standard C++ array, with range checking. Elements can be deleted through the array, but it does not have ownership.

Dynamic arrays

Class name	Description
TArray<T>	An array interface that provides a Count() and At() function only. Its purpose is to allow all array types to be represented in a generic way. Elements can only be accessed through the interface; they cannot be added or deleted and the array cannot be sorted. This interface is implemented by all of the dynamic and fixed-size array types.
RPointerArray<T>	A pointer array. Supports uniqueness, sorting and searching. May be sorted and searched either by pointer address, or by object pointed to – the latter requires the caller to specify a function that compares template class objects.* Can exercise ownership through ResetAndDestroy().
RArray<T>	An array of elements contained by reference. It can be sorted and searched either by using an integer value stored at a specified offset in the template class, or the caller can specify a function that compares template class objects. The Function is packaged in TLinearOrder or a TIdentityRelation object. It provides specializations for arrays of signed and unsigned integers. It does not own the objects in the array.

`CArrayFixFlat<T>`	An array of fixed size elements that uses a flat buffer for storage. It supports sorting, searching, and uniqueness. Sorting and searching are done using a `TKeyArrayFix` key specification (either a descriptor or an integer stored at a specified offset in the template class). It provides various specializations. It owns the objects in the array.
`CArrayFixSeg<T>`	As `CArrayFixFlat`, but using a segmented buffer.
`CArrayPtrFlat<T>`	An array of pointers to objects, using a flat buffer for storage. It supports sorting, searching, and uniqueness. Sorting and searching are done using a `TKeyArrayFix` key specification. Can exercise ownership through `ResetAndDestroy()`
`CArrayPtrSeg<T>`	As `CArrayPtrFlat`, but using a segmented buffer.
`CArrayVarFlat<T>`	An array of variable size elements that uses a flat buffer for storage. Supports sorting, searching, and uniqueness. Sorting and searching are done using a `TKeyArrayVar` key specification. The element's length is specified when inserting. Provides a specialization for arrays of TAny. It owns the objects in the array.
`CArrayVarSeg<T>`	As `CArrayVarFlat`, but using a segmented buffer.
`CArrayPak<T>`	A packed array of variable size elements. Similar to `CArrayVar` except in its implementation. Provides a specialization for arrays of TAny.
`CArrayPakFlat<T>`	As `CArrayPak`, but using a flat buffer only.

*The Function is packaged in a `TLinearOrder`, or `TidentityRelation` object.

Notes:

- Arrays that use segmented storage (with a 'Seg' suffix) are designed for frequent additions and deletions or where the array could grow to a large size.
- Arrays that use flat storage should be used to hold a limited number of elements, or when insertions and deletions are rare.

- RArrays are faster than the equivalent CArrays, so generally should be used in preference, unless segmented storage is a requirement (segmented storage is not available for RArrays).

- If variable-length items need to be contained in the array, use a CArrayVar (if updates are more frequent) or a CArrayPak (if updates are very infrequent).

- CArrayPak arrays have the advantage over CArrayVar of a smaller memory overhead for each element (they don't store a pointer to each element).

Descriptor arrays

Class name	Description
CDesCArrayFlat	A dynamic array of 16-bit descriptors, using a flat buffer for storage. The array is modifiable; for instance, you can append, insert, and delete elements, but the elements are not modifiable. Supports uniqueness, searching, and sorting. To search and sort, you just need to specify the type of comparison to use.
CDesC8ArrayFlat	As CDesCArrayFlat, but stores 8-bit descriptors.
CDesCArraySeg	As CDesCArrayFlat, but implemented using a segmented buffer for storage.
CDesC8ArraySeg	As CDesCArraySeg, but stores 8-bit descriptors.
CPtrCArray	A dynamic array of 16-bit pointer descriptors using a flat buffer for storage. This is similar to CDesCArrayFlat but it stores an array of TPtrC16s rather than the actual data, so it can be used in preference to avoid duplicating memory. Searching and sorting is done in a similar way to the CArrayFix classes, using a key specification (with zero for the offset). Does not own the array elements.
CPtrC8Array	As CPtrCArray, but stores 8-bit pointer descriptors.

Notes:

- Descriptor arrays can be used to hold any type of descriptor. They can be used in APIs that require an `MDesCArray`, for instance, text list boxes.

- `CDesCArray` classes make a copy of each element added to the array, so afterwards the original descriptor can safely be deleted.

- `CPtrCArray` classes hold pointers to the descriptors, rather than the descriptors themselves, so they use less memory than `CDesCArray` classes, but care must be taken since they can hold invalid data if the descriptor pointed to is deleted.

- The build-independent descriptor arrays (`CDesCArrayFlat`, `CDes-CArraySeg` and `CPtrCArray`) are defined as the 16-bit variants, so, when storing binary data rather than text, you need to explicitly specify the 8-bit variant.

8.1.4 Locale

`e32std.h` includes many locale-related classes, covering time zones, formatting for time, date, and currency, measurement units for short distances (inches vs. cm) and long distances (miles vs. km). In a complete Symbian OS platform, these values are usually set through the control panel and are used to prepare times, dates, currency values, and so on for display.

Locale settings depend on

- *The ROM locale*: for instance, a German machine will include German resource files, aif files and spell check dictionaries.

- *The home city*: controls the current time zone, and in UIQ can be set through the control panel or the Time application. The home city default is ROM-locale dependent.

- *Miscellaneous settings*: other locale settings are entered by the user through the control panel, with ROM-locale dependent defaults.

Language downgrade path

In v7.0, the structure of ROM images was changed so that a single ROM can support more than one locale. This is intended to make it easier for Symbian OS licensees to produce ROMs for closely related locales.

Quick selection table

This table lists the most frequently used collection types, and shows the key distinctions among them.

Class name	Frequency of insertions and deletions	Array length	Access type	Data types
TArray<T>	Never	Any	Random	T
TFixedArray<T, TInt S>	Infrequent	Fixed	Random	T
RPointerArray<T>	Infrequent	Bounded	Random	T*
RArray<T>	Infrequent	Bounded	Random	T
RArray<TInt>	Infrequent	Bounded	Random	TInt
RArray<TUint>	Infrequent	Bounded	Random	TUint
CArrayFixSeg<T>	Frequent	Unbounded	Random	T
CArrayPtrSeg<T>	Frequent	Unbounded	Random	T*
TSglQue<T>	Frequent	Unbounded	Sequential, single direction	T
TDblQue<T>	Frequent	Unbounded	Sequential, both directions	T
CDesCArrayFlat	Infrequent	Bounded	Random	Any 16-bit descriptor
CDesCArraySeg	Frequent	Unbounded	Random	Any 16-bit descriptor
CPtrCArray	Infrequent	Bounded	Random	TPtrC16

In a multiple-locale ROM, most files will be identical for all locales supported, even the language-specific ones, such as resource files (assuming that the languages are closely related). In this case, to keep the ROM size down to a minimum, it makes sense to store a single copy of the common files, and only duplicate the files that are different.

To enable this, the ability to set a language downgrade path was added. The language downgrade path is used when a language-specific file cannot be found for the ROM locale; it specifies which languages can provide an alternative version.

The path can contain up to eight languages. The first one is fixed; it is always the language of the ROM locale. The second, third and fourth languages are customizable; they can be set in the ROM by the licensee, and these may be overridden using `TLocale::SetLanguageDowngrade()`. The remaining four are based on a table of near equivalence that is internal to the OS.

For example, on a phone whose ROM locale is Swiss German, if a Swiss German resource file is missing, the language downgrade path could inform the system to search for an equivalent in Austrian German, and failing that, German. You can enquire the whole downgrade path for a particular locale by calling the `BaflUtils::GetDowngradePath()` function.

8.1.5 Math

In `e32math.h`, the static class `Math` defines a range of standard IEEE 754 double-precision math functions, including a random number generator and all the usual log and trig functions.

8.1.6 Variable Argument Lists

Variable argument lists are supported by macros in `e32def.h`. Their commonest use is in providing parameter lists to descriptor formatting functions such as `TDes::FormatList()`.

Here's how it's done in UIKON. `CEikonEnv::InfoMsg(TInt, ...)` takes a resource ID parameter and a variable list of formatting parameters. The resource ID is used to look up a format string, and the format parameters substitute into the string using (ultimately) a `TDes::AppendFormatList()` function.

To implement this version of `InfoMsg()`, Uikon uses a `VA_START` to get the start of the argument list, and a `VA_LIST` to pass a variable list to a lower-level version of `InfoMsg()`:

```
EXPORT_C void CEikonEnv::InfoMsg(TInt aResourceId, ...)
    {
```

```
    VA_LIST list;
    VA_START(list, aResourceId);
    InfoMsg(aResourceId, list);
    }
```

The lower-level version reads in the resource string, and does the real formatting, using `TDes::FormatList()`:

```
EXPORT_C void CEikonEnv::InfoMsg(TInt aResourceId, VA_LIST aList)
    {
    TEikInfoMsgBuf formatString;
    ReadResource(formatString, aResourceId);
    TEikInfoMsgBuf messageString;
    messageString.FormatList(formatString, aList);
    InfoMsg(messageString);
    }
```

In a function such as `FormatList()`, you would use `VA_ARG` *(list, n)* to get the *n*th argument from a `VA_LIST`.

`VA_START` requires that the parameter before the ... is a value parameter, not a reference parameter. So a function prototype of the form,

```
class CConsoleBase : public CBase
    {
public:
    IMPORT_C void Printf(const TDesC& aFormat, ...);
    ...
```

wouldn't work, since the `aFormat` parameter is a reference. If you're coding format lists, you must use the `TRefByValue` class that implements the necessary C++ magic,

```
class CConsoleBase : public CBase
    {
public:
    IMPORT_C void Printf(TRefByValue<const TDesC> aFormat, ...);
    ...
```

and the implementation of `CConsoleBase::Printf()` starts:

```
EXPORT_C void CConsoleBase::Printf(TRefByValue<const TDesC> aFormat, ...)
    {
    VA_LIST list;
    VA_START(list, aFormat);
    ...
```

If you enjoy C++ puzzles, you will have fun with the definitions of the VA_ macros and TRefByValue<T>. You can find them in e32def.h and e32std.h, with descriptor-related overrides in e32des8.h and e32des16.h.

8.1.7   String Formatting

Although Symbian OS is a GUI system, Printf()-style formatting still has a useful role to play, and appears in a number of classes. We've already seen it in CConsoleBase::Printf() and CEiko-nEnv::InfoMsg(). There's another example of using it in buffers:

```
_LIT(KBufFormatString,"[%S]");
console->Printf(KBufFormatString, &temp);
```

As in C print formatting, the % character is followed by a format character, which interprets the corresponding argument for formatting. Here are the main format types:

Format	Argument type	Interpretation
%d	TInt	Decimal value of 32-bit signed integer.
%e	TReal	Real in scientific notation.
%g	TReal	Real in general format.
%x	TUint	32-bit unsigned integer in hexadecimal.
%s	TText*	String passed as the address of a NULL-terminated string of Unicode characters.
%S	TDesC*	String passed as the address of a descriptor

Beware of the difference between %s (for C strings) and %S (for descriptors). Also note that the descriptor version requires the argument to be a pointer to the descriptor.

The SDK documents the format characters along with its documentation for TDes::Format(). There are many more options, including width and precision specifiers and variable argument positions.

8.1.8   RDebug Class

The RDebug class, defined in e32svr.h, includes a host of functions, of which the most interesting is Print(). If you're debugging under the emulator, RDebug::Print() prints to the debug output window. If you're debugging on target hardware, RDebug::Print() uses a serial port, which you can connect to a PC, using a terminal emulator to view the debug output. The port can be set using HAL::Set(), specifying EDebugPort and the port number, or if you are using the eshell text shell, using the debugport command.

This kind of 'print debugging' is useful when a log of activity is handy, or when you have no access to the debugger. The trade-off is that you need a serial port spare on both your target hardware and your PC.

8.2 C Standard Library

STDLIB is the Symbian OS implementation of the standard C library. It delivers standard C functions, which are in general thin layers over corresponding Symbian OS functions.

This means, on the one hand, that you can use almost all your favorite C APIs, from strlen() and malloc() to fopen() and quicksort(), and on the other, that you can usually guess the behavior of functions such as User::QuickSort() because they're there to support standard library functions.

The Symbian OS C standard library was written to support the Java Virtual Machine (JVM) port, which uses Sun's C source code for the JVM with only minor alterations. By this measure alone, the standard library is well tested and powerful. It delivers POSIX-compliant C APIs layering over the Symbian OS user library, file server, and sockets server.

One of STDLIB's design goals was to enable C-language modules to be ported from other systems – particularly those written for a POSIX-like environment and using the standard C library. The next few pages should give you an idea about which parts of your original C code can be ported and which should be replaced by Symbian OS equivalents.

8.2.1 Porting Issues

Compliance with Ansi C/POSIX

STDLIB is not intended to be a complete implementation of all the POSIX standards. Some functions are not fully implemented, because there is no direct equivalent in the C++ libraries. The most important of these are select( ), signal( ), exec( ), and fork( ), whose implementations are at odds with the Symbian OS run-time programming model.

In the case of fork() and exec(), the solution generally involves replacing that part of the code with the Symbian OS equivalent, although there is no way for the called program to inherit open files from the parent – some redesign will be required if this is critical. There is more discussion of this later.

Because STDLIB is layered on the C++ libraries, some functions more closely resemble the equivalent C++ library functions than the standard C library behavior. For example, the behavior of `printf()` follows the equivalent `TDes::Format()` functionality, so that when converting floating point values into strings, the precision is ignored, and the thousands separator is read from the Symbian OS locale.

Console versus GUI apps

The `eshell.exe` program, supplied with the SDK, provides a DOS-style command-line interface instead of the standard graphical one. This can be useful during program development, but it is not suitable for typical end users. Perhaps the major challenge of porting to Symbian OS is that the mobile phones it runs on really demand programs with graphical front-ends are launched in the standard way for the phone.

Symbian OS applications are formed from sets of `.app` and `.dll` files rather than `.exe` files. `.app` files, which are the core DLL associated with each application, do not have a `main()` function. Instead, they require a `NewApplication()` function that returns a `CApaApplication` object – a bootstrap with which a `CApaDocument` is created.

By definition, these and other application object classes must be written in C++, so it can be seen that at least some C++ *must* be used to write any application. Also, although you have some flexibility on how you actually do it, your program will have to process key, pointer, and other events the way Symbian OS expects them to be handled. You can't just port an application; the strategy is to change it while preserving as much of the original as possible.

Global data

One of the consequences of Symbian OS applications being formed from `.app` and `.dll` files is that you can't have writable static data. There are ways around this, which are discussed later in this section, but they all involve some rewriting.

Stack size

One limitation you might come across is if the original program places large variables on the run-time stack – if it uses large arrays as local variables, for example. The standard Symbian OS C++ programming style is to use pointers to heap-based objects rather than having those objects as local variables, and to reflect this the default stack is comparatively small at 8k. On target hardware, this size is fixed for applications, although for executables, you can increase the stack size; this is an option in the .mmp file. A much better solution though would be to change the code so that heap memory is used instead. Heap memory associated with a program grows as required, so requesting too much stack is wasteful.

Error handling

There is something of a mismatch between the Symbian OS approach to error messages and that of traditional C programs. As an essentially graphical environment, Symbian OS uses *alerts* to notify the user about errors, so the C code should return error values instead of printing error messages. Properly written code running in the C++ framework will handle them, often by leaving, resulting in an alert being displayed. Note that the alert will display a generic message based on the error value alone, ignoring the context.

Data types

A problem you may run into involves data types. Under C++, Symbian OS provides a whole range of predefined types such as TInt, TInt16, etc. Unfortunately, these are not available from C, as the e32def.h file includes C++ specific code. Where the use of such integer types is critical in C code, most modules have a config.h file or similar, in which the types used can be given specifically. Copying the relevant parts from e32def.h is one possible approach.

Other considerations

Before your program exits, you should call CloseSTDLIB( ), so that any buffers will be deallocated. If you don't do this on the emulator, you'll get a panic, as happens whenever memory is not deallocated. (This may also happen if your C code doesn't recover any of the heap memory it has requested.)

Changes may need to be made for other non-language reasons, for instance:

Symbian OS phones have limited memory and file space compared to larger machines. You may have to consider how efficiently files are stored by your applications.

The screen is likely to be small. To make effective use of the screen available, you will need to consider at least some of the standard Symbian OS tricks, such as only displaying the menu when required and making only sparing use of separate windows.

8.2.2 Porting Multithreaded Programs

Most programs written for Symbian OS use a single thread where programs written for other platforms would use a process, and active objects (AOs) where other programs would use multiple threads. Ported code that assumes multiple threads with shared resources, for instance, a file or a window, will need to be changed, because in Symbian OS, resources are typically not sharable between threads.

There are two ways around this problem. You can either try to rewrite the program using AOs or you can use STDLIB in such a way that sockets and files *can* be shared between threads. Both of these techniques are discussed here.

Background active object processing

Using this technique, your program needs to be rewritten so that it uses a single thread and asynchronous processing (in which instead of completing immediately, functions issue requests that are completed later).

Interactive applications need to carry out background processing for long-running tasks, so that the main application can still respond to events. Otherwise, the program could effectively seize up while the processing is carried out. Rather than use one thread for event handling and a separate thread for background processing, you can use a single thread where event handling and background processing are handled by **active objects** (AOs).

Asynchronous processing using AOs working together with an **active scheduler** (AS) forms the heart of most, if not all, Symbian OS applications. AOs are covered in more detail in Chapter 17, but they're such a useful and powerful tool that another quick look won't hurt. The basic scheme is shown in Figure 8.2.

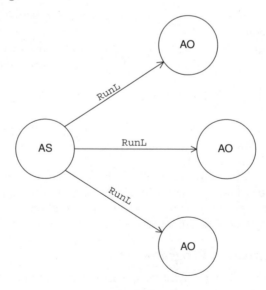

Figure 8.2

Essentially, AOs form an abstract C++ class that encapsulates asynchronous function calls: they contain a status variable, a flag to say they are waiting for action, and a `RunL()` function that's called to handle the completion of whatever 'system call' was issued – together with an extra function `DoCancel()` that's called to cancel an outstanding request. The other part of the system is the AS object, of which there is usually one per program (or one per thread). The AS performs the important job of waiting on the semaphore and calling the appropriate AOs.

To a large extent, this gives us a concurrent execution environment in which AOs are invoked when there is something to be done. The proviso is that it only works when `RunL()` commands do not enter long loops: it is essential that `RunL()` functions execute fairly quickly and return, because the program cannot perform any other functions while they are operating – this is the downside of their non-preemptive semantics.

You can use AOs for background processing. Consider the following code sequence, which can be called from inside an AO:

```
TRequestStatus* status = &iStatus;
User::RequestComplete(status, KErrNone);
SetActive();
```

This construct bypasses the idea of a system call indicating completion of an asynchronous event – it's saying, 'Run me as soon as possible.' Using this sequence, you can get a very similar effect to using threads.

Using this technique will involve some rewriting; for instance, your engine may need to become an active object, and you may need an observer class that tells the main application that the background processing has finished.

One possible problem with converting to the AO/AS scheme is if you want to wait for input on sockets, the console, or similar. If you use synchronous calls, which wait until any input is available, then these will block the whole thread – remember that you can only switch to another active object when you exit the current `RunL()` call. With C++ library calls, the solution would be to use the asynchronous version of the library call, but at face value there are no asynchronous implementations within STDLIB. To help with this, there are overloaded versions of the `ioctl()` call with an extra `TRequestStatus&` parameter. These can be used in a variation on a `select()` call, where you can be told when data is available on a single file descriptor, or when data has been written. For more information, see the STDLIB documentation and the header file `estlib.h`.

CPosixServer

STDLIB runs quite differently depending on whether a CPosixServer object exists or not. If you need one of these, it must be created before you start to use STDLIB, or the system will get confused. If it exists, then all file and socket operations are redirected to the thread that owns the CPosixServer object. Because all files are owned by a single thread, Symbian OS is quite happy with this arrangement – indeed, it is essentially oblivious to the fact, as all it sees is the interthread communication. To create the CPosixServer object, you should use InstallPosixServerActiveObject( ) or Spawn-PosixServerThread() – the latter creates a special thread for the server object, and is probably preferred. Once created, the server object is largely invisible to the program and works behind the scenes. On the other hand, if STDLIB is used without a CPosixServer object having been created, then all resources are owned by the thread that opened the file, socket, and so on, and cannot be accessed by other threads.

There are disadvantages with using the POSIX server. Firstly, it adds an additional client-server transaction each time you access a shared resource and secondly, this approach only applies to STDLIB function calls; you will need to do some extra work if your ported code calls back to C++ code.

The standard recommendation is that you always try to rewrite a thread-based program using the active object technique. Because active objects are non-preemptive, there are no problems with mutual exclusion, and no problems with objects being shared. However, removal of threads completely is not always so simple.

In practice, you will encounter two situations:

- The original program consisted of a sequential engine, with perhaps a second thread being used for interface work. In that case, it is generally more efficient *not* to create a CPosixServer and to use STDLIB in 'direct' mode. Indeed, if you introduce several, separate engines into a program – CODECs for different sound formats, for example – then you still need not use a CPosixServer object, as the various engines won't be sharing files.

- If the original engine was designed in a multithreaded manner, then you will almost certainly have to use a CPosixServer. Remember, though, that you can only share STDLIB objects, so you're totally dependent on the functionality it provides you.

The background active object processing technique is more memory-efficient, more predictable, and almost always preferred – assuming you have the choice.

8.2.3   Global Data

As was mentioned earlier, writable global data is not allowed in Symbian OS DLLs. This can cause problems for ported code. In practice, there are several approaches for dealing with this, but they involve some rewriting:

- Thread local storage (TLS)
- Explicit This
- Global variables masterclass

Thread local storage

Thread Local Storage is the closest thing there is to global variables in Symbian OS. For each DLL, it is possible to register a pointer to a dynamically created struct or class object, and then to look it up whenever it is required. The basis of the technique is that having done this, all global variables are remapped to data fields within the structure. In fact, this is on a per thread, per DLL basis, with each thread potentially having its own globals. (Indeed, this is how the STDLIB implements errno and other global variables so that each thread has its own errno copy.) If you use multithreading, you will have to be careful about this.

The generally recommended way of using TLS is to incorporate OpenL() and Close() functions, and then to call these as the program is created or shut down – typically from the document or app UI classes' constructors and destructors. The OpenL() and Close() functions are then responsible for creating and destroying the global object respectively. A possible implementation is as follows (where Globals is the struct that holds the global data).

```
EXPORT_C void MyEngine::OpenL()
    {
    Globals *globals = STATIC_CAST(Globals*, Dll::Tls());
    if(globals == NULL)
        {
        globals = new(ELeave) Globals;
        Mem::FillZ(globals, sizeof(Globals));
        Dll::SetTls(globals);
        }
    globals->_count += 1;
    }

EXPORT_C void MyEngine::Close()
    {
    Globals *globals = STATIC_CAST(Globals*, Dll::Tls());
    if(globals == NULL)
        return;
```

```
    globals->_count -= 1;
    if(globals->_count == 0)
        {
        delete globals;
        Dll::SetTls(NULL);
        CloseSTDLIB();
        }
    }

Globals* TheGlobals()
    {
    return STATIC_CAST(Globals*, Dll::Tls());
    }
```

Any reference to a global variable x has to be replaced by globals->x. In addition, you need to include the following statement within each function that uses global variables:

```
Globals* globals = TheGlobals();
```

An additional field has been added to the Globals struct: _count. The purpose of this is to allow nested OpenL()/Close() calls – the scheme is sometimes referred to as RegisterL()/Unregister() to reflect this. This is sometimes useful, particularly for unstructured code, or when you only want to create the Globals variable as required in order to save memory. In this particular example, this approach is probably overkill, and it could be replaced with a slightly simpler scheme that dispensed with _count – perhaps adding an assertion within OpenL() that globals is equal to 0.

Another possible TLS scheme is to change TheGlobals() so that the Globals object is created when Dll::Tls() returns 0 – obviously using Dll::SetTls() to store the new value. Although apparently simple, the downside of this technique is all the methods that call TheGlobals() would need to handle out-of-memory events (and TheGlobals() would need to become TheGlobalsL() to reflect this). Always creating from the same place is much simpler.

There are drawbacks with the TLS approach. Firstly, there is a performance penalty: the Dll::Tls() call takes a significantly longer time than a standard pointer lookup (it involves a system call). Secondly, because you can only have one TLS object per DLL, all global and static variables have to be placed within this object. If you have several static variables of the same name, you will have to change their names.

Explicit `this`

In Symbian OS C++ programs, we almost always use data members instead of ordinary variables, with the significant advantage that within member functions, the `this` pointer is implicit. With C programs, of course, this is not so straightforward: there is no implicit `this` pointer, and no method of directly associating functions and data structures. However, that does not stop you doing it explicitly: you can rewrite the original C as 'object-based C' by explicitly defining a `this` pointer.

You declare a structure that contains the global data, store a pointer to it (called `This`) perhaps as a data member of the C++ engine class, and then pass this pointer as an extra parameter to C functions called by the engine that need to access the data.

A disadvantage is that there are potentially many changes to be made: all the functions need to be passed the `This` pointer, except perhaps those that do not call other functions or access global variables. This means that almost all function declarations and calls have to be modified.

On balance, the facts that the original code structure can be largely retained, and that the TLS performance penalty is avoided, tip the balance toward explicit `This` as the preferred technique. However, there are exceptions, and the TLS method should not be dismissed out of hand.

Global variables master class

In this technique, a 'master class' is used to wrap all the C code within a module, so that the original C functions become member functions of this class, and the global variables become data members. In practice, not all C can be so converted.

The number of changes required by this method are likely to be more than for the others previously mentioned – particularly as even functions that are local to a particular file have to be entered into the 'master class', unless they don't use global variables. However, if you bite the bullet and proceed with this technique, there are some distinct advantages. Not least, you can directly employ the predefined Symbian OS types, and you can more easily make calls to the C++ libraries if necessary.

8.2.4   Conclusion

Porting C modules *is* possible. However, you have to be very clear which parts of the original program should be ported, and which will need to be replaced by Symbian OS equivalents. The general approach to porting therefore is not to attempt to port whole programs, but to port the essential parts of the code, (normally the calculation/storage parts of an application, which may themselves need some changes) with the interactive modules written using Symbian OS-oriented C++.

8.3 Summary

This chapter finishes off the section of the book that introduces basic Symbian OS development techniques and APIs. In it, we've seen

- static functions from the `User` and `Math` classes,
- user library support for locales,
- dynamic buffers, flat and segmented, for arrays and rich text,
- collection classes, including arrays and lists,
- the Symbian OS C standard library to support porting,
- how to use variable argument lists,
- debugging with `Printf()`-style formatting of strings and the `RDebug` class.

In the next chapter, we provide a walk-through of a complete Symbian OS application.

9

Stand-alone Applications

In Chapter 4, I described a minimal GUI application that simply showed how the UI framework and app UI fit together. After a further four heavy programming chapters, it's time to go through another application that builds on these ideas. I'll start with a single-player version of the Battleships game and then expand things further with a two-player version.

I also take the opportunity to introduce three new ideas:

- How to program a user interface that doesn't assume a particular screen size. This is important if you want to target your application at multiple user interfaces, as it will help speed up your porting efforts.
- Persistence or file-based applications. This explains the role of the document class in managing how an application interacts with the file system.
- The view architecture, which allows the operating system, and other applications, to communicate directly with an application's views.

Finally, this chapter is a foundation for understanding the Battleships application. I'll later discuss the application in the context of communications and system programming and high-level design. In those chapters, I'll cover the GUI only in passing, so this chapter provides us with an opportunity to get all the GUI aspects together in one place.

9.1 The Game of Battleships

The classic game of battleships is played on paper and works like this. Each player has an 8×8 grid, on which is laid out a fleet of ships, made up as follows:

- one battleship (B), four squares long
- two cruisers (C), each three squares long

Symbian OS C++ for Mobile Phones.   Edited by Richard Harrison
© 2003 John Wiley & Sons, Ltd   ISBN: 0-470-85611-4

- three destroyers (D), each two squares long
- four frigates (F), each one square.

The ships can be aligned horizontally or vertically. Ships may not touch each other, even at the corners. Neither player knows the layout of the ships on the other player's board.

The players take turns. For each turn, one player calls a square to the other, say "B7". The other player responds with "Miss"if that square contained nothing; otherwise "Cruiser"(or another ship type) for a square that was hit. The player whose turn it is then marks the result on his version of the opponent's game board and also marks any squares that, because of the adjacency rules, cannot possibly contain a ship. The winner, of course, is the person who is first to sink their opponent's fleet entirely.

In Solo Ships, the single-player version that we start with, the application lays out the ships and the player has to locate and sink the fleet in as few guesses as possible.

9.2 Overview

Here's a screenshot of Solo Ships in action (Figure 9.1):

Figure 9.1

It's a pretty Spartan display, with no button bar and no player status view – but that's all you need for this game.

The interesting feature of this application is that it's moderately complex: not just another "Hello World!" application; it has 9 classes with a total of 96 functions. That's small in comparison with the full Battleships program (76 classes) and tiny in comparison with Symbian operating system (OS) as a whole. So, as we approach Solo Ships, we'll begin to get an idea of the techniques we use to understand an object-oriented system at a higher level.

9.2.1 Program Structure

In Chapter 4, I introduced the structure of `hellogui`, which had four classes – the minimum for a practical GUI program. `soloships` (you'll find the code in `\scmp\soloships\`) uses these four classes, and a further five, (shown in Figure 9.2) to build up a complete application:

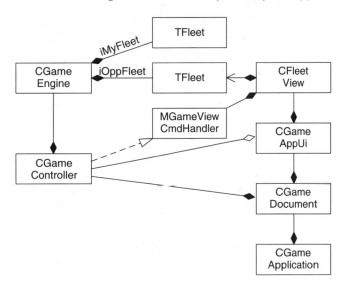

Figure 9.2

On the right, you can see the familiar GUI application classes – application, document, app UI, and app view.

On the left, you can see the classes that represent the application's persistent data: the controller and engine. The engine has two fleet objects: `iMyFleet` stores the actual layout of the fleet of ships the player is trying to sink; `iOppFleet` stores what the player has discovered so far about the fleet (by making shots at it). Each fleet object has a number of ships in it, represented by `TShip`, a class not shown on the diagram.

'Engine' is a synonym for 'model' in the model-view-controller (MVC) sense. Inside Symbian, these words are used interchangeably.

The application's view displays the target fleet, and handles input (the user's shots). The `MGameViewCmdHandler` interface class is used to communicate from the view to the controller.

Note that the document class owns only the objects that hold the application data (controller, engine, fleets), while the app UI owns only the object that displays it (view). This separation will be essential when we come to add persistence to the program.

9.2.2   The Engine

Solo Ships is a simple, one-player game in which the user plays against a randomly generated layout of battleships. However, the application has been designed with flexibility in mind so it can be expanded to a full, multiplayer game in future: Solo Ships and the full Battleships game have the same engine code.

As mentioned earlier, the engine has two fleet objects:

- One fleet object (termed the 'opponent' fleet) is initialized to be entirely empty at the beginning of the game. This is the fleet displayed in the application's view, and it represents what the player has discovered so far about the target fleet that they are trying to sink.
- The real information about the target fleet (termed 'my fleet') is contained in the other fleet object. This is initialized randomly at the beginning of the game.

Each time the user shoots at a square, the controller asks the 'my fleet' object what is at that square, and updates the 'opponent fleet' object appropriately. This is then reflected in the view. So, during the game, the player gets to know more and more about the opponent fleet until – when he's sunk it entirely – he knows everything about it.

As described above, 'my fleet' appears to be a rather odd name in this context. This is because it's carried over from the full Battleships game in which the 'my fleet' object really does represent the player's fleet, which, the opponent, on another phone, is trying to sink.
Also, note that the program refers to a shot at a particular square as a 'hit request'. A square marked as 'hit' is one that has been fired at: it doesn't mean that a ship at that square has been successfully struck.

9.2.3   The Controller

In the MVC paradigm, we saw the following:

- The model contains the application data – in Solo Ships, most of the model is contained in the game engine, though the view's zoom state is also model data.
- The view draws the model in a way that is meaningful to the human user.

- The controller updates the model, and requests view redraws, in response to various kinds of event – including user interactions.

The CGameController class implements controller functionality. It owns the model (CGameEngine, and an iZoomFactor member), and ensures that the model state is saved to file. Finally, the controller handles all events that cause the model (and therefore the view) to change. In Solo Ships, there are two types of events:

- User interactions from the fleet view – handled by the controller as the implementer of the MGameViewCmdHandler interface.
- Commands from the app UI – most app UI CmdXxxL() functions are prechecked by the app UI, and then implemented by some controller function.

For many applications, a separate controller class like this isn't needed. It's OK to put many of the controller functions inside the app UI or the model, provided that all the persistent data is inside the model.

9.3   Engine Classes

We've seen about as much of the high-level view as we can realistically take in. Now it's time to get down to some more detail. Let's start with the engine, which is declared in engine.h and implemented in engine.cpp.

There are three classes (as shown in Figure 9.3):

Class	Description
CGameEngine	The engine itself, which includes two fleets (iMyFleet and iOppFleet), and some utility functions.
TFleet	A fleet. The same class is used to represent 'my fleet', and to represent the 'opponent fleet'.
TShip	A ship. A ship object contains a data member indicating whether it's a Battleship (the largest ship), Cruiser, Destroyer, or Frigate (the smallest).

In UML, that's

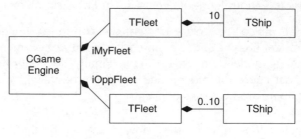

Figure 9.3

There's already an interesting design decision here. The two fleets are in some ways quite different:

- The engine knows everything about `iMyFleet`: the only thing that changes throughout the game is the squares that have been hit by my opponent.
- The engine initially knows nothing about `iOppFleet`: it finds out more throughout the game as it gets reports about the results of hits.

Because of this, I wondered whether to use two different classes, one for each type of fleet, with a base class to represent their properties in common. But there is so much in common that eventually I decided not to do this. Each fleet object uses the same `TFleet` class, though some functions are intended only for use for the 'my fleet' object, and others are intended only for the 'opponent fleet' object.

Having functions that shouldn't be called seems like a messy design, but in this case a 'pure' design seemed even messier. I would need three fleet classes in the engine – and probably in the `CFleetView` too. For such a small application, it didn't seem worth it. Let's look at the individual engine classes more closely.

9.3.1 The Ship Class

The ship class is the most fundamental:

```
class TShip
    {
public:
    enum TShipType
        {
        EUnknown, ESea,
        EBattleship, ECruiser, EDestroyer, EFrigate
        };
public:
    TShip(); // Default constructor for no initialization
    TShip(TShipType aType);  // Initialize type, length, and remaining
```

```
    // Persistence
    void ExternalizeL(RWriteStream& aStream) const;
    void InternalizeL(RReadStream& aStream);
public:
    // Initialized members
    TShipType iType;          // Type
    TInt iLength;             // Length - determined from type
    TInt iRemaining;          // Remaining unit squares

    // Calculated members
    TInt iStartX, iStartY;    // Start position
    TInt iDx, iDy;            // Orientation vector
    };
```

A ship's first attribute is its type, iType, which may be battleship, cruiser, destroyer, or frigate. The type governs the length – 4, 3, 2, or 1 squares, respectively.

The relationship between type and length is encapsulated in the TShip::TShip(TShipType) constructor: you specify a valid ship type, and the constructor sets the length. There's also a default constructor, which is needed so that arrays of TShips can be constructed in TFleet: the default constructor sets the ship type to unknown and the length to zero.

A ship has a starting square and a direction vector, for example, a battleship starting at (3, 4) with a direction vector of (0, 1) will cover squares (3, 4), (3, 5), (3, 6), and (3, 7). Clearly, the same battleship could also be described by a starting square of (3, 7) and a direction vector or (0, −1) – it doesn't matter.

The above information about ships is used during hit detection to work out whether a hit request affects a particular ship (in iMyFleet). When a hit is registered, I decrement the iRemaining count, which is initially set to the length of the ship. I can then tell when the ship has been entirely sunk.

The ship-type enumeration is worth a second look. TShip::TShipType is passed to and returned from many functions in the engine API. The ship type is an enumerated type, including constants not only for the four real ship types, but artificial values for unknown and sea. These are there to ensure that the ship-type enumeration is compatible with the square states in TFleet.

TShip::TShipType is represented in the engine's APIs and is used throughout the higher-level code that plays Solo Ships and, indeed, Battleships. The TFleet square states are private to TFleet. The requirement to keep TFleet's square states in sync with TShipType is a little awkward: initially I judged it tolerable in this small design, but on reflection I'm guilty of microoptimization here – see the comment below.

Another small-time design practice is that I don't hide the data members of TShip *terribly well: in some contexts, I would have taken more trouble to make data members private, and to provide public accessors. Again, that didn't seem worth the trouble here.*

There's nothing too difficult about TShip's implementation: check out engine.cpp for details.

9.3.2   The Fleet Class

A fleet is an 8 × 8 grid of squares containing up to 10 ships. Here's the declaration:

```
class TFleet
    {
public:
    // Square state, similar to TShip::TShipType
    enum TSquareState
        {
        EUnknown, ESea,
        EBattleship, ECruiser, EDestroyer, EFrigate,
        EHit=0x80
        };
public:
    // Setup
    TFleet();
    void SetupBlank();
    void SetupRandom();
    void SetMyFleet();
    void SetOppFleet();

    // Interrogators
    TBool IsMyFleet() const;
    TBool IsOppFleet() const;
    TBool IsShip(TInt aX, TInt aY) const;
    TShip::TShipType ShipType(TInt aX, TInt aY) const;
    TBool IsSea(TInt aX, TInt aY) const;
    TBool IsKnown(TInt aX, TInt aY) const;
    TBool IsHit(TInt aX, TInt aY) const;

    // Complicated interrogators
    TShip& ShipAt(TInt aX, TInt aY) const; // Use for my fleet only
    TShip* PossibleShipAt(TInt aX, TInt aY) const; // Use even if fleet
        unknown
    TInt RemainingSquares() const; // Fleet squares that haven't been
        destroyed
    TInt RemainingShips() const; // Ships that haven't been totally
        destroyed
    TInt SquaresHit() const; // Squares hit (sea or otherwise)

    // Change
    void SetSea(TInt aX, TInt aY); // Say a square is sea
    void SetShipType(TInt aX, TInt aY, TShip::TShipType aShipType);
    void SetHit(TInt aX, TInt aY); // Hit a square - must be known when
        it's hit
```

```
    void TestWholeShip(TInt aX, TInt aY); // Test whether hit has sunk
      a ship
    void SetSeaAroundHit(TInt aX, TInt aY); // Set sea around a hit square

    // Persistence
    void ExternalizeL(RWriteStream& aStream) const;
    void InternalizeL(RReadStream& aStream);
private:
    // Setup
    TBool TryPlaceShip(TInt aShipIndex, TShip::TShipType aShipType);
    void PlaceShip(TInt aShipIndex, TShip aShip);

    // Square accessors
    const TSquareState& Square(TInt aX, TInt aY) const;
    TSquareState& Square(TInt aX, TInt aY);
private:
    TSquareState iSquares[64];
    TBool iMyFleet; // Whether my fleet (which I know all about) or
      opponent's (which I don't)
    TShip iShips[10]; // All the ships we know about
    TInt iKnownShips; // Number of ships known to be in fleet
    TInt iRemainingShips; // How many ships haven't been destroyed
    TInt iRemainingSquares; // How many squares haven't been destroyed
    TInt iSquaresHit; // How many squares (including sea) have been hit
    TInt64 iRandomSeed; // Seed for random-number generation during setup
    };
```

This is much more complex, and much more interesting.

You can think of a fleet as either an 8 × 8 grid of squares, each of which has something on it or as a collection of ships, which happen to occupy squares. These two perspectives are both important for different purposes. I represent the squares perspective with iSquares and related counters and the ships perspective with iShips and related counters.

We also noted above that the fleet supports different functions depending on whether it's a 'my fleet' or an 'opponent fleet' object.

Let's have a closer look at the fleet class from these different perspectives.

A grid of squares

The fleet is an 8 × 8 grid of squares, implemented by iSquares[64] and accessed by the private functions:

```
// Square accessors
const TSquareState& Square(TInt aX, TInt aY) const;
TSquareState& Square(TInt aX, TInt aY);
```

These implement the standard C++ pattern for getting and setting a value in a container. You can write

```
Square(1, 3) = EBattleship;
```

which invokes the non-const version of Square() to set the state, or you can write

```
state = Square(3, 5);
```

which invokes the double-const version of Square() to get the state.

The state of a square, represented by the TSquareState enumeration is unknown, sea, or one of the four ship types. This state can be ORed with a flag, 0 × 80, to indicate that the square has been hit.

> *As I said earlier, I'm guilty of microoptimization here. On reflection, a better way to do this would be to have two squares arrays, one containing TShipTypes and the other containing TBools indicating whether the square had been hit. That would cost me 256 bytes of RAM per TFleet, would probably save me some bit-manipulation code in the TFleet member functions, and would make the square state show up nicely in the debugger as the enumerated constant values. It would be possible to rearrange TFleet to do the right thing, and yet to change its internalizer and externalizers to preserve the same external format.*

In keeping with the way the fleet is coded, there are several functions that interrogate particular squares identified by (x, y) coordinate – ShipType(), IsSea(), IsKnown(), IsHit(), and so on. There are also functions to change the state, such as SetHit(), SetSea(), and SetShipType().

A collection of ships

The second way to think about the fleet is as a collection of ships, implemented by iShips[10]. If the fleet is 'my fleet', these ships are initialized by SetupRandom(), from which point there are always 10 ships. If the fleet is an 'opponent fleet', these ships are cleared by SetupBlank() and then, as I hit complete ships in my opponent's fleet, this is detected and ships are added to iShips until, if I win the game, there are 10 whole ships there.

In keeping with this way of thinking about the fleet, there are functions that place ships, detect the ship at a particular square, work out whether a particular square is part of a whole ship, and so on.

My fleet

When a TFleet represents my fleet, I must first set it up in preparation for a game, and then track the effects of my opponent's hits on it.

The functions to do this are

Function	Description
`TShip& ShipAt(TInt aX, TInt aY) const`	Returns the ship at square (aX, aY), when it is known that there is a ship there.
`TShip* PossibleShipAt(TInt aX, TInt aY) const`	Returns a pointer to the ship at square (aX, aY) if there is one, otherwise a null pointer.
`void SetupRandom()`	Sets ships in a random configuration on the sea.

`PossibleShipAt()` scans through `iShips[]` for each ship. For each ship, it walks from the starting square along the ship and tests whether the square is the one whose coordinates were passed to the function. If so, a pointer to the ship is returned. If no match is found, a null pointer is returned. `ShipAt()` simply calls `PossibleShipAt()`, asserts that the result is non-null, and returns the ship as a reference rather than a pointer.

`SetupRandom()` is an interesting function with some Symbian OS-specific code. The approach it takes is as follows:

- To place a ship, work out its starting coordinate and direction and see if it can be placed on sea: if it does so, mark the ship squares appropriately, and mark all surrounding squares as sea, so that no ships can be placed there. If it does not (because there is already a ship there, or in adjacent squares), then try a different starting point and direction, up to 20 times.
- To place an entire fleet, try to place each of its 10 ships. If one of the ships could not be placed, even after 20 attempts, then judge the layout to be impossible and start again with the whole fleet. Try this up to 10 000 times.

This is a brute-force algorithm, but it works nicely enough in practice. The code for `SetupRandom()` shows how to seed and use random number generators in Symbian OS.

```
void TFleet::SetupRandom()
    {
    // Try to place each ship
    User::After(1);
```

```
    TTime now;
    now.HomeTime();
    iRandomSeed = now.Int64();

// Now try placing, up to 10,000 times
TInt shipsPlaced = 0;
for(TInt attempts = 0; attempts < 10000; attempts++)
    {
    SetupBlank();   // Blank everything
    shipsPlaced = 0; // No ships placed yet
    for(TInt ship = 0; ship < 10; ship++) // Try placing 10
        {
        TShip::TShipType shipType =
            ship < 1 ? TShip::EBattleship :
            ship < 3 ? TShip::ECruiser :
            ship < 6 ? TShip::EDestroyer :
            TShip::EFrigate;

        // If couldn't place, do another attempt
        if(!TryPlaceShip(ship,shipType)) break;

        shipsPlaced++; // One more ship placed
        }
    if(shipsPlaced == 10)
        break; // All ships placed - break
    }

// Check whether we placed all ships, or ran out of attempts
__ASSERT_ALWAYS(shipsPlaced == 10, Panic(EShipsNotAllPlaced));
iKnownShips = 10;

// Set remaining squares to sea
for(TInt x = 0; x < 8; x++)
    for(TInt y = 0; y < 8; y++)
        if(Square(x,y) == EUnknown)
            Square(x,y) = ESea;
}
```

The random number generator requires a 64-bit seed. The generator is seeded at the top of this code. As is conventional, we seed the generator by taking the latest value of the system timer:

```
User::After(1);
TTime now;
now.HomeTime();
iRandomSeed = now.Int64();
```

Incidentally, this is also the way to find out the current time in your local time zone. Check out the SDK for more details of TTime *and see UIQ's (or other Symbian OS UI's) stock controls, including time-and-date editors and clocks, for ways to edit or display the time.*

The first time I wrote this code, which lays out two fleets so that two players can use the same game program, I forgot to include the `User::After(1)` statement. As a result, the random number generators in both game engines picked up the same value of the system time as the initial seed, and therefore generated identical fleet layouts – which made for a fairly boring game!

`User::After(1)` is guaranteed to wait a single clock tick – 1/64 second on a real Symbian OS phone, 1/10 second on the emulator. So by coding `User::After(1)` before getting the seed time, I am guaranteed to get a seed value that has not been used before. The quality of these streams is easily good enough for applications such as Battleships layout.

`TryPlaceShip()` includes some code to use a random number:

```
// Select starting coordinates
TInt coord1 = TInt(Math::FRand(iRandomSeed)*(8 - ship.iLength));
 // 0..8-length
TInt coord2 = TInt(Math::FRand(iRandomSeed)*8);
 // 0..7

// Sanity check in case Math::FRand() returned 1
if(coord1 + ship.iLength == 8 || coord2 == 8)
    continue;
```

This code uses `Math::FRand()` to produce a floating-point number between 0 and 1 and to update the seed for use next time `Math::FRand()` is called. It's not entirely clear from the SDK whether `Math::FRand()` can produce the value 1, so as a precaution I check this after generating the random number.

The approach I've taken to random number generation is good enough for Battleships and for many similar game-type applications. You might want to think harder if random number generation was a really important feature of your game. `Math::Random()` generates integer random numbers, which are more efficient than using floating-point. `Math::FRand()` uses a linear congruential generator with well-chosen constants, while `Math::Random()` derives randomness from hardware events. Note that this won't be good enough for some applications (notably some types of encryption), but for Battleships, the facilities I've chosen are good enough.

The opponent's fleet

Here are some interesting `TFleet` functions relating to the opponent's fleet:

Functions	Description
`void TestWholeShip (TInt aX, TInt aY)`	Knowing that square (aX, aY) contains a ship and has just been hit, this function scans adjacent squares to see whether the whole ship has been hit and, if so, adds a new ship to the fleet's `iShips[]` array.
`void SetSeaAroundHit (TInt aX, TInt aY)`	Knowing that square (aX, aY) has just been hit and contains a ship, this function marks all adjacent squares as sea. Starts by marking diagonally adjacent squares. Then, if the entire ship has been found, marks all its surrounding squares.
`void SetupBlank()`	Sets the 8 × 8 grid to contain blank squares.

The implementations of `TestWholeShip()` and `SetSeaAround-Hit()` involve a lot of careful coding, but nothing very Symbian OS-specific. Check the source for details.

9.3.3   The Game Engine Class

The `CGameEngine` itself acts as a container for 'my fleet' and the 'opponent fleet', and contains query functions to indicate the state of play:

```
class CGameEngine : public CBase
    {
public:
    // Setup
    CGameEngine();

    // Reset
    void Reset();

    // Interrogate
    TBool IsWon() const;
    TBool IsLost() const;
    TBool IsStarted() const;
    TBool IsMyTurn() const;

    // Set up
    void SetFirstPlayer();
```

```
    void SetSecondPlayer();
public:
    TFleet iMyFleet;
    TFleet iOppFleet;
    TBool iFirstPlayer;
    };
```

CGameEngine doesn't require a ConstructL() because its members are all T objects, so construction cannot leave. As it doesn't have a destructor either, arguably it doesn't need to be derived from CBase. However, I've made it a C class, and have treated it as a C class throughout my code, so that the code works as a good basis for other classes without having to convert it from T to C and rethink everything related to cleanup. Using the C class is also useful to ensure zero initialization of all CGameEngine member data.

9.4 The View Class

The view class, CFleetView, has two main roles:

- to display the current state of the opponent fleet,
- to process the key and pointer events that allow the player to shoot at the fleet.

You can find the declaration for CFleetView in view.h, and its source in view.cpp. Here's the declaration:

```
class CFleetView : public CCoeControl
    {
public:
    // Construct/destruct/setup
    ~CFleetView();
    void ConstructL(const TRect& aRect);
    void SetController(TFleet& aFleet, MGameViewCmdHandler& aCmdHandler);
    // Zoom
    void SetZoomL(TInt aZoomFactor);
    TInt GetZoom() const;
    // Cursor
    void SetCursorOff();
    void SetCursor(TInt aX, TInt aY);
    TBool CursorOn() const;
    void GetCursor(TInt& aX, TInt& aY) const;
    // Incremental drawing
    void DrawTilesNow() const;
    void DrawBordersNow() const;
private:
    // From CCoeControl
    void Draw(const TRect&) const;
```

```
    void HandlePointerEventL(const TPointerEvent& aPointerEvent);
    TKeyResponse OfferKeyEventL(const TKeyEvent& aKeyEvent,TEventCode
      aType);
    // Auxiliary draw functions
    void DrawOutside() const;
    void DrawBorders() const;
    void DrawHorizontalBorder(const TRect& aRect) const;
    void DrawVerticalBorder(const TRect& aRect) const;
    void DrawTiles() const;
    void DrawTile(TInt aX, TInt aY) const;
    // Cursor movement
    void MoveCursor(TInt aDx, TInt aDy);
private:
    TFleet* iFleet;
    MGameViewCmdHandler* iCmdHandler;
    // Cursor
    TBool iCursorOn;
    TInt iCursorX;
    TInt iCursorY;
    // Scale to use when calculating drawing stuff
    TInt iZoomFactor;
    // Precalculated drawing stuff
    CFont* iBorderFont;
    CFont* iTileFont;
    TRect iBoardRect; // Board area
    TRect iTopBorder; // All of top border
    TRect iBottomBorder; // All bottom border
    TRect iLeftBorder; // Left border, excluding top and bottom
    TRect iRightBorder; // Right border, excluding top and bottom
    TInt iTileSize; // Side of tile (1/8th board area)
    TInt iBorderSize; // Size of border
    };
```

The first group of public functions (commented as 'Construct/destruct/ setup') is called by the app UI to create the view. The remaining public functions are for use by the controller, to tell the view to respond to user commands or engine data changes.

The first group of private functions (commented as 'From CcoeControl') overrides the control base class functions to draw the view, and to handle pointer and key events, respectively. The remaining private functions implement various parts of drawing the view.

The private data stores parameters passed into the view when it is set up, and data, such as fonts and coordinates, is required to draw the view. iFleet is the model and iCmdHandler is the controller in the MVC sense. So this control works as the V part of a well-structured MVC trio. The fleet view allows a cursor to be displayed, controlled by iCursorOn and (x, y) coordinates. Finally, the view can be zoomed to a size that is (iZoomFactor/1000) times the default size (more on this later).

9.4.1 View Construction

The view's main setup function is ConstructL():

```
void CFleetView::ConstructL(const TRect& aRect)
    {
    // window setup
    CreateWindowL();
    SetRect(aRect);
    // set cursor
    SetCursor(0,0);
    // set zoom factor
    SetZoomL(1000);
    // activate control as ready for drawing
    ActivateL();
    }
```

This begins by creating a new window, and then setting its rectangle (relative to the screen) to the client area rectangle passed as a parameter. From this point, I use Rect() to find coordinates *relative to the window*. This rectangle is smaller than the screen area, as it doesn't include the areas for menus, toolbars, status bars, and so on.

SetCursor(0,0) sets the board cursor to the top-left square. Set-ZoomL(1000) then sets the initial zoom factor to one-to-one, using code that we'll look at closely in the next section. This code calculates all the rectangles used for drawing and pointer event handling. I then activate the app view window, so that it will respond to draw requests and input events.

To complete setup, the view needs to know what fleet object to draw and who to tell about input events. The controller is the object to supply this information, which it does by calling SetController().

```
void CFleetView::SetController(TFleet& aFleet, MGameViewCmdHandler&
  aCmdHandler)
    {
    iCmdHandler=&aCmdHandler;
    iFleet=&aFleet;
    }
```

This simply stores the pointers to the fleet object and the interface, implemented by the controller that handles the player's shots at the fleet.

9.4.2 Drawing the View

Many of the CFleetView data members contain sizes that are used for drawing the view. Shortly, I'll cover how those data members are initialized. First, though, I'll show the drawing code. The Draw() function draws the view in three stages:

```
void CFleetView::Draw(const TRect&) const
    {
    DrawOutside();
```

```
    DrawBorders();
    DrawTiles();
    }
```

DrawOutside() whites out the region outside the board, so that – as required by the Draw() contract – the entire control is drawn.

DrawBorders() draws the top, bottom, left, and right border around the sea area:

```
void CFleetView::DrawBorders() const
    {
    DrawHorizontalBorder(iTopBorder);
    DrawHorizontalBorder(iBottomBorder);
    DrawVerticalBorder(iLeftBorder);
    DrawVerticalBorder(iRightBorder);
    }
```

DrawTiles() simply draws all 64 tiles in the sea area: we'll take a closer look at it below.

Drawing outside the board

DrawOutside() is straightforward, as clearing out the space between two rectangles is sufficiently common that there is utility function to do it:

```
void CFleetView::DrawOutside() const
    {
    CWindowGc& gc=SystemGc();
    gc.SetPenStyle(CGraphicsContext::ENullPen);
    gc.SetBrushStyle(CGraphicsContext::ESolidBrush);
    gc.SetBrushColor(KRgbWhite);
    DrawUtils::DrawBetweenRects(gc, Rect(), iBoardRect);
    }
```

SystemGc() gets a graphics context that the function will use to draw to the control. The graphics context holds the settings to use when drawing, which, in this case are solid white for the drawing brush and null for the drawing pen. It then uses a handy utility function DrawUtils::DrawBetweenRects() to fill the area between the board and the control's boundaries. iBoardRect contains the board's rectangle, and Rect() contains the control's rectangle.

Drawing the borders

Here's DrawHorizontalBorder() – remember that this function gets called twice by DrawBorders(), once with the top border area as a parameter, and once with the bottom border area.

```
void CFleetView::DrawHorizontalBorder(const TRect& aRect) const
    {
    CWindowGc& gc = SystemGc();

    // Draw corners - in fact, whole border
    gc.SetBrushStyle(CGraphicsContext::ESolidBrush);
    gc.SetBrushColor(KRgbBlack);
    gc.SetPenStyle(CGraphicsContext::ENullPen);
    gc.DrawRect(aRect);

    // Draw letters
    gc.SetPenStyle(CGraphicsContext::ESolidPen);
    gc.SetPenColor(KRgbWhite);
    _LIT(KBorderLetters,"ABCDEFGH");
    gc.UseFont(iBorderFont);
    for(TInt i = 0; i < 8; i++)
        {
        TRect rect(
            aRect.iTl.iX+iBorderSize+i*iTileSize, aRect.iTl.iY,
            aRect.iTl.iX+iBorderSize+(i+1)*iTileSize, aRect.iBr.iY
        );
        TPtrC text = KBorderLetters().Mid(i, 1); // Get letter
          at position i
        TInt baseline = rect.Height()/2 + iBorderFont->AscentInPixels()/2;
        gc.DrawText(text, rect, baseline, CGraphicsContext::ECenter);
        }
    gc.DiscardFont();
    }
```

Firstly, I set up the brush to solid black, and then draw the rectangle for the border. Then I draw the border letters.

Looking at it, this loop is less efficient than it might be. I could have set a null brush prior to the loop to avoid repainting the text. I could have saved some multiplication by initializing the rectangle for the A label and then using a simple addition to move along the *x*-coordinates for each iteration, rather than doing the calculation from scratch each time. I have to admit, I wrote this code in a hurry. Then again, it's not the most critical code, so I'm not too worried.

It might also be better to get the label string out into a resource file, rather than hard-coded in the C++ code here. This one isn't an obvious localization issue: it's not clear that the ABCDEFGH string needs to be translated for other Latin-based alphabets, or even for non-Latin-based alphabets. I took the easy option here, for no better reason than that it was easy.

In fact, it wasn't initially obvious whether I should retain the lettered and numbered borders at all. You don't need them, technically, to play the game – as the computer looks after all the coordinates for you. It turns out that in practice it's useful to have them, because players like to talk about the state of the game and without the numbers they would have to invent their own coordinate system for this purpose.

The code for `DrawVerticalBorder()` is essentially the same as this.

Drawing the tiles

`DrawTiles()` simply calls `DrawTile()` 64 times, once for each tile:

```
void CFleetView::DrawTiles() const
    {
    CWindowGc& gc = SystemGc();
    gc.UseFont(iTileFont);
    for(TInt x = 0; x < 8; x++)
        for(TInt y = 0; y < 8; y++)
            DrawTile(x, y);
    gc.DiscardFont();
    }
```

This sets up the font used to draw the tiles, and then draws each tile in turn. Originally, I put the `UseFont()` and `DiscardFont()` around each `DrawText()` call in my `DrawTile()` function below. This means that `UseFont()` and `DiscardFont()` are being called many times more than they need to be – a waste, because this is a surprisingly expensive call. When I took them out of the loop, as shown above, it speeded up redraw performance dramatically.

`DrawTile()` itself works in stages:

- First, determine the on-screen rectangle to use for the tile.

- Then, determine the brush color, depending on whether the tile is unknown (white), known (blue), or hit, that is, shot at (red).

- If the square is a ship (not sea or unknown), determine a character to use to represent it (depending on the ship type), a pen color to draw in (depending on whether it has been hit or not), and then draw the character in the square.

- If the square is not a ship, then just draw the rectangle with the brush color and a null pen.

- If the square is the cursor square, draw the cursor.

Here it is: first, determine the rectangle:

```
void CFleetView::DrawTile(TInt aX, TInt aY) const
    {
    CWindowGc& gc = SystemGc();
    TRect rect(iBoardRect.iTl.iX + aX*iTileSize,
               iBoardRect.iTl.iY + aY*iTileSize,
               iBoardRect.iTl.iX + (aX + 1)*iTileSize,
               iBoardRect.iTl.iY + (aY + 1)*iTileSize);
```

This uses `iBoardRect` as the rectangle for the entire grid, and `iTileSize` (a `TInt`) as the size of a tile. The calculation is simple.

Then, determine the brush color:

```
// Set background color depending on whether known, hit or otherwise
gc.SetBrushStyle(CGraphicsContext::ESolidBrush);
if(!iFleet->IsKnown(aX, aY))
    gc.SetBrushColor(KRgbWhite);
else if(iFleet->IsHit(aX, aY))
    gc.SetBrushColor(KRgbDarkRed);
else gc.SetBrushColor(KRgbCyan);
```

No problems here. Next, I decide whether I need to draw a letter (for part of a ship) or a blank square (for anything else):

```
// Draw either plain square or text
if(iFleet->IsShip(aX, aY))
    {
    // Set pen color depending on whether hit or not
    gc.SetPenStyle(CGraphicsContext::ESolidPen);
    if(iFleet->IsHit(aX, aY))
        gc.SetPenColor(KRgbYellow);
    else
        gc.SetPenColor(KRgbBlack);

    // Set character depending on ship and ship type
    TPtrC text;
    _LIT(KShips,"BCDF");
    text.Set(KShips().Mid(iFleet->ShipType(aX, aY)
      - TShip::EBattleship, 1));
    // Draw the square
    TInt baseline=rect.Height()/2 + iTileFont->AscentInPixels()/2;
    gc.DrawText(text, rect, baseline, CGraphicsContext::ECenter);
    }
else // No ship
    {
    gc.SetPenStyle(CGraphicsContext::ENullPen);
    gc.DrawRect(rect);
    }
```

I was very keen to minimize waste here, so to draw a plain square I use `DrawRect()` rather than the more expensive `DrawText()`.

Just as with the borders, there is a localization issue here: I've chosen to use the string BCDF to represent Battleship, Cruiser, Destroyer, and Frigate. In other languages, these ships doubtless have different names, and perhaps the string should be localized. It's a pretty lousy representation, even in English.

Ideally, I would like to use a graphical representation rather than letters. But a graphical representation for a partial Battleship wouldn't look that good, unless it understood the position of the tile relative to the whole ship. And if that information was given away on the display, it would make the game too easy. So I've stuck with the BCDF string in the C++ source code. I can always change things if needed.

Finally, I highlight the cursor square:

```
// Special border if it's the cursor square
if(iCursorOn && aX == iCursorX && aY == iCursorY)
    {
    gc.SetPenStyle(CGraphicsContext::ESolidPen);
    gc.SetBrushStyle(CGraphicsContext::ENullBrush);
    gc.SetPenColor(KRgbBlack);
    gc.DrawRect(rect);
    rect.Shrink(1,1);
    gc.SetPenColor(KRgbWhite);
    gc.DrawRect(rect);
    }
}
```

The cursor is simply a black line around the outside of the tile and a white line around the inside of that. The cursor had to look good whatever the color of the square underneath and around, whatever the contrast capabilities of the display and at whatever zoom level. And yet the cursor should not altogether hide the square underneath.

It took me quite a few iterations to get a scheme that was acceptable. Unusually, the scheme I finally chose – for good aesthetic and functional reasons – was also very simple to execute in C++ code.

Redrawing the view

Draw() is called by the control framework when the windowing system (the Window server) determines that the control needs redrawing, for example, when a window on-top moves away and exposes the control. The other circumstance in which a program draws to a view is to update it to reflect a change in the program's data: for example, in the case of Solo Ships, to show that a square has been hit. This is called **application-initiated redrawing**.

Sometimes the data will have changed so much that an application will need to redraw the entire view. Often, however, only a small amount of data will have changed, and only part of the view needs changing. This approach, called **incremental redrawing**, is a good idea, as, for views of any complexity, redrawing the entire view is likely to be slow enough to be noticeable to the user. The downside of incremental redrawing is that it can be complex to program. For each view that you write, you will need to determine the appropriate trade off between performance/appearance (requiring more incremental redrawing), and program complexity (requiring less incremental redrawing).

For Solo Ships, some experimentation found that redrawing the whole view produced noticeable display flicker and so was not satisfactory.

Using incremental redrawing to update just the tiles, however, produced an acceptable effect. This meant I didn't have to go to the next level of complexity of incrementally redrawing individual tiles.

The function to update all tiles on the board is `DrawTilesNow()`. It's called whenever a fleet gets a hit, or the cursor moves from one square to another.

```
void CFleetView::DrawTilesNow() const
    {
    Window().Invalidate(iBoardRect);
    ActivateGc();
    Window().BeginRedraw(iBoardRect);
    DrawTiles();
    Window().EndRedraw();
    DeactivateGc();
    }
```

`DrawTiles()` actually draws the tiles, as we've already seen. But now it's sandwiched between some unfamiliar functions. These are discussed in detail in Chapter 11, but for now I'll just say a little about this standard pattern used for incremental redraws. The calls through `Window()` are required because the Window server needs to know when and where applications are drawing in order to coordinate all their display activities. Similarly, the control framework needs to know when its graphic context is being used: the application indicates this through the `ActivateGc()` and `DeactivateGc()` calls.

9.4.3 Handling Events

We can now take a brief look at how the view handles pointer and key events. Input event handling is discussed in detail in Chapter 12.

The view receives key events by implementing the `CCoeControl` base class function `OfferKeyEventL()`.

```
TKeyResponse CFleetView::OfferKeyEventL(const TKeyEvent& aKeyEvent,
  TEventCode aType)
    {
    if (aType!=EEventKey)
        return EKeyWasNotConsumed;
    if (
        aKeyEvent.iCode==EQuartzKeyFourWayLeft ||
        aKeyEvent.iCode==EQuartzKeyFourWayRight ||
        aKeyEvent.iCode==EQuartzKeyFourWayUp ||
        aKeyEvent.iCode==EQuartzKeyFourWayDown
        )
        {
        // move cursor
```

```
        if (aKeyEvent.iCode==EQuartzKeyFourWayLeft)
            MoveCursor(-1,0);
        else if (aKeyEvent.iCode==EQuartzKeyFourWayRight)
            MoveCursor(1,0);
        else if (aKeyEvent.iCode==EQuartzKeyFourWayUp)
            MoveCursor(0,-1);
        else if (aKeyEvent.iCode==EQuartzKeyFourWayDown)
            MoveCursor(0,1);
        // redraw board
        DrawTilesNow();
        return EKeyWasConsumed;
        }
    else if (aKeyEvent.iCode==EQuartzKeyConfirm)
        {
        if (iFleet->IsKnown(iCursorX, iCursorY))
            iEikonEnv->InfoMsg(R_GAME_ALREADY_KNOWN);
        else
            iCmdHandler->ViewCmdHitFleet(iCursorX, iCursorY);
        return EKeyWasConsumed;
        }
    return EKeyWasNotConsumed;
    }
```

The function examines what type of key was pressed: for UIQ, the hardware navigation and CONFIRM keys have special key codes that I test for here. The navigation keys result in the cursor being moved, and the board redrawn to show the new cursor position. A CONFIRM key event indicates the user shooting at a square. If the square has been shot at before (IsKnown() is true), then it calls the controller's ViewCmdHitFleet() function so that the hit request can be processed.

Similarly, the view receives pointer events by implementing the CCoeControl base class function HandlePointerEventL().

```
void CFleetView::HandlePointerEventL(const TPointerEvent& aPointerEvent)
    {
    // check whether we're interested
    if (
        aPointerEvent.iType==TPointerEvent::EButton1Down &&
        iBoardRect.Contains(aPointerEvent.iPosition)
        )
        {
        // identify the tile that was hit
        TInt x=(aPointerEvent.iPosition.iX-iBoardRect.iTl.iX)/
          iTileSize;
        TInt y=(aPointerEvent.iPosition.iY-iBoardRect.iTl.iY)/
          iTileSize;
        // move cursor if necessary
        TBool iCursorMoved=(x!=iCursorX || y!=iCursorY);
        SetCursor(x,y);
        // hit square unless it's already known
        if (iFleet->IsKnown(x,y))
            {
```

```
                iEikonEnv->InfoMsg(R_GAME_ALREADY_KNOWN);
                if (iCursorMoved)
                        DrawTilesNow();
                }
        else
                iCmdHandler->ViewCmdHitFleet(x,y);
        }
    }
```

The function responds to a tap on a screen (a TPointerEvent::EButton1Down event) by translating the event's *x* and *y* coordinates into the coordinates of the relevant square on the board. It then uses these coordinates to both set the cursor position and to make a hit request. The code for these tasks is very similar to that in OfferKeyEventL().

9.5   Scaling and Zooming

Now we've seen everything in CFleetView except how the size of the items in the view is worked out.

SetZoomL() sets up all the precalculated rectangles, sizes, and fonts required to (re)draw a fleet view, and to find the tile associated with a pointer event. Here are the relevant members of (shown in Figure 9.4) CFleetView:

```
CFont* iBorderFont;
CFont* iTileFont;
TRect iBoardRect;      // Board rectangle (without borders)
TRect iTopBorder;      // All of top border
TRect iBottomBorder;   // All bottom border
TRect iLeftBorder;     // Left border, excluding top and bottom
TRect iRightBorder;    // Right border, excluding top and bottom
TInt iTileSize;        // Side of tile (1/8th board)
TInt iBorderSize;      // Size of border
```

How do we calculate these? Well, firstly I have a concept of the board in my mind that says how big it should be:

- Each tile should be 1/3 inch high
- The border width should be 1/6 inch
- Therefore, the total board size is 3 inches square ($8 \times 1/3 + 2 \times 1/6$).

I have encoded these definitions into some constant declarations toward the top of view.cpp:

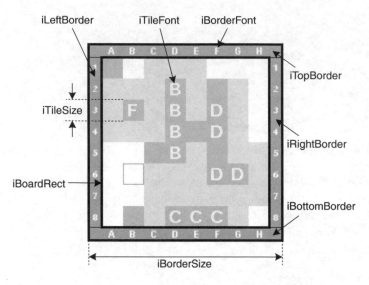

Figure 9.4

```
const TInt KTileSizeInTwips = 480;                              // 1/3"
const TInt KBorderSizeInTwips = 240;                            // 1/6"
const TInt KIdealBoardSizeInTwips = 8 * KTileSizeInTwips +
                                    2 * KBorderSizeInTwips;     // 3"
```

This introduces a new technical term: **twips**. A twip is one-*twenti*eth of a *p*oint, and a point is 1/72 inch. There are, therefore, 1440 twips per inch. In gdi.h, you'll find this number as a constant: you'll also find twips per cm, and twips measurements for common paper sizes including Letter and A4.

Anyway, since the definition of a twip isn't likely to change anytime soon, I don't mind coding constants in the style above.

All programs that require an absolute unit of measurement for paper-type distances should use twips. A twip is small enough that you can express small distances in integers without much loss of precision. A twip is large enough that you can express quite large distances in a 32-bit signed integer – 37 km, or over 130 000 sheets of Letter paper.

Drawing to graphics contexts is done in *pixels*, so I need to convert the board measurements from twips into pixels, taking into account the following facts:

- A scale factor that converts the 3 inches I need for the board into the size, in inches, of the smallest dimension of the fleet view.

- An amendment of this scale factor to take the view's zoom factor into account.

I can then convert the size of the tile, and the border, again by converting inches into pixels. I can also work out the size of the fonts I'll need for the tiles (3/4 of the tile size) and the border (3/4 of the border width).

I do all these calculations in `SetZoomL()`. Let's have a look at the code: on the way we can see how to convert between pixels and inches, how to do zooming, and how to allocate fonts.

Firstly, I do nothing if the zoom factor hasn't changed:

```
void CFleetView::SetZoomL(TInt aZoomFactor)
    {
    // Check we're doing something useful
    if(iZoomFactor == aZoomFactor)
        return;
    iZoomFactor = aZoomFactor;
```

Then, I find the smallest board dimension, in pixels:

```
// Find available size in pixels
TInt boardSize = Rect().Width() < Rect().Height() ? Rect().Width() :
                                                     Rect().Height();
```

Then, I start converting into twips:

```
// Calculate board size in twips, and hence scale factor
TInt boardSizeInTwips =
    iCoeEnv->ScreenDevice()->HorizontalPixelsToTwips(boardSize);
boardSizeInTwips = (boardSizeInTwips * iZoomFactor) / 1000; // Zoom
TInt scaleFactor = (boardSizeInTwips * 1000) / KIdealBoardSizeInTwips;
boardSize =
    iCoeEnv->ScreenDevice()->HorizontalTwipsToPixels(boardSizeInTwips);
```

The `HorizontalPixelsToTwips()` function converts a given number of twips into pixels. This is a device-specific thing: I use the screen device to give me the right answer. Then, I scale the board size (in twips) by the zoom factor to get the board size I will really be using.

Next, I calculate a scale factor: the ratio of the board size I will be using (in twips) to the ideal board size (in twips). I will be applying the same scale factor to the tiles and border, individually.

For both the scale factor and the zoom factor, I use the number 1000 to indicate one-to-one. This means I can do a scaling calculation using code such as

```
scaled_number = (old_number * scale) / 1000;
```

with integer values. This is preferable to doing the calculations in floating-point numbers. This kind of 1000-based scaling is used elsewhere in Symbian OS graphics programming, as we'll see when we look at the `TZoomFactor` class, in Chapter 15. Watch out, though, for two possible sources of error:

- Always multiply by the scale first, and then divide by 1000 – otherwise, you will lose precision.

- Ensure that the number to be scaled, combined with the scale factor, is less than the maximum possible integer – otherwise, you'll get an overflow.

The number 1000 was chosen to be sufficiently large to enable reasonable precision, but sufficiently small to reduce the risk of overflow. With a zoom factor of ten to one, and a twips value the size of a Letter paper, scaling calculations will reach 15 840 (twips high) × 10 000(ten-to-one scale) = 158 400 000, which still leaves a good deal of room before you hit the 2.1 billion maximum signed integer value.

Finally, I calculate a board size in pixels by calling the opposite function to the one I had before – `HorizontalTwipsToPixels()`.

I am assuming that pixels on the screen are square. This assumption is pretty safe.

Next, I calculate the real tile and border sizes, in twips, by applying the same scale factor as was used for the board (I'll be using these values later, to calculate the point sizes of fonts):

```
// Tile and border sizes also
TInt tileSizeInTwips = (KTileSizeInTwips*scaleFactor) / 1000;
TInt borderSizeInTwips = (KBorderSizeInTwips*scaleFactor) / 1000;
```

Then, I have to calculate the sizes, in pixels, of the tiles and border.

```
// Calculate tile and border sizes in pixels, ensuring even distribution
iBorderSize =
    iCoeEnv->ScreenDevice()->HorizontalTwipsToPixels(borderSizeInTwips);

iTileSize = (boardSize-iBorderSize*2)/8; // 8th remaining, rounding down
TInt innerSize = iTileSize*8; // Whole size of inner region
iBorderSize = (boardSize-innerSize) / 2; // Adjust border size again
boardSize = innerSize + iBorderSize*2; // Final board size
```

The algorithm I use here is carefully chosen:

- I get the border size in pixels using twips-to-pixels mapping.
- I get the size of a tile (`iTileSize`) by subtracting twice the border size from the whole-board size, and then dividing by eight.
- It so happens that the tile-size calculation will probably introduce some kind of rounding error. As a result, when I multiply the tile size by eight again to produce `innerSize`, I may be as many as seven pixels short of the inner size you would get if you subtracted twice the border size from the board size.
- I therefore recalculate the border size, as half the remainder after subtracting the inner size from the board size – at worst, this gives me a one-pixel rounding error.
- I finally recalculate the board size, as the inner size plus twice the border size.

This might seem like an odd sequence, but you have to take rounding errors in pixel calculations seriously, and this is the way to do it. Now I know the sizes of everything in pixels, it's easy (if tedious) to calculate the rectangles I'll need.

```
// Precalculate actual rectangles for everything
iBoardRect = TRect(0, 0, boardSize, boardSize);
iTopBorder = TRect(0, 0, boardSize, iBorderSize);
iBottomBorder = TRect(0, iBorderSize + innerSize, boardSize, boardSize);
iLeftBorder = TRect(0, iBorderSize, iBorderSize, iBorderSize + innerSize);
iRightBorder = TRect(iBorderSize + innerSize, iBorderSize,
                     boardSize, iBorderSize + innerSize);
iBoardRect = TRect(iBorderSize, iBorderSize,
                   iBorderSize + innerSize, iBorderSize + innerSize);

// Offset everything to center properly
TPoint offset(Rect().iTl.iX + (Rect().Width() - boardSize) / 2,
              Rect().iTl.iY + (Rect().Height() - boardSize) / 2)
iBoardRect.Move(offset);
iTopBorder.Move(offset);
iBottomBorder.Move(offset);
iLeftBorder.Move(offset);
iRightBorder.Move(offset);
iBoardRect.Move(offset);
```

Finally, I need to allocate fonts: an `iBorderFont` for the border, and an `iTileFile` for the tiles. I use essentially the same algorithm in both cases:

```
// Get small font for drawing border
TFontSpec specBorder(_L("Arial"), (borderSizeInTwips * 3) / 4);
specBorder.iFontStyle.SetStrokeWeight(EStrokeWeightBold);
CFont* borderFont = iCoeEnv->CreateScreenFontL(specBorder);
```

```
if(iBorderFont)
    iCoeEnv->ReleaseScreenFont(iBorderFont);
iBorderFont = borderFont;

// Larger font for drawing tiles
TFontSpec specTile(_L("Arial"), (tileSizeInTwips*3) / 4);
specTile.iFontStyle.SetStrokeWeight(EStrokeWeightBold);
CFont* tileFont = iCoeEnv->CreateScreenFontL(specTile);
if(iTileFont)
    iCoeEnv->ReleaseScreenFont(iTileFont);
iTileFont = tileFont;
}
```

Firstly, I get a `TFontSpec`, in which I specify a font name (Arial in both cases) and size in twips (3/4 of the tile size or border width). I then set the font spec to include the bold attribute.

I call `CreateScreenFontL()` to ask the CONE environment to create a screen font for me, with those attributes – or to leave, if it cannot. Assuming this worked, I replace the previous font with the one I was allocated and release the previous font using `ReleaseScreenFont()`. You need to release the fonts when the program has finished using them, which can be done in the destructor:

```
CFleetView::~CFleetView()
    {
    iCoeEnv->ReleaseScreenFont(iTileFont);
    iCoeEnv->ReleaseScreenFont(iBorderFont);
    }
```

This code has introduced us to many of the issues in size-independent on-screen graphics:

- Absolute units are measured in twips, 1/1440 of an inch.

- You use a device to convert between pixels and twips. Although separate functions are provided for horizontal and vertical measurements, it's a fair bet that pixels, especially on screens, are square.

- You can get at the screen device with `iCoeEnv->ScreenDevice()`.

- In pixel calculations, rounding errors are real, so you have to calculate the sizes of elements in your view carefully to minimize the effects of rounding errors.

- You can get a `CFont*`, needed for GC `UseFont()` functions, by creating a `TFontSpec` with a font name, size in twips, and any other attributes you need, and then using `iCoeEnv->CreateScreenFontL()` to allocate a font matching this as nearly as possible.

In Chapter 15, I'll be explaining these issues in a lot more detail. For now, we have good proof that it works, because by zooming the view, you can see that it scales nicely to a wide variety of sizes.

9.6   The Controller

The controller is at the heart of Solo Ships and – in a much larger form – at the heart of Battleships also. Every update to the model goes through the controller – whether the update originated in the app UI, the app view, or (in the case of Battleships) in a packet coming in on a communications link.

In addition, the controller acts on behalf of the document class as the owner of the application's persistent data. This means that the controller can be changed throughout the lifetime of the application.

Here's the declaration of `CGameController`, in `controller.h`:

```
class CGameController : public CBase,
                        public MGameViewCmdHandler
    {
 public:
    // Construct/destruct
    static CGameController* NewL();
    static CGameController* NewL(const CStreamStore& aStore,
                                 TStreamId aStreamId);
    void SetAppView(CGameAppView* aAppView);
    ~CGameController();
    // Persistence
    TStreamId StoreL(CStreamStore& aStore) const;
    // State
    inline TBool IsMyTurn() const;
    inline TBool IsFinished() const;
    // Game control
    void Reset();
    // Zooming
    void ZoomInL();
    void ZoomOutL();

 private:
    enum TState { EMyTurn, EFinished };

 private:
    // Construct/restore
    void ConstructL();
    void RestoreL(const CStreamStore& aStore, TStreamId aStreamId);
    // Stream persistence
    void ExternalizeL(RWriteStream& aStream) const;
    void InternalizeL(RReadStream& aStream);
    // from MGameViewCmdHandler
    void ViewCmdHitFleet(TInt aX, TInt aY);

 private:
```

```
    CGameAppView* iAppView;
    CGameEngine* iEngine;

private:
    // Cached pointers and values
    CEikonEnv* iEnv;
    // Private persistent state
    TState iState;
    // Zoom for internalizing
    TInt iZoomFactor;
    };
```

The controller knows the current game state: it's either EMyTurn or EFinished (in the full Battleships game, there are nine states). There are utility functions for interrogating the state.

Public functions for construction and persistence allow a caller to create a new default document (NewL() on its own), or a new document restored from a file (NewL() with a CStreamStore parameter), or to store the document (StoreL()).

SetAppView() is called after the controller has been constructed to link it in to the app view, which the controller updates after any model updates.

The controller implements three functions that are executed in response to app UI-originated commands: Reset() starts a new game, while ZoomInL() and ZoomOutL() cycle through zoom states. The controller implements one function in response to view originated commands: ViewCmdHitFleet().

We'll look at construction and persistence later. For now, let's look at how the controller implements commands originated in the app UI and view.

9.6.1 Accessing the GUI Environment

The controller needs to use the GUI environment class for such things as information messages. Its second-phase constructor includes the line

```
iEnv = CEikonEnv::Static();
```

which gets the environment pointer from thread-local storage (TLS) and caches it in a handy pointer iEnv. Since TLS accesses are much slower than pointer accesses, this is normal practice for classes that use CEikonEnv.

You don't need to do this for CCoeControl- or CCoeAppUi-derived classes, since they already contain a pointer to an environment object. The pointer is a protected member variable, iCoeEnv. As the pointer type of the variable is in fact CCoeEnv (the base class for CEikonEnv), a cast is required to use it. A shortcut is provided by a handy #define in eikdef.h:

```
#define iEikonEnv (STATIC_CAST(CEikonEnv*, iCoeEnv))
```

This #define isn't good C++ programming style. Among other unde-sirable effects of this definition, I *can't* call my own pointer to the environment iEikonEnv. That's why, following common practice, I chose to call it iEnv instead.

9.6.2 Zooming

Here's the code for ZoomInL():

```
void CGameController::ZoomInL()
    {
    TInt zoom = iZoomFactor;
    zoom =
        zoom < 250 ? 250 :
        zoom < 350 ? 350 :
        zoom < 500 ? 500 :
        zoom < 600 ? 600 :
        zoom < 750 ? 750 :
        zoom < 850 ? 850 :
        zoom < 1000 ? 1000 :
        250;
    iAppView->SetZoomL(zoom);
    iAppView->DrawNow();
    iZoomFactor = zoom;
    }
```

The controller supports seven levels of zoom here, all fairly arbitrarily chosen. The algorithm is easy to extend to any number of zoom states. Because I wanted to persist the zoom state, it must be a property of the controller: the zoom state in the app view is simply a cached version of the controller's value. I use iZoomFactor to cache the zoom state (it isn't set until the end of the function, so that the controller is not in an inconsistent state if SetZoomL() leaves).

When I have calculated the new zoom state, I call SetZoomL() in the app view to implement the setting, and then DrawNow() to update the view. ZoomOutL() is handled in a similar way.

ViewCmdHitFleet() shows how the controller ties together the two fleets provided by the engine, so that the engine's iOppFleet is what I see on the screen, while the engine's iMyFleet is the real data for the 'opponent's' fleet.

```
void CGameController::ViewCmdHitFleet(TInt aX, TInt aY)
    {
```

```
__ASSERT_ALWAYS(IsMyTurn(), Panic(EHitFleetNotMyTurn));
__ASSERT_ALWAYS(!(iEngine->iOppFleet.IsKnown(aX, aY)),
    Panic(EHitFleetAlreadyKnown));

// Hit fleet
iEngine->iOppFleet.SetShipType(aX, aY, iEngine->iMyFleet.ShipType
  (aX, aY));
iEngine->iOppFleet.SetHit(aX,aY);


// Update view
iAppView->DrawTilesNow();


// If game is won, transition to finished
if(iEngine->IsWon())
    {
    iState = EFinished;
    iEnv->InfoMsg(R_GAME_CONGRATULATIONS);
    }
}
```

Firstly, the controller asserts that it has been called in the right circumstances. The conditions asserted should always be true, but just in case I didn't implement OfferKeyEventL() properly in the view, or because of some other problem I didn't think of – those are usually the worst kinds of problem! I make this assertion so the program can quickly panic if it gets called in the wrong circumstances.

We'll see when we get to the Battleships version of the controller, with nine states and *many* functions that are only valid in certain states, that these assertions are extremely useful.

I use two simple lines to transfer the knowledge of the real fleet from 'my fleet' to the 'opponent's fleet', and to say I hit that square:

```
iEngine->iOppFleet.SetShipType(aX, aY, iEngine->iMyFleet.ShipType
  (aX, aY));
iEngine->iOppFleet.SetHit(aX,aY);
```

If I hit a ship, the engine takes care of ensuring that surrounding squares, on the opponent's fleet, are marked as sea. After this, I update the opponent's fleet view using DrawTilesNow().

Finally, I check whether this means I've won the game. If so, I write an information message to say so, and set the state to finished. In real Battleships, the real complexity in the whole game arises from the fact that this line,

```
iEngine->iOppFleet.SetShipType(aX, aY, iEngine->iMyFleet.ShipType
                              (aX, aY));
```

won't work. Instead of simply doing an object look-up, I have to send
a message to the real opponent, wait for the response, and meanwhile
allow the user of this game to close the file, temporarily or permanently
abandon the game, resend the message in case it got lost, and so on.

9.7 The App UI

Now that we've reviewed the engine, view, and controller, we can return
to the app UI. Back in Chapter 4, I described the app UI as if it was the
main class in a GUI application. In a way, that's true: the entire menu tree
of any GUI application is handled through the app UI, and in a typical
large application, that amounts to a lot of commands.

But we now have another perspective on the app UI: it is just another
source of events to be handled by the controller. This isn't an incompatible
statement; it's just a different perspective. Here's the app UI declaration
in appui.h:

```
class CGameAppUi : public CEikAppUi
    {
public:
    void ConstructL();
    ~CGameAppUi();
private:
    // From CEikAppUi
    void HandleCommandL(TInt aCommand);
    void HandleModelChangeL();
    // Commands
    void CmdStartL();
    void CmdZoomInL();
    void CmdZoomOutL();

private:
    // Uses
    CGameController* iController;
    // Has
    CGameAppView* iAppView;
    };
```

There are no surprises in the command-handling framework. Cmd-
ZoomInL() and CmdZoomOutL() are handled by passing them straight
to the controller:

```
void CGameAppUi::CmdZoomInL()
    {
    iController->ZoomInL();
    }
void CGameAppUi::CmdZoomOutL()
    {
    iController->ZoomOutL();
    }
```

CmdStartL() checks to see if a game is already in progress, and queries the player if so:

```
void CGameAppUi::CmdStartL()
    {
    // User-friendly check
    if(iController->IsMyTurn())
        {
        if(!iEikonEnv->QueryWinL(R_GAME_QUERY_ABANDON))
            return;
        }
    iController->Reset();
    iAppView->DrawTilesNow();
    }
```

If the game had finished anyway, or if the user confirmed that they really did want to start a new game, then the app UI asks the controller to reset, and gets the app view to redraw.

Back in Chapter 4, I introduced the resource file as being something quite heavily associated with the app UI. However, Solo Ships doesn't have a toolbar or any dialogs, so its resource file is not enormous. Here it is (minus #includes and suchlike):

```
NAME SHIP

...


RESOURCE RSS_SIGNATURE { }

RESOURCE TBUF { buf="Battleships"; }

RESOURCE EIK_APP_INFO
    {
    menubar=r_game_menubar;
    hotkeys=r_game_hotkeys;
    }

RESOURCE HOTKEYS r_game_hotkeys
    {
    control=
```

```
        {
        HOTKEY { command=EEikCmdExit; key="e"; },
        HOTKEY { command=EEikCmdZoomIn; key="m"; },
        HOTKEY { command=EGameCmdStart; key="n"; }
        HOTKEY { command=EEikCmdZoomOut; key="o"; }
        };
    }


RESOURCE MENU_BAR r_game_menubar
    {
    titles=
        {
        MENU_TITLE { menu_pane=r_game_file_menu;
                     txt=STRING_r_game_file_menu; },
        MENU_TITLE { menu_pane=r_game_view_menu;
                     txt=STRING_r_game_view_menu; }
        };
    }

RESOURCE MENU_PANE r_game_file_menu
    {
    items=
        {
        MENU_ITEM { command=EGameCmdStart;
                    txt= STRING_r_game_EGameCmdStart; },
        MENU_ITEM { command=EEikCmdExit;
                    txt= STRING_r_game_EEikCmdExit; }
        };
    }


RESOURCE MENU_PANE r_game_view_menu
    {
    items=
        {
        MENU_ITEM {command=EEikCmdZoomIn;
                    txt=STRING_r_game_EEikCmdZoomIn; },
        MENU_ITEM {command=EEikCmdZoomOut;
                    txt=STRING_r_game_EEikCmdZoomOut;}
        }
    };

RESOURCE TBUF r_game_reset { buf=STRING_r_game_reset; }
RESOURCE TBUF r_game_already_known { buf=STRING_r_game_reset; }
RESOURCE TBUF r_game_query_abandon { buf=STRING_r_game_query_abandon; }
RESOURCE TBUF r_game_congratulations { buf=STRING_r_game_congratulations; }
```

There are only four hotkeys (zoom in and out, new-game, and exit), and two menus, with two options each (new-game and exit, zoom in and zoom out). In addition, there are three strings for use in info-messages and one for use in a query dialog.

Released UIQ applications should not have an Exit command, though it can be useful in debug builds for checking against memory leaks. But I'm not following the UIQ style guide very closely for this application.

The other app UI functions are all related to persistence. I've been saving up that topic for a section of its own, so now's the time to tackle it.

9.8   Persistence

A key decision for the user interface designs that run on Symbian OS is whether to expose the file system to the end user or not. Typically, the designs that run on communicator/PDA machines (Psion PDAs, Nokia 9200 family) do allow users to interact directly with files, while designs for phones (UIQ, Nokia Series 60) don't. A directly exposed file system can confuse users, and gives an experience more like using a PC than a phone.

At a certain level, whether the file system is exposed or not is irrelevant to a program's data storage: in either case, a file system exists, and is where persistent data is stored. But when designing how your application interacts with the user, this is a crucial consideration. UIQ applications that allow the user to select an item (what in the PC world would be called a 'document') to work on, such as the Jotter application, each must invent their own means of presenting the available items, and allowing the user to pick one.

The application framework, historically first developed for PDAs, was designed with an expectation that the file system would be exposed, and therefore comes with functionality that enables shell programs to interact with suitably-written applications to load, save, and switch between files. An application that follows the required rules is called a **file-based application**. The chief requirement for a file-based application is that each running instance must have exactly one external document associated with it. Depending on how the application is launched from the shell, the framework can instruct to the application to either load a specified existing document or to create a new empty document. The framework can also instruct an already running application to switch to using another document. Three familiar cases from the Windows world are impossible:

- A blank document called, say, `Document1`, which isn't yet associated with a file. This isn't allowed because it complicates the UI when closing the application.

- No document at all – just the File and Help menus. This isn't allowed, and isn't really needed, because application startup is fast enough anyway.

- Multiple documents, which you can cycle around using the Window menu. However, some UI designs allow multiple open instances of an application.

Despite the lack of an exposed file system, UIQ applications can still be file-based, although in a slightly simplified way. A file-based UIQ application doesn't open different files: whenever it's run, it always opens the same document file. Solo Ships works like this, so we can now look at how to implement this approach.

9.8.1 Solo Ships as a File-based Application

Solo Ships supports three document-related operations:

- Run the application and create a new default document: this only occurs the first time that the application is run.
- Run the application and open the existing document.
- Exit the application and save the data to the document.

Fortunately for the programmer, the application framework handles most of this. As an application programmer, you have to implement the following:

- A function to store the document data.
- A function to read the stored document data.
- A function to set the document to an initial, default, state.
- A C++ destructor for the document.

9.8.2 Store and Restore

The application framework needs to share an understanding with applications about how the application document is saved to file. For this reason, file-based applications aren't free to use whatever file format they wish, but must use a structured format set by the framework. In Symbian OS, structured files are called **stores**, and the elements within a store are called **streams**. An application document's store has three streams, as shown below in Figure 9.5:

The store consists of:

- A special stream called a **stream dictionary**. This provides an index to the other streams in the store.
- A stream that stores a small amount of data to identify the application to which the store relates.
- A stream that contains the document data.

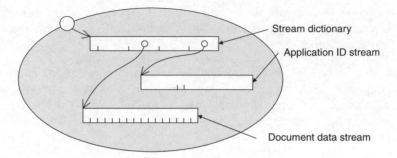

Stream dictionary

Application ID stream

Document data stream

Figure 9.5

The framework handles creating a file store with the appropriate structure, and sets the application identifier stream. You have to supply code for storing and restoring the document data from the third stream. Stores are discussed in detail in Chapter 13.

Document responsibilities

CGameDocument contains the basic functions you need to store and restore document data. Here's StoreL():

```
void CGameDocument::StoreL(CStreamStore& aStore,
                           CStreamDictionary& aStreamDict) const
    {
    TStreamId id = iController->StoreL(aStore);
    aStreamDict.AssignL(KUidExample,id);
    }
```

The framework calls this function, passing objects that represent the store, and the stream dictionary.

This code calls iController->StoreL() to store the controller's data to the stream, and returns the stream ID of the stream so created. The next line makes an entry in the stream dictionary: it records that the data for identified application (KUidExample, the application UID), is stored in the stream with the specified ID.

There's also a corresponding RestoreL():

```
void CGameDocument::RestoreL(const CStreamStore& aStore,
                             const CStreamDictionary& aStreamDict)
    {
    // New controller initialized from store
    TStreamId id = aStreamDict.At(KUidExample);
    CGameController* controller = CGameController::NewL(aStore, id);
    delete iController;
    iController = controller;
    }
```

This time, you look up the ID of the document data stream in the dictionary and restore from it. One possibility would be to call `iController->RestoreL()` and overwrite the data in the existing controller object. An alternative, as shown here, is to construct an entirely new controller object by using `CGameController::NewL(aStore, id)`. After I've constructed the new one successfully, I delete the old one, replacing it with the new controller.

Applications may store more than one stream using the stream dictionary provided. Many applications use this to store different kinds of data.

Storing the controller data

The document just passed the buck to the controller. Here's how the controller stores its data:

```
TStreamId CGameController::StoreL(CStreamStore& aStore) const
    {
    RStoreWriteStream stream;
    TStreamId id = stream.CreateLC(aStore);
    iEngine->ExternalizeL(stream);
    ExternalizeL(stream);
    stream.CommitL();
    CleanupStack::PopAndDestroy(); // stream
    return id;
    }
```

The idea here is to create a stream, write both the engine and any controller data to it, and then close the stream. Symbian OS programs conventionally calls functions that write an object's data to a stream `ExternalizeL()`.

The engine's `ExternalizeL()` function looks like this:

```
void CGameEngine::ExternalizeL(RWriteStream& aStream) const
    {
    aStream << iMyFleet;
    aStream << iOppFleet;
    aStream.WriteUint8L(iFirstPlayer);
    }
```

It externalizes the engine's two fleet objects, and the flag that records whose turn it is.

The first two lines show a neat idiom to use when working with streams. Symbian OS overloads the operator `<<` for streams, so that it calls `ExternalizeL()` on the argument, hence the naming convention. So the first line in the above function could have been equivalently, if less concisely, written as

```
iMyFleet.ExternalizeL(aStream);
```

So, the function called is in fact this:

```
void TFleet::ExternalizeL(RWriteStream& aStream) const
    {
    for(TInt i = 0; i < 64; i++)
        aStream.WriteUint8L(iSquares[i]);
    aStream.WriteUint8L(iMyFleet);
    for(i = 0; i < 10; i++)
        aStream << iShips[i];
    aStream.WriteInt8L(iKnownShips);
    aStream.WriteInt8L(iRemainingShips);
    aStream.WriteInt8L(iRemainingSquares);
    aStream.WriteInt8L(iSquaresHit);
    }
```

It writes the state of the 64 game squares, the ships, and some flags. The `<<` idiom is used again, this time to call the ship class's `ExternalizeL()` function:

```
void TShip::ExternalizeL(RWriteStream& aStream) const
    {
    aStream.WriteUint8L(iType);
    aStream.WriteInt8L(iLength);
    aStream.WriteInt8L(iRemaining);
    aStream.WriteInt8L(iStartX);
    aStream.WriteInt8L(iStartY);
    aStream.WriteInt8L(iDx);
    aStream.WriteInt8L(iDy);
    }
```

The ship data is just a series of 8-bit integers. In all these functions, we don't expect the integer values to be more than 8 bits, so we specify this is what we want to store. This saves some space compared to storing a full 32 bits for each integer.

Finally, the controller itself externalizes some extra persistent data: namely, its state and the current zoom factor:

```
void CGameController::ExternalizeL(RWriteStream& aStream) const
    {
    aStream.WriteUint8L(iState);
    aStream.WriteInt32L(iZoomFactor);
    }
```

Restoring the controller data

We saw that the document's `RestoreL()` uses the controller's restoring `NewL()` function that takes `CStreamStore` and `TStreamId` arguments, which is coded as follows:

```
CGameController* CGameController::NewL(const CStreamStore& aStore,
                                      TStreamId aStreamId)
    {
    CGameController* self = new(ELeave) CGameController;
    CleanupStack::PushL(self);
    self->RestoreL(aStore, aStreamId);
    CleanupStack::Pop();
    return self;
    }
```

`RestoreL()` is private, like `ConstructL()`, so that it can't be accidentally called by `CGameController`'s clients. Here it is:

```
void CGameController::RestoreL(const CStreamStore& aStore, TStreamId
  aStreamId)
    {
    iEnv = CEikonEnv::Static();
    RStoreReadStream stream;
    stream.OpenLC(aStore,aStreamId);
    iEngine = new(ELeave) CGameEngine;
    iEngine->InternalizeL(stream);
    InternalizeL(stream);
    CleanupStack::PopAndDestroy(); // stream
    }
```

The first task is to get a GUI environment pointer. That's needed by the controller, whether it's constructing from scratch or restoring from a document. Next, it creates a new engine object and requests it to initialize itself from the specified stream. As you would expect, `InternalizeL()` is the conventional name for functions that read data from a stream. The controller's and engine's `InternalizeL()` functions are pretty well the reverse of the `ExternalizeL()` functions already seen: they add little that's new, so I'll move quickly on.

9.8.3   Creating a Default Document

When the application is opened with a new file (which for UIQ is only when the application is started for the first time), it can't restore from anything, so instead it needs a new default document. The framework creates the actual new document file: the folder for this is device-specific (in the standard UIQ emulator it's `C:\Documents\<app-name>`), while the default document name is read from the first TBUF in the resource file ('Solo Ships' in this case).

The application class has the responsibility of setting up the application appropriately for a default state:

```
CApaDocument* CGameApplication::CreateDocumentL()
    {
    CGameDocument* doc = new(ELeave) CGameDocument(*this);
    CleanupStack::PushL(doc);
    doc->ConstructL();
    CleanupStack::Pop();
    return doc;
    }
```

The document second-phase constructor creates a new controller:

```
void CGameDocument::ConstructL()
    {
    iController = CGameController::NewL();
    }
```

This uses the conventional `NewL()`, which constructs a default controller, rather than restoring one from file.

9.8.4 App UI and the Document

Finally, we need to link the app UI to the document. Most importantly, the command handler for `EEikCmdExit` which, in all our applications until now, has just been called `Exit()`, now includes `SaveL()`:

```
void CGameAppUi::HandleCommandL(TInt aCommand)
    {
    switch (aCommand)
        {
    case EEikCmdExit:
        SaveL();
        Exit();
        break;

    ...
        }
    }
```

This calls `CEikAppUi::SaveL()`, which ensures that the framework saves the document data (by calling `CGameDocument::RestoreL()`).

As we've seen, I implemented document restore by creating a new controller and deleting the old one. This means that the app UI's pointer to the controller then becomes invalid. To help the app UI cope with this type of circumstance, the framework defines a `HandleModelChangeL()` function, which it calls when a new document is loaded. I implement it here to get an up-to-date value for the pointer to the controller.

```
void CGameAppUi::HandleModelChangeL()
    {
    // Change pointers to new objects
    iController = (STATIC_CAST(CGameDocument*, Document()))->
      iController;
    iAppView->SetController(iController);
    }
```

First, I use the document to find out my new controller, and then I pass on this information to the app view. Without this, you won't keep the app UI and app view up-to-date when the model changes.

9.9 Two Player Battleships

You'll have gathered from the discussion so far that through the application, document, and app UI frameworks the OS knows a lot about the running applications, and that it can have quite fine control of their operations. The exception so far seems to have been the application views: they've just been ordinary control objects, about which the operating system has no special knowledge or control. To fill this gap, Symbian OS v6 introduced the **view architecture**, which allows the operating system and other applications to communicate directly with an application's views.

I'll demonstrate how to use the view architecture by extending Solo Ships into a two-player game. In that version, there'll be two different views for each player, one that shows the opposition fleet, and one that shows the player's own fleet. Along the way, I'll also briefly look at making views more interesting by using bitmap graphics and sound.

9.9.1 View Architecture

One reason not to have introduced the view architecture until now is that applications do not have to use it. Of the current user interface designs, UIQ recommends that applications douse it, while Series 60 allows its use, though encapsulated in Series 60 – specific classes. Its systematic use in UIQ helps the user easily complete tasks that may require more than one application. For example, in the UIQ Contacts application, you can tap on a contact with an e-mail address, and the view will switch to the messaging application, with a new e-mail ready to be written to that contact as shown in Figure 9.6.

Figure 9.7 shows what is happening in architectural terms.

Figure 9.6

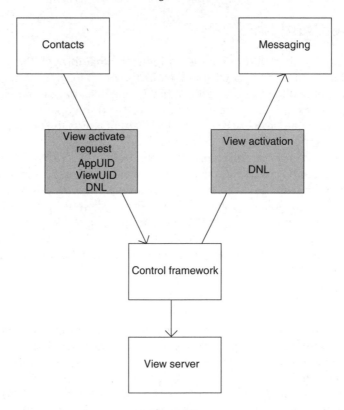

Figure 9.7

'Contacts' issues a view activation request through the control framework. It specifies the application and view that it wants to activate. Optionally, it can supply a message to pass to the target view. Because such messages are typically used to link applications, they are called Dynamic Navigational Link (DNL) messages. Applications that can receive such messages must publish header files that define their message IDs and data formats. In our example case, the UIQ Messaging application defines a DNL to create a new e-mail with a specified recipient.

If the target application (Messaging in this case) is not running, the framework starts it. The next step is that the application registers its views with the control framework. In the background, this record of the available views in the system is managed by a system thread, the View Server. Finally, the framework activates the appropriate application view, and if a DNL is being sent, passes it to the view to process.

As can be seen, architecturally, this is quite a complex process. Fortunately, the frameworks make the task of the application programmer quite simple. I'll look next at what I had to do to implement the required changes for Two Player Battleships.

9.9.2 Views in Two Player Battleships

The Two Player Battleships (project name `tp-ships`) game is for two players using a single device. The players take turns trying to sink each other's fleets. I designed its user interface with the following rules:

- Each player will need a view of his own fleet, so as to know his opponent's progress in attacking it, and a view of the target fleet, which will behave in much the same way as in Solo Ships. The UIQ screen is too small to show both these easily at once, so I decided to allow a player to switch between 'my' and 'opposition' fleet views as he wished.
- A player should not be able to access the views for the other player (as that shows the true positions of the opponent's fleet!).
- Between turns, while the device is handed between the players, there should be a neutral display, which shows nothing of significance. I decided that a player would end his turn by activating this 'hider' view; and that the next player would start his turn by choosing to view his own fleet or his target.

This gives a total of five views (2 × 'my fleet', 2 × 'opposition fleet' + hider view) for the application.

9.9.3 Fleet Views

The main task was to modify Solo Ship's fleet view class. It ended up looking like this

```
class CFleetView : public CCoeControl, public MCoeView
    {
public:
    ~CFleetView();

    // Cursor
    void SetCursorOff();
    void SetCursor(TInt aX, TInt aY);
    TBool CursorOn() const;
    void GetCursor(TInt& aX, TInt& aY) const;

    // Incremental drawing
    void DrawTilesNow() const;

    // Sound
    void ExplSound();
    void MissSound();
    void SunkSound();

protected:
    // Construct
    CFleetView(CFleetViewData& aFleetViewData, TFleet& aFleet);
    void ConstructL();

    // From CCoeControl
    void Draw(const TRect&) const;

    // From MCoeView
    TVwsViewId ViewId() const;
    void ViewActivatedL(const TVwsViewId& aPrevViewId, TUid
      aCustomMessageId, const TDesC8&
aCustomMessage);
    void ViewDeactivated();

    // aAxiliary draw functions
    void DrawOutside() const;
    void DrawBorders() const;
    void DrawHorizontalBorder(const TRect& aRect) const;
    void DrawVerticalBorder(const TRect& aRect) const;
    void DrawTiles() const;
    virtual void DrawTile(TInt aX, TInt aY) const;

    // Cursor movement
    void MoveCursor(TInt aDx, TInt aDy);

protected:
    // Data common to all the fleet views
    CFleetViewData& iData;
    // View specific data
    TFleet& iFleet;
    // Cursor
    TBool iCursorOn;
    TInt iCursorX;
    TInt iCursorY;
    TUid iViewUid; // UID of view
    };
```

The main point is that the class now implements the abstract interface MCoeView. This is what turns the class from being an ordinary control into being usable by the view architecture.

Also, note the following:

- The constructor and ConstructL() functions are now protected rather than public. This indicates that CFleetView is now a base class rather than a concrete class. I found that the 'my fleet' and 'opposition fleet' views behaved differently enough to be their own classes. They still have most functionality in common, though, and this is provided by CFleetView.

- It has far fewer data members than the original CFleetView, but it has a new data member iData.

- It has three public functions for generating sound effects.

Implementing MCoeView

MCoeView's documentation in some SDKs omits some functions, and as the class is quite cryptic, it's worth looking at its entire declaration:

```
class MCoeView
    {
public:
    virtual TVwsViewId ViewId() const=0;
private:
    virtual void ViewActivatedL(const TVwsViewId& aPrevViewId,TUid
        aCustomMessageId,const
TDesC8& aCustomMessage)=0;
    virtual void ViewDeactivated()=0;
protected:
    IMPORT_C virtual TVwsViewIdAndMessage ViewScreenDeviceChangedL();
private:
    IMPORT_C virtual void ViewConstructL();
protected:
    IMPORT_C virtual TBool ViewScreenModeCompatible(TInt
        aScreenMode);
private:
    friend class CCoeViewManager;
    IMPORT_C virtual void MCoeView_Reserved_2();
    };
```

A derived view class always implements the first three functions. Note that the declaration uses a subtlety of C++ access control: ViewActivatedL() and ViewDeactivated() are private, so you can't call them, but are purely virtual, so you must implement them. The purpose of making them private is to allow only one internal control framework class, CCoeViewManager, declared as a friend, to call these functions.

The first function, `ViewId()`, is implemented to return a view identifier, consisting of the application's UID, and a specific UID for the view. In `CFleetView` this is implemented as

```
TVwsViewId CFleetView::ViewId() const
    {
    return TVwsViewId(KUidTpShips, iViewUid);
    }
```

`KUidTpShips` is the application UID and `iViewUid` is the view UID, which is set in the derived my fleet or opposition fleet constructor.

`ViewActivatedL()` is the most interesting of the functions. It is called by the framework when the view is activated. It has parameters for the ID of the view previously activated so that you can go back to it if you need to, and the ID and data for a DNL message. `CFleetView` doesn't process any DNL messages, so its implementation is simple:

```
void CFleetView::ViewActivatedL(const TVwsViewId& /*aPrevViewId*/, TUid
/*aCustomMessageId*/,const TDesC8& /*aCustomMessage*/)
    {
    Window().SetOrdinalPosition(0);
    (static_cast<CGameAppUi*>(iEikonEnv->EikAppUi()))->
        SetActiveView(*this);
    }
```

The first line is standard for `ViewActivated()` implementations: it brings the control's window to the front, that is, causes it to be displayed. The second line is specific to this application. My app UI caches a pointer to the currently active view: `SetActiveView()` updates this cache. Though not needed here, another common operation in this function is to change the menu (or other screen furniture) to be specific to the view.

`ViewDeactivated()` is the complement to `ViewActivatedL()`. If there's something that needs to be done before another view comes to the front, it can be done here. Often there isn't, as is the case with `CFleetView`:

```
void CFleetView::ViewDeactivated()
    {
    }
```

`ViewConstructL()` is called by the framework just before the very first activation of a view. It allows a view's second-phase construction (i.e. the stage at which resources are allocated) to occur only when a view is needed (which may be never). This is a good idea if you have many views, especially if they are expensive in memory.

The final two `MCoeView` virtual functions, `ViewScreenDeviceChangedL()` and `ViewScreenModeCompatible()`, are more specialist. They allow the framework to test the view's ability to handle a change to the screen device (such as a change in the physical display area) and to handle a particular screen mode. The default UIQ design does not require these functions to be implemented.

> *The View Architecture was originally motivated by the needs of the Ericsson R380 smartphone, which could either be closed, in which case a small screen area was displayed, or flipped open, which revealed a larger display. Applications dynamically responded to the screen changing.*

View registration

The final view architecture-related task is to tell the control framework that the views exist. This is done by `CEikAppUi`'s `RegisterViewL()` function. As a control can easily get access to the app UI object, I chose to wrap up view registration as part of view construction.

```
void CFleetView::ConstructL()
    {
    CreateWindowL();
    SetRect(iData.iClientRect);
    SetCursor(0,0);
    // Add to the registered views
    iEikonEnv->EikAppUi()->RegisterViewL(*this);
    iEikonEnv->EikAppUi()->AddToStackL(*this, this);
    // Set ready for drawing
    ActivateL();
    }
```

The alternative of calling `RegisterViewL()` from within the app UI itself, after having created the view, is as good.

Complementarily, `CEikAppUi`'s `DeregisterView()` function must be called before a registered view is destroyed. I do this in the view destructor.

```
CFleetView::~CFleetView()
    {
    iEikonEnv->EikAppUi()->RemoveFromStack(this);
    // Remove from registered views
    iEikonEnv->EikAppUi()->DeregisterView(*this);
    }
```

My fleet and opposition fleet classes

To specialize `CFleetView` to become a 'my fleet' class is simple. I just added a public factory function and a constructor. The constructor takes a flag, `aP1`, indicating to which player (arbitrarily called player1 and player2) the view belongs. This flag is used to set a different view UID for each.

```
CMyFleetView* CMyFleetView::NewL(CFleetViewData& aFleetViewData,
    TFleet& aFleet, TBool aP1)
    {
    CMyFleetView* self = new (ELeave) CMyFleetView(aFleetViewData,
      aFleet, aP1);
    CleanupStack::PushL(self);
    self->ConstructL();
    CleanupStack::Pop(self);
    return self;
    }

CMyFleetView::CMyFleetView(CFleetViewData& aFleetViewData,
    TFleet& aFleet,
    TBool aP1)
    :CFleetView(aFleetViewData, aFleet)
    {
    if (aP1)
        iViewUid = KP1MyViewUID;
    else
        iViewUid = KP2MyViewUID;
    }
```

Note `CMyFleetView` has no functions to handle pointer or keyboard input: the only thing that the user can do is to look at it.

`COppFleetView`, the opposition fleet class, is similar, but has the functions, as defined in Solo Ships, to handle input events.

We now have two concrete view classes, each of which will be instantiated twice, making four objects in all. These objects will need to share a significant amount of data: the zoom factor, and all the data calculated from it, for border areas, tile size, and so on will be the same. C++ would allow me to make these static (i.e. class) members, but Symbian OS does not allow such writable static data. The alternative that I adopted was to encapsulate the data in a class `CFleetView-Data`, and give each view a pointer to the same, single instance of that class.

9.9.4 Hider View

The final view in the application is the hider view, which is displayed between turns. I thought I'd display something prettier than a blank screen, and chose a bitmap of a real battleship in action. The second-phase constructor creates a bitmap object (`CFbsBitmap`), and loads the picture from the application's bitmap store (you'll see how to create such a store in Chapter 14). The view's `Draw()` function simply bit-blit's the bitmap to the view.

Thanks to the U.S. Navy's Naval Historical Center for the 1906 photograph of the USS Connecticut.

```
void CHiderView::ConstructL(const TRect& aRect)
    {
    CreateWindowL();
    SetRect(aRect);
    // Create and load a bitmap
    iHideBitmap = new (ELeave) CFbsBitmap;
    // Get bitmap store--drive independent
    TFileName mbmName = iEikonEnv->EikAppUi()->Application()->
      BitmapStoreName();
    User::LeaveIfError(iHideBitmap->Load(mbmName,0));
    ActivateL();
    }

void CHiderView::Draw(const TRect& /*aRect*/) const
    {
    SystemGc().BitBlt(TPoint(0,0),iHideBitmap);
    }
```

9.9.5 View Test Program

To test that all my views work as expected, I wrote a little test application, `tp-viewtest`. This just defines a menu with options to activate the various views. To active a view, it calls `CEikAppUi`'s `ActivateViewL()` function, passing the target view ID. This is `tp-viewtest`'s menu command handler:

```
void CAppUi::HandleCommandL(TInt aCommand)
  {
 _LIT8(KViewCmdData1, "P1 My");
 _LIT8(KViewCmdData2, "P1 Opp");
 _LIT8(KViewCmdData3, "P2 My");
 _LIT8(KViewCmdData4, "P2 Opp");

 switch (aCommand)
    {
    case ECmdViewPlayer1MyShips:
        ActivateViewL(TVwsViewId(KUidTpShips,KP1MyViewUID),
          KCmd1UID, KViewCmdData1);
        break;
    case ECmdViewPlayer1OppShips:
        ActivateViewL(TVwsViewId(KUidTpShips,KP1OppViewUID),
          KCmd2UID, KViewCmdData2);
        break;
    case ECmdViewPlayer2MyShips:
        ActivateViewL(TVwsViewId(KUidTpShips,KP2MyViewUID),
          KCmd1UID, KViewCmdData3);
        break;
    case ECmdViewPlayer2OppShips:
        ActivateViewL(TVwsViewId(KUidTpShips,KP2OppViewUID),
          KCmd2UID, KViewCmdData4);
        break;
    case ECmdViewHider:
        ActivateViewL(TVwsViewId(KUidTpShips,KHiderViewUID));
```

```
            break;
    case EEikCmdExit:
            CEikAppUi::Exit();
            break;
    }
}
```

If you build and run `tp-viewtest`, and chose a menu option, you'll see the system start up Two Ships, and activate the appropriate view (though activating a player 2 view when it's player 1's turn, or vice versa, doesn't leave Two Ships in a sensible state).

For demonstration purposes, I decided to also pass DNL messages in the view activation requests. In the `ActivateView()` calls, `KCmd<num>Uid` is the ID of the message to pass, and `KViewCmdData<num>` the message data (a simple string in this case). To show the messages being received, I coded an alternative version of `CFleetView::ViewActivatedL()` in `tp-ships`:

```
//#define _DEBUGVIEWS_
#ifdef _DEBUGVIEWS_
 void CFleetView::ViewActivatedL(const TVwsViewId& aPrevViewId, TUid
  aCustomMessageId, const
TDesC8& aCustomMessage)
    {
    Window().SetOrdinalPosition(0);
    (static_cast<CGameAppUi*>(iEikonEnv->EikAppUi()))->
        SetActiveView(*this);
    // display the view switch information
    if (aCustomMessageId.iUid)
        {
        TBuf<200> buf,buf2;
        buf.Format(_L("ID: %x: "), aCustomMessageId.iUid);
        buf2.Copy(aCustomMessage);
        buf.Append(buf2);
        iEikonEnv->InfoMsg(buf);
        }
    }
#else
```

This version formats the message ID and data into a string `buf` and displays it in an information message. If you want to see this in action, uncomment the line `#define_DEBUGVIEWS_`, and rebuild `tp-ships`.

Originally, I used a modal dialog rather than an InfoMsg to display the message information. This caused a panic as waiting for the user to confirm the dialog caused a view server time-out.

9.9.6 Sound Effects

A view is primarily about displaying things, but it can be about interacting with the user through sounds too. Symbian OS can play audio files of various formats (gsm 6.10,. au,. wav,. wve and raw audio data), as well as tones and streaming PCM audio.

For `tp-ships`, I wanted to play three. wav files for hitting a ship, missing a ship, and sinking a ship, respectively. I encapsulated all the sound playing functionality in a class `CSoundEffects`. Its second-phase constructor sets up a `CMdaAudioPlayerUtility` object to play each file:

```
_LIT(KExplSoundFile,"expl.wav");
_LIT(KMissSoundFile,"miss.wav");
_LIT(KSunkSoundFile,"sunk.wav");

void CSoundEffects::ConstructL()
    {
    // Get path to sound files
    TParsePtrC parse(CEikonEnv::Static()->EikAppUi()->Application()->
      AppFullName());
    TPtrC appPath = parse.DriveAndPath();
    ConstructSoundPlayerL(appPath, KExplSoundFile, iExplSoundPlayer);
    ConstructSoundPlayerL(appPath, KMissSoundFile, iMissSoundPlayer);
    ConstructSoundPlayerL(appPath, KSunkSoundFile, iSunkSoundPlayer);
    }

void CSoundEffects::ConstructSoundPlayerL(const TDesC& aPath, const
  TDesC& aName,
    CMdaAudioPlayerUtility*& aPlayer )
    {
    TFileName name = aPath;
    name.Append(aName);
    aPlayer=CMdaAudioPlayerUtility::NewFilePlayerL(name,
      iSoundPlayerCallBack);
    }
```

As well the name of the file to play, `CMdaAudioPlayerUtil-ity::NewFilePlayerL()` requires an `MMdaAudioPlayerCall-back` callback object that tells the client when the file is ready to play (`MapcInitComplete()`), or when playing is complete (`Mapc-PlayComplete()`).

To play a sound, a fleet view object calls `CSoundEffects::PlaySound()`. This cancels any sounds that are already playing, selects the right `CMdaAudioPlayerUtility` object to use, and calls its `Play()` method:

```
void CSoundEffects::PlaySound(TSound aSound)
    {
    if (!iSoundPlayerCallBack.iOK) return;
    CancelSounds();
    switch (aSound)
        {
        case EExplosion:
```

```
            iCurrentSoundPlayer = iExplSoundPlayer;
            break;
    case EMiss:
            iCurrentSoundPlayer = iMissSoundPlayer;
            break;
    case ESunk:
            iCurrentSoundPlayer = iSunkSoundPlayer;
            break;
    default:
            return;
    };
iCurrentSoundPlayer->Play();
iSoundPlayerCallBack.iPlaying = ETrue;
}
```

9.10 Summary

In this chapter I've described a larger-scale GUI application that shows some important aspects about writing real application code:

- How to support persistence
- How to write drawing code that can zoom and use different display sizes
- How to use the view architecture.

The full version of Battleships will add to the framework established by these programs. Battleships adds

- A little more GUI functionality, especially dialogs
- A lot more communications and system programming.

10

Dialogs and Concrete Controls

In Chapter 4, you saw how commands get to `HandleCommandL()` from the application UI's basic interaction resources: the toolbar, menus, and shortcut keys. In our simple `hellogui` application, we handled those commands pretty trivially – either by displaying an info-message or by quitting the program.

In real applications, many commands are handled using a **dialog**; perhaps half of the effort required to program the GUI of a professional application is involved with dialog programming.

In this chapter, I'll introduce the design requirements for dialogs as seen by the user and (only just below the surface) by the programmer. I'll then move on to use some of the simple dialogs from example projects in this book to illustrate the essentials of dialog programming. Then, I'll do a lightning tour through the dialog framework's main APIs, stock controls for inclusion in dialogs, and some standard dialogs. The majority of stock controls and dialogs are generic, but this chapter includes some that are specific to UIQ. In general, each UI will have its own specific controls and dialogs in addition to, and in some cases replacing those supplied by the generic UI layer (Uikon). However, the issues raised here are applicable to all Symbian OS-based UIs.

10.1 Introducing Dialogs

Many readers are familiar with the way dialogs work and have expectations about what dialogs ought to do on the basis of how they work in Windows. Windows and Uikon are designed for different types of hardware and different end users, however, so their dialog designs are different. I'll point out the differences as I go along to help you understand where

Symbian OS C++ for Mobile Phones.   Edited by Richard Harrison
© 2003 John Wiley & Sons, Ltd   ISBN: 0-470-85611-4

Uikon is coming from, and how to use it to deliver the best experience to the users of your applications.

I'll use three dialogs to show you the kinds of things you can do with them.

10.1.1   A Query Dialog

General query dialogs tend not to feature in some UIs such as UIQ, but where they do occur, they tend to be tailored to the specific context. Here's a typical example:

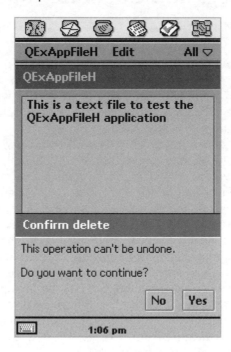

Figure 10.1

I got this by selecting a file in the QExAppFileH example from the UIQ C++ SDK and then pressing the *Delete* button on the resulting dialog. The query is asking me whether I want to continue with the delete operation and I have to answer 'No' or 'Yes'.

The most important point about this dialog is that it's a straightforward query, with a Yes/No answer. Compare it with a similar dialog on Windows as shown in Figure 10.2.

I got this by typing *Ctrl+F4* (a shortcut for File | Close) in Word, as I was typing the previous paragraph. The question being asked here is, 'Do you want to save changes to 2-dialogs.rtf?' I get three options. Consider how new users would respond:

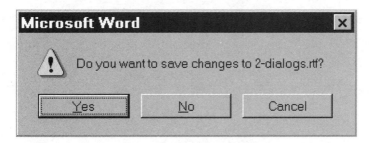

Figure 10.2

- Yes: that ought to save the changes – but it says nothing about exiting – in any case, the data is safe.

- No: if the users are smart Windows users, they'll realize that 'No' means, 'No, I don't want to save changes when closing the file.' If they haven't used Windows before, they'll say, 'No, I didn't ask for my file to be saved, I'll save it later when I've finished.' They select 'No', and their document disappears! They have lost all their changes, even though that's not what they wanted.

- Cancel: a smart Windows user realizes that cancel means, 'No, don't save changes, but keep the file open.' If he's an average user, he won't understand how 'Cancel' can be an answer to the question that was posed.

Almost every Windows user I know, who isn't a professional programmer has lost data because of this Windows dialog.

This illustrates an important aspect of all programming for Symbian OS: write for inexperienced computer users, and make it clear what's going to happen when they select a menu or dialog option. The Symbian OS way is to ask a straight yes/no question in which the expectations and consequences of each possible answer are very clear.

If you're a Windows user, you'll notice a couple of other things about the query dialog:

- Yes and No are swapped around compared to the Windows way of doing things. Yes (and OK) buttons always go on the right of horizontal button lists. Then most people can press these buttons without obscuring the rest of the dialog – all people operate from below the screen, and most people are right-handed.

- There's a title bar that you can move up and down by dragging with the pen.

- There's no notion of a currently focused button and no underscored letters to indicate accelerator keys – more on that later.

10.1.2   A Single-page Dialog

Here's a standard single-page dialog provided by UIQ:

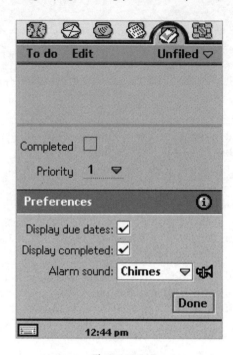

Figure 10.3

I got this by going to my To do list, tapping on the Edit menu item and then tapping on the Preferences item in the menu pane.

One thing is immediately striking about this dialog: all its controls are in a vertical list. Labels are right justified and controls are left justified so that controls and labels are centered neatly on a vertical line. This is a key design decision. It makes many things easier for both user and developer:

- Navigation from field to field is easy: just use the pointer, or the up and down navigation keys.

- Dialog layout is easy: the only difficulty is establishing the width of the controls. In UIQ, for example, all dialogs are exactly 208 or 240 pixels wide (depending on the configuration), no matter how wide their controls.

- The width of the dialog limits the width of a control: if the Alarm sound field had a very long item in its choice list, then this would be truncated and the last three displayable characters would be replaced by an ellipsis to make it fit.

- Because layout is so easy, you don't have to specify the pixel coordinates of each field in the dialog definition resource. In consequence, there is no need for a GUI-based resource builder.

This dialog only has three controls. You can have more, but you need to make sure that they don't overflow the screen vertically. In UIQ, for example, the screen height is 320 pixels, and any dialog that has more than 8 controls becomes a scrollable dialog, with scroll bars automatically added on the right hand side. In UIQ, and in many UIs, it is not a good style to use scrollable dialogs. This has an important impact on dialog design:

- If you need more lines, you can use multipage dialogs (see below).
- Don't create monster dialogs that offer a bewildering set of choices to the user. Experiment with a number of alternatives before committing a design to code – and even then, be prepared for further change.

In UIQ, focus is not indicated except for text fields and numeric fields. For text fields, the flashing cursor indicates the location of focus. For numeric fields, a highlighted background indicates the location of focus.

Finally, note that buttons can't be focused: OK is *always* on the *Confirm* hardware key. Without a keyboard, other buttons are selected with a pointer.

10.1.3 A Multipage Dialog

A sample multipage dialog taken from the UIQ Agenda application can be seen in Figure 10.4.

I got this dialog by selecting the Preferences menu item on the Edit menu pane in Agenda. This allowed me to customize my Agenda view, such as the number of viewed hours in a day. However, there are more details that do not fit.

This is a multipage dialog. I can tap with the pointer on either of the page tabs. Tapping with the pointer on the tab marked Alarm gives me a page on which I can change alarm-related options such as the alarm sound.

The button array on the bottom right is associated with the entire dialog – not with each page. It's bad style to change it when the pages change.

10.1.4 Cue Text

You should make every effort to ensure that the meaning of the controls in your dialogs is transparently obvious to most users. Be prepared to work hard at this: you'll need to choose text and functionality with care, order lines and pages in your dialogs sensibly, and be particularly careful about

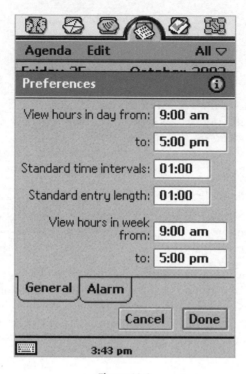

Figure 10.4

options and initial setup. With some thought, you can often produce an application that needs little help with text of any kind.

Sometimes, though, it's not possible – or perhaps not practical – to achieve this ideal. One useful tool when that happens is **cue text** in dialogs. An example from Battleships is shown in Figure 10.5.

I count the need for this cue text as an indication that Battleships still has a way to go in terms of usability. Why should the *users* have to decide on such technicalities if their machines are already in infrared contact, say? Why can't we use some protocol to sort this out? In the case of the SMS protocol, it's much clearer that one person has to make the first call and the other person has to receive it; why couldn't I have chosen words like 'Call' and 'Wait for call' that made that more obvious?

As a justification for needing this dialog, communications is generally complicated by nature and communications setup is especially awkward. Ultimately, the end user pays real money for communications services and makes choices about the level of service. This means that they have to be able to control these choices – whether they like it or not. However hard we work to make these things as easy to control as possible, these options will always be with us, and cue text in dialogs will, therefore, always have a role somewhere.

Figure 10.5

10.1.5 Controls

Each line in a dialog is a captioned control with two or three components:

- a caption, to tell the user what the line is for;
- a control, to allow something to be displayed and/or edited;
- a tag, used by some controls to indicate measurement units, for example, inches or centimeters.

There may be variations in the screen estate occupied by a dialog. In UIQ, for example, dialogs are always full screen width; they may be less than full screen height and are bottom-aligned above the status bar.

Controls allow users to enter data. UIQ, for example, provides 42 stock controls that can be incorporated into dialogs. In the dialogs above, we've already seen text editors, time editors, check boxes, sound selectors, number editors, and choice lists. Knowing the controls available to you is a key aspect of dialog programming. You can also add your own controls.

Sometimes in this book I've used the word 'field' as a user-friendly synonym for 'control'. I'll always use 'control' when referring to programming.

10.1.6 Dialog Processing

Dialogs should help the user to enter valid data and, if it's not valid, should point out the error as early and as helpfully as possible. This works in various ways:

- For some controls, you can specify validity criteria: for numeric editors, for instance, you can specify a range.
- For some controls, you can't enter invalid data anyway: a checkbox can either be checked or unchecked.
- After a control such as a choice list or check box has changed its value, you can override a dialog virtual function that responds by optionally changing the values of other controls. So, if a check box controlling whether an item has an alarm is changed, the fields specifying the alarm information are turned on or off.
- When focus moves from one line to another, you may also wish to do some validation and to change some other fields.
- When OK (or DONE, or another button) is selected, you can do whatever processing you like.

10.1.7 Modality

Dialogs in Symbian OS are modal. While a dialog is being displayed, you can barely see the app view underneath, so there's no point in either allowing the user to do anything with the application while the dialog is active, or in reflecting the dialog-controlled changes instantly in app views.

10.1.8 Summary

Dialogs are culturally similar to their cousins in other systems such as Windows. There are, however, many differences, designed to make life easier both for end users and for programmers.

In this chapter, I'll be covering how to program dialogs in four stages:

- *Simple dialog programming*: getting to grips with the resource file and APIs for a simple dialog with just two text editor fields.
- *Stock controls*: since dialog fields normally use stock controls, the stock controls' resource file and C++ APIs are the most important, and certainly the biggest set of APIs you'll need when programming with dialogs.
- More of CEikDialog's API, showing the functions you can use for processing at the dialog level.
- Some standard dialogs, such as the query window.

This chapter will form the barest introduction to dialog programming. I'll highlight the major issues and take you quickly through the possibilities supported by the framework so that simply by reading the book you can get a taste of what's possible.

Again note that while we use UIQ as an example UI, the issues raised here apply to all UIs based on Symbian OS.

A whole book could be written to cover the topics addressed by this chapter, and indeed the SDKs contain comprehensive documentation. If you want to program real projects, you'll need to use the SDK appropriate to your UI to get the information you need.

In the next two chapters, I'll explain how you can write your own controls. You'll also be able to see, as you read those chapters, how the architecture of the control framework and the window server has been influenced by the requirements of dialogs.

10.2   Some Simple Dialogs

Here's a simple dialog, taken from Chapter 13's streams application, in action:

Figure 10.6

This shows the typical elements that comprise a single-page dialog:

- a title
- OK and Cancel buttons
- a vertical list of controls – in this case, two controls, each with a caption.

The basic techniques involved in dialog programming are

- constructing the dialog: this is done using resource files
- initializing each control when the dialog is first displayed
- checking individual controls, and the dialog as a whole, for validity
- getting information from each control, and kicking off some action when OK (or another button) is pressed.

The good news is that you *don't* have to perform the complicated processing that could be required, say, for keyboard handling and character drawing within a text control. That's already done for you by the text editor control, which we'll study later on.

Here's the code from `streams.cpp` that launches the Write file dialog. It's a command handler function, `CmdWriteFileL()`, called directly from `HandleCommandL()` in the case for `EStreamsExampleCmdWriteFile`:

```
void CExampleAppUi::CmdWriteFileL()
    {
    // Use a dialog to get parameters and verify them
    CEikDialog* dialog = new(ELeave) CExampleWriteFileDialog(this);
    if(!dialog->ExecuteLD(R_EXAMPLE_WRITE_FILE_DIALOG))
        return;

    // Write file under a trap
    TRAPD(err, WriteFileL());
    if(err)
        {
        delete iText;
        iText = 0;

        // Don't check errors here!
        iCoeEnv->FsSession().Delete(*iFileName);
        iAppView->DrawNow();
        User::Leave(err);
        }

    // Update view
    iAppView->DrawNow();
    }
```

Processing is in three stages:

- A dialog is used to set up values for iFileName and iText.

- Some code from the stream store API is used to open iFileName and write iText to it.

- The application updates the view to reflect the data that has changed.

This is a taste of **model-view-controller programming** (MVC), which I'll cover in more detail in the next chapter. For the moment, however, our interest is in the first lines of code in this function, which construct and run a dialog. Here they are again:

```
CEikDialog* dialog = new(ELeave) CExampleWriteFileDialog(this);
if(!dialog->ExecuteLD(R_EXAMPLE_WRITE_FILE_DIALOG))
    return;
```

This rather terse pattern is used to launch *every* dialog. First, a dialog object of type CExampleWriteFileDialog is allocated and C++-constructed. The app UI is passed as a parameter to the constructor, so that the dialog can later get at the app UI's data members – the whole point of the dialog is to update iText and iFileName.

Next, a single line is used to

- second-phase construct the dialog from a resource ID, in this case, R_EXAMPLE_WRITE_FILE_DIALOG

- run the dialog (that's what you'd expect something called ExecuteXxx() to do)

- destroy the dialog after it's run (the D in ExecuteLD() means 'destroy when finished')

- leave if there are any resource allocation problems or other environment errors (the L in ExecuteLD())

- return ETrue if the user ended the dialog with OK

- return EFalse if the user ended the dialog with Cancel

The if statement distinguishes between the return ETrue and return EFalse cases. If Cancel was pressed, then the return statement is taken, to prevent the rest of the WriteFileL() function being called to actually write the file.

As you program more dialogs, you'll soon come to appreciate the simplicity and regularity of these conventions.

10.2.1 Resource File Definition

The dialog was defined in a resource labeled R_EXAMPLE_WRITE_FILE_ DIALOG. Here it is, from `streams.rss`:

```
RESOURCE DIALOG r_example_write_file_dialog
    {
    title = "Write file...";
    buttons = R_EIK_BUTTONS_CANCEL_OK;
    flags = EEikDialogFlagWait;
    items =
        {
        DLG_LINE
            {
            prompt="File name";
            id=EExampleControlIdFileName;
            type=EEikCtFileNameEd;
            control=FILENAMEEDITOR
                                {};
            },
        DLG_LINE
            {
            type = EEikCtEdwin;
            prompt = "Text";
            id = EExampleControlIdText;
            control = EDWIN { width = 25; maxlength = 256; };
            }
        };
    }
```

Simply put, the dialog has a title, standard OK and Cancel buttons, some flags, and some items. In keeping with guidelines for positioning buttons, the Cancel/OK buttons are indicated by the resource name R_EIK_BUTTONS_CANCEL_OK.

Almost all dialogs should be coded with the line flags = EEikDialogFlagWait, which makes the dialog modal.

Regrettably, this is not the default. The default behavior is that your application can continue executing while the dialog is displayed. This isn't quite the same as a nonmodal dialog. Nonwaiting dialogs are typically used for activities like progress monitoring, in which the application is busy while the dialog is being displayed. Because the application is busy, it doesn't accept user input, so a nonwaiting dialog is effectively modal.

The body of the dialog is a vertical list of controls, each of which has

- a caption or prompt, such as Text;
- an ID, such as EExampleControlIdText;

- a type, such as `EEikCtEdwin`, which may have some initialization data of a format corresponding to the type, such as `EDWIN { width = 25; maxlength = 256; }`.

Don't use too many controls. As far as C++ is concerned, there is no limit, but in UIQ, for example, more than eight won't fit nicely onto a screen only 320 pixels high. If you code too many controls, the dialog becomes scrollable, and the user will have to scroll it to see all the controls. The dialog begins to overwhelm users with too much choice, making it hard to use. In some UIs, having too many controls will cause the dialog to overflow the screen, making it effectively unusable.

The prompt serves to identify its purpose to the user, while the ID identifies the control to the programmer. Later, we'll see that control IDs are used by C++ programs to specify the controls whose values they want to set or read. Like command IDs, control IDs are defined in the application's `.hrh` file, so they can be accessed both by resource file definitions and C++ programs.

One of the controls used here, `EEikCtEdwin`, is an edit window; the `EDWIN` resource `STRUCT` is required to initialize such a control. In this example, I specify the size of the control (25 characters) that affects the dialog layout, and the maximum length of the data (256 characters).

10.2.2 Dialog Code

The base class for all dialogs is `CEikDialog`. Any dialog you write in your application will derive from `CEikDialog`, and it will typically implement at least two member functions – one for initializing the dialog and one for processing the OK key (or the DONE key that is often used in UIQ).

'Read-only' dialogs, for displaying application data, need only implement the initialization function. Ultra-trivial dialogs, initialized entirely from resource files, needn't even implement the initialization function. More complex dialogs can implement many functions besides the two shown below: we'll return to this later on. All `CEikDialog` virtual functions have a do-nothing default implementation: you only override them if necessary.

The following code extract shows the declaration of `CExampleWrite-FileDialog`, from **streams.cpp**:

> *That's right:* `streams.cpp`, *not* `streams.h`. *I've treated dialogs as being private to the app UI, so they were not given their own header.*

The C++ constructor takes whatever parameter is necessary to connect the dialog to the outside world – in this case, my app UI, since it's this dialog's job to set its `iFileName` and `iText` members.

```
class CExampleWriteFileDialog : public CEikDialog
    {
public:
    CExampleWriteFileDialog(CExampleAppUi* aAppUi);

private:
    // From CEikDialog
    void PreLayoutDynInitL();        // Initialization
    TBool OkToExitL(TInt aKeycode);  // Termination

private:
    CExampleAppUi* iAppUi;
    };
```

On reflection, this isn't actually a very good encapsulation of the interface: I should really have passed references to the iFileName and iText members to make it clear that the dialog is intended to alter them and nothing else.

Initialization is performed by PreLayoutDynInitL().The 'prelay-out' part of the name means that the data you put into the dialog here will influence its layout – dialogs are laid out automatically to incorporate the optimum size of controls for the initialization data supplied.

Here's PreLayoutDynInitL():

```
void CExampleWriteFileDialog::PreLayoutDynInitL()
    {
    CEikFileNameEditor* fnamed=STATIC_CAST(CEikFileNameEditor*, Control
                                    (EExampleControlIdFileName));
    fnamed->SetTextL(iAppUi->iFileName);
    CEikEdwin* edwin=STATIC_CAST(CEikEdwin*, Control
                            (EExampleControlIdText));
    edwin->SetTextL(iAppUi->iText);
    }
```

This simply sets the edit windows to the existing values in iFile-Name and iText. Controls are identified by their ID, as specified in the id= line in the resource file definition. Control() returns a CCoeControl* type and this must be cast to the actual control type that it represents.

OK is handled by OkToExitL(). In fact, pressing any of the dialog buttons – *except* Cancel – will result in a call to this function. The function extracts values from the controls and, if everything is OK, returns ETrue, causing the dialog to be dismissed. If there's a problem (if, say, the value for iFileName isn't actually a valid filename), then OkToExitL() may either leave or return EFalse, which will continue the dialog.

OkToExitL() is more complicated because it has to check the validity of the requested operation before returning control to the app UI:

```
TBool CExampleWriteFileDialog::OkToExitL(TInt /* aKeycode */) //
  termination
  {
  // Get file name
  CEikEdwin* edwin = STATIC_CAST(CEikEdwin*,
                          Control(EExampleControlIdFileName));
                  // NB Cast as a CEikEdwin because only base
                  // class functionality required.
  HBufC* fileName=edwin->GetTextInHBufL();

  // Check it's even been specified
  if(!fileName)
      {
      TryChangeFocusToL(EExampleControlIdFileName);
      iEikonEnv->LeaveWithInfoMsg(R_EIK_TBUF_NO_FILENAME_SPECIFIED);
      }
  CleanupStack::PushL(fileName);

  // Check it's a valid filename
  if(!iCoeEnv->FsSession().IsValidName(*fileName))
      {
      TryChangeFocusToL(EExampleControlIdFileName);
      iEikonEnv->LeaveWithInfoMsg(R_EIK_TBUF_INVALID_FILE_NAME);
      }

  // Get the text string
  edwin = STATIC_CAST(CEikEdwin*, Control(EExampleControlIdText));
  HBufC* text = edwin->GetTextInHBufL();
  if(!text)
      text = HBufC::NewL(0);
  CleanupStack::PushL(text);

  // Ensure the directories etc. needed for the file exist
  TInt err = iCoeEnv->FsSession().MkDirAll(*fileName);
  if(err != KErrNone && err != KErrAlreadyExists)
      User::Leave(err);

  // Check whether it's going to be possible to create the file for
     writing
  RFile file;
  err = file.Create(iCoeEnv->FsSession(), *fileName, EFileWrite);
  if(err != KErrNone && err != KErrAlreadyExists)
      User::Leave(err); // No need to close file, since it didn't open

  // Check whether the user wants to replace the file, if it already
     exists
  if(err == KErrAlreadyExists)
      {
      if(iEikonEnv->QueryWinL(R_EIK_TBUF_FILE_REPLACE_CONFIRM))
          User::LeaveIfError(file.Replace(iCoeEnv->FsSession(),
                            *fileName, EFileWrite));
      else
          iEikonEnv->LeaveWithInfoMsg(0); // Let user try again
```

```
    }
file.Close();

// Finished with user interaction: communicate parameters and return
delete iAppUi->iFileName;
iAppUi->iFileName = fileName;
delete iAppUi->iText;
iAppUi->iText = text;
CleanupStack::Pop(2); // text, fileName
return ETrue;
}
```

On a dialog with only Cancel and OK buttons, there's no point in checking the key code: it can only be the code for OK. For that reason, I've commented out the `aKeyCode` parameter name. On a dialog with other buttons you would also have to check for those key codes.

The processing sequence is as follows:

- *Get the filename*: check that it's nonempty, and check that it's a valid filename.
- *Get the text*: if it's empty, turn it into a zero-length string (rather than no string at all), because my file format requires that a string be written, even if it's an empty one.
- Make sure the directory exists.
- Check whether the user wants to overwrite an existing file.
- Store the values from the dialog.
- Return.

Any of the checks or file operations in this sequence could fail, and the dialog is carefully coded to ensure that any such failure is entirely cleanup-safe. A `User::Leave()` from `OkToExitL()` is trapped and ensures that the dialog doesn't exit. There are three types of leave from within this code:

- `iEikonEnv->LeaveWithInfoMsg`(*resource-id*): this prints an info-message and then leaves with a special error code that causes no error message to be displayed. This is the recommended option for leaving when you detect an error in user input. Use the info-message to identify the error. Rely on cleanup to delete temporary variables such as `fileName` and `text` that have been pushed to the cleanup stack.
- `User::Leave()` with a genuine error code, such as the error code from opening a file: the UI will display the error message appropriate for that code.
- `iEikonEnv->LeaveWithInfoMsg(0)`: this leaves without giving any message at all, but cleans up temporary variables. I use this

variant when the user opts not to overwrite an existing file: it's obvious what the 'error' is at this point, so the user doesn't need to be told again.

The code above also shows how to use a query window. In this case, I want to check whether the user wants to overwrite an existing file. `iEikonEnv->QueryWinL()` takes a resource ID indicating the question to ask, and returns `ETrue` for a Yes answer, `EFalse` for No.

In the `PreLayoutDynInitL()` code, the control for capturing the file name was cast as a `CEikFileNameEditor`. In `OkToExitL()` above, it's cast as a `CEikEdwin`. This is ok because `CEikFile-NameEditor` is derived from `CEikEdwin` and we only need to use the functionality provided by the base class.

10.2.3 Read-only Dialogs

The Battleships program also has some useful dialog examples. The settings display simply shows fields without allowing them to be edited. Here it is:

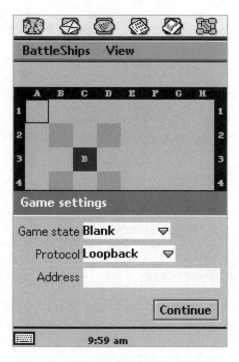

Figure 10.7

The Continue button is deliberately ambiguous: you can use this button on a dialog that conveys either good news (where OK would have been just as good) or bad news (where it would be an insult to the users to tell them that it was OK).

Here's the resource definition from `\scmp\battleships\battleships.rss`:

```
RESOURCE DIALOG r_game_settings_dialog
    {
    title = "Game settings";
    buttons = R_EIK_BUTTONS_CONTINUE;
    flags = EEikDialogFlagWait;
    items =
        {
        DLG_LINE
            {
            type = EEikCtChoiceList;
            prompt = "Game state";
            id = EGameControlIdState;
            control = CHOICELIST { array_id = r_game_state; };
            itemflags = EEikDlgItemNoBorder | EEikDlgItemNonFocusing;
            },
        DLG_LINE
            {
            type = EEikCtChoiceList;
            prompt = "Protocol";
            id = EGameControlIdProtocol;
            control = CHOICELIST { array_id = r_game_gdp_protocols; };
            itemflags = EEikDlgItemNoBorder | EEikDlgItemNonFocusing;
            },
        DLG_LINE
            {
            type = EEikCtEdwin;
            prompt = "Address";
            id = EGameControlIdOtherAddress;
            control = EDWIN { width = 25; maxlength = 100; };
            itemflags = EEikDlgItemNoBorder | EEikDlgItemNonFocusing;
            }
        };
    }
```

The `buttons = R_EIK_BUTTONS_CONTINUE` line brings in the standard Continue button from the resource file.

Each `DLG_LINE` has `itemflags = EEikDlgItemNoBorder | EEikDlgItemNonFocusing`. Nonfocusing lines can't be edited. There's not much point in having a border around something that can't be edited, so we turn that off too.

This dialog contains two choice lists and an editor. The choice lists are indicated by `type = EEikCtChoiceList`, and a `CHOICELIST` struct to initialize the control. The struct specifies an array of text items, one for each choice. The editor is only used to display a value, although in the dialog shown above, this is blank. Here's `r_game_gdp_protocols`:

```
RESOURCE ARRAY r_game_gdp_protocols
    {
    items=
        {
        LBUF { txt = "Loopback"; },
        LBUF { txt = "Infrared"; },
        LBUF { txt = "SMS"; }
        };
    }
```

The C++ code for a display-only dialog doesn't need an `OkToExitL()` – just a `PreLayoutDynInitL()`. Here it is:

```
void CGameSettingsDialog::PreLayoutDynInitL()
    {
    // Game state
    SetChoiceListCurrentItem(EGameControlIdState, iController->State());

    // Protocol
    SetChoiceListCurrentItem(EGameControlIdProtocol,
                            iController->Gsdp().GetGdpProtocol() - 1);

    // Other address
    TBuf<KMaxGsdpAddress> address;
    iController->Gsdp().GetOtherAddress(address);
    SetEdwinTextL(EGameControlIdOtherAddress, &address);
    }
```

The calls to `SetChoiceListCurrentItem()` set an index into the array of items, which is assumed to be zero-based. That works well for the game state (which uses a zero-based enumeration) but not for the GDP protocol (which uses a one-based enumeration, with zero as an invalid value), so I have to subtract one from the GDP protocol enumeration value.

10.2.4   Simple Dialog Processing

The startup dialog in Battleships, which we saw earlier, includes a couple of other interesting features:

- cue text, to indicate the purpose of the Start mode line
- the address field, which is displayed only if necessary – that is, only if you're initiating, and using a networked protocol that requires addresses.

Here's the resource file definition:

```
RESOURCE DIALOG r_game_initiate_dialog
    {
    title = "Start first game";
    buttons = R_EIK_BUTTONS_CANCEL_OK;
    flags = EEikDialogFlagWait;
    items =
        {
        DLG_LINE
            {
            type = EEikCtChoiceList;
            prompt = "Protocol";
            id = EGameControlIdProtocol;
            control = CHOICELIST { array_id = r_game_gdp_protocols; };
            },
        DLG_LINE
            {
            type = EEikCtChoiceList;
            prompt = "Start mode";
            id = EGameControlIdStartMode;
            control = CHOICELIST { array_id = r_game_start_mode; };
            },
        DLG_LINE
            {
            type = EEikCtLabel;
            control = LABEL
                {
                standard_font = EEikLabelFontAnnotation;
                txt = "One player must Initiate, the other must Listen";
                };
            },
        DLG_LINE
            {
            type = EEikCtEdwin;
            prompt = "Address";
            id = EGameControlIdOtherAddress;
            control = EDWIN { width = 25; maxlength = 100; };
            },
        DLG_LINE
            {
            type = EEikCtChoiceList;
            prompt = "First move";
            id = EGameControlIdFirstMovePref;
            control = CHOICELIST { array_id = r_game_first_move; };
            }
        };
    }
```

The cue text is indicated by:

```
        DLG_LINE
            {
            type = EEikCtLabel;
```

```
control = LABEL
    {
    standard_font = EEikLabelFontAnnotation;
    txt = "One player must Initiate, the other must Listen";
    };
},
```

This is a control without a prompt or control ID (neither are needed). The control is a label. You should always use the annotation font for cue text labels.

The other controls use an EDWIN and CHOICELISTs, which we've already met. The derived dialog class shows the functions we use to control the visibility of the address field:

```
class CGameInitiateDialog : public CEikDialog
    {
public:
    CGameInitiateDialog(CGameController* aController);

private:
    // From CEikDialog
    void PreLayoutDynInitL();         // Settings on dialog launch
    TBool OkToExitL(TInt aKeycode);   // Action when OK pressed

    // Listen to changing selections
    void HandleControlStateChangeL(TInt aControlId);

    // Show only necessary controls
    void ShowRelevantControls();      // Called when needed
private:
    CGameController* iController;
    };
```

As well as PreLayoutDynInitL() and OkToExitL(), I have implemented another CEikDialog framework function, HandleControlStateChangeL(), which gets called with the relevant control ID whenever the value of any of the choice list fields in the dialog is changed. I also have a function called ShowRelevantControls() that I use to show or hide the address depending on the settings of other controls.

Here's the implementation of ShowRelevantControls():

```
void CGameInitiateDialog::ShowRelevantControls()
    {
    CEikChoiceList* choicelist1 =
    static_cast<CEikChoiceList*>(Control(EGameControlIdStartMode));

    CEikChoiceList* choicelist2 =
    static_cast<CEikChoiceList*>(Control(EGameControlIdProtocol));
```

```
// Get start mode: 1=listen, 0=initiate
TInt listen = choicelist1->CurrentItem();

// Get protocol
TUid protocol = TUid::Uid(choicelist2->CurrentItem());

// Show other address if mode=initiate, and protocol is networked
TBool otherAddressNeeded = !listen && ProtocolRequiresAddress
    (protocol);
MakeLineVisible(EGameControlIdOtherAddress, otherAddressNeeded);
}
```

The idea is to test the value of the GDP protocol and start mode choice lists, and work out whether another address is needed. I then call `MakeLineVisible()` – a `CEikDialog` library function – to make the address line visible only if it is needed.

I call `ShowRelevantControls()` from two places – `PreLayout-DynInitL()` and `HandleControlStateChangeL()`. The first call is the last line of `PreLayoutDynInitL()` after I've set up all the controls:

```
void CGameInitiateDialog::PreLayoutDynInitL()
    {
    CEikEdwin* edwin;

    // Protocol
    CEikChoiceList* choicelist =
    static_cast<CEikChoiceList*>(Control(EGameControlIdProtocol));

    TUid protocol = iController->GetGdpProtocol();

    // Start mode;
    if(protocol == KGdpLoopbackUid)
        choicelist->SetCurrentItem(0);
    if(protocol == KGdpSmsUid)
        choicelist->SetCurrentItem(1);
    if(protocol == KGdpBluetoothUid)
        choicelist->SetCurrentItem(2);

    // Other address
    _LIT(KBlank," ");
    edwin=STATIC_CAST(CEikEdwin*, Control(EGameControlIdOtherAddress));
    edwin->SetTextL(&KBlank);

    // First-move preferences
    // Show relevant controls
    ShowRelevantControls();
    }
```

`HandleControlStateChangeL()` calls `ShowRelevantControls()` only if there was a change in the protocol or start mode lines:

```
void CGameInitiateDialog::HandleControlStateChangeL(TInt aControlId)
    {
    if(aControlId == EGameControlIdProtocol ||
       aControlId == EGameControlIdStartMode)
        ShowRelevantControls();
    }
```

CEikDialog provides other framework functions that you can implement to detect relevant events in the dialog and its controls. It also provides other library functions that you can implement to control the dialog.

10.3   Dialog APIs

The simple dialogs we've seen are enough to show us that CEikDialog offers a large range of framework functions that you can override and library functions that you can use to provide specific dialog processing.

In this section, our lightning-quick tour of dialogs continues with an overview of these APIs.

10.3.1   Resource Specifications

Let's start with a closer look at the resource specification for dialogs. You define a dialog using the DIALOG resource STRUCT. Its members are:

Member	Description
title = <string>	The title of the dialog, displayed at the top
flags = <bitmask>	Optional bitmask of flags governing attributes of the dialog. Defaults to 0, but for nearly all dialogs you should specify EEikDialogFlagsWait.
buttons = <resource>	A resource defining the buttons to use in the dialog. Default is no buttons; many dialogs use R_EIK_BUTTONS_CANCEL_OK.
items = <list>	A comma-separated list of dialog items
pages = <resource>	A resource defining the pages in a multipage dialog. Don't specify this if you only want a single-page dialog.

The flags for the dialog as a whole are specified in the `flags` member of the `DIALOG` structure. Flag bit values are defined by `EEik-DialogFlagXxx` constants in `eikdialg.hrh`. Typical dialogs specify `EEikDialogFlagWait`; other flags control button positioning (right or bottom), whether there is a title, and several others.

The flags for a dialog *line* are specified in the `flags` member of the `DLG_ITEM` structure. Bit values for these flags are defined by `EEikDlgItemXxx` constants in `eikdialg.hrh`, and they allow you to specify that there should be a separator after this item, that the control doesn't take focus, that it has no border, and so on.

10.3.2 Adding Buttons

If you want to code more buttons than the standard `R_EIK_BUTTONS_CANCEL_OK`, use code such as:

```
DLG_LINE { buttons = r_example_buttons_test_cancel_ok; ... }

   RESOURCE DLG_BUTTONS r_example_buttons_test_cancel_ok
   {
   buttons=
       {
       DLG_BUTTON
           {
           id = EExampleBidTest;
           button = CMBUT { txt = "Test"; };
           hotkey = 'T';
           },
       DLG_BUTTON
           {
           id = EEikBidCancel;
           button = CMBUT { txt = "Cancel"; };
           hotkey = EEikBidCancel;
           },
       DLG_BUTTON
           {
           id = EEikBidOk;
           button = CMBUT { txt = "OK"; };
           hotkey = EEikBidOk;
           }
       };
   }
```

This causes a Test button to be included in the buttons for the dialog. You will need to test whether Test or OK has been pressed in your `OkToExitL()` function.

If you want to display buttons vertically to the right of the dialog lines, use dialog flag `EEikDialogFlagButtonsRight`.

10.3.3 Basic Functions

CEikDialog provides a rich API for derived dialog classes. As usual, I'll summarize the functions here, and leave you to hunt around in the documentation and headers for more.

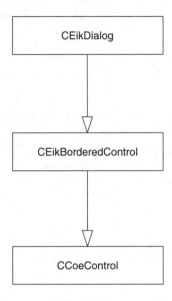

Figure 10.8

Many of the functions in the dialog API are either convenience functions that add to the API's size without really adding to its complexity, or else they're for unusual requirements such as constructing dialogs dynamically without using a resource file. Leaving them aside, we get to a manageable set of framework and library functions. Here are some of the most important.

Framework functions

The following virtual functions are called (see over) when various events happen during dialog processing. If you want to handle those events, override the default function provided by CEikDialog. In each case, the default function does nothing.

Sometimes, you want to do some processing when the user has finished working with one of the dialog's lines. Changes in time and date editors don't get reported using HandleControlStateChangeL(), so you need to intercept the event when the user changes the focused line.

The solution is to provide an implementation for LineChangedL(). The dialog automatically calls this function when the focused line

Function	Description
`virtual void PreLayoutDynInitL();`	Called prior to layout so you can initialize the dialog. Set control values here if you want them to influence sizing and layout.
`virtual void PostLayout DynInitL();`	Called after layout so you can initialize the dialog. Set control values here if you *don't* want them to influence sizing and layout.
`virtual void SetInitial CurrentLine();`	Called during initialization to allow you to set the current line. Implement this if you want to override the default (which is the top line).
`virtual TBool OkToExitL(TInt aButtonId);`	Called when a button is pressed, but not for the Cancel button, unless you specify this in dialog flags. Return `ETrue` if it's OK to exit the dialog; `EFalse` otherwise. Leave if you want to report an error: this is interpreted as not being OK to exit.
`virtual void Handle-ControlStateChangeL (TInt aControlId);`	Called when the state of the given control changes, as reported by an `EEventStateChanged` event sent by the control to its observer. Certain controls, such as choice lists, call this when their state changes. Implement this function if you wish to change dialog settings – or those of other controls – in response to state changes.
`virtual SEikControlInfo CreateCustomControlL (TInt aControlType);`	Called during dialog construction when the control type indicated in the resource file is not recognized by the control factory. You must implement this if your dialog struct specifies a control type ID not recognized by the control factory – otherwise dialog construction will panic. `SEikControlInfo` is a struct containing a pointer to the new control, flags, and the resource ID of optional text after the control.

changes, and passes the ID of the control to which the focus is moving. You can find out the ID of the control that currently has the focus by calling `IdOfFocusControl()`. For example:

```
CExampleMyDialog::LineChangedL(TInt aControlId)
    {

    TInt id = IdOfFocusControl();

    // Now do my stuff
    ...
    }
```

An older technique is to override `PrepareForFocusTransi-tionL()`, which is virtual, more or less by accident. It has a nontrivial default implementation, but note that it does not include a parameter specifying the line that is currently focused. Start your overridden function with

```
CExampleMyDialog::PrepareForFocusTransitionL()
    {
    CEikDialog::PrepareForFocusTransitionL();
    TInt id = IdOfFocusControl();

    // Now do my stuff
    ...
    }
```

Call the base class implementation first; this leaves if the control with the current focus isn't valid, and is unable to lose focus. Then call `IdOfFocusControl()` to get the control ID that's currently focused. Then decide what you need to do, and do it.

Library functions

You can call the following library functions from the framework functions listed above:

Function	Description
`CCoeControl* Control(TInt aControlId) const;`	Get a pointer to the control whose ID is specified: panic if the ID doesn't exist.
`CCoeControl* ControlOrNull(TInt aControlId) const;`	Get a pointer to the control whose ID is specified: return 0 if the ID doesn't exist.

`CEikLabel* ControlCaption(TInt aControlId) const;`	Get the caption associated with a control whose ID is specified.
`void SetLineDimmedNow(TInt aControlId, TBool aDimmed);`	Dim or undim a line. Lines should be dimmed if it is not currently meaningful to select them.
`void MakeLineVisible(TInt aControlId, TBool aVisible);`	Make the control on a line visible or invisible (but don't change the visibility of its caption).
`void MakeWholeLineVisible (TInt aControlId, TBool aVisible);`	Make a whole line visible or invisible, both the caption and control.
`TInt IdOfFocusControl() const;`	Get the ID of the control with focus – that is, the control currently being used
`void TryChangeFocusToL(TInt aControlId);`	Calls `PrepareForFocusLossL()` on the currently focused control and, if this doesn't leave, transfer focus to the control whose ID is specified. This is the way to change focus: the control with the focus should only refuse the request if its state is invalid.

10.4   Stock Controls for Dialogs

From the dialog code above, we've seen that a basic part of dialog programming is using stock controls in your dialogs. The general techniques for using stock controls are:

- specify a control type in your `DLG_LINE` struct, using `type =`
- specify initialization data for the control in your `DLG_LINE` struct, using `control =` and an appropriate resource `STRUCT`
- do further control initialization from `PreLayoutDynInitL()` or `PostLayoutDynInitL()`
- extract values from the control when needed, in `OkToExitL()` or other dialog processing functions

- do other things, such as controlling the control's visibility, using dialog library functions.

UIQ provides 42 stock controls that you can use in dialogs. Other UIs may have more or less depending on the UI design. In this section, I'll give a lightning-fast tour of those controls, including the resource STRUCTs you use to initialize them, and their C++ classes.

Here are the stock control classes, sorted by base class:

```
CEikHorOptionButtonList          CEikBorderedControl
                                   CEikCalendar
CEikLabeledButton                  CEikClock
                                   CEikComboBox
CQikSlider                         CEikProgressInfo
                                   CEikSecretEditor
CQikVertOptionButtonList           CEikWorldSelector
                                   CQikIpEditor
CQikSoundSelector                  CEikScrollBar
                                     CEikArrowHeadScrollBar
CQikTabScreen                      CQikToolbar
                                   CQikScrollableContainer
CQikTabScreenPage                CQikTTimeEditor
                                   CQikDateEditor
CQikNumericEditor                  CQikDurationEditor
  CQikFloatingPointEditor          CQikTimeEditor
  CQikNumberEditor                 CQikTimeAndDateEditor
                                 CEikButtonBase
CEikAlignedControl                 CEikCheckBox
  CEikImage                        CEikOptionButton
  CEikLabel                        CEikCommandButtonBase
                                     CEikTwoPictureCommandButton
                                     CEikTextButton
                                     CEikBitmapButton
                                     CEikCommandButton
                                       CEikMenuButton
                                 CEikChoiceListBase
                                   CQikColorSelector
                                   CEikChoiceList
                                 CEikEdwin
                                   CEikGlobalTextEditor
                                     CEikRichTextEditor
                                 CEikListBox
                                   CEikHierarchicalListBox
                                   CEikTextListBox
                                     CEikColumnListBox
```

All these classes are derived ultimately from `CCoeControl`. The few classes at the beginning of the table, along with `CEikAligned-Control` and `CEikBorderedControl`, are derived directly, while all other controls on the diagram are derived indirectly.

The C++ SDK includes information and examples for all the controls above, plus others that aren't intended for direct inclusion into dialogs.

To give you a very quick taster, here's a tour of the most important controls mentioned in the table above. You can either guess the function of the other controls based on their name, or look up the details in the appropriate SDK as and when you need them.

10.4.1   Buttons

```
CEikBorderedControl
CEikButtonBase
 CEikCheckBox
 CEikOptionButton
 CEikCommandButtonBase
  CEikTwoPictureCommandButton
  CEikTextButton
  CEikBitmapButton
  CEikCommandButton
  CEikMenuButton
```

A **command button** has text and/or graphics, and reports events to some observer. The button bar in the Agenda uses command buttons.

Option buttons cooperate with each other to allow the user to choose from a small (up to four) number of options. In a dialog, option buttons should be combined in a **horizontal option button list**, so that all option buttons appear on a single line. In other systems, a horizontal option button list would be called a radio button group.

A **check box** allows you to enable or disable an option using a check mark.

10.4.2   Lists

```
CEikBorderedControl            CEikBorderedControl
 CEikChoiceListBase             CEikListBox
  CQikColorSelector              CEikHierarchicalListBox
  CeikChoiceList                 CEikTextListBox
                                 CEikColumnListBox
```

Choice lists allow you to select from a large number of options. These controls are normally displayed in a small area that allows you to see one option at a time. If you tap on the control, a vertically scrolling window pops up that displays as many options as possible in the given screen size. Specific examples of choice lists include a basic selector for text items, a color selector, and a selector for sounds.

For some lists, the popup paradigm is inappropriate: it's better to display the whole list in its own (potentially large) area on the screen. UIQ, for example, uses **list boxes** derived ultimately from `CEikListBox` to do this. List boxes can display data in rows that scroll down, or they can produce a 'snaking' multicolumn effect. Rows themselves that have multiple columns and hierarchical lists that can be expanded and collapsed are also available.

The list box APIs are generalized to support any kind of data, although `CEikTextListBox` is a useful specialization for simple text lists. The other list classes are more complex.

10.4.3 Editors

CQikNumericEditor	**CEikBorderedControl**
CQikFloatingPointEditor	CEikComboBox
CQikNumberEditor	CQikIpEditor
	CEikSecretEditor
	CQikTTimeEditor
	CQikDateEditor
	CQikDurationEditor
	CQikTimeEditor
	CQikTimeAndDateEditor
	CEikEdwin
	CEikGlobalTextEditor
	CEikRichTextEditor

Numeric input is supported by **numeric editors** whose base class is `CQikNumericEditor`. A numeric editor can edit integers and real (floating point) numbers, with checks on range.

Time and date editors, whose base class is `CQikTTimeEditor`, allow times and dates to be displayed and changed. Tapping on these editors causes a popup to be displayed. For a time editor, the popup is in the form of a digital clock. Hours, minutes and the a.m./p.m. value can be changed independently by simply tapping on the leaves displayed in the popup. For a date editor, the popup is in the form of a calendar.

Basic text input is supported by the **edit window** or **edwin**, which simply edits a string. A **combo box** is a hybrid editor and choice list: you can pick a preset value from the choice list, or enter a new value.

Edwins edit plain text – just characters, without formatting. The derived global text editor adds support for global formatting – providing the same paragraph and character formatting for the whole text. The further derived rich text editor adds support for formatting paragraphs and characters individually, and adds support for embedded objects and pictures. These classes reuse the Symbian OS classes `CGlobalText` and `CRichText`, and they make it possible to deliver very sophisticated editing in a simple application. However, the APIs to these classes are wide, and the more sophisticated edwin classes also have a reputation of being hard to use.

Other editors include a **secret editor** (for passwords and the like).

10.4.4 Using Controls in Dialogs

Now that we've seen what stock controls can go into a dialog, we need to know how to specify them in a resource file `DLG_LINE` struct. `DLG_LINE` has the following members:

Member	Description
`prompt = <string>`	The caption for the control.
`id = <number>`	A numeric ID that can be used to retrieve the control. Use a constant defined in your application's `.hrh` file.
`itemflags = <bitmask>`	Optional bitmask of flags governing attributes of the line, such as whether to put a horizontal line below it. Defaults to 0.
`type = <number>`	A numeric ID indicating the control class to construct. See below for more information about this.
`control = <struct>`	Further initialization data, mapped by a `STRUCT` appropriate for the particular control type.
`trailer = <string>`	Optional trailer text, not usually specified. May be up to 40 characters.

The control in each dialog line is constructed by the **control factory**. The ID you specify to the `type` member of the `DLG_LINE` indicates to the control factory, which control to construct. A new control is constructed, and its `ConstructFromResourceL()` function is then called to read the data specified by `control =` in order to initialize itself. (`ConstructFromResourceL()` is virtual in `CCoeControl` and implemented by any derived class that can be constructed from resource file data.)

As a dialog programmer, then, you have to know the type IDs and resource `STRUCT`s associated with each control. The IDs are enumerated in two locations. IDs for common Uikon controls are enumerated in

TEikStockControls in eikctrls.hrh. The UIQ specific IDs are enumerated in TQikStockControls in qikstockcontrols.hrh. The complete list in class hierarchy order is given on pages 310 and 311.

The table lists the header files where you can find the C++ API for the control. Also, check out the following:

- eikon.rh and Qikon.rh contain the resource STRUCT and the resource STRUCT members you can use (almost all controls have initialization data that can be conveyed from a STRUCT)

- eikon.hrh and Qikon.hrh contain the flag definitions (many controls support one or more flags).

10.4.5   Accessing Controls

You can access any control in a dialog using the Control() function: specify the ID you use to identify the control in the resource file and you will get a CCoeControl*. You can then cast this pointer to the type of control you know it really is, and access member functions such as getters, setters, and many others:

```
CEikEdwin* edwin = STATIC_CAST(CEikEdwin*,
                    Control(EExampleControlIdFileName));
HBufC* fileName = edwin->GetTextInHBufL();
```

This pattern is *very* common in dialogs' PreLayoutDynInitL(), HandleControlStateChangeL(), and OkToExit() functions.

10.4.6   Custom Controls in Dialogs

Dialogs aren't limited to using stock controls. You can add your own controls into a dialog as well. To do so, you'll first need to understand about writing controls – which is what the next two chapters are about.

Then, you'll need to do the following:

- For each dialog in which you want to include a custom control, implement its CreateCustomControlL() function.

- In the resource file for the dialog, specify a type = that isn't used by the control factory.

- In your CreateCustomControlL(), test for the relevant type, and construct an SEikControlInfo appropriate for your control.

- Define a resource file STRUCT in a .rh file associated with your control, specifying the member names and types for the resource initialization data.

Class	Header	ID	STRUCT
CEikHorOptionButtonList	eikhopbt.h	EEikCtHorOptionButtList	HOROPBUT
CEikLabeledButton	eiklbbut.h	EEikCtLabeledButton	LBBUT
CQikSlider	QikSlider.h	EQikCtSlider	QIK_SLIDER
CQikVertOptionButtonList	QikVertOptionButtonList.h	EQikCtVertOptionButtonList	QIK_VERTOPBUT
CQikSoundSelector	QikSoundSelector.h	EQikCtSoundSelector	(none)
CQikTabScreen	QikTabScreen.h	EQikCtTabScreen	QIK_TABSCREEN
CQikTabScreenPage	QikTabScreen.h	EQikCtTabScreenPage	QIK_TABSCREENPAGE
CQikFloatingPointEditor	QikFloatingPointEditor.h	EQikCtFloatingPointEditor	QIK_FLOATING_POINT_EDITOR
CQikNumberEditor	QikNumberEditor.h	EQikCtNumberEditor	QIK_NUMBER_EDITOR
CEikImage	eikimage.h	EEikCtImage	IMAGE
CEikLabel	eiklabel.h	EEikCtLabel	LABEL
CQikDateEditor	QikDateEditor.h	EQikCtDateEditor	QIK_DATE_EDITOR
CQikDurationEditor	QikDurationEditor.h	EQikCtDurationEditor	QIK_DURATION_EDITOR
CQikTimeEditor	QikTimeEditor.h	EQikCtTimeEditor	QIK_TIME_EDITOR
CQikTimeAndDateEditor	QikTimeAndDateEditor.h	EQikCtTimeAndDateEditor	QIK_TIME_AND_DATE_EDITOR
CEikCalendar	eikcal.h	EEikCtCalendar	CALENDAR
CEikClock	eikclock.h	EEikCtClock	CLOCK
CEikComboBox	eikcmbox.h	EEikCtComboBox	COMBOBOX
CEikProgressInfo	eikprogi.h	EEikCtProgInfo	PROGRESSINFO
CEikSecretEditor	eikseced.h	EEikCtSecretEd	SECRETED
CEikWorldSelector	eikwsel.h	EEikCtWorldSelector	WORLD_SELECTOR

CQikIpEditor	QikIpEditor.h	EQikCtIpEditor	QIK_IP_EDITOR
CEikScrollBar	eikscrlb.h	EEikCtScrollBar	(none)
CEikArrowHeadScrollBar	eikscrlb.h	EEikCtArrowHeadScrollBar	(none)
CQikToolbar	QikToolbar	EQikToolbar	QIK_TOOLBAR
CQikScrollableContainer	QikScrollableContainer.h	EQikCtScrollableContainer	(none)
CEikCheckBox	eikchkbx.h	EEikCtCheckBox	(none)
CEikOptionButton	eikopbut.h	EEikCtOptionButton	(none)
CEikTwoPictureCommandButton	eikcmbut.h	EEikCtTwoPictureCommandButton	(none)
CEikTextButton	eikcmbut.h	EEikCtTextButton	TXTBUT
CEikBitmapButton	eikcmbut.h	EEikCtBitmapButton	BMPBUT
CEikCommandButton	eikcmbut.h	EEikCtCommandButton	CMBUT
CEikMenuButton	eikmnbut.h	EEikCtMenuButton	MNBUT
CQikColorSelector	QikColorSelector.h	EQikCtColorSelector	QIK_COLOR_SEL
CEikChoiceList	eikchlst.h	EEikCtChoiceList	CHOICELIST
CEikEdwin	eikedwin.h	EEikCtEdwin	EDWIN
CEikFileNameEditor	eikfsel.h	EEikCtFileNameEd	FILENAMEEDITOR
CEikFolderNameEditor	eikfsel.h	EEikCtFolderNameEd	FOLDERNAMEEDITOR
CEikGlobalTextEditor	eikgted.h	EEikCtGlobal TextEditor	GTXTED
CEikRichTextEditor	eikrted.h	EEikCtRichTextEditor	RTXTED
CEikTextListBox	eiktxlbx.h	EEikCtListBox	(none)
CEikColumnListBox	eikclb.h	EEikCtCoListBox	(none)

- In your control, implement `ConstructFromResourceL()` to read initialization data from a resource `STRUCT`.

Have a look at the **Examples**\... example source code that includes an example of creating custom controls.

10.5   Standard Dialogs

There are a number of convenient standard dialogs. Perhaps the most convenient are the alert and query dialogs.

10.5.1   Alerts

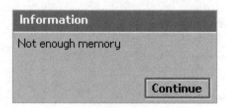

Figure 10.9

An **alert** displays a title saying 'Information', one or two lines of text, and a button labeled 'Continue'. The UI environment constructs a ready-made alert dialog that you can invoke with `iEikonEnv->AlertWin()`, specifying either one or two string parameters. As the dialog is preconstructed, you can never run out of memory when using `AlertWin()`. Indeed, alert dialogs are used to indicate error conditions, including out-of-memory!

The Continue button is carefully chosen. How often have you seen error messages on your PC such as, 'The system detected an unrecoverable error – OK'? This is a frequent source of user annoyance, so the UI uses Continue, which is exactly what will happen when you press the button.

I don't know where the technical term 'alert' came from. Alerts preallocate their resources using a technique you can use yourself, if you want to – paradoxically, it's called **sleeping dialogs**. Check out the appropriate UI SDK for more information.

10.5.2   Queries

A query dialog enables a minimal form of interaction: you can use one to ask a simple Yes/No question. As we saw earlier in this chapter, the `streams` application contains an example:

Figure 10.10

```
if(err == KErrAlreadyExists)
    {
    if(iEikonEnv->QueryWinL(R_EIK_TBUF_FILE_REPLACE_CONFIRM))
        User::LeaveIfError(file.Replace(iCoeEnv->FsSession(),
                            *fileName, EFileWrite));
    else
        iEikonEnv->LeaveWithInfoMsg(0); // Let user try again
    }
```

You run a query by specifying a string in a resource file that will be used as a question. I happened to be able to find a resource string in the Symbian OS source, for this purpose: often, you'll need to write your own. The query dialog has a Yes button and a No button; `iEikonEnv->QueryWinL()` returns `ETrue` if Yes is pressed, or `EFalse` otherwise.

As I pointed out at the beginning of this chapter, the art is to ask truly Yes/No questions – not to use the puzzling Windows-style Yes/No/Cancel.

Actually, the real art is to avoid queries altogether, if you can. The example above is not *really* the Symbian OS way: very few users want to throw away data. Applications running on Symbian OS don't ask; they save and exit quietly.

Note that query dialogs *aren't* sleeping dialogs, so the process of constructing and executing a query *can* leave. A query dialog is a special case of a `CEikInfoDialog`.

10.5.3 Other Standard Dialogs

UIs provide many other standard dialogs; UIQ, for example, includes dialogs to

- add, edit and delete categories
- enter and change a password
- set the current date and time
- set the options for formatting dates and times.

Additionally, many of the more sophisticated controls include dialogs of their own – edwins, for example, include dialogs for find, replace, and options for replace.

10.6 Summary

In this chapter, I've introduced a lot of topics, answered the main questions, and left a *lot* of detail unanswered. You've now seen

- what dialogs can do
- the basic shape of dialog programming, in both resource files and C++
- the framework, library and convenience functions in `CEikDialog`
- the stock controls offered by a typical UI for inclusion in dialogs
- some standard dialogs.

There is much more information in the UI specific SDKs, including documentation, source code and examples about:

- stock controls
- important `CEikDialog` functions
- how to implement custom controls for dialogs
- programming multipage dialogs.

11

Graphics for Display

In the previous two chapters, we've begun to get familiar with the GUI and we've got about as far as we can without a better understanding of Symbian OS graphics. In Chapter 4, I passed over hellogui's app view's Draw() member without much comment and, although we saw some specific examples of drawing to an application's views in Chapter 9, most of the other drawing – in button bars, menus, dialogs and standard controls – has been done by the Uikon and UIQ frameworks.

Now it's time to look at graphics in more detail. In this chapter, I'll take you through the things you need to know for on-screen drawing and in the next, the way that graphics support user interaction based on keyboard and pointer devices.

In this chapter we'll study:

- *Drawing basics*: how to get graphics on screen, working with basic example code.

- *Using the* CGraphicsContext *API*: this fundamental class contains the drawing functions.

- *The model-view-controller paradigm (MVC)*: this is a key concept for both drawing and interaction.

- *Flicker-free drawing*: how to update the screen without producing visible flicker.

- *Screen sharing*: how to share the screen using **windows** (RWindow) and **controls** (CCoeControl) and outline the implications of sharing on drawing code.

- *Special effects*: some of those supported by the Symbian OS graphics system.

Symbian OS C++ for Mobile Phones. Edited by Richard Harrison
© 2003 John Wiley & Sons, Ltd   ISBN: 0-470-85611-4

11.1   Drawing Basics

GUIs present many more opportunities for displaying data than a console program does. Even in a program as simple as 'Hello World', you face these issues:

- What font should you use?
- What colors should you use for foreground and background?
- Where should you put the text?
- Should you set the text off in some kind of border or frame?
- How big is your screen and how much of it do you get to draw the text?

Whichever way you look at it, you *have* to make these decisions, so the part of your program that says 'Hello world!' will inevitably be bigger than the corresponding part of a text-mode program. Here, once again, is `CHelloGuiAppView::Draw()` from `HelloGui_AppView.cpp`:

```
void CHelloGuiAppView::Draw(const TRect& /*aRect*/) const
    {
    CWindowGc& gc = SystemGc();
    gc.Clear();
    TRect rect = Rect();
    rect.Shrink(10, 10);
    gc.DrawRect(rect);
    rect.Shrink(1, 1);
    const CFont* font = iEikonEnv->TitleFont();
    gc.UseFont(font);
    TInt baseline = rect.Height() / 2 + font->AscentInPixels() / 2;
    gc.DrawText(*iHelloWorld, rect, baseline, CGraphicsContext::ECenter);
    gc.DiscardFont();
    }
```

That's 11 lines of code, where one would have been enough in `hellotext`. The good news, though, is that you are *able* to make decisions you need, and it's relatively easy to write the code to implement whatever you decide. The `Draw()` example illustrates the essentials of drawing:

- Draw your graphics to a **control**. In the example, `CHelloGuiAppView` is derived from `CCoeControl`.
- Use the `CGraphicsContext` API to draw the graphics themselves.

11.1.1   Controls

From the perspective of a Symbian OS application programmer, *all* drawing is done to a **control**. A control is a rectangular area of screen that occupies all or part of a **window**. The base class for all controls is `CCoeControl`, which is defined by the CONE component in Symbian OS.

Look at this display in Figure 11.1:

Figure 11.1

As it happens, this screen includes two windows – the app view and the button bar. The app view is a single control, while the button bar comprises several controls such as

- a **container** for the whole button bar, which is a **compound control**;
- **component controls**, including the four buttons.

This application already allows us to make some generalizations about controls:

- For an application, a control is the basic unit of GUI interaction: a control can do any sensible combination of drawing, pointer handling, and key handling.
- A window is the basic unit of interaction for the system: controls always use all or part of a window.
- Controls can be compound: that is, they can contain component controls. A compound control is sometimes known as a container.

In this chapter, we'll see the role that controls (and windows) play in drawing. We'll continue to use the `hellogui` and Battleships applications as examples. In the next chapter, we'll look more closely at key- and pointer-based interaction.

11.1.2   Walking through `Draw()`

In Symbian OS, all drawing is done through a **graphics context** (GC). In this section, we'll take a closer look at the example function `CHelloGui-iAppView::Draw()` to see how the GC is used.

Getting the graphics context

The `CHelloGuiAppView::Draw()` function begins by getting hold of a GC using `SystemGc()`, a function in `CCoeControl`:

```
CWindowGc& gc = SystemGc();
```

All graphics context classes are derived from `CGraphicsContext`. Each derived class – such as `CWindowGc` here – is used for drawing on a particular graphics device (in our example, a window) and implements *all* the functionality specified by the base class, plus (and optionally) some extra functionality appropriate for the device in question. Therefore, we can clear the screen through the graphics context:

```
gc.Clear();
```

Graphics contexts are a common notion in the world of computer graphics. Windows uses a 'device context'; Java uses a `Graphics` object.

Drawing a rectangle

The next three lines of code conspire to draw a rectangular border ten pixels in from the edge of the app view's area on the screen:

```
TRect rect = Rect();
```

`CCoeControl::Rect()` gives the coordinates of the rectangle occupied by the control from within which it's called, in this case the app view. The coordinates are given relative to the window that the control uses. The coordinates used by `CWindowGc` drawing functions must also be relative to the window, so this is convenient.

```
rect.Shrink(10, 10);
```

This use of `Shrink()` makes the rectangle 10 pixels smaller than the control's rectangle on every side – top, right, bottom, and left. `TRect` contains many utility functions like this, as we'll see later.

```
gc.DrawRect(rect);
```

This draws a rectangle using the default graphics context settings. These settings specify

- that the pen creates a black, one pixel wide, solid line: this causes the boundary of `rect` to be drawn in black;
- that the brush is null, which means that the rectangle is not filled.

You can rely on the default GC configuration being set up prior to your `Draw()` function. Don't waste your time setting things that are guaranteed to be the default anyway.

Drawing the text

Now we are going to draw the text, centered in the rectangle. For good measure, we start by shrinking the rectangle by one pixel each side so that we can afford to white it out without affecting the border we have just drawn:

```
rect.Shrink(1, 1);
```

Then, we get a font from the Uikon environment:

```
const CFont* font = iEikonEnv->TitleFont();
```

This is our first encounter with a `CFont*`. In Chapter 15, we'll have a careful look at how to get a font of a desired face, size, bold/italic attributes, and so on. To avoid these issues right now, I just used a title font from the Uikon environment – it's the font used on the title bar of dialog boxes and it's suitably bold and large.

It's not enough just to have a pointer to the font; we must also tell the graphics context to use it.

```
gc.UseFont(font);
```

This `UseFont()` lasts for all subsequent text-drawing functions – until another `UseFont()` is issued or until `DiscardFont()` is called.

Now we need to draw the text, centered in the `rect` rectangle:

```
TInt baseline = rect.Height() / 2 + font->AscentInPixels() / 2;
gc.DrawText(*iHelloWorld, rect, baseline, CGraphicsContext::ECenter);
```

This `DrawText()` function conveniently draws text with the GC's pen and font settings and the entire rectangle area with the current brush settings. Horizontal justification is specified by its final parameter, which we specify here as `CGraphicsContext::ECenter` to indicate that the text should be horizontally centered.

Vertical justification

Oddly, `DrawText()` *doesn't* handle vertical justification for you, so you have to calculate the baseline yourself. Fortunately, the algorithm is simple and (unlike horizontal justification) it doesn't depend on the text of the string. You have to specify the baseline in pixels *down* from the top of the rectangle so start with half the height of the rectangle (measured from its top downwards) and then add half the font's ascent, as in Figure 11.2.

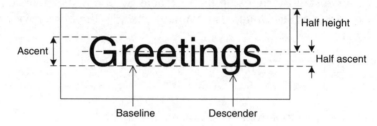

Figure 11.2

At this point, we've drawn the text so we're finished with the font and must discard it:

```
gc.DiscardFont();
```

> **It is important to discard the font at this stage in order to avoid memory leak.**

And there you have it. Our drawing code has drawn both graphics (a box) and text (a string in a box).

11.2   The `CGraphicsContext` API

As I stated above, all concrete graphics context classes are derived from `CGraphicsContext`, which offers a rich API for device-independent drawing. Let's pause briefly to describe this API. Its main features in UML are shown in Figure 11.3:

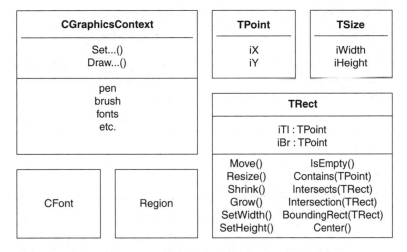

Figure 11.3

CGraphicsContext contains the main drawing functions and is defined in gdi.h. All drawing is done using the current pen, brush, and font settings and is clipped to the currently set clipping region. The pen, brush, font, and clipping region settings, therefore, provide context for graphics functions – hence, the name of the class.

You can only set GC settings. There is no class for pen, brush and so on, and you can't interrogate a GC to find out its current settings. You can keep a GC if you need to keep its settings, and you can reset a GC with a single function call if you need to throw all the settings away.

This section discusses the main features of CGraphicsContext, including:

- using coordinate classes to specify position
- setting the GC features, such as pen, brush, and so on
- using the drawing functions.

The drawing functions are illustrated by many examples throughout this book, and CGraphicsContext is thoroughly documented in the Symbian OS UIQ C++ SDK.

11.2.1 Coordinate Classes

Graphics are drawn to a device whose coordinate system is defined in pixels. Each point on the device has an (x, y) coordinate, measured from an origin at the top left of the device, with *x* coordinates increasing toward the right and *y* coordinates increasing downwards.

In Chapter 15, we'll see how pixel coordinates are related to real-world units such as inches or centimeters. In this chapter, we'll concentrate on pixels and screen-oriented graphics.

2D supporting classes for points, rectangles, sizes, and regions are defined in e32std.h:

- TPoint contains iX and iY coordinates
- TRect contains two points, iTl for top left, and iBr for bottom right
- TSize contains iWidth and iHeight dimensions.

These classes are equipped with a large range of constructors, operators, and functions to manipulate and combine them, but they make no attempt to encapsulate their members. You don't have to use get/set functions to access the (x, y) coordinates of a point and so on. In truth, there would be very little point in doing so – the representations of these objects are genuinely public.

The two points that define Trect can be interpreted by specific graphics implementations in different ways. One common interpretation is to place the top-left point inside the rectangle, and the bottom-right point is just outside it, as shown in Figure 11.4:

Figure 11.4

This definition makes some things easier, such as calculating the size, because you simply subtract the *x* and *y* coordinates of the top-left from the bottom-right point. It also makes other things harder, such as rubber-banding calculations for interactively drawing a rectangle, because you have to add (1, 1) to the pointer coordinates to include the bottom-right corner correctly. If TRect is defined to include its bottom-right corner, it simply makes different things easier and different things harder. It's important to remember that the definition of the rectangle depends on the specific graphics interpretation.

Rectangles should be normalized, so that iTl coordinates are never greater than corresponding iBr coordinates. If you perform a calculation on a TRect that might violate this condition, call Normalize() to clear things up, by swapping the *x* coordinate values and/or *y* coordinate values as necessary.

Region-related classes

Several region-related classes are also defined in e32std.h. These define a region of arbitrary shape as the union of several rectangles. The region classes are used extensively by the window server, but only in specialized application programs.

A region can potentially have very many rectangles so that the region classes, in general, can allocate resources on the heap. They are heavily optimized so that, if only relatively few rectangles are needed to define the region, then no heap-based allocation is necessary. So, while points, rectangles, and sizes are simple, T classes that are easy to allocate anywhere and pass around in client-server call's regions require more careful management and like C classes, need to be deleted or cleaned up when no longer required.

11.2.2   Setting up the Graphics Context

CGraphicsContext holds several important items of context for drawing functions:

- Pen
- Brush
- Font
- Current position
- Origin
- Clipping Region
- Justification.

Pen

The **pen** defines draw modes (color and style). These are used for drawing lines, the outlines of filled shapes, and text.

Draw mode options include Boolean operations on pixel color values – probably the only useful ones are solid (use the color specified), null (don't draw) and XOR with white (invert), which can be useful for cursor selection, rubber banding, and so on.

Style options include solid, dotted, dashed, and also pen width. However, the BITGDI that draws screen graphics (that you saw in Chapter 9) doesn't support combinations of style and pen width – that is, it can't do thick dotted lines.

Use the `SetPenColor()`, `SetPenStyle()`, and `SetPenSize()` member functions to control the pen. By default, the pen is black, solid, and one pixel thick.

Brush

The **brush** defines fill and background color or pattern.

The brush can be null, solid, a hatching pattern, or a bitmap. For hatching and bitmaps, you can set an offset so that pattern fills on adjacent drawing primitives abut each other without odd edge effects. Use `SetBrushStyle()`, `SetBrushColor()`, `SetBrushOrigin()`, `SetBrushPattern()`, and `DiscardBrushPattern()` to control brush settings; defaults are null brush, zero origin.

Font

The **font** defines the font to be used for drawing text.

You specify it by passing a `CFont*` to `CGraphicsContext`. We'll cover fonts properly in Chapter 15, but for now you note that the CONE environment has one font (`iCoeEnv->NormalFont()`), while the Uikon environment contains several (`iEikonEnv->TitleFont()`, `LegendFont()`, `SymbolFont()`, `AnnotationFont()`, and `Dense-Font()`). Use `UseFont()` to set a font, `DiscardFont()` to say you no longer wish to use that font, and `SetUnderlineStyle()` and `Set-StrikethroughStyle()` to set algorithmic enhancements to the font in use.

By default, *no* font and *no* algorithmic enhancements are in use – you'll get panicked if you try to draw text without a font in use. You can find more information about the use of fonts in Chapter 15.

Current position

Current position is set by `MoveTo()` and various `DrawXxxTo()` member functions, and moved by `MoveBy()` and corresponding `DrawXxxBy()` functions. It is also affected by `DrawPolyLine()`. The `XxxBy()` functions support relative moving and drawing. By default, the current position is at (0, 0).

Origin

The **origin** defines the offset from the device origin that will be used for drawing and you can use `SetOrigin()` to control it. By default, the origin is (0, 0).

Clipping region

The **clipping region** defines the region to which you want your graphics to be clipped.

You can specify a simple rectangle or a region that may be arbitrarily complex. Use `SetClippingRect()` to set a rectangular clipping region and `CancelClippingRect()` to cancel it. By default, no clipping region (other than the device limits) applies.

Justification

Specialized **justification** settings for a variant of `DrawText()` can be set, although it's best not to call these directly from your own code. Instead, use the FORM component in Symbian OS to create text views for you.

Use `Reset()` to set all contexts to default values.

11.2.3   Drawing Functions

Once you've set up the GC to your liking, there are numerous ways to draw to the screen. All GC functions are virtual so they can be implemented in derived classes. Furthermore, all GC functions are designed to succeed and so don't return anything (in C++ declarations, they return `void`). This requirement is such that multiple GC commands can be batched into a single message and sent to a server for execution – this would not be possible if any GC command had a return value.

Points and lines

You can plot a single point or draw an arc, a line, or a polyline. These functions all use the current pen; here are their declarations in `gdi.h`:

```
virtual void MoveTo(const TPoint& aPoint) = 0;
virtual void MoveBy(const TPoint& aVector) = 0;
virtual void Plot(const TPoint& aPoint) = 0;

virtual void DrawArc(const TRect& aRect,
                     const TPoint& aStart,
                     const TPoint& aEnd) = 0;
virtual void DrawLine(const TPoint& aPoint1,
                      const TPoint& aPoint2) = 0;
virtual void DrawLineTo(const TPoint& aPoint) = 0;
virtual void DrawLineBy(const TPoint& aVector) = 0;
virtual void DrawPolyLine(const CArrayFix<TPoint>* aPointList) = 0;
virtual void DrawPolyLine(const TPoint* aPointList,
                          TInt aNumPoints) = 0;
```

Note that line drawing (including arcs and the last line in a polyline) excludes the last point of the line. As with the specification of `TRect`, this is a mixed blessing: sometimes it makes things easier, sometimes harder. If the last pixel was plotted automatically, it would be harder to unplot it in the cases in which this behavior was not desired. However, it's easy enough to fix tricky cases by using `Plot()`.

Check out the SDK for the interpretation of `DrawArc()` parameters.

`DrawPolyLine()` starts at the current cursor position set with `MoveTo()`, any `XxxTo()` or `XxxBy()` function, or `DrawPolyLine()`. Effectively, `DrawPolyLine()` uses `DrawLineTo()` to draw to every point specified.

Filled-outline shapes

You can draw several filled-outline shapes: a pie slice, ellipse, rectangle, rectangle with rounded corners, or a polygon. These functions use the pen and/or the brush. Use pen only to draw an outline. Use brush only to draw the shape. Use both to draw an outlined shape.

Here are the functions:

```
virtual void DrawPie(const TRect& aRect,
                     const TPoint& aStart,
                     const TPoint& aEnd) = 0;
virtual void DrawEllipse(const TRect& aRect) = 0;
virtual void DrawRect(const TRect& aRect) = 0;
virtual void DrawRoundRect(const TRect& aRect,
                     const TSize& aCornerSize) = 0;
virtual TInt DrawPolygon(const CArrayFix<TPoint>* aPointList,
                     TFillRule aFillRule = EAlternate) = 0;
virtual TInt DrawPolygon(const TPoint* aPointList,
                     TInt aNumPoints,
                     TFillRule aFillRule = EAlternate) = 0;
```

The `DrawPie()` parameters are essentially the same as for `DrawArc()`.

`DrawPolygon()` connects and fills all the points specified and just as `DrawLine()`, no relative drawing is used or needed. Self-intersecting polygons may be drawn, in which case the fill rule parameter specifies the behavior for regions of even enclosure parity. Check the SDK and the `grshell` example for details.

Bitmaps

You can draw a bitmap either on the scale of 1 : 1 or stretched to fit a rectangle you specify. Here are the functions:

```
virtual void DrawBitmap(const TPoint& aTopLeft,
                        const CFbsBitmap* aSource) = 0;
virtual void DrawBitmap(const TRect& aDestRect,
                        const CFbsBitmap* aSource) = 0;
virtual void DrawBitmap(const TRect& aDestRect,
                        const CFbsBitmap* aSource,
                        const TRect& aSourceRect) = 0;
```

Use the same-size variant for high-performance blitting of GUI icons and the stretch-blit variants for device-independent view code supporting on-screen zooming or printing.

See Chapter 15 for more on drawing bitmaps.

Text

You can draw text in the current font. Here are the functions for doing so:

```
virtual void DrawText(const TDesC& aString,
                      const TPoint& aPosition) = 0;
virtual void DrawText(const TDesC& aString,
                      const TRect& aBox,
                      TInt aBaselineOffset,
                      TTextAlign aHoriz = ELeft,
                      TInt aLeftMrg = 0) = 0;
```

The first (and apparently simpler) function uses the GC's justification settings, but you shouldn't call it yourself. Instead, use FORM if you need to handle properly laid out text. For general use, use the `TRect` variant that clips the text to the specified rectangle, and paints the rectangle background with the current brush.

A graphics context has no default font and if you call a text-drawing function without a previous call to `UseFont()` in effect, you get a panic.

The panic is particularly ugly when you're drawing to a `CWindowGc` because all window-drawing functions are batched together and sent to the window to be executed later. The window server doesn't detect the absence of a `UseFont()` until the buffer is executed by which time there is no context information about where the panic occurred. Always remember to use `DiscardFont()` to discard the font after use and so avoid a memory leak.

11.3   Drawing and Redrawing

In a GUI program, all drawing is done to controls, which form all or part of a screen window as we saw with `hellogui`'s `CHelloGuiApppView::Draw()`:

- The derived control class's `Draw()` function is called when drawing is required.
- `Draw()` gets a graphics context using `SystemGc()`.
- It draws into the area defined by its `Rect()` function.

But it's a bit more complicated than that. Your control must not only **draw** its content but must also **redraw** it when it changes, or when the system requires a redraw.

System-initiated redraws occur when:

- the window is first constructed;
- the window, or part of it, is exposed after having been obscured by some other application or a dialog box.

Application-initiated redraws occur when:

- the application changes the control's content and wants these changes to be shown in an updated display
- the application changes the drawing parameters (such as color, scrolling, or zoom state), and wants these changes to be shown in an updated display.

In addition, there are various other circumstances – partly system-initiated, partly application-initiated – in which redrawing must occur. For example, dismissing a dialog is application-initiated, but the redrawing of the controls underneath comes as a system-initiated request.

To understand redrawing properly, we have first to review the **model-view-controller** (MVC) paradigm. This is a good way to think about GUI systems and using MVC concepts makes the following discussions much easier.

11.3.1   The Model, View, and Controller Pattern

On inspection you'll notice that `hellogui`'s `CHelloGuiApppView::Draw()` function assumes that the data we need is already available. It doesn't interrogate a database or ask the user for the string to draw – rather, it just uses the data that's already there in the `iHelloWorld` member of the control.

This is a standard paradigm in graphics: draw functions simply draw their model data; they don't change anything. If you want to change something, you use another function and then call a draw function to reflect the update. In fact, this pattern is so common that it has a name: **model-view-controller**, often abbreviated simply to MVC:

- The **model** is the data that the program manipulates: in the case of 'Hello World', it's the string text.

- The **view** is the view through which a user sees the model: in the case of 'Hello World', it's the `CHelloGuiAppView` class.

- The **controller** is the part of the program that updates the model and then requests the view to redraw in order to show the updates. In 'Hello World', there are no updates and, therefore, there is no controller. When we come to the Battleships application, the controller will be quite sophisticated because updates may be generated either through user interaction or through events from the other player.

A strict MVC structure is shown in Figure 11.5:

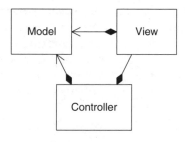

Figure 11.5

From the diagram:

- The model is an independent entity; if the program is file-based, the model very often corresponds with the program's persistent data.

- The view *uses* the model since its job is to draw it.

- The controller coordinates the model and view updates, so it *uses* both the model and the view.

Depending on the program design, either of the *uses* relationships above may be upgraded into a stronger *has* relationship. The MVC pattern dictates the *uses* relationships, but doesn't force the *has* relationships, specifically.

Blurring the MVC distinction

In some programs, the distinctions between model, view, and controller are cleanly reflected by boundaries between classes. But there are many reasons for the boundaries to become blurred in practice.

The 'Hello world!' program is so simple that there's no point in making such fine distinctions – its CHelloGuiAppView contains both the model and the view and there is no controller at all. Battleships, on the other hand, is complicated enough to use this structure and greatly benefits from it. However, it turns out that the model (in the MVC sense) isn't the same as the CGameEngine engine class; it also includes some aspects of the CGameController. The view has to display things from both these, so the boundary between MVC model and controller is not quite the same as that between the C++ engine and controller classes.

Battleships has a more than one view to display two fleets and the overall game status. The model for each part of this view is different. A sophisticated application uses the MVC pattern again and again – at the large scale for the whole application and at a smaller scale for each interaction within it. You could say that even the button bar has a model (defined by the resource file definitions that construct it) and a controller (somewhere in the Uikon application framework).

The MVC paradigm can become particularly blurred when giving feedback to some kinds of user interaction – navigation, cursor selection, animation, or drag-and-drop (which admittedly is rare in Symbian OS, though it does exist, for example when resizing grid columns in a Symbian OS spreadsheet). Nonetheless, it remains extremely useful and you can use it to think about the design of many Symbian OS controls and applications.

Words used in MVC

I should finish this section with a word on nomenclature. A 'control' in Symbian OS is not usually a 'controller' in the MVC sense, which is precisely why the word 'controller' is not used. A Symbian OS control does, however, usually contain pure MVC view functionality: its Draw() function draws a model without changing it.

In Symbian OS literature, the word 'view' is used for a control or some drawing/interaction code to highlight the fact that the 'view' is entirely separate from the 'model'. A good example is ETEXT and FORM, Symbian OS rich text components:

- ETEXT is a model without views
- FORM provides views but has no model.

We use 'app view' in this sense, while the model is often contained in the document class. In Battleships, I use 'fleet view' and 'player status view' as the names of my controls, because the fleet and player status 'model' data is kept in separate classes.

Often, in Symbian OS literature, the word 'model' is used for application data that can be saved to file.

11.3.2 The `Draw()` Contract

Symbian OS controls use the `Draw()` function to implement MVC view functionality. `CCoeControl::Draw()` is defined in `coecntrl.h` as

```
IMPORT_C virtual void Draw(const TRect& aRect) const;
```

A derived class will override this virtual function to draw – or redraw – its model. In the rare cases in which this function is not overridden, there's a default implementation that leaves the control blank.

> Because `CCoeControl::Draw()` is *strictly* an MVC view function, it should not update the model. It is therefore `const`, and nonleaving. Your `Draw()` implementation *must not leave*.

This is another reason `CGraphicsContext` functions return `void`: if they could fail, `Draw()` could fail also.

Redraw handling

System-initiated redraw handling starts in the window server, which detects when you need to redraw part of a window. In fact, it maintains an **invalid region** on the window, and sends an event to the application that owns the window, asking it to redraw the invalid region. CONE works out the control that intersects the invalid region and converts the event into a call to `Draw()` for all affected controls. A system-initiated redraw must redraw the model *exactly* as the previous draw.

Application-initiated redraw handling starts (by definition) in the application. If you update a model and need to redraw a control, you can simply call its `DrawNow()` function. `DrawNow()` is a nonvirtual function in `CCoeControl` that:

- tells the window server that the control is about to start redrawing,

- calls `Draw()`,

- tells the window server that the control has finished redrawing.

In theory, then, you don't need to code any new functions in order to do an application-initiated redraw. You can simply call `DrawNow()` so that your `Draw()` function is called in turn.

Where to draw

It's possible that only a part of your control will need to be drawn (or redrawn). To understand this, you need to distinguish between the four regions shown in Figure 11.6:

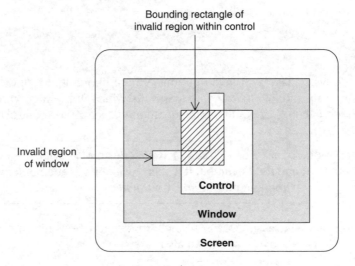

Figure 11.6

Your control is part of a window. The window server knows about the window and knows the regions of the window that are invalid – that is, the parts that need to be redrawn. Your `Draw()` function must draw the entire invalid region, but it must *not* draw outside the boundary of the control.

The window server will clip drawing to the invalid region – which is clearly bounded, in turn, by the boundary of the window itself.

> **But if your control doesn't occupy the entire window, *you* are responsible for ensuring that your redraw doesn't spill beyond the boundaries of the control.**

Often, this turns out to be not too onerous a responsibility: many controls such as buttons and the various sections of the Battleships application screen, draw rectangles, lines, and text are guaranteed to be inside the control's boundary in any case.

In the few cases in which this doesn't happen, you can issue a `SetClippingRect()` call to the graphics context that ensures that future drawing is clipped to the control's rectangle. Here's an example, from the `drawing` example developed in Chapter 15:

```
aGc.SetClippingRect(aDeviceRect);
aGc.SetPenColor(KRgbDarkGray);
aGc.DrawRect(surround);
```

This is necessary because `surround` could have been bigger than `aDeviceRect`, which is the region of the control that this code is allowed to draw into. You can cancel this later, if you wish, with `CancelClip-pingRect()`, but since `CGraphicsContext::Reset()` does this anyway and `Reset()` is called prior to each control's `Draw()`, you don't need to do this explicitly from a control.

How to draw

Naturally, you can draw using the system GC and its member functions.

> **You can assume that the GC was reset before `Draw()` was called. Don't reset it yourself and don't set colors and options that you don't need.**

Avoiding wasteful redraws

Drawing outside the invalid region is technically harmless (because such drawing will be clipped away by the window server whether it's inside your control's boundaries or not), but it's potentially wasteful. You may be able to save time by confining your drawing activity to the invalid region; the trade-off is that you will have to do some testing to find out what you must draw and what you do not need to draw.

That's the purpose of the `TRect` passed to your `Draw()` function; it is the bounding rectangle of the invalid region. If you wish, you can use this to draw (or redraw) only the part of the control within the passed `TRect`. It will be worth doing this if the cost of testing is outweighed by the savings from avoiding irrelevant drawing.

In practice, very few controls gain much by confining their redraw activity entirely to the bounding rectangle – it's simpler and not much slower to redraw the whole control. As a result, the majority of controls are coded to ignore the bounding rectangle that's passed. If you're writing a control that *does* use the `TRect`, remember that you still have to obey the contract to cover the entire invalid region within the boundary of your control, and nothing outside your control. You may still have to set a clipping region to ensure this – the system doesn't set one for you.

> *Early in Symbian OS development, we passed the invalid region (rather than its bounding rectangle) to* Draw(). *This turned out to be more trouble than it was worth. Regions are data structures of arbitrary size, which are much harder to pass around than* TRects, *but we had to pass them whether they were needed or not – and they usually weren't. As a compromise, we passed the bounding rectangle of the invalid region.*

Breaking the const and leave rules

In quite rare circumstances, you may need to do some nondraw processing in Draw(). This could happen, for instance, if your view is very complicated and you're doing lazy initialization of some of the associated data structures in order to minimize memory usage.

In this case, you may need to allocate memory during Draw() to hold the results of your intermediate draw-related calculations, and this allocation could cause a leave. In addition, you'll want to use a pointer to refer to your newly allocated memory, perhaps in the control. This requires you to change the pointer value, which would violate the const-ness of Draw().

The solution, in this case, is to use casting to get rid of const-ness and to put your resource-allocating code into a leaving function that gets called from a TRAP() within Draw(). You also have to decide what to draw if your resource allocation fails.

11.4   Flicker-free Redraw

So far, we have suggested that you only need Draw() to do all your application's drawing. Draw() gets called when necessary for system-initiated redraws, and you can use DrawNow() to call Draw() for application-initiated redraws. However, there's a big problem with this simplistic approach: it makes applications impossible to use.

Firstly, it makes them slow, because you do too much redrawing in response to the most trivial updates. Secondly, it makes them ugly, because the draw-everything approach usually causes unacceptable flicker while the display is drawn and redrawn.

> **The art of graphics, in general, is making it look pretty. The art of on-screen graphics is redrawing quickly and without flicker – and knowing when to stop optimizing.**

We'll use the Battleships views to show the most important considerations involved here. Many Symbian OS applications are more complicated than Battleships and take these considerations much further than I have done. In some cases, such as layout and updating rich text, the logic is hideously complicated. Symbian OS provides a single, reusable component for this purpose – FORM – so that application authors don't have to invent their own.

The most important cases for updating the views are as follows:

- When we start the game, bring the game to the foreground, or reload a game, we have to draw everything.

- When most events take place in the game, we have to redraw the status view.

- When something is hit, we have to reflect it in a fleet view – either my fleet or the opponent's fleet.

- When the cursor is moved (on the opponent's fleet view), we have to move the highlight quickly from the old cursor location to the new one.

Let's examine each of these in turn.

11.4.1 Drawing Everything

The draw-everything situations are fairly easy. The status view is very quick to draw and I don't need to do anything special. I had more trouble with the fleet view, pictured in Figure 11.7:

Figure 11.7

I started out with code that looked something like this:

```
void CFleetView::Draw(const TRect&) const
    {
    DrawBoard();
    DrawBorders();
    DrawTiles();
    }
```

This code:

- draws a black square over the region of the board including both the border area and the sea area,
- draws the letters and numbers for the top, bottom, left, and right borders,
- draws the 64 tiles in the sea area.

This is a classic flickery-draw function, in which the backgrounds are drawn first and then overpainted by the foreground. It looks especially bad towards the bottom right of the sea area, because there is a significant delay between the first function call (which painted the whole board black) and the last one (which finally painted the 64th tile).

Whiting out the background

There is another problem, which I'll demonstrate in Chapter 15, because I had not whited out the background area between the board and the edge of the control. I could have tackled that easily enough using, say,

```
void CFleetView::Draw(const TRect&) const
    {
    ClearBackground();
    DrawBoard();
    DrawBorders();
    DrawTiles();
    }
```

but that would have made the flicker even worse.

> **The general solution to flicker problems is to avoid painting large areas twice.**

And so to my code in its present form:

```
void CFleetView::Draw(const TRect&) const
    {
    DrawOutside();
    DrawBorders();
    DrawTiles();
    }
```

This code

- whites out the area of the control between the board rectangle and the border of the control – it doesn't touch the board area itself,

- draws the whole top, bottom, left, and right borders – without affecting the sea area,

- draws each of the 64 tiles in the sea area.

My new draw-border code draws the border background and then overpaints it with the letters or numbers, which is a potential source of flicker. But the border is small and the time interval between drawing the background and overpainting the eighth letter or number is too short to notice any flicker.

Likewise, the code I use to draw each tile starts by drawing the tile with its letter and then, if it's the cursor tile, overpaints the cursor. Again, this is OK – it happens so quickly that no one notices.

Don't overpaint on a large scale

This example emphasizes the point about the general rule for avoiding flicker: don't overpaint on a large scale. In some circumstances, redraws need to be optimized much more than I've done here. You can use many techniques for optimizing drawing to eliminate flicker:

- Draw all the interesting content first – that is, draw the tiles, then the borders, and then the legend. This means that the things the user is interested in get drawn first.

- Optimize the drawing order so that the tile at the cursor position is drawn first. Again, this is what the user is most interested in.

- Draw subsequent tiles in order of increasing distance from the cursor tile, rather than scanning row-by-row and column-by-column.

- Use active objects to allow view drawing to be mixed with user interaction – cursor movement or hit requests, for example – so that the application becomes responsive immediately.

- Draw to an off-screen bitmap and bitblitt that bitmap to the screen.

Each level of increased redraw optimization adds to program complexity. Fortunately, none of this was necessary for the fleet view. In some Symbian OS application views, however, these techniques make the difference between an application that is pleasant to use and one that can barely be used at all. The Agenda year view, for instance, would use all the techniques mentioned above.

11.4.2   Status View Update

The status view update didn't need any optimization, even though the status view draw function appears to be quite complicated with lots of detailed coordinate calculations, font selection, and string assembly.

The status view actually benefited from the buffering performed by the graphics system. As I mentioned above, drawing commands are buffered and only sent from the client application to the window server when necessary. They are executed very rapidly indeed, by the window server – typically, within a single screen refresh interval. This is too fast for a user to notice any flicker.

The status view update uses only around 10 draw function calls, which probably all fit within a single buffer and so are executed all together. If the status view had been more complicated (which it would have been, had I used a suitably professional graphic design), then it might have been more flicker-prone and I would have had to take more precautions when redrawing it.

> **In any professional application, the aesthetics of a view are more important than the ease with which that view can be programmed.**

In this book, I've paid enough attention to aesthetics to make the points I need to make, but no more. I don't really think any of my graphics are satisfactory for serious use and the status view is a prime example. In a real application, it would have to be better and if this meant the redraw code would need optimizing, then that would have to be done.

Good status views are particularly demanding. On the one hand, a rich status view conveys very useful information to the user. On the other hand, the user isn't looking at the status view all the time and it *must not* compromise the application's responsiveness. For these reasons, status views are often updated using background active objects.

A good example of a status view from the Symbian OS standard application suite would be the toolband at the top of a Word view. Of most interest to us here is that it shows the font, paragraph formatting, and other information associated with the current cursor position. Its implementation is highly optimized using background active objects and a careful drawing order so that document editing is not compromised at all.

11.4.3   Hit Reports

When a hit report comes in from the opponent's fleet, the fleet view is updated to show the affected tile. Calling `DrawNow()` would have done the job, but it would have involved drawing the board and its borders, which is slow and completely unnecessary as these could not possibly have changed.

Looking for a better approach, I considered redrawing only the tiles that were affected by the hit. These are as follows:

- The tile that was hit.
- If that tile was a ship, then the squares diagonally adjacent to it (provided they're on the board, and provided they haven't already been hit), because we now know that these tiles must be sea.
- If the tile was the final tile in a ship, then we know that *all* the tiles surrounding the ship must be sea, so we have to redraw them.

It turns out that working out exactly the tiles that are affected and doing a minimal redraw is nontrivial – though we could do it if it was really necessary. Instead, I decided that I would redraw all the tiles. The code would be quick enough and wouldn't cause perceived flicker because there would be no change to tiles that weren't affected. I wrote a `DrawTilesNow()` function to do this:

```
void CFleetView::DrawTilesNow() const
    {
    Window().Invalidate(iSeaArea);
    ActivateGc();
    Window().BeginRedraw(iSeaArea);
    DrawTiles();
    Window().EndRedraw();
    DeactivateGc();
    }
```

This function contains the logic needed to start and end the drawing operation and, in the middle, the same `DrawTiles()` function that I use to draw the board in the first place. During system-initiated redraw, the window server preparation is handled by the CONE framework. During application-initiated redraw, we have to do it ourselves before we can call `DrawTiles()`.

The `DrawXxxNow()` pattern

> You can easily copy this `DrawXxxNow()` pattern for any selective redraws in your own applications.

It's useful to pause to note a few rules about application-initiated redraw here:

- Application-initiated redraw is usually done using a function whose name is `DrawXxxNow()`.

- A `DrawXxx()` function (without the `Now`) expects to be called from within an activate-GC and begin-redraw bracket, and to draw to an area that was invalid.

- A simple `DrawXxxNow()` will invalidate activate-GC, begin-redraw, call `DrawXxx()`, and then end-redraw and deactivate-GC.

- A more complex `DrawXxxNow()` function may need to call many `DrawXxx()` functions.

- You should avoid calling multiple consecutive `DrawXxxNow()` functions if you can because this involves (typically) wasteful invalidation, activate-GC, and begin-redraw brackets.

- You *must*, in any case, avoid calling a `DrawXxxNow()` function from within an activate-GC/begin-redraw bracket, since it will cause a panic if you repeat these functions when a bracket is already active.

Later, I'll explain what the activation and begin-redraw functions actually do.

Mixing draw and update functions

> **Don't mix (view-related) draw functions with (model-related) update functions.**

For example, don't specify a function such as `MoveCursor()` to move the cursor *and* redraw the two affected squares. If you write all your model-update functions to update the view as well, you won't be able to issue a sequence of model updates without also causing many wasted view updates. The crime is compounded if your view update after, say, `MoveCursor()` is not optimized so that it updates the whole view.

Instead, make `MoveCursor()` move the cursor *and nothing else*. You can call lots of model-update functions like this, calling an appropriate `DrawXxxNow()` function to update the view only when they have all executed. After a *really* complicated sequence of model updates, you might simply call `DrawNow()` to redraw the entire control.

If you *must* combine model updates with redrawing, make it clear in your function name that you are doing so – `MoveCursorAndDrawNow()`, for example. Then your users will know that such functions should not be called during optimized update processing.

11.4.4 Cursor Movement

Cursor movement is highly interactive, and must perform supremely. When writing the application, I was prepared to optimize this seriously if necessary, and that would not have been difficult to do. When you move the cursor, by keyboard or by pointer, at most two tiles are affected – the old and new cursor positions. It would have been easy to write a function to draw just the two affected tiles.

But it turned out to be unnecessary. Early in development, I experimented with `DrawTilesNow()`, which draws all 64 tiles. That turned out to be fast enough and sufficiently flicker-free.

In more demanding applications, cursor movement can become very highly optimized. A common technique is to invert the affected pixels so that no real drawing code is invoked at all – all you need to know is which region is affected and use the logical operations of the GDI to invert the colors in the affected region. However, although this technique can be very fast, it needs careful attention to detail:

- Color inversion is good for black and white, but for color or more subtle shades of gray, it doesn't always produce visually acceptable results.

- You still have to be able to handle system-initiated redraws, which means that you must be able to draw with the inverted color scheme on the affected region. It's insufficient simply to draw the view and then to invert the cursor region. This would produce flicker precisely in the region in which it is least acceptable. You must draw the view and cursor in one fell swoop.

- In fact, you have to combine system-initiated redraws with very high application responsiveness so that the cursor can move even while a redraw is taking place. This simply amplifies the difficulties referred to, above.

In general, cursor-movement optimization is nontrivial. In almost every PC application I've used (including the word processor I'm using to write this book), I've noticed bugs associated with cursor redrawing.

It's the age-old lesson again: reuse existing code if you can and don't optimize unless you have to. If you do have to optimize, choose your technique very carefully.

11.5 Sharing the Screen

Until now, I've covered the basics of drawing and in many cases I've had to tell you to do something without explaining why – for instance, the `ActivateGc()` and `BeginRedraw()` functions in `DrawTilesNow()`.

Now it's time to be precise about how windows and controls work together to enable your application to share the screen with other applications and to enable the different parts of your application to work together.

Symbian OS is a full multitasking system in which multiple applications may run concurrently. The screen is a single resource that must be shared among all these applications. Symbian OS implements this sharing using the **window server**. Each application draws to one or more **windows**; the window server manages the windows, ensuring that the correct window or windows are displayed, exposing and hiding windows as necessary, and managing overlaps (Figure 11.8).

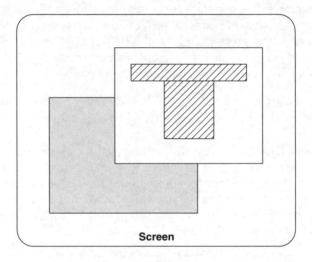

Screen

Figure 11.8

An application must also share the screen effectively between its own components. These components include the main application view, the button bar, and other ornaments: dialogs, menus, and the like. An application uses **controls** for its components. Some controls – dialogs, for instance – use an entire window, but many others simply reside alongside other controls on an existing window. The buttons on a button bar behave this way, as do the fleet views in the main application view of Battleships.

11.5.1 CONE

Every GUI client uses CONE, the control environment, to provide the basic framework for controls and for communication with the window server in Figure 11.9:

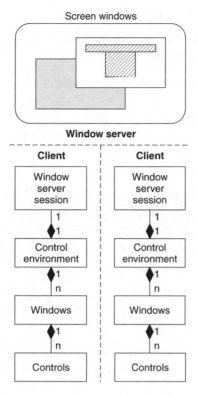

Figure 11.9

The window server maintains the windows used by all applications. It keeps track of their (x, y) positions and sizes, and also their front-to-back order, which is referred to as a z coordinate. As windows are moved and their z order changes, parts of them are exposed and need to be redrawn. For each window, the window server maintains an **invalid region**. When part of a window is invalid, the window server creates a redraw event, which is sent to the window's owning application so that the application can redraw it.

Every application is a client of the window server (we'll be describing the client-server framework in detail in Chapter 18). Happily, though, it's not necessary to understand the client-server framework in enormous detail for basic GUI programming because the client interface is encapsulated by CONE.

CONE associates one or more controls with each window and handles window server events. For instance, it handles a redraw event by calling the Draw() function for all controls that use the window indicated and fall within the bounding rectangle of the invalid region.

11.5.2   Window-owning and Lodger Controls

I introduced you to the concept of controls at the start of this chapter. There are two types of control:

- A control that requires a whole window is called a **window-owning control**.
- A control that requires only *part* of a window, on the other hand, is a **lodger control** or (more clumsily) a non-window-owning control.

Consider the dialog in Figure 11.10:

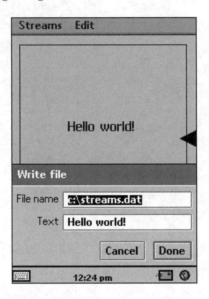

Figure 11.10

It has a single window, but 12 controls, as shown in Figure 11.11.

Advantages of lodgers

Although a window can have many controls, a control has only one window. Every control, whether it is a window-owning control or a lodger, ultimately occupies a rectangle on just one window, and the control draws to that rectangle on that window. A control's window is available via the `Window()` function in `CCoeControl`. There are certain advantages in using lodgers:

- *Reduced traffic*: Lodgers vastly reduce the client-server traffic between an application and the window server. Only one client-server message is needed to create an entire dialog since it includes only one

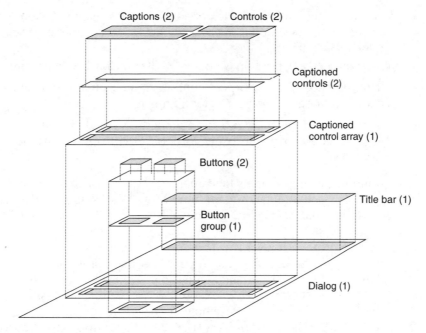

Figure 11.11

window. Only one event is needed to redraw the whole dialog, no matter how many of its controls are affected. Dialogs are created and destroyed frequently in application use, so these optimizations make a significant difference.

- *Reduced overheads*: Lodgers also reduce the overheads associated with complex entities such as a dialog because controls are much more compact in memory than windows.

- *Less processing*: Lodgers have less demanding processing requirements. Windows may move, change *z* order, and overlap arbitrarily. Lodgers at peer level on the same window never intersect and they only occupy a subregion of their owning window or control. This makes the logic for detecting intersections much easier than that required for the arbitrarily complex regions managed by the window server.

When you need a window

All these factors improve the system efficiency of Symbian OS, compared to a scenario with no lodger controls. In order to take advantage of these features, most controls should be coded as lodgers, but there are a few circumstances in which you *need* a window:

- When there is no window to lodge in – this is the case for the application view.

- When you need shadows, as described later in this chapter. Shadows are used by dialogs, popup menus, popup list-boxes, menu panes, and the menu bar.

- When you need a backed-up window – we'll come back to these later.

- When you need to overlap peer controls in an arbitrary way – not according to lodger controls' stricter nesting rules.

- When you need the backup-behind property (see below), which is used by dialogs and menu panes to hold a bitmap of the window *behind* them.

Being window-owning is a fairly fundamental property of a control. There isn't much point in coding a control bimodally – that is, to be *either* a lodger *or* to be window-owning. Decide which it should be and commit to it.

On the other hand, only small parts of your control's code will be affected by the decision. So, if you find out later that (for instance) your control that was previously a stand-alone app view now has a window to lodge in, then you should be able to modify your control quite easily.

For instance, in the `drawing` example in Chapter 15, the `CExample-HelloControl` class adapts `hellogui`'s `CHelloGuiAppView` to turn it into a lodger. The class declaration changes from,

```
class CHelloGuiAppView : public CCoeControl
    {
public:
    static CHelloGuiAppView* NewL(const TRect& aRect);
    ~CHelloGuiAppView();
    void ConstructL(const TRect& /*aRect*/);
private:
    void Draw(const TRect& /* aRect */) const;
private:
    HBufC* iHelloText;
    };
```

to:

```
class CExampleHelloControl : public CCoeControl
    {
public:
    static CExampleHelloControl* NewL(const CCoeControl& aContainer,
                                      const TRect& aRect);
    ~CExampleHelloControl();
. . .

private:
    void ConstructL(const CCoeControl& aContainer, const TRect& aRect);
```

```
private: // From CCoeControl
    void Draw(const TRect&) const;
private:
    HBufC* iText;
. . .
    };
```

The essential change here is that I have to pass a `CCoeControl&` parameter to the control to tell it which `CCoeControl` to lodge in.

The construction changes from

```
void CHelloGuiAppView::ConstructL(const TRect& aRect)
    {
    CreateWindowL();
    SetRectL(aRect);
    ActivateL();
    iHelloWorld = iEikonEnv->AllocReadResourceL(R_HELLOGUI_TEXT_HELLO);
    }
```

to

```
void CExampleHelloControl::ConstructL(const CCoeControl& aContainer,
                                      const TRect& aRect)
    {
    SetContainerWindowL(aContainer);
    SetRect(aRect);
    iView=CExampleHelloView::NewL();
    iText=iEikonEnv->AllocReadResourceL(R_EXAMPLE_TEXT_HELLO_WORLD);
    iView->SetTextL(*iText);
    . . .
    ActivateL();
    }
```

Instead of calling `CreateWindowL()` to create a window of the right size, I call `SetContainerWindowL()` to register myself as a lodger of a control on an existing window.

11.5.3 Compound Controls

There needs to be some structure in laying out lodger controls such as those in the Battleships Start first game dialog, or indeed in the Battleships app view. That discipline is obtained by using **compound controls**: a control is compound if it has one or more **component controls** in addition to itself.

- A component control is contained entirely within the area of its owning control.

- All components of a control must have nonoverlapping rectangles.
- A component control does not have to be a lodger, it can also be window-owning. In the majority of cases, however, a component control is a lodger.

To indicate ownership of component controls to CONE's framework, a compound control must implement two virtual functions from `CCoeControl`:

- `CountComponentControls()` indicates how many components a control has – by default, it has zero, but you can override this.
- `ComponentControl()` returns the *n*th component, with *n* from zero to the count of components minus one. By default, this function panics (because it should never get called at all if there are zero components). If you override `CountComponentControls()`, you should also override this function to return a component for each possible value of *n*.

Here is a generic example implementation of these functions. Most Symbian OS applications use enums for their controls like

```
enum
        {
        EMyFirstControl,
        EMySeconfControl,
        EAmountOfControls
        }
```

This enables you to simply return `EAmountOfControls` in the `CountComponentControls`. This ensures that you do not forget to change your return value when you add or remove controls over time:

```
TInt anyExampleAppView::CountComponentControls() const
    {
    return EAmountOfControls;
    }

CCoeControl* anyExampleAppView::ComponentControl(TInt aIndex) const
    {
    switch (aIndex)
        {
    case 0: return EMyFirstControl;
    case 1: return EMySeconfControl;
    case 2: return EAmountOfControls;
        }
    return 0;
    }
```

A dialog is also a compound control with typically only a single window. A dialog has an unpredictable number of component controls, so instead of hardcoding the answers to CountComponentControls() and ComponentControl() as I did above, CEikDialog uses a variable-sized array to store dialog lines and calculates the answers for these functions.

11.5.4 More on Drawing

Drawing to a window is easy for programs but involves complex processing by Symbian OS as you can see in Figure 11.12:

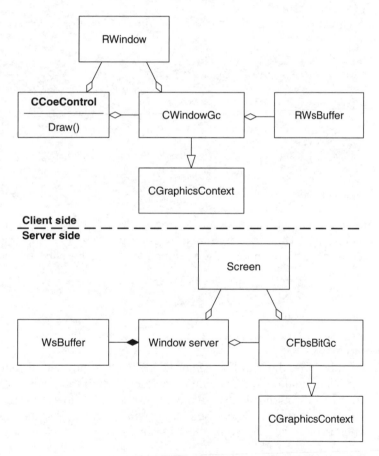

Figure 11.12

On the client side, an application uses a CWindowGc to draw to a window. CWindowGc's functions are implemented by encoding and storing commands in the window server's client-side buffer. When the buffer is full, or when the client requests it, the instructions in the buffer are all sent to the window server, which decodes and executes them by drawing directly onto the screen, using a `CFbsBitGc` – a `CGraphicsContext`-derived class for drawing onto bitmapped devices. Prior to drawing, the window server sets up a clipping region to ensure that only the correct region of the correct window can be changed, whatever the current state of overlapping windows on the screen. The window server uses the BITGDI to 'rasterize' the drawing commands.

The client-side buffer, which wraps several window server commands into a single client-server transaction, significantly speeds up system graphics performance.

We can now explain the `DrawTilesNow()` function that we saw earlier:

```
void CFleetView::DrawTilesNow() const
    {
    Window().Invalidate(iSeaArea);
    ActivateGc();
    Window().BeginRedraw(iSeaArea);
    DrawTiles();
    Window().EndRedraw();
    DeactivateGc();
    }
```

This is a member function of `CFleetView`, which is derived from `CCoeControl`. The central function is `DrawTiles()`, but this is bracketed by actions necessary to function correctly with the window server.

Invalidating

First, we use an `Invalidate()` function to invalidate the region we are about to draw.

Remember that the window server keeps track of all invalid regions on each window and clips drawing to the total invalid region. So before you do an application-initiated redraw, you must invalidate the region you are about to redraw, otherwise nothing will appear (unless the region happened to be invalid for some other reason).

Activating the graphics context

Then, CONE's system graphics context must be activated. If you take a look at `coecntrl.cpp`, you'll find that `CCoeControl::Activate-Gc()` is coded as:

```
EXPORT_C void CCoeControl::ActivateGc() const
    {
    CWindowGc& gc = iCoeEnv->SystemGc();
    if(iContext)
        iContext->ActivateContext(gc, *iWin);
    else
        gc.Activate(*iWin);
    }
```

The usual case just executes gc.Activate() on the control's window, telling the window server's client interface to use CONE's system GC to start drawing to it. The function also resets the window GC to use default settings. I'll explain the other case later on.

Beginning and ending the redraw

Immediately before drawing, we tell the window server we are about to begin redrawing a particular region. And immediately after redrawing, we tell the window server that we have finished. When the BeginRedraw() function is executed by the window server, it has two effects:

- The window server sets a clipping region to the intersection of the invalid region, the region specified by BeginRedraw(), and the region of the window that is visible on the screen.

- The window server then marks the region specified by BeginRedraw() as *valid* (or, more accurately, it subtracts the begin-redraw region from its current invalid region).

The application's draw code must then cover every pixel of the region specified by BeginRedraw(). If the application's draw code includes an explicit call to SetClippingRegion(), the region so specified is intersected with the clipping region calculated at BeginRedraw() time.

When the application has finished redrawing, it calls EndRedraw(). This enables the window server to delete the region object that it allocated during BeginRedraw() processing.

Concurrency

You're probably wondering why the window server marks the region as valid at *begin* redraw time rather than *end* redraw. The reason is that Symbian OS is a multitasking operating system. The following theoretical sequence of events shows why this protocol is needed:

- Application A issues begin-redraw. The affected region is marked valid on A's window.
- A starts drawing.

- Application B comes to the foreground, and its window overwrites A's.
- B is terminated, so that A's window is again exposed.
- Clearly, A's window is now invalid. The window server marks it as such.
- A continues redrawing, and issues end-redraw.

At the end of this sequence, the region of the screen covered by the reexposed region of A's window is in an arbitrary state. If the window server had marked A's window as valid at end-redraw time, the window server would not know that it still needs to be redrawn. Instead, the window server marks A's window as valid at begin-redraw time so that, by the end of a sequence like this, the window is correctly marked invalid and can be redrawn.

You might think this sequence of events would be rare, but it *is* possible, so the system has to address it properly.

Redrawing

You should now find it pretty easy to understand how redrawing works. When the window server knows that a region of a window is invalid, it sends a redraw message to the window's owning application, specifying the bounding rectangle of the invalid region. This is picked up by CONE and handled using the following code:

```
EXPORT_C void CCoeControl::HandleRedrawEvent(const TRect& aRect) const
    {
    ActivateGc();
    Window().BeginRedraw(aRect);
    Draw(aRect);
    DrawComponents(aRect);
    Window().EndRedraw();
    DeactivateGc();
    }
```

This code has exact parallels to the code we saw in `DrawTilesNow()`: the activate and begin-redraw brackets are needed to set everything up correctly. However, CONE doesn't need to call `Invalidate()` here because the whole point of the redraw is that a region is already known to be invalid. In fact, if CONE did call `Invalidate()` on the rectangle, it would potentially extend the invalid region, which would waste processing time.

Inside the activate and begin-redraw brackets, CONE draws the control using `Draw()` and passing the bounding rectangle. Then, CONE draws every component owned by this control using `DrawComponents()`, which is coded as follows:

```
void CCoeControl::DrawComponents(const TRect& aRect) const
    {
    const TInt count = CountComponentControls();
    for(TInt ii = 0; ii < count; ii++)
        {
        const CCoeControl* ctrl = ComponentControl(ii);
        if(!(ctrl->OwnsWindow()) && ctrl->IsVisible())
            {
            TRect rect;
            const TRect* pRect = (&aRect);
            if(!(((ctrl->Flags()) & ECanDrawOutsideRect))
                {
                rect = ctrl->Rect();
                rect.Intersection(aRect);
                if(rect.IsEmpty())
                    continue;
                pRect = (&rect);
                }
            ResetGc();
            ctrl->Draw(*pRect);
            ctrl->DrawComponents(*pRect);
            }
        }
    }
```

CONE simply redraws every visible lodger component whose rectangle intersects the invalid rectangle and then its components in turn. CONE adjusts the bounding invalid rectangle appropriately for each component control.

> *CONE also makes an allowance for a rare special case: controls that can potentially draw outside their own rectangle.*

Default settings are assured here: the original call to `ActivateGc()` set default settings for the window-owning control that was drawn first; later calls to `ResetGc()` ensure that components are drawn with default settings also.

The loop above *doesn't* need to draw window-owning components of the window-owning control that received the original redraw request. This is because the window server will send a redraw message to such controls in any case, in due time.

You can see again here how lodger components promote system efficiency. For each component that is a lodger (instead of a window-owning control), you avoid the client-server message and the 'activate' and 'begin-redraw' brackets. All you need is a single `ResetGc()`, which occupies a single byte in the window server's client-side buffer.

Support for flicker-free drawing

As an application programmer, you should be aware of two aspects of the window server that promote flicker-free drawing.

Firstly, the window server clips drawing down to the intersection of the invalid region and the begin-redraw region, so if your drawing code tends to flicker, the effect will be confined to the area being necessarily redrawn.

You can exploit this in some draw-now situations. Imagine that I wanted to implement a cursor-movement function, but didn't want to alter my `DrawTiles()` function. I could write a `DrawTwoTilesNow()` function that accepted the (x, y) coordinates of two tiles to be drawn, enabling me to calculate and invalidate only those two rectangles. I could then activate a GC and begin-redraw the whole tiled area, calling `DrawTiles()` to do so. The window server would clip drawing activity to the two tiles affected, eliminating flicker anywhere else. It's a poor man's flicker-free solution, but in some cases, it might just make the difference.

Secondly, the window server's client-side buffer provides useful flicker-free support. For a start, it improves overall system efficiency so that everything works faster and flickers are therefore shorter. Also, it causes drawing commands to be batched up and executed rapidly by the window server using the BITGDI and a constant clipping region. In practice, this means that some sequences of draw commands are executed so fast that, even if your coding flickers by nature, no one will ever see the problem, especially on high-persistence LCD displays. The key here is to confine sequences that cause flicker to only a few consecutive draw commands so that they all get executed as part of a single window server buffer.

Finally, and most obviously, the use of lodger controls helps here too because it means the window server buffer contains only a single `ResetGc()` command between controls, rather than a whole end bracket for redraw and GC deactivation, followed by a begin bracket for GC activation and redraw.

11.5.5   Backed-up Windows

In the window server, a standard window is represented by information about its position, size, visible region, and invalid region – and that's about all. In particular, no memory is set aside for the drawn content of the window, which is why the window server has to ask the application to redraw when a region is invalid.

But in some cases, it's impractical for the application to redraw the window, for instance, if it's:

- an old-style program that's not structured for event handling, and so can't redraw;
- an old-style program that's not structured in an MVC manner, has no model, and so can't redraw, even if it can handle events;

- a program that takes so long to redraw that it's desirable to avoid redraws if at all possible.

A program in an old-style interpreted language such as OPL is likely to suffer from all these problems.

In these cases, you can ask the window server to create a **backed-up window**; the window server creates a backup bitmap for the window and handles redraws from the backup bitmap without sending a redraw event to the client application.

The backup bitmap consumes more RAM than the object required to represent a standard window. If the system is running short on memory, it's more likely that creation of a backed-up window will fail, rather than creation of a standard window. If it does fail, the application will also fail, because requiring a backed-up window is a fairly fundamental property of a control. If you need backup, then you need it. If you can code proper redraw logic of sufficient performance, then you don't need backup.

Code that is designed for drawing to backed-up windows usually won't work with standard windows because standard windows require redraws, which code written for a backup window won't be able to handle.

On the other hand, code that is good for writing to a standard window is usually good for writing to a backed-up window; although the backed-up window won't call for redraws, there's no difference to the application-initiated draw code. The only technique that won't work for backed-up windows is to invalidate a window region in the hope of fielding a later redraw event – but this is a bad technique anyway.

Standard controls such as the controls Uikon offers to application programmers are usually lodger controls that are designed to work in standard windows. Such lodger controls will also work properly in backed-up windows, unless they use invalidation in the hope of fielding a later redraw. All Uikon stock controls are designed to work in both windows.

`CCoeControl`'s `DrawDeferred()` function works on a standard window by invalidating the window region corresponding to the control. This causes a later redraw event. On a backed-up window, this won't work, so in that case `DrawDeferred()` simply calls `DrawNow()`:

```
void CCoeControl::DrawDeferred() const
    {
    ...
    if(IsBackedUp())
        DrawNow();
    else
        Window().Invalidate(Rect());
    ...
    }
```

11.6   `CCoeControl`'s Support for Drawing

Now is a good time to summarize the drawing-related features of `CCoeControl` that we've seen so far.

First and foremost, a control is a rectangle that covers all or part of a window. All concrete controls are (ultimately) derived from the abstract base class `CCoeControl`. Various relationships exist between controls, other controls, and windows:

- A control can own a window, or be a lodger.

- A control may have zero or more component controls: a control's components should not overlap and should be contained entirely within the control's rectangle.

- A control is associated with precisely one window, whether as the window-owning control, or as a lodger.

- All lodgers are components of some control (ultimately, the component can be traced to a window-owning control).

- Component controls do not have to be lodgers; they can also be window-owning (say, for a small backed-up region).

Controls contain support for drawing, application-initiated redrawing, and system-initiated redrawing:

- Applications request controls to draw using the `DrawNow()` function.

- The window server causes controls to draw when a region of the control's window becomes invalid.

- In either case, `Draw()` is called to handle the drawing.

- Functions exist to provide access to a GC for use on the control's window, to activate and deactivate that GC and to reset it.

Here are the main functions and data members associated with the above requirements.

11.6.1   Control Environment

Each control contains a pointer to the control environment, which any control can reach by specifying `iCoeEnv` (protected) or `ControlEnv()` (public):

```
class CCoeControl : public CBase
    {
public:
    ...
    inline CCoeEnv* ControlEnv() const;
    ...
protected:
    CCoeEnv* iCoeEnv;
    ...
    };
```

There are four ways in which you can access the control environment:

- From a derived control or app UI class, including your own application's app UI, you can use iCoeEnv to get at the CcoeEnv.
- If you have a pointer to a control or app UI, you can use its public ControlEnv() function.
- If you have access to neither of these things, you can use the static function CCoeEnv::Static(), which uses thread-local storage (TLS) to find the current environment.
- Since TLS isn't particularly quick, you can also store a pointer somewhere in your object for faster access, if you need to do this frequently.

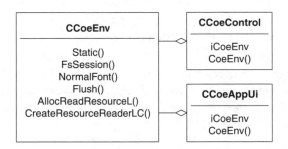

Figure 11.13

The control environment's facilities include the following:

- Access to the basic GUI resources: window server session, window group, screen device, and graphics context.
- A permanently available file server session, available via FsSession().
- A normal font for drawing to the screen (10-point Arial), available via NormalFont().

- A Flush() function to flush the window server buffer and optionally wait a short period.
- Convenience functions for creating new graphics contexts and fonts on the screen device.
- Support for multiple resource files and many functions to read resources (see Chapter 7).

See the definition of CCoeEnv in coemain.h for the full list.

11.6.2　Window-owning and Lodging

A control may be either window-owning or a lodger. A window-owning control *has-a* window: a lodger simply *uses-a* window (Figure 11.14).

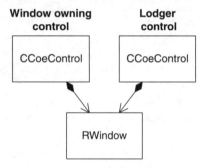

Figure 11.14

Either way, throughout the lifetime of a control, an iWin member points to a drawable window. The drawable window may be either standard (RWindow) or backed-up (RBackedUpWindow) – RDrawableWindow is a base class for both these.

You can call a CCoeControl function from those listed below during the second-phase constructor of a concrete control class to indicate whether it's window-owning or a lodger.

The functions for specifying and testing the window are:

```
class CCoeControl : public CBase
    {
public:
    ...
    IMPORT_C virtual void SetContainerWindowL(const CCoeControl&
      aContainer);
    IMPORT_C void SetContainerWindow(RWindow& aWindow);
    IMPORT_C void SetContainerWindow(RBackedUpWindow& aWindow);
    ...
```

```
    inline RDrawableWindow* DrawableWindow() const;
    ...
    IMPORT_C TBool OwnsWindow() const;
    IMPORT_C TBool IsBackedUp() const;
    ...
protected:
    ...
    inline RWindow& Window() const;
    inline RBackedUpWindow& BackedUpWindow() const;
    IMPORT_C void CloseWindow();
    IMPORT_C void CreateWindowL();
    IMPORT_C void CreateWindowL(const CCoeControl* aParent);
    IMPORT_C void CreateWindowL(RWindowTreeNode& aParent);
    IMPORT_C void CreateWindowL(RWindowGroup* aParent);
    IMPORT_C void CreateBackedUpWindowL(RWindowTreeNode& aParent);
    IMPORT_C void CreateBackedUpWindowL(RWindowTreeNode& aParent,
                                        TDisplayMode aDisplayMode);
    ...
protected:
    CCoeEnv* iCoeEnv;
    ...
private:
    RDrawableWindow* iWin;
    ...
    };
```

The `CreateWindowL()` functions cause a new window – either standard or backed-up – to be created.

The `SetContainerWindow()` functions tell the control to use an existing standard window or backed-up window. This should be used by controls that are themselves components of a control associated with the same window. `SetContainerWindowL()` tells the control to lodge in an existing control – and hence, ultimately, to use an existing window.

This function is both virtual and potentially leaving. That's not the best design in Symbian OS: really, it should be neither. You can guarantee that this function won't leave if it's not overridden, so try to think of this function as not having been designed to be overridden. A few classes in Uikon use it for purposes that could be achieved by other means.

11.6.3 Components

A control can have any number of component controls, from zero upwards. Here are the component-control functions:

```
class CCoeControl : public CBase
    {
public:
    ...
```

```
IMPORT_C TInt Index(const CCoeControl* aControl) const;
...
IMPORT_C virtual TInt CountComponentControls() const;
IMPORT_C virtual CCoeControl* ComponentControl(TInt aIndex) const;
...
};
```

If you want to implement a container control, you can store controls and use any data structure you want. You override `CountComponent-Controls()` to indicate how many controls you have and `Compo-nentControl()` to return the control corresponding to each index value, starting from zero.

As we saw earlier, by default, `CountComponentControls()` returns zero, and `ComponentControl()` panics. These functions work as a pair, so make sure you override them both consistently.

`Index()` searches through the component controls one by one to find one whose address matches the address passed. If none is found, `Index()` returns `KErrNotFound`, which is defined as −1.

The `CCoeControl` base class does not dictate how component controls should be stored in a container.

If your container is a fixed-purpose container such as the Battleships application view, which contains just three components, then you can use a pointer to address each component, hardcode `Count-ComponentControls()` to return 3, and use a switch statement in `ComponentControl()`.

On the other hand, if your container is a general-purpose container such as a dialog, you may wish to implement a general-purpose array to hold your component controls.

11.6.4   Position and Size

You can set a control's position and size. Here are the declarations related to position and size:

```
class CCoeControl : public CBase
    { m
public:
    ...
    IMPORT_C void SetExtentL(const TPoint& aPosition, const TSize& aSize);
    IMPORT_C void SetSizeL(const TSize& aSize);
```

```
    IMPORT_C void SetPosition(const TPoint& aPosition);
    IMPORT_C void SetRectL(const TRect& aRect);
    IMPORT_C void SetExtentToWholeScreenL();
    ...
    IMPORT_C TSize Size() const;
    IMPORT_C TPoint Position() const;
    IMPORT_C TRect Rect() const;
    IMPORT_C TPoint PositionRelativeToScreen() const;
    ...
    IMPORT_C virtual void SizeChangedL();
    IMPORT_C virtual void PositionChanged();

    IMPORT_C void SetCornerAndSizeL(TCoeAlignment aCorner, const TSize&
        aSize);
    IMPORT_C void SetSizeWithoutNotificationL(const TSize& aSize);
    ...
protected:
    ...
    TPoint iPosition;
    TSize iSize;
    ...
    };
```

Position and size are stored in iPosition and iSize. You can interrogate them with Position(), Size(), or Rect() and change them with SetExtentL(), SetPosition(), SetSizeL(), and SetRectL().

Changing the size of a control could, in rare cases, cause memory to be allocated, which could fail – so all functions that change size are potentially leaving. SetPosition() does not change size so it cannot leave.

- When a control's size is changed, its virtual SizeChangedL() function is called.

- A position change is notified by PositionChanged().

- SetExtentL() calls SizeChangedL() but not Position-Changed() – so think of SizeChangedL() as always notifying size change, and potentially notifying position change.

- You can use SetSizeWithoutNotificationL() to prevent SizeChangedL() being called.

- You can set and interrogate position relative to the owning window and set the size to the whole screen. SetCornerAndSizeL() aligns a control's rectangle to one corner of the whole screen.

Merely resizing a control should not cause extra resources to be allocated, except in the rare kinds of control which might need to allocate resources in Draw(). In this case, you should take the same action: trap any leaves yourself.

11.6.5   Drawing

Functions relevant for drawing include:

```
class CCoeControl : public CBase
    {
public:
    ...
    IMPORT_C virtual void MakeVisible(TBool aVisible);
    ...
    IMPORT_C virtual void ActivateL();
    ...
    IMPORT_C void DrawNow() const;
    IMPORT_C void DrawDeferred() const;
    ...
    IMPORT_C TBool IsVisible() const;
    ...
protected:
    ...
    IMPORT_C void SetBlank();
    ...
    IMPORT_C CWindowGc& SystemGc() const;
    IMPORT_C void ActivateGc() const;
    IMPORT_C void ResetGc() const;
    IMPORT_C void DeactivateGc() const;
    IMPORT_C TBool IsReadyToDraw() const;
    IMPORT_C TBool IsActivated() const;
    IMPORT_C TBool IsBlank() const;
    ...
private:
    ...
    IMPORT_C virtual void Draw(const TRect& aRect) const;
    ...
    };
```

Use the functions as follows:

- You have to activate a control using ActivateL() as the final part of its second-phase construction. Assuming that by the time ActivateL() is called, the control's extent is in place and its model is fully initialized makes the control ready for drawing. You can use IsActivated() to test whether ActivateL() has been called.

- You can set a control to be visible or not – Draw() is not called for invisible controls.

- IsReadyToDraw() returns ETrue if the control is both activated and visible.

- SetBlank() is an obscure function that only affects controls that don't override Draw(). If you don't SetBlank(), then CCoeControl::Draw() does nothing. If you *do* SetBlank(), then CCoeControl::Draw() blanks the control.

- We have already seen that `Draw()` is the fundamental drawing function. `DrawNow()` initiates the correct drawing sequence to draw a control and all its components.

- `DrawDeferred()` simply invalidates the control's extent so that the window server will send a redraw message, causing a redraw later. This guarantees that a redraw will be called on the control at the earliest available opportunity, rather than forcing it now.

- `ActivateL()`, `MakeVisible()`, and `DrawNow()` recurse as appropriate through component controls.

- `SystemGc()` returns a windowed GC for drawing. `ActivateGc()`, `ResetGc()`, and `DeactivateGc()` perform the GC preparation functions needed for redrawing.

> **Always use these functions, rather than directly calling `SystemGc.Activate(Window())`. It's more convenient and it allows control contexts to be supported properly.**

11.7 Special Effects

The window server provides many useful special effects to application programs. These include

- Shadows
- Backed-up-behind Windows
- Animation
- Use of debug keys
- Using a control context
- Scrolling.

We'll examine each of these in turn.

> **Note that the availability of these special effects depends on the implementation that you are using. UIQ for example, does not use shadows; instead it fades the background.**

11.7.1 Shadows

Shadows can be used in many circumstances – behind dialogs, behind menus, behind popup choice lists and so on. Not all Symbian OS implementations use shadows. UIQ, for example, fades the background behind a window and so does not implement shadows.

You have to specify that you want a window to cast a shadow and say how 'high' the window is. The shadow actually falls on the window(s) behind the one that you specify to cast shadows. To implement a shadow when it is cast, the window server asks the BITGDI to dim the region affected by the shadow. To maintain a shadow even when the window redraws, the window server executes the application's redraw command buffer twice:

- Firstly, it uses a clipping region that *excludes* the shadowed part of the window and uses the BITGDI to draw using normal color mapping.

- Secondly, it uses a clipping region for only the shadowed parts, and puts the GC it uses for BITGDI drawing into shadow mode.

This causes the BITGDI to 'darken' all colors used for drawing in the affected region. The net result is that the drawing appears with shadows very nicely – without affecting the application's drawing code at all.

Shadows are implemented in dialogs and the like by calling the `AddWindowShadow()` function in `CEikonEnv`:

```
void CEikonEnv::AddWindowShadow(CCoeControl* aWinArea)
  {
  aWinArea->DrawableWindow()->SetShadowHeight(LafEnv::ShadowHeight());
  }
```

Shadow height is therefore dependent on the GUI customization.

11.7.2 Backing Up Behind

Backed-up-behind windows maintain a copy of the window *behind* them, so that when the backed-up-behind window is dismissed, the window behind can be redrawn by the window server without invoking application redraw code.

This effect is used for menu panes: it speeds up the process of flicking from one menu pane to another immensely. It's also used for dialogs, where it speeds up dialog dismissal.

When a backed-up-behind window is created, a big enough bitmap is allocated to backup the entire screen area that the window *and its shadow* are about to cover. The screen region is copied into this backup bitmap and then the window is displayed. When the window is dismissed, the backup bitmap is copied back onto the screen. The net effect is that the application doesn't have to redraw at all so that window dismissal is very quick.

The backup-behind code is clever enough to update the backup bitmap when, for instance, a dialog is moved around the screen.

The backup-behind code, however, is not an essential property of the window, but an optimization. So it gives up when it runs out of memory, when the window behind tries to redraw, or when another window with backup-behind property is placed in front of the existing one. Then an application redraw is needed, after all. This doesn't have any effect on application code – just on performance.

11.7.3 Animation

Sometimes, you want to do some drawing where timing is an essential feature of the visual effect. This isn't really something that fits well into the MVC paradigm, so we need special support for it.

One kind of animation is used to give reassuring cues in the GUI.

- If you select OK on a dialog – even using the keyboard – the OK button goes down for a short time (0.2 s, in fact) and then the action takes place
- If you select a menu item using the pointer, the item appears to flash briefly, in a subtle way, before the menu bar disappears and your command executes.

In both cases, this animation reassures you that what you selected actually happened. Or, just as importantly, it alerts you to the possibility that something you *didn't* intend actually happened. Either way, animation is an extremely important cue; without it, the Symbian OS GUI would feel less easy to use.

> As it happens, animation isn't the only potential clue that something happened Sound, such as key or digitizer clicks, can be useful too.

This animation is achieved very simply. In the case of the dialog button animation, for instance,

- draw commands are issued to draw the button in a 'down' state;
- the window server's client-side buffer is flushed so that the draw commands are executed;
- the application waits for 0.2 s;
- the dialog is dismissed and the relevant action takes place – in all probability, this will cause more drawing to occur.

The key point here is the flush-and-wait sequence. CONE provides a function to implement this, `CCoeEnv::Flush()`, which takes a time interval specified in microseconds. The following code, therefore, implements the flush and the wait:

```
iCoeEnv->Flush(200000);
```

The flush is vital. Without it, the window server might not execute your draw commands until the active scheduler is next called to wait for a new event – in other words, until the processing of the current key has finished. By that time, the dialog will have been dismissed so that your draw commands will execute 'flicker-free' – just at the point when some flicker would have been useful!

Don't use this command to wait for longer than about 0.2 s; it will compromise the responsiveness of your application if you do so. The entire application thread, with all its active objects, is suspended during this 0.2-s wait. If you need animation to occur on a longer timescale, use an active object to handle the animation task. See Chapter 17 for a detailed description of active objects.

Even animation using active objects isn't good enough for some purposes because active objects are scheduled non-preemptively and particularly in application code, there is no guarantee about how long an active object event handler function may take to run. Some animations have to keep running, or else they look silly. Examples include the flashing text cursor, the analog, or digital clocks on the bottom of your application button bar, or the Uikon busy message that appears when your program is running a particularly long event handler that (by definition) would prevent any other active object from running.

These animations run *as part of the window server* in a window server **animation DLL**. They are required to be good citizens of the window server and to have *very* short-running active objects.

11.7.4   Uikon Debug Keys

As we saw in Chapter 6, debug builds of Uikon (in other words, on the emulator) allow you to use various key combinations to control and test your application's use of memory. You can also use the following key combinations to control your program's drawing behavior:

Key	Effect
Ctrl+Alt+Shift+M	Creates a 'mover' window that you can move all around the screen, causing redraws underneath it.
Ctrl+Alt+Shift+R	Causes the entire application view to redraw.
Ctrl+Alt+Shift+F	Causes the window server client API to enable auto-flush, so that draw commands are sent to the window server as soon as they are issued, rather than waiting for the buffer to fill or for a program-issued `Flush()` function.

Try this and watch things slow down: it can be useful for naked-eye flicker testing.

It's also handy for debugging redraws. You can step through drawing code and see every command produce an instant result on the emulator. You'll need a big monitor to see both the emulator and Visual Studio at the same time – but that's a hardware problem!

Ctrl+Alt+Shift+G	Disables auto-flush.

Remember that these settings only apply to the current application's Uikon environment.

11.7.5 Control Context

The `ActivateGc()`, `DeactivateGc()`, and `ResetGc()` functions in a control normally pass directly through to window server functions: `gc.Activate(*iWin)`, `gc.Deactivate(*iWin)`, and `gc.Reset()`.

These functions and the way that CONE calls them when recursing through component controls, guarantee that your GC is properly reset to default values before each control's `Draw()` function is called.

In some cases, you don't want your GC to reset to system default values; instead, you want to set the default values to something decided by the control. For this purpose, you can use a **control context**, an interface that overrides GC activate and reset behavior. See the SDK and examples for further details.

11.7.6 Scrolling

The window server supports scrolling but once again, this feature may not be available in all UI implementations. If it is available, you can ask for a region of a window to be scrolled, up, down, left, or right. This results in

- some image data being lost,
- the window server moving some image data in the requested direction,
- an area of the window becoming invalid.

The (new) invalid area is exactly the same size as the amount of (old) image data that was lost.

Scrolling is clearly always application-initiated. After a scroll, you should `DrawXxxNow()` – you don't need to invalidate the area that was invalidated by the scroll. You must ensure that the new drawing precisely abuts onto the old data without any visible joins.

11.8   Summary

In this chapter, I've concentrated on graphics for display – how to draw and how to share the screen. More specifically, we've seen the following:

- The `CGraphicsContext` class and its API for drawing, getting position, and bounding rectangles.
- The MVC pattern, which is the natural paradigm for most Symbian OS applications. MVC updates and redraws work well with standard `RWindows`. For non-MVC programs, or programs whose updates are particularly complex, backed-up windows may be more useful instead.
- How the work is shared between controls and windows.
- The use of compound controls to simplify layout.
- The difference between window-owning and lodger controls.
- Application- and system-initiated drawing.
- Techniques for flicker-free drawing and how CONE helps you achieve this.
- The activate-GC and begin-redraw brackets in `DrawXxxNow()`-type functions.
- `CCoeControl`'s functions for drawing.
- How to animate, cast a shadow, scroll, and backup behind.

Graphics also provides the fundamental mechanism for user interaction, which is the subject of the next chapter.

12

Graphics for Interaction

In the last chapter, we saw the way that Symbian OS uses the MVC pattern, and how to draw in the context of controls and windows. In this chapter, I'll cover how to use controls to enable user interaction with your programs.

In theory, it's easy. Just as controls provide a virtual `Draw()` function for drawing, they also provide two virtual functions for handling interaction: `HandlePointerEventL()` for handling pointer events, and `OfferKeyEventL()` for handling key events. You simply work out the coordinates of the pointer event, or the key code for the key event, and decide what to do.

Basic interaction handling does indeed involve knowing how to use these two functions, and I'll start the chapter by describing them. I'll also show that interaction is well suited to MVC: you turn key and pointer events into commands to update the model, and then let the model handle redrawing. However, this description of key and pointer event handling is really just the tip of the interaction iceberg. There are plenty of other issues to deal with:

- Key events are normally associated with a cursor location or the more general concept of **focus** – but not all key events go to the current focus location, and some key events cause focus to be changed.

- A 'pointer-down' event away from current focus usually causes focus to be changed. Sometimes, though, it's not good to change focus – perhaps because the currently focused control is not in a **valid** state, or because the control in which the event occurred doesn't accept focus.

- Focus should be indicated by some visual effect (usually a cursor, or a highlight). Likewise, temporary inability to take focus should be

Symbian OS C++ for Mobile Phones.   Edited by Richard Harrison
© 2003 John Wiley & Sons, Ltd   ISBN: 0-470-85611-4

indicated by a visual effect (usually some kind of dimming, or even making a control invisible altogether).

- The rules governing focus and focus transition should be easy to explain to users – better still, they shouldn't need to be explained at all. They should also be easy to explain to programmers.

The big issues in interaction are intimately related: namely drawing, pointer handling, key handling, model updates, component controls, focus, validity, temporary unavailability, ease of use, and ease of programming. During early development of Symbian OS, we rewrote the GUI and the control environment twice to get these issues right – the first time for the user, the second time for the programmers.

Explaining these things is only slightly easier than designing them. Here goes, as smoothly as I can.

12.1 Key, Pointer, and Command Basics

As mentioned above, the basic functions that deal with interaction are `HandlePointerEventL()` and `OfferKeyEventL()`. To describe how they work, I'll start with the latter's implementation in `COppFleetView`, the interactive heart of the Battleships app view.

12.1.1 Handling Key Events

Here's how `COppFleetView` handles key events generated via the keypad:

```
TKeyResponse COppFleetView::OfferKeyEventL(const TKeyEvent& aKeyEvent,
                                           TEventCode aType)
    {
    if(aType!=EEventKey)
        return EKeyWasNotConsumed;
    if(aKeyEvent.iCode   ==  EQuartzKeyFourWayLeft ||
       aKeyEvent.iCode   ==  EQuartzKeyFourWayRight ||
       aKeyEvent.iCode   ==  EQuartzKeyFourWayUp ||
       aKeyEvent.iCode   ==  EQuartzKeyFourWayDown)
        {
        //Move cursor
        if(aKeyEvent.iCode == EQuartzKeyFourWayLeft)
            MoveCursor(-1, 0);
        else if(aKeyEvent.iCode == EQuartzKeyFourWayRight)
            MoveCursor(1, 0);
        else if(aKeyEvent.iCode == EQuartzKeyFourWayUp)
            MoveCursor(0, -1);
        else if(aKeyEvent.iCode == EQuartzKeyFourWayDown)
            MoveCursor(0, 1);
```

```
        // Redraw board
        DrawTilesNow();
        return EKeyWasConsumed;
        }
    else if(aKeyEvent.iCode == EQuartzKeyConfirm)
        {
        if (iFleet.IsKnown(iCursorX, iCursorY))
            iEikonEnv->InfoMsg(R_GAME_ALREADY_KNOWN);
        else
            iData.iCmdHandler.ViewCmdHitFleet(iCursorX, iCursorY);
        return EKeyWasConsumed;
        }
    return EKeyWasNotConsumed;
    }
```

OfferKeyEventL() is called on a control if the framework thinks that it ought to be offered the chance of handling a key event (by the end of this chapter you'll know what that means). If necessary, a control can indicate that it doesn't take keys; I do this elsewhere in the Battleships GUI using methods that I'll explain later in this chapter.

We should note the following points about the function:

- I don't have to check whether I should have been offered the key event; I can rely on the framework offering it to me only if I am supposed to have it.

- I don't have to 'consume' the key. I return a value (EKeyWas-Consumed or EKeyWasNotConsumed) to indicate whether I consumed it.

The function starts by determining the kind of key event. There are three possibilities – hardware key down, standard key, or hardware key up. Here, I'm only interested in standard key events indicated by the value EEventKey being passed in the second parameter of the function. I can handle the key in one of the three ways:

- I can ignore it if it's not a key that I recognize or want to use at this time.

- I can handle it entirely internally, within the control, by (say) moving the cursor, or (if it were a numeric editor) changing the internal value and visual representation of the number being edited by the control.

- I can generate some kind of command that will be handled outside the control, like the hit-fleet command that's generated when I press the *Confirm* key when on a new tile, or (if this were a choice list item in a dialog) an event saying that the value displayed in the choice list had been changed.

For issuing commands from `COppFleetView`, I call `iData.iCmdHandler.ViewCmdHitFleet()`, specifying the coordinates of the tile to hit. `iCmdHandler` is of type `MGameViewCmdHandler&`, an interface class that I implement in `CGameController`. I'll show the definition of the interface below.

Those are the main issues in key event handling. But you also need to know some of the details about how to crack key events and key codes. The first parameter to `OfferKeyEventL()` is of type `TKeyEvent&`, a reference to a window server key event that's defined in `w32std.h`:

```
struct TKeyEvent
    {
    TUint iCode;
    TInt iScanCode;
    TUint iModifiers;    // State of modifier keys and pointing device
    TInt iRepeats;       // Count of auto repeats generated
    };
```

The legal values for `iCode`, `iScanCode`, and `iModifiers` are in `e32keys.h`.

- Use values from the `TKeyCode` enumeration to interpret `iCode`. The values are `<0x20` for nonprinting Unicode characters, `0x20-0xf7ff` for printing Unicode characters, and `>=0xf800` for the usual function, arrow, menu, and other keys found on PC keyboards. Although Symbian OS phones may have no keyboard, many of these key events may still be generated using a front end processor (**FEP**).

- Use `TStdScanCode` and `iScanCode` for the extremely rare cases when scan codes are of interest. The scan codes are defined by the historical evolution of IBM PC keyboards since 1981. They originally represented the physical position of a key on the 81-key IBM PC keyboard, and have evolved since then to support new keyboards and preserve compatibility with older ones. Since keyboards and keypads used on Symbian OS phones are rather different, scan codes have limited value.

- Use `TEventModifier` to interpret the bits in `iModifiers`. This enables you to test explicitly for *Shift, Ctrl,* and other modifier keys, where supported. These are of particular interest in combination with navigation keys.

UIQ defines some additional key codes in `quartzkeys.h`. These represent the events generated by the two or four direction keys on the keypad, and the *Confirm* key – they are the key events that are consumed by `COppFleetView::OfferKeyEventL()`, above.

TEventCode is the type of window server event that is being handled; it can be one of EEventKey, EEventKeyDown, or EEventKeyUp. For ordinary event handling, most controls are interested in EEventKey.

If key down *is* relevant to your application, then the auto repeat count in the TKeyEvent might also be of interest. It tells you how many auto repeats you have missed since you handled the last key event.

12.1.2 Handling Pointer Events

Here's how COppFleetView handles pointer events:

```
void COppFleetView::HandlePointerEventL(const TPointerEvent&
    aPointerEvent)
    {
    // Check whether we're interested
    if(aPointerEvent.iType == TPointerEvent::EButton1Down &&
        iData.iSeaArea.Contains(aPointerEvent.iPosition))
        {
        // Identify the tile that was hit
        TInt x=(aPointerEvent.iPosition.iX-iData.iSeaArea.iTl.iX)/
            iData.iTileSize;
        TInt y=(aPointerEvent.iPosition.iY-iData.iSeaArea.iTl.iY)/
            iData.iTileSize;
        // Move cursor if necessary
        TBool iCursorMoved=(x!=iCursorX || y!=iCursorY);
        SetCursor(x,y);
        // Hit square unless it's already known
        if (iFleet.IsKnown(x,y))
            {
            iEikonEnv->InfoMsg(R_GAME_ALREADY_KNOWN);
            if (iCursorMoved)
                DrawTilesNow();
            }
        else
            iData.iCmdHandler.ViewCmdHitFleet(x,y);
        }
    }
```

Like OfferKeyEventL(), HandlePointerEventL() is called on a control if the framework thinks the pointer event ought to be handled. Generally, that means the control is not dimmed or invisible, and it's the control with the smallest area that surrounds the coordinates of the pointer event. Unlike OfferKeyEventL(), HandlePointerEventL() requires the control to handle the pointer event or ignore it. The event is not offered to be consumed optionally; it will not be offered to any other control. That distinction is reflected in both the name (HandlePointerEventL()) and the return type (void).

As with keys, I can choose to handle the pointer event in one of three ways:

- *I can ignore it*: I do this if it's not a pointer-down event, or if the event occurs outside the sea area on the game board.

- *I can handle it internally, in this case by moving the cursor*: I do this here if the pointer event happens in a tile that has already been uncovered.

- *I can generate a command to be handled outside the control*: I generate a hit-fleet command if the pointer event occurs in a tile that hasn't already been uncovered.

I'm using the pointer events to generate exactly the same command as I was with the key events – a hit request. So, I handle these commands using the same MGameviewCmdHandler interface.

The TPointerEvent object passed to HandlePointerEventL() is defined in w32std.h as

```
struct TPointerEvent
    {
    enum TType
        {
        EButton1Down,
        EButton1Up,
        EButton2Down,
        EButton2Up,
        EButton3Down,
        EButton3Up,
        EDrag,
        EMove,
        EButtonRepeat,
        ESwitchOn,
        };
    TType iType;              // Type of pointer event
    TUint iModifiers;         // State of pointing device and associated
                                buttons
    TPoint iPosition;        // Window co-ordinates of mouse event
    TPoint iParentPosition;  // Position relative to parent window
    };
```

You can check for the following:

- Button 1 down (pen down).

- Button 1 up (pen up).

- *Other buttons*: you don't get these on pen-based devices.

- *Drag*: is rarely used in Symbian OS but can be useful.

- *Move*: you will *never* get this on pen-based devices. Devices supporting move also have to support a mouse cursor (which the window server does), control entry/exit notification, and changing the cursor to indicate what will happen if you press button 1. That's all hard work and adds little value to Symbian OS phones.

- *Button repeat*: this is generated when a button is held down continuously in one place – it's just like keyboard repeat. It's most useful for controls such as scroll arrows or emulated on-screen keyboards.

- *Switch-on*: some phones can be switched on by a tap on the screen.

- *Modifiers*: state of the pointer event modifiers, including *Shift, Ctrl,* and *Alt* keys (where available) – using the same values as for key event modifiers. For instance in a text editor, *Shift* and tap is used to make or extend a selection.

12.1.3 Turning Events into Commands

Whether I interact with the control using the keypad or the pointer, I ultimately generate commands that should be handled by a different part of the program. For getting commands from one to the other, I use an interface class. Here it is:

```
class MGameViewCmdHandler
    {
public:
    virtual void ViewCmdHitFleet(TInt aX, TInt aY) = 0;
    };
```

```
class CFleetViewData : public CBase
    {
public:
    ...
    CFleetViewData(CSoundEffects& aSoundEffects,
        MGameViewCmdHandler& aCmdHandler, const TRect& aRect);
    ...
private:
    // Command handler
    MGameViewCmdHandler& iCmdHandler;
    ...
    };
```

```
class CFleetView : public CCoeControl, public MCoeView
    {
    ...
protected:
    CFleetView(CFleetViewData& aFleetViewData, TFleet& aFleet);
    ...
    CFleetViewData& iData;
    TFleet& iFleet;
    ...
    };
```

```
class COppFleetView : public CFleetView
    {
public:
    static COppFleetView* NewL(CFleetViewData& aFleetViewData,
        TFleet& aFleet, TBool aP1);
private:
    // construct
    COppFleetView(CFleetViewData& aFleetViewData,
        TFleet& aFleet, TBool aP1);
    // from CCoeControl
    void HandlePointerEventL(const TPointerEvent& aPointerEvent);
    TKeyResponse OfferKeyEventL(const TKeyEvent& aKeyEvent,TEventCode
        aType);
    };
```

And here it is in Figure 12.1:

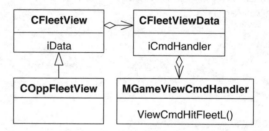

Figure 12.1

The interface class is a useful way of encapsulating the interaction that's happening between the control and the rest of the program. In the MVC sense, the rest of the program is the controller; the control doesn't care about its internal details provided that it handles the hit-fleet command correctly. Likewise, the controller doesn't care whether the command originated with a key event, a pointer event, or some other kind of event – it just wants to get called at the right time with the right parameters.

This pattern is repeated throughout Uikon and it's a good one to adopt for your programs. As a further illustration, think about the way commands reach HandleCommandL(). The menu bar uses an MEik-MenuObserver for this, and doesn't otherwise care about the large API of the CEikAppUi class. The button bar uses the MEikCommandObserver interface. Similarly, command buttons use an MCoeControlObserver interface, which the button bar implements by converting button events into application commands.

12.2   Interaction in Dialogs

From the preceding discussion, it should be apparent that it's not handling pointer and key events that makes programming an interactive graphics framework complex. Rather, it's handling the relationships between different parts of the program and the visual feedback you give in response that can cause difficulty.

The dialogs in Figure 12.2, from the UIQ control panel, show many of the issues involved. By studying them, we'll get some way of understanding how all of the separate aspects of interaction relate to one another. That will help you not only to make better use of the dialog framework but also to write good app views and to understand many aspects of the CCoeControl API.

Figure 12.2

In this dialog, the first line is an edit window, below which is a secret editor, another edit window, a choice list, and a numeric editor.

12.2.1   User Requirements

Here are some user requirements, which I'll express in nontechnical terms. (This is what you're supposed to do with user requirements: it helps you to see things clearly, and besides, users never give their requirements in technical terms.)

Firstly, the user needs to understand what's going on. They can, because

- the dialog in which the action is taking place is in front of everything else and the window that was previously in the foreground is now faded (you can see fading in Figure 12.3),
- the field in which the action is taking place is highlighted.

Secondly, the user should not be able to enter invalid data through the dialog. That implies the following:

- If the user presses Done, when the value in the numeric editor is outside the valid range, the dialog should complain that the value is invalid, and should not cause the usual action associated with Done (in this case, saving the e-mail account settings) to happen. Instead, the dialog should stay active, with the highlight on the Smaller than field, resetting it to the nearest valid value.
- If the user tries to use the pointer to select a different dialog line from the one containing the invalid data, the dialog should reset the value and notify the user. If the dialog allowed the current control to become invalid, there could be many invalid controls by the time the user pressed Done, and it would be difficult to know where to start correcting them.
- When an invalid field value is reset, the user should be notified.
- If the user presses Cancel, the dialog should disappear without any validation.

When the Schedule button is pressed, the following dialog appears:

Figure 12.3

The first line here is a checkbox that allows you to select whether to schedule downloads at all; the remaining lines are dimmed unless this is selected. The next two lines are a vertical option button list, followed by three check boxes, paired with time editors.

A third requirement is that the user should not be able to enter data that is inconsistent with other settings in the dialog. So

- if the user unchecks the Schedule download box, then the remaining fields in this dialog should be dimmed or may even disappear altogether, so that they can't be selected with the pointer;

- if the Set interval button is selected, the Set time button should be deselected, and the remainder of the dialog should change appropriately (in this case, the time editors are replaced by a choice list for selecting time intervals);

- if the user sets two Check for new messages at values to be the same, as in Figure 12.3, then the second value should be dimmed when the user moves away from the line on the dialog or presses Done. It makes no sense to schedule two downloads to take place at the same time.

12.2.2 Some Basic Abstractions

On the basis of this list of user requirements, you can see some abstractions beginning to take shape:

- The highlight (or in text editors, the cursor) is actually a visual indication of **focus**. The control that currently has focus receives the majority of key events; it therefore ought to know whether it has focus or not, and draw any necessary highlight accordingly. (Not all key events go to the control with focus: in dialogs, pressing the hardware *Confirm* key causes the default button – usually Done – to be pressed, rather than going to the highlighted control.)

- A control needs to be able to *refuse* interactions such as pointer events. **Invisible** or **dimmed** controls should certainly refuse interactions. A control should know whether it is dimmed and thereby draw itself in a suitable way.

- A control needs to be able to say whether it is in a **valid** state, and to respond to queries about its state.

- If a control's **state** changes, it needs to be able to report that to an **observer** such as the dialog, so that the dialog can handle any knock-on effects. (For example, if the checkbox changes to indicate no scheduled downloads, the dialog must make it impossible to select the other controls.)

These are the user requirements. Throughout the rest of this chapter, I'll be describing how the GUI framework makes it possible for you meet them.

12.2.3   Programmer Requirements

If ease of use matters to end users, it certainly matters to programmers, who will have some requirements on the way all the ideas raised above should hang together:

- It should be possible to invent new dialogs with rich functionality and to implement the validation rules with sufficient ease that programmers will want to use these facilities to deliver helpful and usable dialogs.

- It should be possible to use *any* control in such dialogs – not only the stock controls provided by Uikon and UIQ but also any new control that you invent and wish to include in a dialog. (Not *all* controls are designed to be included in dialogs. I don't see any need to include my fleet view in a dialog.)

- Given the number of different ideas here, it should be possible to write code that supports only the things you require for a particular control, without having to worry about implementing things that are unnecessary. Furthermore, you need to be confident that you've decided to include only those things that need including and to exclude only those things that need excluding.

12.2.4   Compound Controls

I began to introduce the idea of compound controls in relation to drawing. It turns out that compounding also makes it very much easier to implement general-purpose containers such as dialogs. Returning once again to our e-mail account dialog, the visible controls on it, from the top, are as follows:

- A dialog, containing a title bar, a help icon, a captioned control array, a button (**Schedule**), a page selector, and a button group.

- A captioned control array with five captioned controls, including a numeric editor with up and down arrows.

- A page selector with three tabs and left and right scroll arrows.

- A button group with two buttons (**Cancel** and **Done**).

That's 21 controls altogether, all lodging in a single window.

For another example, see the **Write file** dialog in Chapter 11, which had two captioned controls and two buttons, for a total of 12 controls.

It should be obvious that the dialog isn't the only container here. The captioned control array, and some controls themselves, are also important containers with their own responsibilities, as we'll see. What we think of as 'the controls' in the dialog (the five controls in the captioned control array we were talking about above) are more complicated than they appear, and account for only about a quarter of the controls actually on this dialog page.

12.2.5   Key Distribution and Focus

Here's a simplified account of how a dialog processes `OfferKeyEventL()` (for the full truth, check out `CEikDialog::OfferKeyEventL()` in the SDK).

In UIQ, the primary method of text input is the FEP. UIQ provides two FEPs: a handwriting recognition method and an on-screen virtual keyboard; others may be written by third parties. Both FEPs receive pointer events and translate them into key events, which are passed to the application. Whether the key event is generated by a FEP or by a real keyboard shouldn't make any difference to your app. If you are creating a new control that can receive text input, you need to override `CCoeControl`'s virtual `InputCapabilities()` method. This tells the FEP whether it can accept text input, and if so, which types of input are supported. The default implementation in `CCoecontrol` returns `TCoeInputCapabilities::EnIone`.

At any one time, precisely one of the controls in the captioned control array has focus. That means the line is either highlighted, or displays a cursor, and is the recipient of 'most' key events. When the dialog is offered a key event, it handles some special cases itself (for instance, it offers *Confirm* to the dialog buttons), but otherwise it offers the key to the currently focused general-purpose control.

> A **general-purpose control** is one that can be used both in dialogs and app views. Incidentally to make a control intended for dialogs usable in app views isn't always difficult, and it's a good thing to aim for. But to make a control intended for app views usable in a dialog is rarely necessary, and you shouldn't try to do so without good reason.

There are plenty of special cases (such as support for the page selector), but this description is enough to illustrate the role of focus, and also to begin to show why keys are offered, and why they are not always consumed.

It's important to give a clear visual indication of focus and all the components work together to achieve this:

- The dialog is the topmost window: it has focus merely by being there.
- The buttons and the title bar can never receive focus so they don't need to change their drawing code in response to whether they're focused or not.

- The general-purpose control designed to live in a dialog should show the cursor, if it has one, when focused, and not otherwise. Many editors include some kind of cursor, or even the window server's flashing text cursor to indicate that they have focus.

12.2.6　Dimming and Visibility

There are two ways to indicate that a control in a dialog cannot receive focus:

- *Make it invisible*: You can make the entire line – prompt and control – invisible. The control environment has full support for this action; invisible controls are omitted from redraw and pointer handling, and dialogs will not allow invisible lines to be given focus.

- *Dim it*: If you dim a control, it will be omitted from pointer handling. Dialogs will not allow a dimmed line to gain focus. However, the control environment will not omit a dimmed control from redrawing; rather, the implementer of the control has to code Draw() to dim the control explicitly.

The good thing, supposedly, about dimming is that the user can still see the control and the value it contains, even though they can't change it. But actually, dimming is a nuisance; the writer of the control has to add support in Draw() (using logical colors can help with this – see Chapter 15 for more information). In any case, unavailable options might be entirely meaningless, in which case you don't want to dim the control, you want to hide it altogether.

Some dialogs in UIQ use a cute compromise; when you call Make-LineVisible(EFalse), the dialog dims the prompt and makes the general-purpose control invisible. The user gets the right effect, the general-purpose control doesn't have to support dimming explicitly, and there's no attempt to display meaningless values.

The relevant functions for dimming are found in CCoeControl:

```
class CCoeControl : public CBase
    {
public:
    ...
    IMPORT_C virtual void SetDimmed(TBool aDimmed);
    ...
    IMPORT_C TBool IsDimmed() const;
    ...
    };
```

If you need to, honor IsDimmed() in your Draw() function.

12.2.7 Validation

While a control has focus and the user is editing it, it may become temporarily invalid – numeric editors are an example of this. However, when focus is taken from a control – for a change in dialog line, or because **Done** was pressed – it's important that the control should be valid.

UIQ calls `PrepareForFocusLossL()` on any control from which it is about to remove focus. The default `CCoeControl` implementation of this function is empty: there is no need to override it for a control whose internal state can never be invalid (for example, a choice list, a button, a checkbox, or a text editor). If your control *could* be invalid, however, you should implement this function to check the current validity of the control. If it *is* invalid, you should:

- issue some kind of message (such as an **info-message**) to inform the user that the control is invalid,
- reset it to the nearest valid value,
- if resetting the value is not possible (for instance the user may have entered text in a numeric editor), you should leave, usually with error code `KErrNone`. This informs the dialog that the control is invalid and prevents the dialog from changing the current line, continuing with **Done** button processing, and so on.

As we saw in the discussion of the `streams` example in Chapter 10, you can issue a single call to `iEikonEnv->LeaveWithInfoMsg()` to display an info-message and leave without displaying Uikon's standard alert dialog.

12.2.8 Control Observers

Compounding produces a hierarchy of controls that's strictly related to coordinates. Components are contained within their container's extent and peer controls don't overlap. As we've seen, that greatly simplifies drawing, and you can probably guess that it makes pointer event processing easier too.

Draw and pointer event processing operate from top to bottom in the hierarchy, that is from the dialog down to the controls.

Controls also report events such as whether their state changed. Event reporting usually goes *up* the hierarchy – for instance, from a control contained in a dialog, to the dialog, so that you can handle the event with the dialog's `HandleControlStateChangeL()` function.

For this reason, many systems handle a chain of events by passing them up their window ownership hierarchy.

However, making the event reporting hierarchy the exact opposite of the compounding hierarchy turns out to be very awkward.

> **A key design decision in Symbian OS was to avoid fixing any associa-tion between the observer of an event and the container control. The observer does not have to be the container control, or a control in the containment hierarchy; it doesn't even have to be a control at all.**

In Symbian OS, each control contains a member called `iObserver` that it can use to report various general-purpose events required by all controls that live in dialogs:

```
class CCoeControl : public CBase
    {
public:
    ...
    IMPORT_C void SetObserver(MCoeControlObserver* aObserver);
    IMPORT_C MCoeControlObserver* Observer() const;
    ...
protected:
    ...
    IMPORT_C void ReportEventL(MCoeControlObserver::TCoeEvent aEvent);
    ...
private:
    ...
    MCoeControlObserver* iObserver;
    ...
    };
```

`MCoeControlObserver` is an interface that can be implemented by any class that wishes to observe controls. It is defined in `coecobs.h`, and has just one member function:

```
class MCoeControlObserver
    {
public:
    ...
    virtual void HandleControlEventL(CCoeControl* aControl,
                                     TCoeEvent aEventType) = 0;
    };
```

If you're writing a control, you can call `ReportEventL()` to report events to your observer (if you have one), which will give the observer an opportunity to do something about them.

The available event types are defined in `MCoeControlObserver:: TCoeEvent`. They are:

```
enum TCoeEvent
    {
    EEventRequestExit,
```

```
EEventRequestCancel,
EEventRequestFocus,
EEventPrepareFocusTransition,
EEventStateChanged,
EEventInteractionRefused
};
```

We also saw this earlier in the chapter, with the `MGameViewCmdHandler` interface and `COppFleetView` class.

State changed

If you're implementing a general-purpose control, the only useful event to report is `EEventStateChanged`. If you report this event from within a dialog, the dialog framework will call `HandleControlStateChangeL()`, which the dialog implementer can use to change the values or visibility of other controls in the dialog. That's the function the `Schedule download` dialog uses to make the settings visible or invisible in response to the checkbox changing.

For some controls, state-change reporting is optional. Edit windows, for example, only report state changes if they're asked to – otherwise, every key press (except for navigation keys) would cause a state change.

You should certainly not report a state change until you have reached a valid state – it would be inappropriate, for instance, to report a state change while the number in a numeric editor is invalid.

Container behavior

If you're implementing a container such as a dialog, you should make yourself the observer of your contained controls and handle the following three events that are generated by the control environment in specific circumstances:

Event Type	Description
`EEventInteraction-Refused`	Pointer-down on a dimmed control. UIQ dialogs handle this with the virtual function `HandleInteractionRefused()`, whose default implementation issues an infoprint saying the control is not available. You can override this in a derived dialog to give a better-tailored message. ("You can't select this because you haven't installed x, y, z.")

`EEventPrepare-` `FocusTransition`	Pointer-down on a focusable control that is not focused. UIQ dialogs call `PrepareForFocusLossL()` in response to this on the currently focused control, so that that control can validate – and leave, if necessary. If the currently focused control leaves, processing of the pointer event will abort, so that focus is not transferred.
`EEventRequestFocus`	A pointer event has been handled on a nonfocused control, which now needs focus. Permission to change focus has been granted by the handler for the prepare-focus transition event above, so the container must now execute the focus change.
	UIQ dialogs handle this message by unfocusing the currently focused control and focusing the control associated with the pointer event.

If you implement a general-purpose container, you should handle these three messages with semantics similar to those used by UIQ dialogs. The messages are all generated by `CCoeControl::ProcessPointerEventL()`, and you should not generate them yourself.

If the container is the observer, why not just say 'the container' and have an ownership hierarchy as in some other systems? Firstly, because controls can be used in containers which don't need to observe them. Secondly, because the dialog isn't actually the imme-diate container of, say, a choice list control – the dialog contains captioned controls, and in turn a captioned control contains the choice list. But the dialog is a direct observer of the choice list.

Other events

The `MCoeControlObserver` interface is designed for the general-purpose requirements of contained controls (reacting to changes in state) and containers (responding to interaction refused, prepare focus transition, and request focus events).

Special-purpose events should be handled by special-purpose inter-faces, such as `MGameViewCmdHandler` from `COppFleetView`, or `MEikMenuObserver` from UIQ menus.

12.2.9 Containers

UIQ dialogs are just one example of a container: not every control you write needs to go into a dialog. Many controls will be designed for use in your app view instead, or perhaps for some other kind of container. Your container may be general purpose (such as a dialog, which can contain an arbitrary number of controls of any type) or special purpose (such as an application's view).

If your container is general purpose, you should use design patterns similar to those used by dialogs for handling validation, focus, changes of state, dimming, and so on, but for navigating around the container, views and dialogs use different design patterns.

UIQ supports a keypad with either two (up and down) or four (up, down, left, right) direction keys and a *Confirm* key. In dialogs, the pen is the only way to navigate between controls, so that pressing the direction keys has no effect; in views, the direction keys can be used for navigation. So, if your control is intended for an app view, it will be offered all the direction key events, but it should only consume them when necessary. (If your control is intended for a dialog, this is not an issue because direction key events are not consumed by the dialog.)

Most views use the left and right keys to navigate around the view or between views. For instance, in views that use tabs, the view normally uses the horizontal direction keys to move between the tabs.

So, if your control is used in a view, it should only consume the direction keys if it really needs to. By doing this, the *view* will be able to consume the events instead.

When you tap on a UIQ choice list, it displays a vertical (potentially scrolling) list of the items. The popout is modal; it captures the pointer and consumes all keys offered to it. You have to tap outside the choice list or select an item in the list to pop the list back in again – then you can use the rest of the dialog or view.

Similarly, if you tap on a date control, you get a popup calendar for a whole month, which you can navigate without worrying about the dialog or view as a container. Only when you tap to select a date does the calendar pop back in again.

12.3 Key Processing Revisited

Now we've seen:

- what a simple control does with a key event in its `OfferKeyEventL()` member function;
- the idea of focus, which means the location where most key events go, and which is indicated visually by a cursor, or some kind of highlight;

- how, in a dialog, keys are handled either by the dialog itself, or by the currently focused control;
- how a control may or may not consume a key offered to it. For instance, most controls in views don't consume direction keys; this allows the view to consume them instead, for navigation around the view.

This puts us in a better position to understand the full picture of key distribution as handled by the window server and the control environment together.

When a key is pressed, it is initially passed from the keyboard driver to the window server. If the key is a window server hotkey, then it causes an action associated with the hotkey, and it is not specifically routed to any application. Window server hotkeys include *Ctrl+Alt+Shift+K*, which kills the process associated with the foreground window.

However, the window server routes most key events to the application in the foreground, where they are detected by CCoeEnv::RunL() and passed on to CCoeAppUi::HandleWsEventL(), which identifies the event as a key event, and then CCoeControlStack::OfferKeyL(). The control environment's **control stack** is responsible for offering keys to controls that can then handle them further.

If the key event was generated by the on-screen virtual keyboard FEP, the window server is not involved – the FEP simulates a key event, passing it directly to CCoeAppUi::HandleWsEventL(), as though it had come from the window server. The handwriting recognition FEP is slightly different – it generates key events and passes them to the window server, which adds them to its event queue. In both cases though, the result is the same – the key event ends up on the control stack.

In the UIQ environment, there are typically five types of controls on the control stack:

- *The debug keys control*: this is an invisible control that consumes all *Ctrl+Alt+Shift+* keys and associates a particular action with most of them. For instance, *Ctrl+Alt+Shift+R* causes the current screen to be completely redrawn. If the debug keys control doesn't consume the key, it gets offered further down the stack.
- *Any FEP that needs to intercept key events*: most FEPs own a control that is added to the control stack at a higher priority than any visible controls. This ensures that the FEP receives first refusal of all key events, allowing it to consume and optionally process them and commit its output to the focused control (if the control supports text input). In UIQ, the virtual keyboard FEP needs to be on the control

stack – when it is active, it consumes all key events, except those it generated itself; the handwriting recognition FEP on the other hand does not consume any key events, so is not on the control stack.

- *Any dialogs that might be active*: the key is offered to the topmost active dialog, which will consume the key. Only if no dialog is active will the key be offered further down the stack.

- *The menu bar*: in UIQ, the menu bar does not need to process shortcut keys defined by the application, but in some other UIs, it does. If so, when the menu bar is invisible, it will consume only the shortcut keys and the key that causes it to be displayed. When the menu bar is visible, it will consume all keys offered to it. The fact that the menu bar is lower down in the control stack than *any* dialog prevents the menu bar from being invoked when a dialog is active.

- *Any app views*: an application view should usually ignore keys (and therefore not consume them) if it is invisible; otherwise it should handle and consume them.

For an application with three views (one of which is active), and two active nested dialogs, the control stack is shown in Figure 12.4:

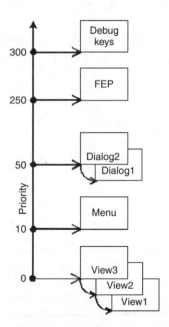

Figure 12.4

Key events are offered down the control stack, firstly in order of priority (highest to lowest), and secondly in stack order (most recently added to least recently added, within each priority). The priorities shown above are defined in an enumeration in `coeaui.h`.

Using this stack structure, Symbian or an application programmer can insert something new into the GUI environment without having to rewrite the control environment or the existing GUI components.

Exactly how a control on the stack handles a key event depends on the control:

- Dialogs are general containers, and we've already seen how they offer keys around their component controls.

- App views are often also containers, and they will use their own logic, including at least some of the patterns used by dialogs to offer keys to component controls.

- The debug keys control is a single control without any components – it either consumes a key, or it doesn't.

- The menu bar, when visible, includes at least two controls – the menu bar and a menu pane – and possibly more, since there may also be cascaded menus. The menu bar offers and handles keys among these controls to implement conventional menu navigation and item selection.

- Most FEPs need to handle key events but how they do this depends on the type of FEP. UIQ's virtual keyboard FEP when active, consumes all key events but the ones it generates itself. FEPs activated by a key press or combination of key presses need to be on the control stack even when they are inactive, but when inactive they would typically not consume any other key events.

We can see again how focus and key handling are usually – but not always – related:

- The window server sends most keys to the application with focus, or to the FEP, if active – but not its own hotkeys.

- The control stack mechanism *usually* results in keys being handled by a control with focus – the topmost dialog, the menu if visible, or the application view. But keys are always offered to the debug keys control, which is always invisible, and if no dialog is showing, keys are always offered to the menu bar, even if it is invisible. In other words, these keys can be handled even though they're not associated with focus.

- Dialogs normally channel keys to the focused line but the direction keys and the *Confirm* key are handled differently.

- The menu bar maintains a focused pane and a focused item, and normally offers keys to them – but shortcut keys, if supported, are handled independently of focus.

The Battleships app view can only send keys to the opponent's fleet view. It does this provided the fleet view can accept input events:

```
TKeyResponse CGameAppView::OfferKeyEventL(const TKeyEvent& aKeyEvent,
                                          TEventCode aType)
    {
    // Let the opponent-fleet view handle the key
    if(iOppFleetView->IsDimmed())
        return EKeyWasNotConsumed;
    else
        return iOppFleetView->OfferKeyEventL(aKeyEvent, aType);
    }
```

It might be a surprise that I test for whether the control is dimmed here, not whether it's focused. If the app view contained more than one interactive control, I would have to use focus, but as it is, focus is not the issue: it's whether the control can handle interaction right now, and that's indicated by the dimmed property. The control environment recognizes dimmed controls, and doesn't pass pointer events to them .

Clearly, a dimmed control can't have focus, but in situations where there is more than one nondimmed control, focus must be used to distinguish between them.

12.3.1 Focus

Focus is supported throughout Symbian OS graphics components. First, the window server associates focus with a **window group**. The window server routes keys to the application whose window group currently has focus. A window group owns all of an application's windows, including app views, menus, dialogs, and any others. The window server sends focus-gained and focus-lost events to applications as their window groups gain and lose focus.

In response to window server focus-changed events, most applications do nothing. There is no point in redrawing their controls to indicate that they have lost focus, since unfocused apps can't be seen on the screen anyway.

The control environment maintains the top focusable control on the control stack, and calls FocusChanged() on it to indicate when focus has changed. Similarly, container controls should also call a contained control's FocusChanged() function when they change the focus given to it.

A control should change its appearance depending on whether it has focus. If you need to indicate focus visually, use IsFocused() in your drawing code to draw an appropriate highlight or cursor. Handle FocusChanged() to change between focus states (for example, to redraw, activate, or deactivate the cursor). FocusChanged() includes a parameter TBool aRedrawNow to indicate whether an immediate redraw is needed.

You can control and interrogate focus with SetFocus(). This does *not* iterate through component controls; each container handles propagation of SetFocus() according to its own requirements.

You can use SetNonFocusing() and related functions to set whether a control allows itself to be focused. This can be a permanent state for some controls (they simply don't handle input), or a temporary state for others (this is analogous to dimming, see below).

To summarize, here are CCoeControl's focus-related functions:

```
class CCoeControl : public CBase
    {
public:
    ...
    IMPORT_C virtual void PrepareForFocusLossL();
    IMPORT_C virtual void PrepareForFocusGainL();
    ...
    IMPORT_C void SetFocus(TBool aFocus, TDrawNow aDrawNow = ENoDrawNow);
    ...
    IMPORT_C TBool IsFocused() const;
    ...
    IMPORT_C void SetNonFocusing();
    IMPORT_C void SetFocusing(TBool aFocusing);
    IMPORT_C TBool IsNonFocusing() const;
    ...
protected:
    ...
    IMPORT_C virtual void FocusChanged(TDrawNow aDrawNow);
    ...
    };
```

12.3.2   The Text Cursor

Focus may be associated with a cursor. The window server provides a text cursor, and there may only be a single cursor on the screen at any one time. By convention, the text cursor must be in the focused control in the focused window. If your control uses the text cursor, you should be sure to implement FocusChanged() so that you can turn the cursor off when you lose focus, and on again when you regain it.

A control that uses the text cursor doesn't have to use it all the time. For instance, the secret editor (for entering passwords) hides it if the user tries to move it.

12.4 Pointer Processing Revisited

The window server ensures that a pointer event gets to the right window, and the control environment framework ensures that it gets to the right control. The event can then be handled by `HandlePointerEventL()`.

12.4.1 Interaction Paradigms

A control should interpret the entire pointer sequence. Two sequences in particular are very common.

Press-and-release is appropriate for many types of buttons:

- *Pen down inside the button*: provide visual feedback by making it obvious that the button is pressed down.

- Pen may stay down and/or be dragged for an arbitrary period.

- *Pen up*: if the pen is outside the button, release it, redraw it in its neutral state, and do nothing. If the pen is inside the button, release it, redraw it, and then do the action associated with the button.

Other related sequences are possible. For instance, UIQ supports buttons that activate as soon as you press them, whose state toggles, or which expand to show a drop-down list when pressed, so that you can drag and release on one of the items. There are many alternatives.

In UIQ, a single tap to select and 'open' is used, where other systems would use double-click or two taps. Pen down on an item selects and opens it – in the case of `COppFleetView`, 'open' means 'request a hit on'; in the case of the Application launcher, 'open' means open the application (or, if its already open, switch to it).

12.4.2 Pick Correlation

Pick correlation means associating the right object with a pointer event. As we saw in `HandlePointerEventL()` for `COppFleetView` at the start of the chapter, this was very easy to do:

- There is only one object type that could have been selected: a sea tile,

- The object is rectangular,

- The object is part of a simple grid.

This meant that selecting the object was a matter of a simple bounds checking and division.

In more complicated cases (such as selecting text in a word processor, or an object in a vector graphics package), pick correlation can be an awkward business, and can involve even more optimization and complexity than is involved in drawing. However, you can use a few handy techniques to make life easier:

- Object orientation makes designing for pick correlation easier, just as it does most things.

- Construct a pick list optimized for easy checking when a pointer event occurs.

It so happens that you can often use the same code to handle both incremental redraw and pick correlation requirements. (You can see this happening quite clearly in `COppFleetView`: the precalculated border and tile positions and sizes are used by both drawing and pick correlation.) The net effect is that your redraw and pick correlation code can be optimized together. Often, optimizing feels like inventing two solutions to solve one problem. In this case, one solution solves two problems, which is nice.

12.4.3   Grabbing the Pointer-down Control

When you drag outside a control – in, say, the press-and-release paradigm – you usually want all your pen events, including the release, to go to the control in which the pen went down. The framework therefore has to remember with which control the pen-down event was associated, and to channel all subsequent drags and the pen-up event to that control.

Symbian OS calls this **pointer grab**, and you'll see grab-related APIs in the window server and the control environment to support it; the window server has to remember the right window, while the control environment has to remember the right control. If you write a control that contains multiple press-and-release type objects, but for various reasons you don't want to implement those objects as component controls (perhaps they are not square), then you will also have to implement grab logic.

12.4.4   Capturing the Pointer

If an application launches an app-modal dialog (that is, most UIQ dialogs), then although the app view window may still be visible above the dialog window, no pointer events should be allowed into the app view. This is **pointer capture**: the dialog captures the pointer.

You can use `CCoeControl::SetPointerCapture()` to capture all pointer events to a control. Clearly, the window server has to be involved also: displayable windows also support a pointer capture API.

Grab and capture can seem confusing. Grab means keeping pointer events associated with the control on which pointer-down event occurred. Capture means preventing pointer-downs being handled outside a particular control.

12.4.5 Getting High-resolution Pointer Events

The window server normally amalgamates drag events so that, after an application has handled the previous pointer event and asks the window server for another event, the window server only tells the application the coordinates to which the pointer has now been dragged.

This is the right thing to do when handling non-time-critical MVC-type interactions. But for time-critical applications, such as handwriting recognition, it's rather awkward. The client-server communication between the window server and application is too slow to handle events at the rate produced by handwriting. Without any special support, the system will effectively sample handwriting at too low a rate, with very poor results:

Figure 12.5

Some Symbian OS engineers refer to this affectionately as 'thruppenny-bitting', after the twelve-sided three-penny coin ('thruppenny bit') that went out of circulation in the UK in 1971. It is enough of a challenge to write a screen device driver that doesn't 'thruppenny-bit', let alone convey all this through the window server, client-server interfaces, handwriting recognition software for Western or Far Eastern alphabets, and finally to a mere application.

To avoid this problem, an application can ask the window server to buffer pointer events, and then receive the whole buffer in a single massive event. The window server will send the buffer when it is full, or immediately after the pen-up event, whichever is sooner.

Controls handle a full pointer buffer with the `HandlePointerBufferReadyL()` function, which deals with the `EEventPointerBufferReady` event. The window server has corresponding supporting APIs.

12.4.6 Processing Pointer Events

We now have all the pieces of the puzzle, so let's run through the processing of a pointer event from start to finish.

Pointer events are initially generated by the digitizer driver and passed to the window server. The pointer event is usually associated with a window whose on-screen region includes the pointer event. However, as we've just seen, pointer grab may be in use to associate the event with the same window that was associated with pointer down, or pointer capture may be in use to reject the event altogether.

The window is a member of a window group and the window group corresponds to an application. The window server sends the event to the relevant application. The control environment's CCoeEnv::RunL() is called: it passes the event on to CCoeAppUi::HandleWsEventL(). This is the same process that's used for the key events, and is also what we saw in the Chapter 4 debugger run through of Hellogui. HandleWsEventL() then identifies it as a pointer event associated with a particular window.

The control environment finds the window-owning control associated with that window and calls its (nonvirtual) ProcessPointerEventL() function. ProcessPointerEventL() calls HandlePointerEventL() to handle the event. It also does some preprocessing, which I'll explain shortly.

HandlePointerEventL() is virtual: when implementing a simple control, you implement it to handle a pointer event, however you wish. If you don't implement this function, you get the default implementation (in the CCoeControl class), which searches through all the visible, non-window-owning component controls to find the one that includes the event coordinates, and then calls ProcessPointerEventL() on that control. This default implementation is good for compound controls and you override it at your peril.

So, the pointer event is ultimately channeled to the right noncompound control, where you can handle it by overriding HandlePointerEventL(). Note, however, that the complications we've already seen in this section affect both ProcessPointerEventL() and HandlePointerEventL():

- The control environment's pointer-processing functions support pointer grab, so that once a pointer is grabbed, all subsequent events, to pointer-up, are channeled to the same control.

- The control environment doesn't support pointer capture: you have to use windows (and window-owning controls) for that.

`ProcessPointerEventL()` implements the event reporting in the container needed for focus transfer between components: it generates interaction-refused events for dimmed controls, a prepare-focus-transition event for a focusable but nonfocused control on pointer-down, and a request-focus *after* `HandlePointerEventL()` has been called, for controls that were not refused focus at prepare-focus-transition time.

If you use the pointer buffer to capture high-resolution pointer event sequences, the control environment handles it with `ProcessPointer-BufferReadyL()`, and you have to handle it with `HandlePointer-BufferReadyL()`.

12.4.7 Customizing Pointer and Key Sounds

The window server provides support for customizing the sounds made when the screen is tapped or a key is pressed. If the phone allows it you can change the sounds by creating a DLL that implements the `CClickMaker` interface. You can then load and enable the DLL using the `RSoundPlugIn` class. When your DLL is enabled, its implementations of `KeyEvent()` and `PointerEvent()` are called whenever key or pointer events occur. The function that lets you find out whether you can unload the phone's current DLL and load your own is

```
TBool RSoundPlugIn::IsLoaded(TBool& aIsChangeable);
```

The return value tells you if one is loaded and the parameter returns whether you are allowed to change it. For more information on the classes and functions involved, see the SDK.

12.5 More on Window Server and Control Environment APIs

Now to review the classes provided by the window server and control environment, and how they affect the application framework.

12.5.1 Application to Window Server Communication

In every Symbian OS GUI application, four main system-provided framework classes are used to ensure that the application can communicate properly with the window server as shown in Figure 12.6:

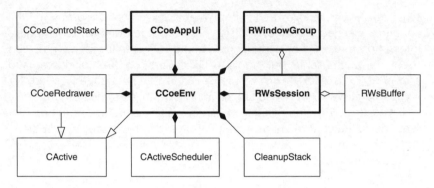

Figure 12.6

These classes are as follows:

Class Name	Definition	Description
RWsSession	w32std.h	Window server session: this provides a client-server session from the application to the window server. All window server classes (window groups and windows) use this session for communication with the window server. The session also owns the application's client-side buffer, in which drawing and window manipulation commands are batched up before being sent to the server for execution.
RWindowGroup	w32std.h	A window group: this is the client-side version of the window at the top of the application's entire window hierarchy. A window group is associated with keyboard focus.
CCoeEnv	coemain.h	The control environment base class: this encapsulates the window server session in active objects, whose RunL() member functions are invoked when events

		are received from the window server. These `RunL()`s analyze the events and call framework functions to handle them. The control environment sets up a cleanup stack for graphics programs. It also contains many useful utility functions.
`CCoeAppUi`	`coeaui.h`	The app UI base class: The essential purpose of the control environment's app UI is to handle the control stack for first-level key event distribution. The app UI also performs some other, more incidental functions.

The control environment provides some prerequisites for cleanup handling (a cleanup stack) and event handling (an active scheduler). It wraps the window server session API in *two* active objects to handle events generated by the window server. As we saw in Chapter 4, all event handling in a GUI program – key, pointer, redraw, and others – takes place under an active-object `RunL()`.

The two active objects are a higher-priority one for user-initiated events and a lower-priority one for redraw events. So, if both a user-input and a redraw event occur while the previous user-input event is being handled, the framework will handle the user-input event first. Otherwise, the use of two active objects makes no difference at all to the application programmer.

There's some interesting design history here. Earlier in the development of Symbian OS, only a single active object was used. It seemed natural, then, to derive CCoeEnv directly from CActive. But then the redraw event stream was separated from the user-input event stream, creating a second active object. The result is that CCoeEnv is-a CActive (for user events), and has-a CActive (for redraw events). If we had designed this from scratch rather than making a late change, we would surely have said CCoeEnv has two active objects. This is a good case study for two of the guidelines I express elsewhere in this book: don't use inheritance where ownership will do, and hide active objects from your interfaces.

12.5.2   Window Types

In the last chapter, we concentrated on the drawing-related interactions between `CCoeControl` and `RWindow` (and, occasionally, `RBackedUp-Window`). In this chapter, we've introduced key processing, which brings in `RWindowGroup` and `CCoeAppUi`. The window server classes involved here are part of a small window class hierarchy, defined in `w32std.h`, and illustrated in Figure 12.7:

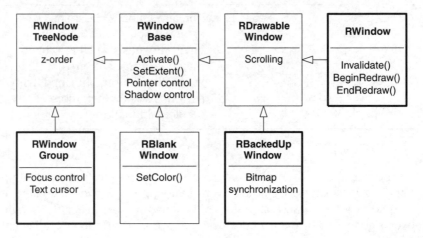

Figure 12.7

For much of the time, you can call these classes' functions through the control environment. Nevertheless, it is useful to understand them because the control environment is not designed to encapsulate the window server. Rather, the control environment provides a convenience layer for lodger controls and compound controls, and for the window server's major functions such as drawing, pointer, and key event handling.

I won't be providing detailed information on these facilities but I'll give enough of an overview that you can understand what's available and find the information you need in the SDK.

Class Name	Description
RWindowTreeNode	Base class for all windows: a node in the tree that defines *z*-order
RWindowGroup	Unit of keyboard focus, and top-level owner of displayable windows
RWindowBase	Base class for all displayable windows

RBlankWindow	Entirely blank window
RDrawableWindow	Base class that defines windows which support drawing
RBackedUpWindow	Backed-up window: window server redraws invalid areas
RWindow	Standard window: application redraws invalid areas

For most application programming, the most important concrete classes are RWindow and RWindowGroup. These are described in more detail in the following two sections.

Because all displayable windows are ultimately owned by a window group, a window group is the top-level node in the tree that defines *z-order*. This means that all windows belonging to an application move back and forth in the *z-order as a group*. We can therefore use the terms 'foreground application' and 'application whose window group has focus' interchangeably.

The window server allows applications to have more than one window group, but the control environment supports only a single window group per application, and this assumption is built into other Symbian OS components as well.

The window server provides other features that aren't supported by the control environment, such as blank windows, and even nonrectangular windows whose shape is defined by a region.

12.5.3 Standard Window

A standard window has functionality inherited from the chain right up to RWindowTreeNode. However, the *interesting* functionality starts with RWindowBase, the base class for all displayable windows. RWindow-Base includes:

- Activate(): you can use this function as the final step in a three-phase construction. The steps are as follows: (1) A default RWindow is just an empty client-side handle. (2) Use Construct() to connect the RWindow to an RWsSession and construct a server-side window. (3) After you have set all the window parameters, you need to use Activate() to display the window and enable it to receive events. In the case of an RWindow, the whole window will be invalid immediately after you activate it, so (unless you proceed immediately to redraw it) you'll get a redraw event.

- Position and size setting functions.
- Pointer control.
- Shadow control.
- Backed-up behind.

These functions can only be associated with a window that has a visible extent on the screen, so that's why they're introduced with `RWindowBase`. `RWindowBase` serves as a base class for blank windows and also for `RDrawableWindows`. If you can draw a window, you can also scroll its contents, so scrolling functions are introduced here.

Finally, we get to `RWindow`. We have already seen the most important functions introduced here: `Invalidate()`, `BeginRedraw()`, and `EndRedraw()`. However, there are a few other interesting functions too:

- `Construct()` requires you to pass an `RWindowTreeNode` to serve as this window's parent. This can be another displayable window or a window group. Window groups are the only kinds of windows that don't need a parent, so window groups are the top-level windows in the window tree – that is, in the z-order.

  `Construct()` also requires you to pass a 32-bit handle. All events relating to this window include the handle. The control environment passes the address of the window-owning control that owns this window: when the control environment fields an event, it simply casts the handle to the address in order to associate it with the correct control.

- Easier-to-use variants of `SetSize()` and `SetExtent()` than those provided by `RWindowBase`.

- Two variants of `SetBackgroundColor()`.

- `GetInvalidRegion()`: with this you can get the exact region that is invalid. If you have a particularly complex and well-optimized program, it enables you to improve on the bounding rectangle passed to a control environment's `Draw()` function. You only need to invalidate and redraw the rectangles that comprise the invalid region, rather than their bounding rectangle, which is passed to the control environment. Few programs are as demanding as this – but for those that are, the facility is there.

Fading is used in Symbian OS to change a window's colors so that other windows stand out. It is implemented by remapping color values to a more limited range, and optionally also making them lighter or darker. For example, in UIQ, when a dialog is displayed, the window that was previously in the foreground is redrawn, faded, and darkened.

In UIQ, the default fading values are zero for black and 190 for white (unfaded windows have values of 0 and 255), although you can override these defaults through the window itself (the functions to do this are defined in `RWindowTreeNode`), or through the graphics context used to draw it.

12.5.4   Window Group

The `RWindowGroup` class's primary role is to handle focus and key handling. Window groups handle focus because they are the top-level nodes in the z-order. The only window group that can reasonably grant focus is the foreground group.

As we have seen, focus-related functions enable you to say that this window group can't take focus, or that it should automatically get focus (and foreground) when a pointer event occurs.

The window server supports a flashing text cursor, whose window, position, shape, and so on, can be controlled through the window group. This is clearly the right place to do it since the cursor is associated with focus.

One implication of this is that there can only be one text cursor per application. An application such as the UIQ agenda displays a flashing cursor in its detail view when the view has focus. But when a dialog is displayed, the view loses focus, and it must stop the cursor flashing. Moreover, it should relinquish control of it, so that it can be used for editing any text fields that might appear in the dialog.

`RWindowGroup` also allows you to configure the way in which key events are generated when a key is held down for a certain length of time. For instance, on some phone keypads, holding down an alphanumeric key rather than releasing immediately it can cause the generated key event to switch between alphabetic and numeric; it may also be sent to a different application from the one currently in the foreground.

The default behavior is for a standard key event to occur on key down, then after the key has been held down for a short time interval (usually a fraction of a second), it auto repeats. You can override this using `RWindowGroup`'s `CaptureLongKey()` function.

You can customize several things. Firstly, when the user presses the key but releases it before the required time interval, the standard key event can occur when the key is pressed, or when it is released. Secondly, if the key **was** held down long enough, a different key event can be generated from the one that was captured, and this may or may not auto repeat. It can also be sent to a different window group from the one with focus. For instance, if the key event was changed from numeric to alphabetic, you might want to send it to the address book, rather than the phone number display.

12.6   The Shell

In any Symbian OS GUI, one special application and two servers are always active:

- The shell (called `Application launcher`, in UIQ), which controls application launching and task switching.

- The Eikon server, which handles off-screen pointer events generated in the sidebar window (UIQ does not implement this) and also looks after other system-wide windows such as the password window.

- The application architecture server, which maintains lists of installed and active applications, and rescans the file system and window group list when necessary in order to keep these lists up to date. The application architecture server maintains notify requests on the file system and the window server so it can rescan when necessary (and so it doesn't rescan when it's not necessary).

Other servers may also be active but these are GUI-specific. For instance, UIQ uses a text input server to load the handwriting recognition FEP and handle switching between FEPs and a memory manager server to perform memory management, for instance, closing down apps running in the background when memory is low.

The shell application works intimately with the application architecture server and Eikon server. The shell interprets events from the application picker and displays all application icons and their captions in the application launcher grouped by category. The application launcher is kept up to date by consulting the list of installed applications maintained by the application architecture server.

12.7   Summary

In this chapter, I've covered the control and window server framework that supports GUI interaction, and shown how you should use this framework. Because everything is interlinked, it's probably the hardest chapter in the book. We've seen

- the key and pointer event types
- how events are turned into commands
- how to handle focus, using window groups
- how the control stack for key event handling routes events to various destinations

- dimming controls and validation
- using observers to preserve the MVC pattern
- the difference between pointer-grab and pointer-capture
- the relationship between the control environment and the window server.

13

Files, Streams, and Stores

In this chapter, we'll cover file and data management for Symbian OS in more detail and discuss the main conventions and idioms for using files and file-related APIs.

The Symbian OS C++ framework includes abstractions to manipulate **files, streams,** and **stores**. Streams do not only exist in the context of files, but also extend in other domains. We use streams in preference to raw file operations not only because they are syntactically more easy to use but also because they offer reuse in manipulating and serializing data to and from objects. Stores are used to complement streams in storing and retrieving externalized data.

While explaining how to use streams and stores to manipulate, retreive and save data to files, I will also discuss accessing the file server, data file placements and media configuration support. Finally, we will discuss the use of **embedded**, **permanent**, and **dictionary** stores.

13.1 File-based Applications

Symbian OS application architecture includes the following:

- Support for applications that don't use native file format: this is done by associating a MIME type with a file, according to industry standards so that, for instance, the type of an HTML file is **text/html**. You can launch an HTML file from the shell and from within the Web browser.

- A system for recognizing nonnative files and application types such as Java applications.

- Viewing attachments such as those on e-mails.

Symbian OS C++ for Mobile Phones. Edited by Richard Harrison
© 2003 John Wiley & Sons, Ltd   ISBN: 0-470-85611-4

There are two important types of file-based application:

- An application such as Word, Sheet, Sketch, or Record is clearly file-based. You can save a Word document into a file and load one from a file.

- An application such as Agenda or Data is clearly file-based, but you don't load and save the whole file at once. Instead, you use the file as a database and you load and save individual *entries* at the same time. For efficiency, you also maintain some index data in RAM.

Figure 13.1 illustrates the difference:

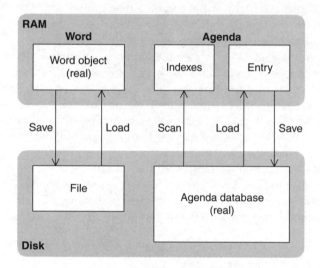

Figure 13.1

There are some similarities between these two types of file-based application – for instance, in either case, you can open the file by select-and-open from the shell in UIS that support this functionality.

However, there are real differences also. In load/save-type applications, we tend to think that the 'real' document is in RAM: the file is just a saved version of the real document. In database-type applications, we tend to think that the 'real' document is the database – each entry is just a RAM copy of something in the database. The disciplines for managing the documents as a whole are, therefore, quite different.

In general, it is meaningful to make load/save-type documents embeddable, but for databases, it is less useful. So, databases cannot be embedded (though they can certainly embed other documents in them).

Finally, there are a few other ways an application handles files:

- An application such as a calculator is clearly *not* file-based. It neither loads nor saves the state of its calculation.
- Similarly, there is no need for many test applications to be file-based.
- An application such as e-mail clearly deals with files: it has several directories and files containing inbox, outbox, other folders, and all the messages. On the other hand, no sensible e-mail program – including those on the PC – would expose these directories and files to the user; it would be damaging for the user to try to manipulate them from the shell (or Windows Explorer).

In all three cases, you can't open an application by opening a *file* belonging to that application from a shell application.

So the idea of **file-based application** has a particular meaning in Symbian OS. Load/save applications such as Word and so on are file-based and potentially embeddable. Database applications such as Agenda are also file-based, but can't be embedded. Applications such as Calc and e-mail aren't file-based.

> **A key implication of this is that Symbian OS native documents are not recognized on the basis of their file extension because you don't get file names in embedded documents.**

So a file that's displayed on the Symbian OS shell view (like on the Psion Revo) as Agenda has no hidden file extension and file save dialogs don't add a default extension for you as happens on Windows. If an Agenda icon appears next to the file, that's because its *internal* format indicates it's an Agenda file.

Strictly, documents are not the same as files. However, it's easy to confuse them because file-based applications are very often called document-based applications in Symbian OS literature.

13.1.1 User and System Files

Usually from the user's perspective, there are just two types of file:

- document files that contain the end user's own data,
- system files that contain frightening and mysterious things, which shouldn't be touched.

For most end users, it's actually better to pretend that there's only *one* type of file – document files – and hide the system files altogether.

Moreover, all system files are given a location in the \System\ folder – on whatever drive, z:, c:, or another. By default, in devices that have a document-exposing shell (and Uikon file-related dialogs), the shell doesn't show the user the \System\ folder or any of its subfolders. Programs, applications .ini files, and even the user data files maintained by programs such as e-mail, should reside somewhere in the \System\ directory so that end users can't find them using a shell-type application.

On devices with a shell, a 'default folder' is maintained for end-user document files. The initial setting for this default is language specific – on an (English) Psion Series 5, it's c:\Documents\. You can change this to any drive and any folder you like. You can also store documents in folders other than the default – either subfolders of the default folder or folders outside the default folder.

For many end users, any kind of file system is a source of unnecessary and worrisome choice. Many Symbian OS phones like the Nokia 7650, Sony Ericsson P800 and Ericsson R380, choose to completely hide the file system from the user interface, with no option to expose it.

In any case, system files are not indicated by the system attribute bit in the file's directory entry. That makes life too awkward when interchanging data with PCs. It's their location in the \System\ tree on some drive that counts.

13.1.2   UIQ Application Data File Placement

Applications have to specify where their data files are kept on the disk. The recommended placements are as follows:

Data files for standard applications should be stored in the path supplied to the application at startup. The standard is \Documents\Appname \appname. This should not be hardcoded into the application or the resource files. All application data files are known by the storage manager server and are either in the default location supplied via the method called:

```
QikFileUtils::MakeAppDocumentFileNameL((TFileName,
  const TUid) or specified in the application's AIF file.
```

Other application housekeeping data such as .ini files are conventionally stored in the path \System\Apps\AppName\.

If the application uses other paths than the ones stated above, the application author has to specify the location by setting the file_ownership _list information in its .aif resource file.

Furthermore, the application architecture makes default assumptions about the capabilities of the application – specifically, that it can't be embedded.

If you want a nice caption and a nice icon or if you want to allow your application to be embedded, you have to code an **application information file** or AIF. In Chapter 14, I'll show you how to create `hellogui.aif` to give an icon and captions to my GUI hello world application.

13.1.3 Summary of File Naming and Location Conventions

Here's a summary of the file types we've met so far, along with the folder and file naming conventions required by the application architecture.

Type	Name	Location
Application document files	Any name	Any folder on any read/write drive. `\Documents\` is an initial default suggestion, in English-language locales
Nondocument application files	Any name	In `\System\` on any drive, according to requirements
Application program	*appname*`.app`	`\System\Apps\`*appname*`\` on any drive (including `z:` for built-in applications)
AIF	*appname*`.aif`	Same as `.app`
Resource file	*appname*`.r??`	Same as `.app`
`.ini` file	*appname*`.ini`	`c:\System\Apps\`*appname*`\` – only `c:` is allowed, so that the `.ini` file is (1) on a read/write drive and (2) always accessible to an application, even if removable media are changed
Application-specific DLLs	`*.dll`	Same as `.app`
Shared DLLs	`*.dll`	`\System\Libs\`, ideally on same drive as `.app`
Shared `.exes`	`*.exe`	`\System\Programs\`, ideally on same drive as `.app`

13.2   Introducing the APIs

Now that we've seen how the file system is used, it's time to look at the main file and data-related APIs. Here they are

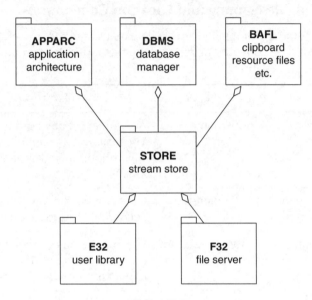

Figure 13.2

The file server provides fundamental file services such as drives, directories, files, and installable file systems, for use by higher-level components. It also provides program-loading facilities required by E32 – which is why the file server is an integral part of the Symbian OS base.

Central to all native Symbian OS data handling is the **stream store**. Objects exchange data between RAM and streams using << and >> operators or `ExternalizeL()` and `InternalizeL()` functions. Stores are just collections of streams. Native file and document formats use file stores and embedded stores. The APIs of the file server and the stream store will form the body of the material in the rest of this chapter.

Also, the clipboard API that is specified in `baclipb.h`, uses stream store technology and a single data file, `clipbrd.dat`, which is shared by all applications. This allows a copying program to put data on the clipboard in multiple formats and a pasting program to select the best available format for its purposes. It is documented in the C++ SDK.

13.3   The File Server

The file server (F32) offers a client API that allows user programs to manipulate drives, directories, and files and to read and write data in files.

F32 uses DOS-like conventions to offer up to 26 devices (drives) identified as a: to z:, a fully hierarchical directory structure and long filenames incorporating almost any character – except those reserved by the file system itself. Directory names are separated by backslashes ('\', as in Windows), rather than by forward slashes ('/', as in UNIX). A period ('.') may be used to indicate an extension; although this has no special meaning to Symbian OS, some applications may assign their own meaning to them. A filename, *including* its drive and directory portion, may be up to 256 characters long.

Like Windows (and unlike UNIX), the file server (VFAT driver) is case preserving, but not case sensitive. In other words, if you create a file called **My File**, all directory operations will return the name My File with the original case. But if you search for My File, my file, MY file, or any other combination, the file My File will be returned. Clearly, this means that you can't have two or more files in the same directory whose names differ only in the cases of some of their letters.

The file server's API has been designed for easy mapping to POSIX APIs. The Symbian OS C standard library is built on top, using FILE and its associated functions to map to RFile and its member functions and so on for other file-related operations.

The main classes in the file server API are

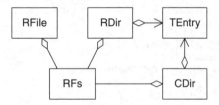

Figure 13.3

An RFs is a session from your program to the file server. You need a session for all file-related operations and in order to be able to use other classes such as RFile, RDir, and CDir – whose purpose I'll explain below.

All these classes are defined in f32file.h.

13.3.1   File Server Sessions

All servers use session-based communication so that a client function such as RFs::MkDir() or RFile::Write() is converted into a message that's sent to the server. The requested function is performed in the server and then any result is passed back to the client. It isn't necessary to understand how servers work in order to use them; all you need to know is that you can't do anything without a connected session – in the case of the file server, a connected `RFs`.

So the pattern for using the file server is

- connect an `RFs` to the file server,
- open a file, specifying which `RFs` to use,
- do what you want to do,
- close the file,
- close the `RFs` using `Close()`, since R objects don't have destructors.

You can open any number of files or directories using a single `RFs`. You can carry an `RFs` as part of your application's object data, open it at the time you open your application, and close it when you finish. As was mentioned in Chapter 6, the CONE environment already has an open `RFs`, so you don't need to create one of your own from a GUI program; just use `iCoeEnv->FsSession()`.

After you have closed the session, you won't be able to call any more member functions on objects that were opened using that session. You should make sure you close and clean up these objects – preferably before you close the session. In any case, when the session is closed, the server will clean up any server-side resources associated with the session.

`RFs` contains many useful file system-related operations:

- manipulating the current directory;
- making, removing, and renaming directories;
- deleting and renaming files;
- changing directory and file attributes;
- notifying changes;
- manipulating drives and volumes;
- peeking at file data without opening the file (used by some file format recognizers);
- adding and removing file systems;
- system functions to control and check the status of the server.

Refer to the SDK for more details.

13.3.2   The Current Directory

Most RFs-related functions are *stateless* – that is, the results of a function don't depend on any functions previously called. That's why you can share CONE's RFs with other libraries used by your application.

RFs has just one item of state: its current directory. When you open an RFs, usually its current directory is set to c:\; which as discussed, will be hidden from users in UIQ and may be just a ramdrive in some devices, or a flash-based drive in others. Indeed, c: is *always* an internal read/write drive on Symbian OS phones. On the Psion Series 5, it was a RAM disk, but on smartphones, it's a flash-based medium, which is (a) persistent with respect to taking the battery out – something phone users do frequently – and (b) slower, especially when handling reads and writes scattered over discontiguous sectors.

You can use SetSessionPath() to change the current directory used by an open RFs. You can use SetDefaultPath() to change the initial current directory for all future RFs objects. The current directory includes the drive as well as directory names. So, unlike DOS, there is no concept of one current directory per drive.

If you manipulate or rely on the current directory, make sure you use your own RFs rather than sharing one with other programs.

13.3.3   Drives, File Systems, and Media

Symbian OS supports at least three different media types:

- The ROM file system that is assigned to z: was built by the ROM builder and is clearly read-only as far as Symbian OS programs are concerned.
- An internal read/write drive that is always assigned to c: On smartphones and communicators, this is a flash-based drive, with flash performance characteristics, on which data persists when all power is removed. Historically (i.e. Series 5 and netBook), the internal drive was a RAM drive, which had the benefit of being much faster, but which needed battery power to maintain even when the device was switched off, and whose entire contents would disappear when all power was removed (such devices had a secondary backup power scheme such as a backup battery – e.g. Psion Series 5 – or some reserved main battery power like the Psion Revo). Symbian OS takes enormous care to ensure that a sector write is atomic (that is, it completes entirely, or it doesn't complete at all). In turn, higher-level components such as STORE and DBMS use formats and protocols that ensure failure to write to the media never results in a corrupt file.
- Removable media such as CF, MMC, MemoryStick, Microdrive or SD cards, on which, if present, the main drive is assigned to d:. Such

technologies require little power to read, but rather more power to write. As battery power drains, it may become impossible to write to such a drive. Again, Symbian OS takes enormous care to ensure that a sector write is atomic.

F32 supports the addition of new file system drivers, either for networked drives or for new media types. Drivers may be added dynamically, without rebooting and without interrupting any connected file server sessions.

Media configuration is an important variable of Symbian OS. For example, the Psion Revo doesn't support removable media and the Nokia 7650 uses a very small flash drive as c: for temporary file storage, whereas its d: drive corresponds to its main internal flash-based media.

However, it's always safe to assume that a Symbian OS phone has a read-only z: drive and a read/write c: drive.

13.3.4 Files

An open file is represented by an RFile object. You can open a file with one of four RFile functions, each of which takes an RFs so that the RFile has a session within which to communicate with the server. The 'open' functions are:

- Open(), which opens an existing file for either reading or writing.
- Create(), which creates a new file for writing.
- Replace(), which deletes an existing file and creates a new one for writing.
- Temp(), which opens a temporary file and allocates a name to it.

When you've opened a file, you can Read() from it into a TDes8 or Write() to it from a TDesC8. You can also Seek() to a position and Flush() any server-side write buffers to the file.

Various access modes are supported: shared read, exclusive write, or shared write. Operations are blocked if they violate the access and sharing modes on a file. You normally specify the access mode when you open the file, although you can change it while the file is open using ChangeMode(). If you're using shared write access, you can use Lock() to claim temporary exclusive access to regions of the file and then UnLock() it later. Moreover, while the file is open, you can change its name using Rename().

Later we will see examples of opening, reading, and writing. If you're interested in more details on the other functions, see the SDK.

13.3.5 Directories

A directory contains files and other directories each of which is represented by a **directory entry** or simply, in file server API language, an **entry**.

The `RDir` class allows you to iterate through all the entries in a directory, while the `TEntry` type is used to contain a single entry.

It's expensive to call the file server once for each directory entry, so `RFs` provides high-level `GetDir()` functions that get more than one entry into a `CDir`.

You can change attributes of directory entries, including the hidden, system, read-only, and archive bits. The only bit with a really unambiguous meaning is read-only. If a file is specified as read-only, then you won't be able to write to it or erase it.

Hidden files are optionally hidden by some higher-level components. System files are optionally hidden by some higher-level components. These attributes are supported for strict compatibility with VFAT, but usage conventions in the PC world are confused and they aren't important in Symbian OS, so it's probably best not to use them. If you want to conceal files from average end users, use the `\System\` folder.

Directory entry timestamps are maintained by Symbian OS in UTC, not local time, so that backup programs that are timestamp-based don't get confused when there's a time zone change.

13.3.6 Cracking Filenames

Code that manipulates filenames is tricky. Potentially, a filename contains four parts, that are

- the drive (a single letter and a colon)

- the path (a list of directories starting and ending with a backslash)

- the filename (if an extension is specified, everything before the final dot; if not, everything after the final backslash)

- the extension (everything after the final dot (after the final backslash)).

> **Don't attempt to manipulate filenames yourself. Use `TParseBase` classes to do it for you.**

Symbian OS offers a concise class hierarchy for manipulating filenames:

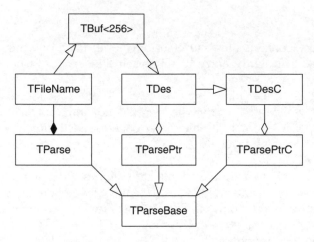

Figure 13.4

The base class for parsing is `TParseBase`. There are three implementations:

- `TParse`, which includes an entire filename and is therefore a large object.
- `TParsePtr`, which functions like a `TPtr` in that it refers to a buffer outside itself that can be changed.
- `TParsePtrC`, which functions like a `TPtrC` in that it refers to a buffer outside itself that cannot be changed.

`TFileName` is simply `TBuf<256>`, a descriptor long enough to contain the longest possible file name.

> **`TFileName` is a big object – 524 bytes. For this reason, you should never allocate or pass a `TFileName` on the stack. Where possible, use an `HBufC*` or some other descriptor type to contain the filename in as little space as possible.**

For instance, if you *know* that a certain filename is only 20 characters long, you could use a `TBuf<20>` to contain it.

The concrete `TParseXxx` classes allow you to query or manipulate a filename:

- `TParsePtrC` uses an existing filename referred to by a const `TDesC&`. It allows you to use all the query functions supported by the `TParseBase` class, which we'll examine shortly.
- `TParsePtr` uses an existing filename referred to by a `TDes&` parameter.

- `TParse` contains the filename as a `TFileName`. This is a big object, so don't use a `TParse` if a `TParsePtr` or `TParsePtrC` will do.

`TParseBase`'s query functions allow you to extract each element of a full filename: the drive, the path, the drive-and-path, the filename, the extension, the name-and-extension, or the full filename (which includes everything). You can ask whether the drive, path, filename, extension, or name-and-extension is present, and also whether the filename is in the root directory of a drive.

Wildcard matching using '`*`' for any string and '`?`' for any individual character is supported. You can ask whether the file specification contains any wildcards. Wildcard matching doesn't treat '`.`' as a special character (as it does on DOS).

If you're using `TParse` or `TParsePtr`, you can use `AddDir()` to add a single directory qualifier onto the end of the path or `PopDir()` to take the final directory qualifier off the path.

`TParse`'s constructor allows you to pass three filenames that are parsed and stored in `TParse`'s internal `TFileName`. The drive, path, name of the file, and extension are taken from the first of the three filenames to specify them, respectively.

If you're specifying a directory name to any `TParse`-related class, you must include the trailing '`\`'.

As usual in C++ source code, be careful to double all your '`\`' characters, as in `_L("c:\\System\\")`.

For more details on all this, see the SDK.

13.4 The `streams` Program

The `streams` program, in `\scmp\streams\`, is a UIQ program that illustrates some aspects of the file server APIs we have mentioned above. It provides a very simple example of how to use the streams API, which I'll describe in the next section.

Some screenshots of the application in action are seen in Figure 13.5.

`streams` is based on `hellogui`. The Edit options in the menu provide three operations

- Write file allows you to specify a filename and some text to write to it. The text is then displayed in the main application view.

- Read file allows you to specify a filename from which text will be read and displayed in the main application view.

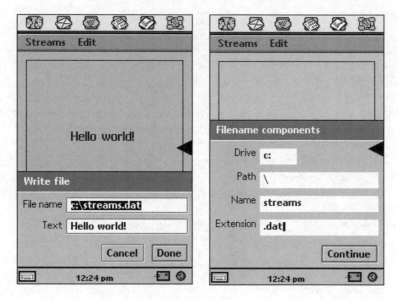

Figure 13.5

- Parse filename uses a `TParse` object to crack the various components of the filename you specified.

The first screenshot shows the **Write file** dialog. In a real UIQ program, you'd never browse or be shown a file using any file-selection dialogs. For the sake of this example, though, we're going down to basics and allowing the user to type in the filename directly. Here's the application structure:

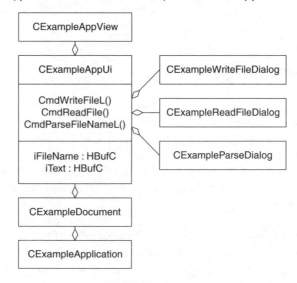

Figure 13.6

Compared with `Hellogui`, this program has

- nontrivial command handling functions for its three main commands,
- dialogs that implement some of the handling of its three main commands.

The main action is in `CExampleAppUi`'s `CmdXxxL()` functions and the corresponding functions in their dialogs' postprocessing functions, titled `OkToExitL()`.

13.4.1 Connecting to the File Server

Originally, I coded `streams` to include an `RFs` object which I opened from the app UI's second-phase constructor – in addition to initializing the `iFileName` and `iText` strings:

```
void CExampleAppUi::ConstructL()
    {
    ...
    iFileName = iEikonEnv->AllocReadResourceL
        (R_EXAMPLE_INITIAL_FILE);
    iText = iEikonEnv->AllocReadResourceL
        (R_EXAMPLE_TEXT_HELLO);
    User::LeaveIfError(iFs.Connect());
    }
```

I closed the `RFs` again in the app UI's C++ destructor – as well as freed the `iFileName` and `iText` strings:

```
CExampleAppUi::~CExampleAppUi()
    {
    iFs.Close();
    delete iText;
    delete iFileName;
    ...
    }
```

But, as we saw earlier, CONE provides its own `RFs` that can normally be used by other components too. So I deleted the `iFs` member, deleted initialization and destruction code for it and changed all references to use `iCoeEnv->FsSession()` instead.

13.4.2 Writing a File

The file writing command handler invokes `CExampleWriteFileDialog` and, after you press the OK button or the *Enter* key, its `OkToExitL()` function is called. It's quite a large piece of code, so I'll take it a step at a time. First, I get the filename:

```
TBool CExampleWriteFileDialog::OkToExitL(TInt /* aKeycode */)
   // termination
   {
   // Get filename
   CEikEdwin* edwin = static_cast<CEikEdwin*H>
      (Control(EExampleControlIdFileName));
   HBufC* fileName = edwin->GetTextInHBufL();
    // Check it's even been specified
    if(!fileName)
        {
        TryChangeFocusToL(EExampleControlIdFileName);
        iEikonEnv->LeaveWithInfoMsg(R_EIK_TBUF_NO_FILENAME_
            SPECIFIED);
        }
   CleanupStack::PushL(fileName);
```

There's nothing obviously file-related in here, but there's some interesting descriptor and cleanup stack code. If this function succeeds, I have an HBufC* filename, which is pushed to the cleanup stack and contains a nonempty string. It can leave for numerous reasons, the most interesting of which is my call to iEikonEnv->LeaveWithInfoMsg(). This leaves, cleans up anything that needs cleaning up (nothing does, as it happens) and notifies the error to the user.

In fact, iEikonEnv->LeaveWithInfoMsg() reads its message from a resource file. To do this, a CONE function that uses CONE's built-in RFs is called. I'll also be using this in the next step, which is to check that the filename is valid:

```
// Check it's a valid filename
if(!iCoeEnv->FsSession().IsValidName(*fileName))
    {
    TryChangeFocusToL(EExampleControlIdFileName);
    iEikonEnv->LeaveWithInfoMsg(R_EIK_TBUF_INVALID_FILE_NAME);
    }
```

I just use IsValidName() to check whether it's valid; if not, I move the focus to the filename control and leave with an info-message. This time, I'm getting value from the cleanup stack. The filename will be popped and destroyed as part of this leave processing.

Now, I get the text to write to the file:

```
// Get the text string
edwin = static_cast(CEikEdwin*>(Control EExampleControlIdText));
HBufC* text = edwin->GetTextInHBufL();
```

```
if(!text)
    text = HBufC::NewL(0);
CleanupStack::PushL(text);
```

I don't really mind if it's blank this time, but for uniformity reasons, I allocate a zero-length string rather than using the zero pointer returned by this dialog API.

There's plenty of leaving code to follow, so I push the text to the cleanup stack before moving on to ensure that all the directories needed for the file exist:

```
// Ensure the directories etc. needed for the file exist
TInt err = iCoeEnv->FsSession().MkDirAll(*fileName);
if(err != KErrNone && err != KErrAlreadyExists)
    User::Leave(err);
```

The code makes a simple MkDirAll() call. This time you can see why RFs functions don't leave with an error code: I'm happy if either the directories didn't exist and they were created successfully, or they did exist and nothing happened. Otherwise, I initiate a leave.

Now things get more delicate. I tentatively create the file, by opening it:

```
// Check whether it's going to be possible to create the
    file for writing
RFile file;
err = file.Create(iCoeEnv->FsSession(), *fileName, EFileWrite);
if(err != KErrNone && err != KErrAlreadyExists)
    User::Leave(err); // No need to close file, since it didn't open
```

RFile::Create() creates a new file if possible. If the file was already there, err contains KErrAlreadyExists: I want to know whether the user really wants to replace this file, and I'll check that in the next step. If the file wasn't there, but was created successfully by RFile::Create(), then err is set to KErrNone. In any other case, I leave.

If I had wanted to replace the file without checking with the user, I would have used RFile::Replace() to open the file. As it is, I want to check with the user:

```
// Check whether the user wants to replace the file,
    if it already exists
if(err == KErrAlreadyExists)
```

```
    {
    if(iEikonEnv->QueryWinL(R_EIK_TBUF_FILE_REPLACE_CONFIRM))
        User::LeaveIfError(file.Replace(iCoeEnv->FsSession(),
                                        *fileName,
                                        EFileWrite));
    else
        iEikonEnv->LeaveWithInfoMsg(0); // Let user try again
    }
```

If the attempt to create the file revealed that it already existed, I use a UIQ query dialog to confirm with the user. If the user insists, then I open the file again with `RFile::Replace()`. Any errors *this* time must be genuine errors, so I don't have to check specific codes – I simply enclose the call with `User::LeaveIfError()`.

But if the user didn't want to replace the file, I leave silently with `iEikonEnv->LeaveWithInfoMsg(0)`. The leave helps to ensure that anything I've pushed to the cleanup stack is popped and destroyed. The 0 indicates that I'm not passing a resource ID of a message, because I don't want a message – none is needed, since the user knows exactly what has happened: they just replied No to a query.

Now I've done all the checks I need to make. I've verified user input and I've verified that I can create the file for writing.

I prefer to separate UI code (like this) from engine code (which actually does things, like writing files), so I don't write the file from this function. Instead, I pass the parameters back to the app UI, which will process them when I return:

```
file.Close();
// Finished with user interaction: communicate parameters
    and return
delete iAppUi->iFileName;
iAppUi->iFileName = fileName;
delete iAppUi->iText;
iAppUi->iText = text;
CleanupStack::Pop(2); // text, fileName
return ETrue;
}
```

I start by closing the file because that's no longer needed here. I then set the `HBufC` pointers in the app UI to refer to the new strings and make sure that whatever was there before is deleted. Finally, I pop both the new string pointers from the cleanup stack because they are stored safely as member variables now.

The scene of processing now returns to my `WriteFileL()` function:

```
void CExampleAppUi::WriteFileL()
    {
    // Create a write stream on the file
    RFileWriteStream writer;
    writer.PushL();                     // Writer on cleanup stack
    User::LeaveIfError(writer.Replace(iCoeEnv->FsSession(),
                                  *iFileName,
                                  EFileWrite));
    // Write the text
    writer  << *iText;
    writer.CommitL();
    // Finish
    CleanupStack::PopAndDestroy(); // Writer
    }
```

Having used an `RFile` to check that this file could be written, I *don't* use `RFile::Write()` to write it. Instead, I'm now using a higher-level class, `RFileWriteStream`, which is derived from `RWriteStream`. Every function you see in the code above, except the `Replace()` function I use to open the stream, is actually a member of the base class. `RWriteStream` provides a rich API including the insertion operator `<<`, which is used to write the text.

Before writing the text, I have to open the write stream, which, in reality means opening the file and then initiating the stream for writing. And before I do that, I use `PushL()` so the write stream can push itself to the cleanup stack.

After writing, I commit the stream data using `CommitL()` and then close the stream using `CleanupStack::PopAndDestroy()`. Committing to a write stream causes buffers to be flushed and sent to the file using `RFile::Write()`. This may fail, which is why `CommitL()` is a leaving function.

There are two good reasons for preferring to write this file through an `RWriteStream` rather than through an `RFile`:

- `RFile::Write()` is inconvenient to use and sometimes positively harmful. It's much safer to use the stream functions.
- `RFile::Write()` sends data *immediately* to the server and `RFile::Commit()` flushes *server-side* buffers to the file. In contrast, `RFileWriteStream` keeps its buffers in memory on your thread's default heap. `RFileWriteStream::CommitL()` flushes *client-side* buffers by doing an `RFile::Write()`. If your pattern of writing activity consists of many operations that write small amounts of data, then writing to the file would involve significantly more client-server messages, which would severely impact performance.

RFileWriteStream includes a range of << functions for all Symbian OS built-in types. As we'll see below, you can code an ExternalizeL() function on any class to allow it to be written out to any type of write stream using <<.

As usual, WriteFileL() is fully error checked. Every function that can leave is handled. Note that << can also leave, but doesn't have an L to indicate it!

WriteFileL() is extremely unlikely to fail in this case: only a few microseconds before calling it, I established that the conditions for writing were good and I'm only writing a small amount of data. But it *could* fail because conditions could change even in those few microseconds and if I was writing a lot of data, it could easily fail with KErrDiskFull.

There is actually no way to ensure that WriteFileL() can't fail, so we have to handle failure in the calling function. This is one of those cases where the cleanup stack doesn't do what we want: we actually need a trap. Here's the code I use to invoke the dialog and to call WriteFileL() under a trap:

```
void CExampleAppUi::CmdWriteFileL()
    {
    // Use a dialog to get parameters and verify them
    CEikDialog* dialog = new(ELeave) CExampleWriteFileDialog(this);
    if(!dialog->ExecuteLD(R_EXAMPLE_WRITE_FILE_DIALOG))
        return;
    // Write file under a trap
    TRAPD(err, WriteFileL());
    if(err)
        {
        delete iText;
        iText = 0;
        iCoeEnv->FsSession().Delete(*iFileName);
            // Don't check errors here!
        iAppView->DrawNow();
        User::Leave(err);
        }
    // Update view
    iAppView->DrawNow();
    }
```

The normal sequence of processing here is that I invoke the dialog (using the terse syntax that was covered in Chapter 10), write the file, and redraw the app view to reflect the new iText value.

If WriteFileL() leaves, I handle it by deleting iText and setting its pointer to 0. I also delete the file. Then I redraw the app view and use User::Leave() to propagate the error so that Uikon can display an error message corresponding to the error code. Test this, if you like, by inserting User::Leave(KErrDiskFull) somewhere in WriteFileL().

The trap handler has one virtue: it lets the user know what's going on. It's still not good enough for serious application use, though, because this approach loses user and file data. It's a cardinal rule that Symbian OS shouldn't do that. A better approach would be to do the following:

- *Write the new data to a temporary file*: if this fails, keep the user data in RAM, but delete the temporary file.

- *Delete the existing file*: this is very unlikely to fail, but if it does, keep the user data in RAM but delete the temporary file.

- *Rename the temporary file*: this is very unlikely to fail, and if it does, we're in trouble. Keep the user data in RAM anyway.

I would need to restructure my program to achieve this; the application architecture's framework for load/save and embeddable documents looks after these issues for you.

For database-type documents, you face these issues with *each entry* in the database. They are managed by the permanent file store class, which we'll encounter below.

13.4.3 Reading it Back

The code to read the data we have written is similar. Firstly, we read the filename from a dialog and then use `OkToExitL()` to do some checking. This time, the code is much easier and I'll present it in one segment:

```
TBool CExampleReadFileDialog::OkToExitL(TInt /* aKeycode */)
    // Termination
    {
    // Get filename
     CEikFileNameEditor* fnamed=static_cast(CEikFileNameEditor*>
                        (Control(EExampleControlIdFileName));
    HBufC* fileName = fnamed->GetTextInHBufL();
    // Check it's even been specified
    if(!fileName)
        {
        TryChangeFocusToL(EExampleControlIdFileName);
        iEikonEnv->LeaveWithInfoMsg(R_EIK_TBUF_NO_FILENAME_SPECIFIED);
        }
    CleanupStack::PushL(fileName);
    // Check it's a valid filename
    if(!iCoeEnv->FsSession().IsValidName(*fileName))
        {
        TryChangeFocusToL(EExampleControlIdFileName);
        iEikonEnv->LeaveWithInfoMsg(R_EIK_TBUF_INVALID_FILE_NAME);
        }
    // Check whether it's going to be possible to create the file
        for reading
    RFile file;
```

```
User::LeaveIfError(file.Open(iCoeEnv->FsSession(), *fileName,
    EFileRead));
file.Close();
// Finished with user interaction: communicate parameters
    and return
delete iAppUi->iFileName;
iAppUi->iFileName = fileName;
CleanupStack::Pop(); // fileName
return ETrue;
}
```

As before, the job of `OkToExitL()` is to check that the user's input is sensible. This function checks

- that a filename has been specified,
- that the filename is valid,
- that it's going to be possible to read the file: I use `RFile::Open()` for this and leave if there was any error.

Assuming all is well, control returns to my command handler that processes it using:

```
void CExampleAppUi::CmdReadFileL()
    {
    ...
    // Create a read stream on the file
    RFileReadStream reader;
    reader.PushL();
        // Reader on cleanup stack
    User::LeaveIfError(reader.Open(iCoeEnv->FsSession(),
                                   *iFileName,
                                   EFileRead));
    // Read the text
    HBufC* string = HBufC::NewL(reader, 10000);
    delete iText;
    iText = string;
    // Finish
    CleanupStack::PopAndDestroy();      // Reader
    ...
    }
```

The code is largely a mirror image of the code I used to write the data:

- I create an `RFileReadStream` object.
- Instead of using a `>>` operator to read the string that I wrote with `<<`, I use `HBufC::NewL(RReadStream&, TInt)`. This function takes a peek into the descriptor I wrote, checks how long it is and allocates

an `HBufC` that will be big enough to contain it (provided it's smaller than maximum length I also pass – in this case, `10 000` characters). It then reads the data.

- I don't need to commit a read stream because I've got all the data I want and I'm not writing anything. So I simply close with `Cleanup-Stack::PopAndDestroy()`.

`RFileReadStream` is a mirror to `RFileWriteStream` and as you might expect, it's derived from `RReadStream`. `RReadStream` contains many `>>` operators and you can add support for reading to a class by coding `InternalizeL()`.

The only real issue in the code above was predicting how long the string would be. `HBufC::NewL(RReadStream&, TInt)` reads the descriptor that was written previously when `*iText` was written. Before allocating the `HBufC`, this function

- checks the length indicated in the stream,

- returns `KErrCorrupt` if the length is longer than the maximum I passed or in various other circumstances where the length in the stream can't be valid.

13.4.4 Parsing Filenames

The `streams` example shows how to use a `TParsePtrC` to crack a filename into its constituent parts. Here's some code to display all four parts of a filename in a dialog:

```
void CExampleParseDialog::PreLayoutDynInitL()
    {
    TParsePtrC parser(*iAppUi->iFileName);
    CEikEdwin* edwin=static_cast(CEikEdwin*>
        (Control(EExampleControlIdDrive));
    edwin->SetTextL(&parser.Drive());

    edwin=static_cast<CEikEdwin*>(Control(EExampleControlIdPath));
    edwin->SetTextL(&parser.Path());

    edwin=static_cast<CEikEdwin*>(Control(EExampleControlIdPath));
    edwin->SetTextL(&parser.Name());

    edwin=static_cast<CEikEdwin*>(Control(EExampleControlIdPath));
    edwin->SetTextL(&parser.Ext());
    }
```

The interesting thing here is that the `TParsePtr` constructor causes `TParsePtr` simply to store a reference to the filename string in `iFile-Name`. Because `TParsePtr` is essentially only a pointer, it uses very little space on the stack. I can then retrieve all its constituent parts using functions like `Drive()`.

For file manipulation, I need to use more space. Here's how I could find the name of my resource file:

```
TFileName appFileName = Application()->AppFullName();
TParse fileNameParser;
fileNameParser.SetNoWild(_L(".rsc"), &appFileName, NULL);
TPtrC helpFileFullName = fileNameParser.FullName();
```

First, I get my application's full name into a `TFileName`. If I installed to `c:`, it's `c:\system\apps\streams\streams.app`. Then I set up a `TParse` to parse the name: its first argument is `.rsc`, the second argument is my application name, and the third argument is null. Finally, I ask the `TParse` to return the full name of my file, which is calculated as above by scanning each of the three parameters, so it's `c:\system\apps\streams\streams.rse.hlp`.

This time, I had to change the data rather than simply pointing to it, so I couldn't use a `TParsePtr`. Having both a `TFileName` *and* a `TParse` on the stack uses a lot of room (over 1k). You need to avoid this except where it's both necessary (as here) and safe (meaning you don't then call many more functions that are likely to have significant stack requirements).

13.4.5   Summary of the File APIs

Symbian OS file APIs contain all the functions you would expect of a conventional file system. We use `RFs`, `TParse`, and `RFile` functions to manipulate the file system, files, and directories. We also use `RFs` to ensure that our client program can communicate with the file server.

But we rarely use `RFile::Write()` or `RFile::Read()` for accessing file-based data. Instead, we usually use streams.

13.5   Streams

In the `streams` example, we got our first sight of the central classes for data management:

- `RWriteStream`, which **externalizes** objects to a stream.
- `RReadStream`, which **internalizes** objects from a stream.

13.5.1 External and Internal Formats

Data stored in program RAM is said to be in **internal format**. Endian-ness, string representations, pointers between objects, padding between class members, and internally calculated values are all determined by the CPU type, C++ compiler and program implementation.

Data stored in a file or sent via a communications link is said to be in **external format**. The actual sequence of bits and bytes matters, including string representation and endian-ness. You can't have pointers – instead, you have to **serialize** an internal object network into an external stream and **de-serialize** when you internalize again. Compression or encryption may also be used for the external format.

> You should distinguish carefully between internal and external formats. Never 'struct dump' (that is, never send your program's structs literally) when sending data to or over an external medium.

For reference, the Symbian OS emulator and ARM platform implementations have only a couple of internal format differences, such as:

- 64-bit IEEE 754 double-precision, floating-point numbers are stored with different endian-ness on ARM and x86 architectures.

- ARM requires that all 32-bit data be 32-bit aligned, whereas x86 does not. Therefore, ARM data structures potentially include padding that isn't present in their x86 equivalents.

13.5.2 Ways to Externalize and Internalize Data

We have two ways to externalize and (implicitly) three ways to internalize:

- You can use insertion and extraction operators: externalize with `stream << object` and internalize with `stream >> object` (remember these operators can leave).

- You can externalize with `object.ExternalizeL(stream)` and internalize with `object.InternalizeL(stream)`.

- You can incorporate allocation, construction, and internalization into a single function of the form `object = class::NewL(stream)`.

There are, in fact, many write stream and read stream classes that derive from `RWriteStream` and `RReadStream`, and access streams stored in different objects. These objects include

- files, as we have just seen;

- memory: a fixed area of memory that's described by a descriptor or a (pointer, length) pair; or an expandable area of memory described by a CBufBase (see Chapter 8);

- stream stores, which I'll describe below;

- dictionary stores, which I'll also describe below.

Some streams exist to perform preprocessing before writing to other streams. An example of this is REncryptStream, which encrypts data before writing it to a write stream, and RDecryptStream, which decrypts data just read from a read stream.

To externalize, you always need an RWriteStream: in the code fragments below, writer could be an object of any class derived from RWriteStream.

To internalize, you always need an RReadStream: in the code fragments below, reader could be an object of any class derived from RReadStream.

<< and >> operators

To externalize a built-in type, you can use <<:

```
TInt32 x;
writer  << x;
TBufC <20> text = KText;
writer  << text;
```

To internalize again, you can use >>:

```
TInt32 x;
reader >> x;
TBuf <20> text;
reader >> text;
```

However, you can't *always* use << and >>. The semantics of TInt specify only that it must be *at least* 32 bits; it may be longer. Furthermore, users may employ TInts to represent quantities that are known to require only, say, 8 bits in external format. As the application programmer, you know the right number of bits and the stream doesn't try to second-guess you. If you write this

```
TInt i;
writer  << i;
```

the stream class doesn't know what to do. You will get a compiler error. If you find yourself in this situation, you can either cast your `TInt` to the type you want to use or use one of the specific write or read functions described below.

> You cannot externalize a `TInt` using << or internalize it using >>. You must choose a function that specifies an external size for your data.

WriteXxxL() and ReadXxxL() functions

If you want to be very specific about how your data is externalized, you can use the `WriteXxxL()` and `ReadXxxL()` member functions of `RWriteStream` and `RReadStream`. Here's some code:

```
TInt i = 53;
writer.WriteInt8L(i);
...
TInt j = reader.ReadInt8L();
```

By doing this, it's clear that you mean to use an 8-bit external format. Here's the complete set of `WriteXxxL()` and `ReadXxxL()` functions:

RwriteStream Functions	RReadStream Functions	<<Type	External Format
WriteL()	ReadL()		Data in internal format
WriteL(Rread Stream&)	ReadL(Rwrite Stream&)		Transfer from other stream type
WriteInt8L()	ReadInt8L()	TInt8	8-bit signed integer
WriteInt16L()	ReadInt16L()	TInt16	16-bit signed integer, bytes stored little-endian
WriteInt32L()	ReadInt32L()	TInt32	32-bit signed integer, bytes stored little-endian
WriteUint8L()	ReadUint8L()	TUint8	8-bit unsigned integer
WriteUint16L()	ReadUint16L()	TUint16	16-bit unsigned integer, bytes stored little-endian
WriteUint32L()	ReadUint32L()	TUint32	32-bit unsigned integer, bytes stored little-endian

WriteReal32L()	ReadReal32L()	TReal32	32-bit IEEE754 single-precision floating point
WriteReal64L()	ReadReal64L()	TReal, TReal64	64-bit IEEE754 double-precision floating point

If you use << and >> on built-in types, it will ultimately call these functions. The '<< type' column shows what Symbian OS data type will invoke these functions if used with the << and >> operators.

Raw data

The WriteL() and ReadL() functions for raw data deserve a closer look. Here are the WriteL() functions, as defined in the header file s32strm.h:

```
class RWriteStream
    {
public:
    ...
    IMPORT_C void WriteL(const TDesC8& aDes);
    IMPORT_C void WriteL(const TDesC8& aDes, TInt aLength);
    IMPORT_C void WriteL(const TUint8* aPtr, TInt aLength);
    ...
//
    IMPORT_C void WriteL(const TDesC16& aDes);
    IMPORT_C void WriteL(const TDesC16& aDes, TInt aLength);
    IMPORT_C void WriteL(const TUint16* aPtr, TInt aLength);
    ...
```

These functions simply write the data specified, according to the following rules:

- WriteL(const TDesC8& aDes, TInt aLength) writes aLength bytes from the beginning of the specified descriptor.
- Without the aLength parameter, the whole descriptor is written.
- The const TUint8* variant writes aLength bytes from the pointer specified.
- The const TDesC16 and const TUint16* variants write Unicode characters (with little-endian byte order) instead of bytes.

RReadStream comes with similar (though not precisely symmetrical) functions:

```
class RReadStream
    {
public:
    ...
    IMPORT_C void ReadL(TDes8& aDes);
    IMPORT_C void ReadL(TDes8& aDes, TInt aLength);
    IMPORT_C void ReadL(TDes8& aDes, TChar aDelim);
    IMPORT_C void ReadL(TUint8* aPtr, TInt aLength);
    IMPORT_C void ReadL(TInt aLength);
    ...
//
    IMPORT_C void ReadL(TDes16& aDes);
    IMPORT_C void ReadL(TDes16& aDes, TInt aLength);
    IMPORT_C void ReadL(TDes16& aDes, TChar aDelim);
    IMPORT_C void ReadL(TUint16* aPtr, TInt aLength);
    ...
```

The problem when reading is to know when to stop. When you're writing, the descriptor length (or the `aLength` parameter) specifies the data length. When you're reading, the rules work like this:

- The `TDes8& aDes` format passes a descriptor whose `MaxLength()` bytes will be read.

- If you specify `aLength` explicitly, then that number of bytes will be read.

- If you specify a delimiter character, the stream will read up to *and including* that character. If the `MaxLength()` of the target descriptor is encountered before the delimiter character, reading stops after `MaxLength()` characters – nothing is read and thrown away.

Like all other `ReadXxxL()` functions, these functions will leave with `KErrEof` (end of file) if the end of file is encountered during the read operation.

You should use these raw data functions with great care. Any data that you externalize with `WriteL()` is effectively struct-dumped into the stream. This is fine provided that the data is already in external format. Be sure that it is!

When you internalize with `ReadL()`, you must always have a strategy for dealing with the anticipated maximum length of data. For example, you could decide that it would be unreasonable to have more than 10 000 bytes in a particular string and so you check the length purportedly given and if you find it's more than 10 000 you leave with `KErrCorrupt`. That's what `HBufC::AllocL(RReadStream&, TInt)` does.

Strings

You'll remember that I used this,

```
writer.iObj  << *iText;
```

to externalize the content of the string in the **streams** program in which `iText` was an `HBufC*`. This doesn't match against any of the basic types externalized using an `RWriteStream::WriteXxxL()` function. Instead, it uses C++ templates to match against an **externalizer** for descriptors that write a header and then the descriptor data.

To internalize a descriptor externalized in this way, if the descriptor is short and of bounded length, you can use `>>` to internalize again:

```
TBuf<20> text;
reader.iObj >> text;
```

But if the length is variable you can internalize to a new `HBufC` of exactly the right length, which is the technique I used in `streams`:

```
iText = HBufC::NewL(reader.iObj, 10000);
```

In either case, the Symbian OS C++ framework uses an **internalizer** for descriptors to reinternalize the data. The internalizer reads the header that contains information about the descriptor's character width (8 or 16 bits) and length (in characters). You get panicked if the character width of the descriptor that was externalized doesn't match the descriptor type to which you're internalizing. The length is used to determine how much data to read.

It's possible to externalize strings using two `WriteL()` functions (one for the length of the data and another for the data itself) and then reinternalize them by reading the length and the data. But it's better to use the `<<` operator to externalize and either `>>` or `HBufC::NewL(RReadStream&)` to internalize, because the code is less difficult, but also, more importantly because you'll get standard Unicode compression (defined by the Unicode consortium) on data read and written this way.

You don't *get this compression when using* `WriteL(TDesC16&)`. *The standard Unicode compression scheme involves state, but* `WriteL()` *is of necessity stateless.*

ExternalizeL() and *InternalizeL()* functions

If you have an object of some class type, you need to write your own functions to enable that object to be externalized and internalized. These functions must have the following prototypes:

```
class Foo
    {
public:
    ...
    void ExternalizeL(RWriteStream& aStream) const;
    void InternalizeL(RReadStream& aStream);
    ...
    };
```

A general template for `operator<<()` ensures that you can externalize a Foo using either this:

```
Foo foo;
foo.ExternalizeL(writer);
```

or this:

```
writer  << foo;
```

A similar template exists for `operator>>()`.

The `ExternalizeL()` and `InternalizeL()` functions are not virtual and there's no implication that Foo is derived from any particular base class or that it has to be a C, T, or R class.

You then have to implement your own code to externalize and internalize the class. Here's some externalizing and internalizing code from my Battleships application:

```
void CGameController::ExternalizeL(RWriteStream& aStream) const
    {
    aStream.WriteUint8L(iState);
    aStream.WriteUint8L(iHaveFirstMovePref);
    aStream.WriteUint8L(iFirstMovePref);
    }
void CGameController::InternalizeL(RReadStream& aStream)
    {
    iState = (TState)aStream.ReadUint8L();
    iHaveFirstMovePref = aStream.ReadUint8L();
    iFirstMovePref = (TFirstMovePref)aStream.ReadUint8L();
    }
```

The patterns here are characteristic:

- The two functions mirror each other closely.
- I know that all my data can be externalized into 8-bit unsigned integers, so I use `WriteUint8L()` to write, and the corresponding `ReadUint8L()` to read.
- I don't use `<<` and `>>` because my internal format for all variables is a 32-bit integer – an enumeration for `iState` and `iFirstMovePref`, and a `TBool` for `iHaveFirstMovePref`.
- I need some casting to convert integers back into enumerations when I read them in.

If your object is more complicated, you can recursively externalize and internalize your member data.

ExternalizeL(), <<, or WriteXxxL()?

We have now seen *three* ways to externalize:

Technique	Application
`writer << object`	`object` may be a built-in integer (but not `TInt`), a real type, a descriptor, or any class with a properly specified `ExternalizeL()` member function.
`writer.WriteXxxL (object)`	`object` must be a suitable built-in type, or descriptor whose contents are to be externalized as-is.
`object.ExternalizeL (writer)`	`object` is of class type with a suitable `ExternalizeL()` function.

Which method should you use?

- If you want to externalize a descriptor *with its header*, then use `<<`.
- If you have to specify the exact length to use for a built-in type and the internal format is either `TInt` or some length that's not what you want to use for the external format, then use a `WriteXxxL()` function.
- If you prefer to save typing, use `<<` in preference to `ExternalizeL()` when dealing with a class type for which an `ExternalizeL()` exists.
- If you are writing a container class that's templated on some type `T`, you know whether `T` will be a built-in type or a class type. Use `<<` and C++ will match against the right function.

This boils down to

- use << if you can
- use specific `WriteXxxL()` functions if you have to.

InternalizeL(), >>, ReadXxxL(), or NewL(RReadStream&)?

For the corresponding question about the best way to internalize, the basic rule is very simple: do the opposite of what you did when externalizing. Here's a complication: when writing a 32-bit integer compactly to a write stream, you could use this:

```
writer  << (TInt8)i;
```

but, when reading, you can't use this:

```
reader >> (TInt8)i;
```

For this reason, it's better to use `WriteInt8L()` and `ReadInt8L()` in both cases so you can easily check the symmetry of your `Internal-izeL()` and `ExternalizeL()` functions.

Another complication is that you can think of internalizing as either an assignment or a construction. For a simple `T` class, assignment is okay,

```
reader >> iFoo;
```

but for a class of any complexity or of variable length, it's better to think of internalizing as a constructor. If you're replacing an existing object, construct the new one by internalizing it and then delete the old one and replace it:

```
CBar* bar = CBar::NewL(reader, other_parms);
delete iBar;
iBar = bar;
```

It uses more memory, but in many cases it's the only practical approach.

13.5.3 Types of Stream

The base `RWriteStream` and `RReadStream` interfaces are implemented by many interesting and useful derived classes that write to and read from streams in different media. Concrete stream types include:

Header File	Class Names	Medium
s32file.h	RFileWriteStream, RFileReadStream	A file. Constructors specify either an open RFile or a file server session and a filename.
s32mem.h	RDesWriteStream, RDesReadStream	Memory, identified by a descriptor.
s32mem.h	RMemWriteStream, RMemReadStream	Memory, identified by a pointer and length.
s32mem.h	RBufWriteStream, RBufReadStream	Memory, managed by a CBufBase-derived dynamic buffer. As new data is written through an RBufWriteStream, the destination CBufBase will be expanded as necessary.
s32std.h	RStoreWriteStream, RStoreReadStream	A stream store of which there is more in the next section. RreadStream constructors specify a stream store and a stream ID. RWriteStream constructors for a new stream specify a stream store and return a stream ID. RWriteStream constructors for modifying an old stream specify a stream store and a stream ID.
s32stor.h	RDictionaryReadStream, RDictionaryWriteStream	A dictionary store. Constructors specify a dictionary store and a UID. See the section near the end of this chapter for more information.
s32crypt.h	REncryptStream, RDecryptStream	Another stream. Constructors specify the host stream, the CSecurityBase algorithm, and a string to initialize the CSecurityBase – effectively, a password.

13.6 Stores

Many other systems provide stream APIs (such as Java and standard C), but Symbian OS goes further – streams do not exist in isolation. The stream system was designed from the outset with file-based stores in mind. Two principal types of store were envisaged:

- Direct file stores where an entire file is written or read in a single operation; when a document in memory is saved the entire file store is written in a single operation and previous file data is erased.

- Permanent file stores that are databases of objects and that support efficient writing, reading, deleting, and indexing of individual objects within the store without ever deleting the entire store itself.

13.6.1 Direct File Stores

Let's consider the file format of the Boss Puzzle, that's delivered with the UIQ SDK. The Boss Puzzle is a single-player game in which you move the tiles around:

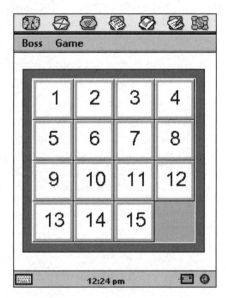

Figure 13.7

If you want to build and launch this yourself, you'll find it in the UIQ C++ SDK in the directory tree headed by \UIQExamples\papers\boss. If you build the engine\v1, view\v3, and quartz\v7 projects for the winscw udeb target, you'll then be able to run the application from the UIQ emulator's application launcher.

I closed the application after taking the screenshot above, then looked at the document file in c:\Documents\Boss. In hex, it looks like this:

```
37 00 00 10 12 3A 00 10    53 02 00 10 EE 4A 28 77 ........    ........
31 00 00 00 01 02 03 04    05 06 07 08 09 0A 0B 0C ........    ........
0D 0E 0F 00 53 02 00 10    20 42 4F 53 53 2E 61 70 ........    .BOSS.ap
70 04 53 02 00 10 14 00    00 00 34 3A 00 10 24 00 p.......    ........
00 00
```

This file consists of the following:

- A 16-byte header containing the file's UIDs and a checksum. The UIDs are `0x10000037` (for a direct file store), `0x10003a12` (for a Uikon document), and `0x10000253` (for a Boss document). The file server generates the header and checksum.

- A 4-byte stream position indicating the beginning of a stream dictionary, which is `0x00000031`.

- The document data that comprises 16 consecutive bytes containing the tile values in the Boss Puzzle. Since the puzzle has just been initialized, these 16 bytes simply contain increasing values 1, 2, 3, 4, ..., 15, 0 (the 0 represents the empty tile at the bottom right of the puzzle).

- An application header indicating the application's UID (four bytes), an externalized descriptor header byte `0x20`, and the name of the application DLL `BOSS.app`.

- The stream dictionary that starts with `0x04` (which is an externalized `TCardinality` for the value 2) to indicate two associations. The first associates the application UID `0x10000253` with the document data at offset `0x00000014`; the second associates the application identifier UID `0x10003a34` with the application identifier stream at offset `0x00000024`.

TCardinality is a class that is used to provide a compact externalization of positive numbers that can potentially have large values, but are usually small. Like in the above example, it is typically used to create an externalized representation of values such as a count of array elements or a descriptor header.
We can picture the file content like this:

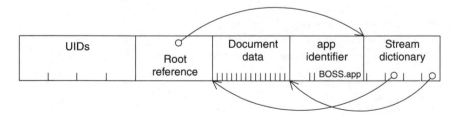

Figure 13.8

This type of file layout is frequently used, and found in many 'load/save' applications – applications that keep their data in RAM, and load and save the whole document only when necessary. The main features of this kind of layout are that seek positions are used to refer to data that has already been written – almost every reference in the file is backwards. The only exception is the reference to the root stream, which happens to be a forward reference from a fixed location early in the file.

In the language of Symbian OS, the document file is a **direct file store**, which is a kind of **persistent store**. The document has three streams, which we can picture like this:

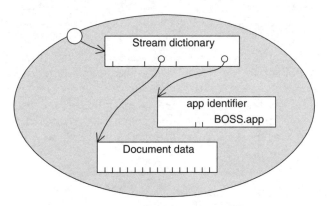

Figure 13.9

The **root stream** is accessible from the outside world. It contains a **stream dictionary**, which in turn points to two other streams. The **application identifier stream** contains information such as the application's DLL name, while the single **document data stream** contains the board layout. To write a file like this, we have to

- create a direct file store with the right name and UIDs. After this store is created, I don't need to know that it is a file store anymore – I just access it through persistent store functions;

- create a stream dictionary that will eventually be externalized onto the root stream;

- create, write, and close the document data stream – save its ID in the stream dictionary;

- create, write, and close the application identifier stream – save its ID in the stream dictionary;

- write the stream dictionary to a stream;

- close the persistent store, setting the stream containing the stream dictionary to be the root stream.

The `dfbosswrite` example does just this. First, the file store is opened and remembered as a persistent store:

```
void CBossWriter::OpenStoreL(const TDesC& aFileName)
    {
    CFileStore* store = CDirectFileStore::CreateLC(iFs,
        aFileName, EFileWrite);
    store->SetTypeL(TUidType(KDirectFileStoreLayoutUid,
                            KUidAppDllDoc,
                            KUidBoss));
    CleanupStack::Pop();        // store
    iStore = store;             // iStore is a CPersistentStore*
    }
```

This creates a file with direct file store layout and the right UIDs. It saves a pointer to the newly opened store in `iStore`, which is a `CPersistentStore*` (a base class of `CDirectFileStore`). Now that the file has been created, we need only use the more generic persistent store functions, as nothing is specific to `CDirectFileStore`.

Then we create the stream dictionary that will be written to the root stream:

```
void CBossWriter::OpenRootDictionaryL()
    {
    iRootDictionary = CStreamDictionary::NewL();
    }
```

Next, we call two functions in turn to write the data streams and store their stream IDs in the stream dictionary:

```
void CBossWriter::WriteDocumentL()
    {
    TStreamId id = iPuzzle.StoreL(*iStore);
    iRootDictionary->AssignL(TUid::Uid(0x10000253), id);
    }
void CBossWriter::WriteAppIdentifierL()
    {
    TApaAppIdentifier ident(TUid::Uid(0x10000253), KTxtBossApp);
    RStoreWriteStream stream;
    TStreamId id = stream.CreateLC(*iStore);
    stream << ident;
    stream.CommitL();
    CleanupStack::PopAndDestroy(); // stream
    iRootDictionary->AssignL(KUidAppIdentifierStream, id);
    }
```

`WriteDocumentL()` calls the Boss engine's `StoreL()` function that creates a stream, externalizes the engine data to the stream, closes the streams, and returns the stream ID. Then it stores that stream ID in the dictionary, associating it with the Boss Puzzle's UID.

`WriteAppIdentifierL()` shows how to create a stream: you use an `RStoreWriteStream`. Calling `CreateLC(*iStore)` creates it and gets its stream ID – and pushes it to the cleanup stack. The stream is then open, so you can write to it using `<<`. After writing, commit it using `CommitL()` and close it using `CleanupStack::PopAndDestroy()`. Finally, as before, we store the association between stream ID and UID in the stream dictionary.

Now we've written the two data streams, we finish by writing the stream dictionary in a new stream, setting that as the root stream of the store, and closing the store:

```
void CBossWriter::WriteRootDictionaryL()
    {
    RStoreWriteStream root;
    TStreamId id = root.CreateLC(*iStore);
    iRootDictionary->ExternalizeL(root);
    root.CommitL();
    CleanupStack::PopAndDestroy(); // root
    iStore->SetRootL(id);
    iStore->CommitL();
    }
```

We use the same technique as before to create the stream and get its ID. Then we externalize the dictionary and commit the stream. Finally, we set the root stream ID and commit the entire store. Shortly after this code, we call the C++ destructor on the `CPersistentStore` that releases all the resources associated with the store.

13.6.2 Embedded Stores

The code above was not specific to a direct file store. It would have worked equally well with any kind of persistent store.

The stream store framework provides an **embedded store** type that is intended specifically for object embedding. Imagine you have a Symbian OS Word document (like in the Nokia 9210 and Psion Series5) that embeds the Boss Puzzle. Here's what the store layout might look like, conceptually, with a Boss document inside the Word document:

Actually, the word processor's store format is a bit more complicated than this, but the simplified version here is enough to explain our point.

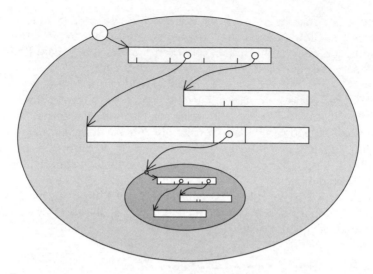

Figure 13.10

The main document is a Word document that uses a direct file store. As with the Boss Puzzle, the Word document has a root stream that is a stream dictionary referring to other streams. One of the streams will contain a stream ID that refers to a stream containing the embedded Boss Puzzle. From the point of view of the embedding store, this is a single stream.

From the point of view of the Boss Puzzle, though, this stream is an embedded store. The streams inside the embedded store are *exactly* as they were inside the direct file store.

The layout of an embedded store is *nearly* the same as a direct file store, but not quite. Embedded stores don't need the leading 16 bytes of UID information required by file stores, so these are omitted. The first four bytes of an embedded store contain the root stream ID. Stream IDs within an embedded store are *stream* seek positions relative to the stream in the embedding store, not file seek positions.

13.6.3   Permanent File Stores

We have now seen two types of store:

- Direct file stores
- Embedded stores.

In these store types, the store adds little value above that of the medium containing it (the medium is either the containing file or the containing stream).

- Stream IDs are seek positions within the medium.

- You can only refer to streams already created.

- You cannot delete a stream after it has been created.

- When you open a new stream, it is impossible to write anything else to any stream that was previously open.

- When you close the store, you cannot later reopen it and change it (except under obscure conditions and with additional constraints).

Despite – in fact because of – these restrictions, the so-called direct-layout store types are simple to work with. They are well suited for load/save-type applications such as the Boss Puzzle or Word. For these applications, the 'real' document is in user RAM and it's saved to file in entirety (or loaded from file) when necessary. When the document is saved, the old file is deleted and a new file is written again from the beginning.

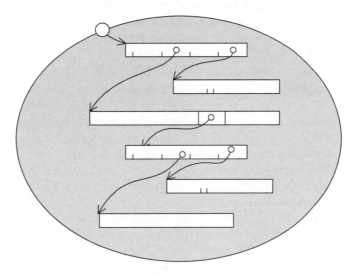

Figure 13.11

For database-type applications, the 'real' document is the data in the database file. An application loads *some* information from the database into RAM to work with it, but it doesn't load the entire database into RAM at once. In fact, it loads and replaces information in the database, a single entry at a time. In effect, for a database application, a *single entry* is like a load/save document, but the database as a whole is a permanent entity.

The stream store provides a store type for databases: the **permanent file store**. In a permanent file store:

- You can delete a stream after it has been created. You can also truncate it, and add to it in any way you like.

- You can keep as many streams open as you like, for either writing or reading. You can interleave writing to many streams (provided that you `CommitL()` between writing different streams' data).

- You can reopen the store after it has been closed and do any manipulations you like.

However, this flexibility comes at a price. Most obviously, there is no correspondence between stream ID and seek position. This relationship is private to the implementation of the permanent file store. Furthermore, you can't guarantee that all data in a stream is contiguous.

> **You have to manage a permanent file store very carefully, just as you have to manage Symbian OS memory. You must avoid 'stream leaks' with the same vigilance as you avoid memory leaks. In fact, you must be even more vigilant because permanent file stores survive even shutdown and restart of your applications and of the Symbian OS phone as a whole. And you must do all this using techniques that are guaranteed even under failure conditions.**

The stream store provides a tool analogous to the cleanup stack for cleaning up write streams that have been half-written, due to an error occurring during the act of writing to a permanent file store. The central class is `CStoreMap` that contains a list of open streams.

As you manipulate streams in a permanent file store, the store will gradually get larger and larger. The stream store provides incremental compaction APIs, so you can gradually compact a store, even while you're doing other work on it.

The permanent file store has been designed to be extremely robust. Robustness was prioritized even higher than space efficiency – though the format is still space-efficient.

The Agenda application uses the permanent file store directly. The DBMS component also uses the permanent file store; most other permanent file store users are indirect users, through the DBMS. For more on using permanent file stores, `CStoreMaps` and so on, see the SDK.

13.7 Types of Store

To summarize what we've seen so far:

- A load/save application uses a `CDirectFileStore` for its main document.

- A load/save application uses a `CEmbeddedStore` when it is embedded.

- A database application uses a `CPermanentFileStore` for its database.

These three store types are part of a small hierarchy:

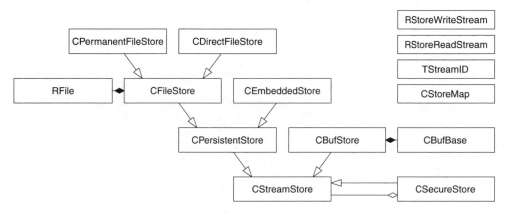

Figure 13.12

The base class for all stores is `CStreamStore`, whose API provides all the functionality needed to create, open, extend, and delete streams; to commit and revert (a kind of rollback) the entire store and to reclaim and compact.

On top of `CStreamStore`, `CPersistentStore` provides one extra piece of functionality: you can designate a single stream as the root stream. This allows you to close the store and, later, open it again. Hence the store is **persistent**; like a file, its existence persists after a program has closed it and even after the program itself has terminated.

The two file store types are derived from `CPersistentStore` via the `CFileStore` class. `CEmbeddedStore` is derived from `CPersistentStore` directly.

A `CBufStore` implements the `CStreamStore` interface in a dynamic buffer in RAM. Such a store is clearly *not* persistent: it cannot survive the destruction of its underlying buffer. `CBufStore` implements the full stream manipulation interface of `CStreamStore`. `CBufStores` are used for undo buffers in some apps including Word.

Finally, `CSecureStore` allows an entire store to be encrypted or decrypted, just as `CSecureWriteStream` and `CSecureReadStream` support encryption and decryption of individual streams.

This class hierarchy uses a useful object-oriented pattern. Derivation in this class hierarchy is based on the distinction between nonpersistent and persistent stores, and between file stores and other types. But the stream manipulation functionality cuts across the hierarchy – a full interface is supported by permanent file stores and buffer stores, while only a partial interface is supported by direct file stores and embedded stores. The only way to support this is to provide the full stream interface in the base class: derived classes implement these functions as needed and return error codes when an unsupported function is called.

Here's a summary of the store types:

File	Name	Purpose
s32stor.h	CStreamStore	Base class, with extend, delete, commit, revert, reclaim, and compact functions – not all of which are available in all implementations.
s32stor.h	CPersistentStore	Adds a root stream to CstreamStore.
s32stor.h	CEmbeddedStore	An embedded store: opens a new one on a write stream or an old one on a read stream.
s32file.h	CFileStore	File-based persistent store. Constructors specify either an RFs and a filename or an already open RFile.
s32file.h	CDirectFileStore	Direct file store. Has a wide variety of constructors supporting all file-related open functions (open, create, replace, and temp). Can also be constructed from an already open RFile.
s32file.h	CPermanentFileStore	Permanent file store. Has a wide variety of constructors.
s32mem.h	CBufStore	Nonpersistent store in a privately owned CbufBase.
s32crypt.h	CSecureStore	Secure store with encrypted streams and the like. Constructors specify host stream store, CSecurityBase encryption algorithm, and an initialization string.

Beware of the following sources of potential confusion:

- Don't get confused between a **persistent store** and a **permanent file store**. A persistent store has a root stream and can persist after you've closed it. A permanent file store is the type of file store that's used by database applications; the database itself is permanent, although its *entries* may be saved, deleted, or replaced.

- Don't use file write streams and file read streams to access file stores; you use them to access files when the file is *not* a file store. *All* store types should be accessed with store write streams and store read streams – most often, you only need to use the write stream and read stream interfaces.

13.8 Dictionary Stores and .ini Files

A persistent stream store includes a stream network in which streams may contain stream IDs that refer to other streams and which has a single root stream.

In contrast, a **dictionary store** contains a list of streams, each of which is accessed using a UID, rather than a stream ID:

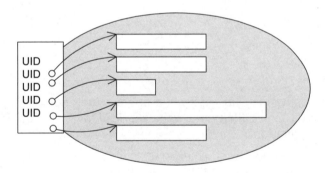

Figure 13.13

There is a small class hierarchy associated with dictionary stores, as seen in Figure 13.14.

You write to a dictionary store using a dictionary write stream that you construct, specifying the dictionary store and a UID. You read from a dictionary store using a dictionary read stream that you open, specifying a dictionary store and a UID.

There is one concrete dictionary store class – the dictionary file store that is used for .ini files. You can open the system .ini file or a named .ini file for an application.

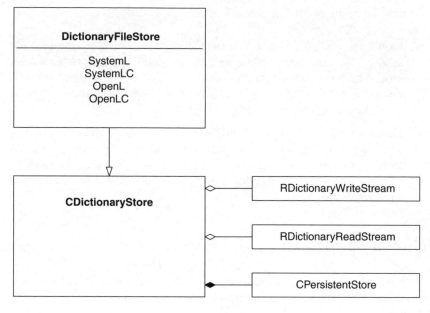

Figure 13.14

You should not keep dictionary stores permanently open. When you need to access a dictionary store:

- *Open it*: if this fails, wait a second or so and retry.
- Open the stream you want.
- Read or write the stream.
- Close the store.

To each application, the application architecture assigns a .ini file, which is a dictionary file store. Again, we need to beware the following sources of confusion:

- Dictionary stores have nothing to do with the stream dictionaries that we saw when looking at application document formats.
- A dictionary store is not a stream store at all.
- Therefore, a dictionary file store is not a file store at all: it is a dictionary store that happens to use a file. Perhaps a better name would have been 'file dictionary store'.

The Application Architecture

In order to make C++ documents embed efficiently, Symbian OS requires that C++ applications must be polymorphic dynamically loadable libraries. So when you launch an embedded document, a new library is effectively loaded *in the same process as the embedding application* – in fact, the embedding application is run in the same thread as the embedded application. This is *much* more efficient than systems requiring interprocess communication, and because the design is simple, it's also more robust.

Virtually everything else in the resulting application architecture flows from these requirements. The application architecture:

- specifies the association between document files and applications

- says how to associate basic information with an application – including an icon, a caption, whether it is file-based, whether it can be embedded etc.

- includes an API that you can interrogate to get lists of all installed applications, all embeddable applications, all running applications etc.

- specifies the location of an application's `.ini` file

One implication is that conventions are needed to detect installed applications and query their capabilities. As we've already seen, an installed application must be in a directory of the form `\system\apps\appname\`, and must be called appname.app.

Symbian OS enables and supports many diverse devices to be built, with varying philosophies and architectures in mind. Since this book is supplied with the UIQ SDK, it is appropriate at this point to focus briefly on the philosophy of the UIQ Application Architecture and its direct effects on the file, stream and store characteristics.

The two components in UIQ that really deal with applications closely are the Application Launcher and Uikon (the application framework underlying UIQ). It follows that these two components are heavily dependent on the application architecture and its APIs.

The philosophy of UIQ, which is reflected in the application and file-handling framework, is that the *system, rather than the end-user, is responsible for managing memory and storage.*In that respect, as far as the interaction with the user goes, UIQ does not ordinarily allow users to close an application, nor does a UIQ device usually allow an end-user to see the file system and thus erase files.

As we have seen, the system implications of the UIQ paradigm are pervasive throughout the system. Although they have no fundamental effect the way that files, streams and stores are handled, they do impose various constraints on application developers. Fortunately, these constraints are not too onerous and you should be able to satisfy them without too much additional effort.

13.10 Summary

In this chapter, we've introduced the application architecture APIs for communicating with the file server, and the Symbian OS framework that deals with streams and stores.

Data management, in Symbian OS, is based around a conventional file system, accessed through the file server APIs. In practice, however, applications usually use streams and stores to read and write file data. The stream and store APIs support the application architecture and deliver compactness in both code and data formats. More specifically, then, we've seen:

- The difference between load/save files that can embed or be embedded and database applications that can embed load/save applications.

- A document has essentially the same structure whether it is embedded or not – and as such it isn't identified by a file extension, but by its internal UID.

- Where an application's .ini file is stored and how it uses a dictionary store.

- The relationship between the file server, the stream store, and the application architecture.

- How to use an RFs object to get a session with the file server, how to use the CONE environment's RFs to save resources, and how RFs functions are stateless.

- Using an RFile and navigating the file system in code.

- How to parse the path of a filename.

- How data and objects can be externalized and internalized into streams.

- How any classes can use the streams and serialization capabilities.

- Using the stream store APIs as a generic way to externalize and internalize data between streams and user RAM.

- When to use which function.

- The structure of a persistent store, with a root stream containing a stream dictionary that points to the other streams of data in the application and in consequence, how an embedded store is just another stream.

- The difference between a direct file store and persistent store.

- How a dictionary store isn't a store at all.

14

Finishing Touches

After several heavy programming chapters, it's time to step back from the C++ and take a look at some of the softer, but equally important, aspects of producing a well-crafted GUI application for Symbian OS smartphones based on UIQ. The title of this chapter is *Finishing Touches*, but that's probably misleading – without these finishing touches, your application will be a non-starter in the mass market.

In this chapter, we'll go through the following improvements to the `HelloGUI` application that was the subject of Chapter 4:

- adding a button bar and buttons to the user interface

- adding an icon and a localizable language caption to be displayed in the Application Launcher

- wrapping up the whole application into a single installable package and certifying it to allow easy delivery and secure installation for end users.

We'll also discuss how to build applications from the command line, using some of the underlying Symbian OS build tools. All these tools and the file formats used are described in greater detail within the Tools and Utilities section of Developer Library supplied on the UIQ SDK.

Finally, we'll look at some of the stylistic aspects to consider when creating GUI applications based on UIQ. The aim of this style guide is also to help maintain consistency between applications produced by different suppliers.

Let's begin by adding the finishing touches to `HelloGUI`. You'll find all the source code in `\scmp\HelloGUIfull`, though all source files are still named `HelloGui.*`.

Symbian OS C++ for Mobile Phones.   Edited by Richard Harrison
© 2003 John Wiley & Sons, Ltd   ISBN: 0-470-85611-4

14.1   Adding Buttons

Adding graphical icons to the button bar of your program can make a big difference to the end user's impression of your application, as well as enhancing usability. The procedure for doing this is fairly straightforward – in fact, all you need to do is to make the following changes:

- Create Windows bitmaps for the icons and make them available to the application at build time.
- Change the resource file for the application to include specifications for the button bar.
- Change the project specification (.mmp) file to enable conversion of your Windows bitmaps to Symbian OS format files during the application build process.
- That's it! You don't need to modify any C++ at all!

The (Figure 14.1) screenshot shows the button bar of the HelloGUI application we're working toward.

Figure 14.1

Like some of the built-in applications available on UIQ, there are graphical icons visible on the button bar of the program itself.

14.1.1   Creating the Bitmaps

The first step of the process is to create Windows bitmaps and convert them into the specific file format used by Symbian OS, called a 'multi-bitmap' file or MBM. The .mbm file, together with an associated .mbg file, is constructed from one or more Windows .bmp files using the tool bmconv (Bitmap Converter), which is called during the main application build process (Figure 14.2).

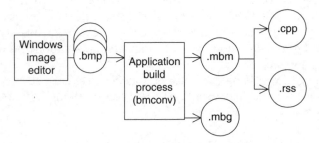

Figure 14.2

The .mbm format is also used when icons are required elsewhere in your application – for example, a splash screen on startup.

You can also use the `bmconv` tool as a stand-alone application, converting `.bmp` and `.mbm` files in both directions, as explained later in *More on the bmconv tool.*

As a Windows programmer, you may find it easier to think of a `.mbm` file as an addition to the application's resource file, and a `.mbg` as an addition to the `.rsg` generated header, which contains symbolic IDs for the resources.

Under Windows, bitmaps and other resources are incorporated directly into resource files. Under Symbian OS, they are treated separately because the tools used to create MBMs and resource files are different. This also permits finer control over the content of these files – for example, some Symbian OS phones may compress resource file text strings to save space, but wish to leave bitmaps uncompressed to avoid performance overheads at display time.

The icon you wish to add to the button bar should be created from two Windows bitmaps:

Firstly, there's the icon itself.

Secondly, there's a mask, which is black in the area of the icon.

Only the regions of the bitmap corresponding to the black regions of the mask are copied onto the button bar. Everything else is ignored – the white areas of the mask are effectively transparent, whatever the color in the corresponding parts of the data bitmap.

The unmasked region of the icon overlays the button underneath. The masked region of the icon is ignored, so that the region of the toolbar button underneath is unchanged (Figure 14.3).

Figure 14.3

As you can see above, when built into a UIQ application, the button's icon has a surrounding border. Because of this, and the fact that the button bar also has additional margins, you cannot create icons higher than **14 pixels** in order for them to be fully visible in a UIQ button bar.

When it comes to color support for the bitmaps, it's important to bear in mind the fact that UIQ supports different color depths for items shown on screen. Symbian OS provides support from 1-bit color up to 24-bit color, and individual phones will vary in terms of the color depth supported by their hardware.

Most current Symbian OS phones support 12-bit color ($2^{12} = 4,096$ colors) on screen. However, there is a trade-off between battery-life and RAM usage. Higher color depth images (and icons) consume more power when displayed, occupy more disk space when stored *and* use more RAM when loaded. For this reason, most UIQ icons use an 8-bit color depth (giving $2^8 = 256$ colors), striking a suitable balance between aesthetics and file size/power consumption. As a general guide, unless you have a specific reason for using more colors, your icons and images should use 256 colors, or fewer, on UIQ.

In general, these bitmap icons are easy to produce and at this stage you don't need to spend too much time on them; simply create them using Paint Shop Pro or another equally suitable graphics package.

For the purpose of adding buttons to our `HelloGUI` example application, we have created three icons with corresponding icon masks and placed them in the source folder within the `HelloGUIfull` example directory. These are named:

- icon and iconmask
- icon2 and icon2mask
- icon3 and icon3mask.

Altogether then, there are **six** bitmaps to be turned into a single `.mbm` file, along with its associated `.mbg`.

14.1.2 Converting the Bitmaps

Now that we have dealt with the bitmaps, the next step is to add the bitmap conversion information to the application build process for `HelloGUI`. You do this by including the following text in the project's `.mmp` file:

```
START BITMAP Hellogui.mbm
```

```
        HEADER
        SOURCEPATH ..\src
        SOURCE c8 icon.bmp
        SOURCE 1 iconmask.bmp
        SOURCE c8 icon2.bmp
        SOURCE 1 icon2mask.bmp
        SOURCE c8 icon3.bmp
        SOURCE 1 icon3mask.bmp
   END
```

The statements have the following meanings:

START BITMAP	Marks the start of the bitmap conversion data and specifies the `.mbm` multibitmap filename.
HEADER	Specifies that a symbolic ID file, `Hellogui.mbg` is to be created (in the `\epoc32\include` folder).
SOURCEPATH	Specifies the location of the original Windows bitmaps.
SOURCE	Specifies the color depth and the names of one or more Windows bitmaps to be included in the `.mbm` Symbian developers conventionally.
END	Marks the end of the bitmap conversion data.

The `.mbm` is generated into the target folder of the project – for a `winscw udeb build`, this is `\epoc32\release\winscw\udeb\z\system\apps\HelloGui`. Standard practice is to specify the `.mbm` name to be the same as that of the `.app` file. As we shall see later, this makes it easier to access the `.mbm` file.

Symbian OS developers conventionally specify only one bitmap file per SOURCE statement and specify each mask file immediately after its corresponding bitmap file. The ordering of the statements does not really matter, but the color depth value does. A value of `c8` specifies 8-bit color (the default for UIQ). The value specified for each mask is 1 – the lowest form of color depth. Since masks contain only black, they need not be stored with a high color depth. By specifying this value you will normally save valuable bytes of storage space for the resulting `.mbm` – and normally also save on RAM usage when the `.mbm` is loaded at runtime.

Depending on circumstances, and particularly if bitmaps are stored in compressed form, using a color depth of 1 may not save on memory usage. Also, if your icon and mask are to be used in a speed-critical operation (an animation, for example) you may want to include both the icon and mask at the same color depth as is used by the phone itself. Since the Window Server will then not need to convert the images between the display mode of the phone and the color-depth in which they are stored, you will sacrifice some storage space and RAM for a bit of extra speed. For general use, however, the above approach is recommended.

The format of the generated `Hellogui.mbg` file is as follows:

```
// HELLOGUI.mbg
// Generated by BitmapCompiler
// Copyright (c) 1998-2001 Symbian Ltd.  All rights reserved.
//

enum TMbmHellogui
    {
    EMbmHelloguiIcon,
    EMbmHelloguiIconmask,
    EMbmHelloguiIcon2,
    EMbmHelloguiIcon2mask,
    EMbmHelloguiIcon3,
    EMbmHelloguiIcon3mask
    };
```

The naming convention here is clear enough: there's an enumeration whose type name includes the MBM filename, and several enumerated constants whose names are made up from the MBM filename *and* the source bitmap filename.

You will need to use the enumerations generated in this file, in the resource file, as described below.

14.1.3   Changing the Resource File

Before you can rebuild your application and convert the bitmaps into a `.mbm` file, you need to define exactly how your icons should appear on the button bar. You do this by referencing the bitmap files and corresponding masks from the application resource file. The information you need to add includes:

- the names of all the icon bitmaps
- the names of all the mask bitmaps
- the name and location of the `.mbm` file that contains them
- a specification of the function of each button
- optionally, where to insert any button text in relation to the bitmap.

The resource file, **HelloGui.rss**, needs the following additional `#include` statements:

```
#include <hellogui.mbg>
#include <qikon.rh>
```

and the button bar is specified within the RESOURCE EIK_APP_INFO
statement, as shown below:

```
RESOURCE EIK_APP_INFO
    {
    hotkeys = r_hellogui_hotkeys;
    menubar = r_hellogui_menubar;
    toolbar = r_example_toolbar;
    }
```

The named button bar is defined in a QIK_TOOLBAR resource,
as follows:

```
RESOURCE QIK_TOOLBAR r_example_toolbar
    {
controls=
        {
        QIK_TBAR_BUTTON // Done image button aligned to right side of bar.
            {
            id=EHelloGuiCmd0;
            flags=EEikToolBarCtrlHasSetMinLength;
            alignment=EQikToolbarRight;
            bmpfile="*";
            bmpid=EMbmHelloguiIcon;
            bmpmask=EMbmHelloguiIconmask;
            },

        QIK_TBAR_BUTTON
            {
            id=EHelloGuiCmd1;
            flags=EEikToolBarCtrlHasSetMinLength;
            bmpfile="*";
            bmpid=EMbmHelloguiIcon2;
            bmpmask=EMbmHelloguiIcon2mask;
            },

        QIK_TBAR_BUTTON
            {
            id=EHelloGuiCmd2;
            flags=EEikToolBarCtrlHasSetMinLength;
            bmpfile="*";
            bmpid=EMbmHelloguiIcon3;
            bmpmask=EMbmHelloguiIcon3mask;
            }

        };

    }
```

In each `QIK_TBAR_BUTTON` struct, the members have the following meaning:

`id`	Specifies the command to be executed when the button is pressed. In this case, we are reusing the commands of the original `HelloGUI` example.
`flags`	Allow you to define various properties for the button – see the `qikon.rh` header file for more options. The default value for UIQ is `EeikTool BarCtrlHasSetMinLength` as used above – this will ensure the buttons are sized in the standard way.
`alignment`	Specifies where on the button bar the button should appear; left, right, or center. Because the default is `EQikToolbarLeft`, you can omit this setting if you do not wish to alter it – for example, in the above definition we specify alignment for the first button only, where we specifically set it to `EQikToolbarRight`.
`bmpfile`	Specifies the location of the multibitmap file. Here we use a special case of "*", which is described in more detail below.
`bmpid`	The enum for the relevant bitmap, defined in the generated `.mbg` file.
`bmpmask`	The enum for the corresponding bitmap mask, also defined in the `.mbg` file.

The asterisk used for each `bmpfile` argument is shorthand (in these circumstances only) for 'my own `.mbm` file'. For this to work, your `.mbm` file must have the same name as your `.app` file and be installed in the same folder. This works for us because we deliberately constructed our `.mbm` to follow these conventions by calling it `Hellogui.mbm` in the project specification file, `HelloGui.mmp`. Under the emulator, the icon is always placed in the same folder as our `.app` file but, as we shall see later, you must be careful to install it in the right place on the target phone.

It's possible to specify other `.mbm` files within the `bmpfile` statement as well, but you have to supply the full file path in order to do so. Some potentially useful ones are the UIQ's own System MBMs – such as `qikon.mbm`, `quartz.mbm` and `eikon.mbm`. These files are located in `Z:\System\Data` and their corresponding `.mbg` files are in `\epoc32\include`. These bitmaps include icons for arrowheads used on scrollbars, bold/italic/underline symbols, various other arrows, application icons, background textures, and more.

14.1.4 Building the Application

After making all the necessary changes to the resource file, you can finally rebuild the `HelloGUI` application using the – by now familiar – `bldmake` and `abld` commands.

From a command prompt, move to the `\scmp\HelloGUIfull\Group` directory and type:

```
abld reallyclean
```

to remove any files that might have already been built from the previous version of `HelloGUI`. Then, type:

```
bldmake bldfiles
```

to create new build files incorporating the new bitmap build and resource information. Finally, use the `abld` command to rebuild the full application to be tested in the emulator, in Figure 14.4.

```
abld build winscw udeb
```

Figure 14.4

You can view the result by launching the emulator in the usual way; either type epoc from the command line, or double-click on epoc.exe within epoc32\release\winscw\udeb.

Figure 14.4 shows what our modified HelloGUI now looks like.

14.1.5 More on the bmconv Tool

The Bitmap Converter tool (bmconv) can also be used as a stand-alone application, either to package bitmaps into a single .mbm file, or to extract bitmaps from a multibitmap file.

To convert bitmaps into a single .mbm file you only need to type a command such as the following:

```
bmconv /h Hellogui.mbg Hellogui.mbm icon.bmp icon2.bmp
    iconmask.bmp icon2mask.bmp
```

This gives a verbose log (that you can choose to suppress by using the /q switch if you wish):

```
BMCONV version 110.
Compiling...
Multiple bitmap store type: File store
Epoc file: Hellogui.mbm

Bitmap file 1    : icon.bmp
Bitmap file 2    : icon2.bmp
Bitmap file 3    : iconmask.bmp
Bitmap file 4    : icon2mask.bmp
Success.
```

To extract .bmp files from a multibitmap file you specify a /u flag after the bmconv command, for example:

```
bmconv /u Hellogui.mbm icon.bmp icon2.bmp iconmask.bmp icon2mask.bmp
```

One useful application of this facility is to capture a screen from the emulator using *Ctrl+Alt+Shift+S*. This results in a .mbm file containing a single entry. Once you have extracted the bitmap, you can display and manipulate it using an editor such as Paint Shop Pro.

You can also view the contents of a .mbm file by specifying a /v flag after the command:

```
bmconv /v Hellogui.mbm
```

To see the full set of supported options, just type bmconv on the command line.

14.2 Adding Application Icons

Now that we've seen how to handle bitmap and resource conversion for adding buttons to a button bar, we can generate an application icon and a better caption for HelloGUI, to be visible from the Application Launcher of your UIQ smartphone.

Here is what the application icon will look like (Figure 14.5):

Hello!

Figure 14.5

As we saw in Chapter 13, the icon and caption, along with some other information about an application's capabilities are contained in an application information file or AIF. Without an AIF, the application uses the default system icon (consisting of two squares on a gray background with a black border), displays a caption identical to its filename, and is assumed not to be associated with any MIME file types. The AIF has the same name as the application, and resides in the same directory – unsurprisingly, its extension is .aif.

An AIF is created with the aid of aiftool.exe and the two main ways of using it are:

- using aiftool as part of the application build process, by adding a statement to the project specification file, as we did with .mbm files earlier
- using aiftool directly.

For the purpose of this book, we'll concentrate on producing the .aif file as part of the application build process and mention the stand-alone use of aiftool only briefly.

There is also a third method, using a GUI application called AIF Builder, which is especially useful for creating .aif files and icons for Java-based applications. More information on using the AIF Builder is available within the Tools & Utilities section of the Developer Library provided on the UIQ SDK.

To change the icon that represents your application in the Application Launcher, you need to perform the following steps:

- Create the icon and corresponding mask
- Create an AIF resource file containing language captions and usage information
- Add the AIF conversion process command to the application build
- Rebuild the application.

The process is illustrated in Figure 14.6:

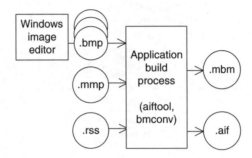

Figure 14.6

14.2.1 Creating the Icon

As when we added icons to the button bar of HelloGUI earlier, a UIQ application icon has to be created from up to four Windows bitmaps, made up of icon and bitmap pairs. As with the button icons, these will then be converted into a multibitmap format during the application build process. There is, however, a difference; the .mbm file is only generated temporarily, and not stored – all the necessary image information is contained in the resulting .aif file.

To create the icons, you can once again use Paint Shop Pro or any other suitable graphics package, but this time the size of the icons is even more important. Ideally, for UIQ applications, you should supply two icons and their corresponding masks. One of the icons, and its mask, should be 20 × 16 pixels and the other icon (plus mask) should be 32 × 32 pixels. The two different icons cater to the different zoom states of the Application Launcher application – the smaller icon will be used in the 'List' view, the larger icon in the 'Icons' view. However, if your icon is not very detailed there is another, easier option. If you create a single icon (plus mask) of 24 × 24 pixels, the Application Launcher will autosize your icon to be suitable for use at all zoom levels. This is the approach we will use for HelloGUI.

As discussed earlier, the standard UIQ color depth is 8-bit or 256 colors – we will create our icons to follow this convention.

You will find the example bitmaps, `hello.bmp` and `hellomask.bmp`, in `\scmp\HelloGUIfull\Src`, and they look like this (Figure 14.7):

Figure 14.7

The AIF itself is defined by inserting the following line into `Hellogui.mmp`:

```
AIF HelloGui.aif ..\src Helloaif.rss c8 hello.bmp hellomask.bmp
```

where the various elements have the following meanings:

`AIF` `HelloGui.aif`	Specifies the start of the AIF specification. The name of the AIF to be generated in the target directory – in the case of a `winscw` udeb build, this will be `\epoc32\release\winscw\udeb\Z\System\apps\HelloGui`
`..\src`	The source directory for the AIF `.rss` and `.bmp` files.
`Helloaif.rss`	The AIF resource file to be used – this file should have a different name to the `HelloGui.rss` file and should be created separately before building your application.
`c8`	Specifies 8-bit color depth, as for `.mbm` files. However, we do not have such fine control over the icon inclusion with AIF files as we did with `.mbm` files – all icons and masks must use a common color depth.
`hello.bmp` `hellomask.bmp`	The names of the bitmaps, in the correct order – icons before masks.

If you provide more than one icon and bitmap pair, the AIF format requires you to list them in ascending size order.

14.2.2   Adding Captions

Before you can rebuild your application once more, you need to create a resource file called `Helloaif.rss` within your source directory. This is where you specify any caption information (in various languages) for the application icon and define how it should be displayed. The format of this file is similar to that of `HelloGui.rss`, but only needs to contain the statements illustrated in this example:

```
// Helloaif.rss
//
// Copyright (c) 1998-2002 Symbian Ltd. All rights reserved.

#include <aiftool.rh>

RESOURCE AIF_DATA
{
caption_list=
  {
  CAPTION { code=ELangEnglish; caption="Hello!"; },
  CAPTION { code=ELangFrench; caption="Bonjour!"; },
  CAPTION { code=ELangSpanish; caption="Ola!"; }
  };
//
app_uid=0x101f74aa;
//
num_icons=1;
}
```

- The `caption_list` struct contains a list of captions in various languages. For our example, the English caption is **Hello!**, which is going to look better on the Application Launcher than **HelloGUI** did.

- The `app_uid` item specifies the application's UID and must be the same as the one that is listed in the project specification file and is returned by the application's `AppDllUid()` function.

- The `num_icons` item specifies the number of icons to be added to the application (not including any masks) which, in this case, is one.

  Your application icon will not be displayed if you don't include the correct UID. As is explained in Chapter 4, you should always check that all the UIDs you have specified in your application match each other.

  It is possible to add more information to this file if necessary (for example, MIME type associations). For more information on the resource file format syntax used by AIF files, see the Tools & Utilities section of the UIQ SDK.

14.2.3 Rebuilding Your Application

Once you have created the bitmap files, included the AIF statement in your project specification file, as well as created an AIF resource file, you are ready to build your application again, using the same commands as described earlier in this chapter. The only difference this time is that `aiftool` is called straight after `bmconv` in order to produce the `.aif` file in your target directory.

When viewing the result in the emulator, you should be able to see the new, more visually appealing icon and localized caption in the Application Launcher (Figure 14.8):

Figure 14.8

14.2.4 More on aiftool

As with `bmconv`, it is possible to use `aiftool` as a stand-alone command line program. This may be useful when you need to produce `.aif` files separately for inclusion in your program at a later stage.

To run `aiftool` on its own, simply type a command of the form:

```
aiftool helloaif helloaif.mbm
```

where the first parameter is the name of the AIF resource file, minus its extension. Don't type the `.rss` extension on the resource file – you'll get a strange error if you do. Assuming that everything goes according to plan, you'll see another rather verbose log:

```
AIF tool
Copyright (C) Symbian 2002
Compiling resource file
        1 file(s) copied.
Copying mbm
        1 file(s) copied.
Running AIF writer
Reading resource file...
Adding icons
Reading icons
Adding captions
Adding capability
Adding data types
Saving
        1 file(s) copied.
```

When this has finished, you'll see that `Helloaif.aif` is in your current directory. Copy it to `\epoc32\release\winscw\udeb\z\system\apps\HelloGui`, and rename it to `HelloGui.aif`.

To avoid the rename, you can build your AIF in a separate directory, so that you can have an AIF specification file called `HelloGui.rss`, which doesn't clash with the application's main resource file.

14.3 Making Your Application Installable

Up to this point, we've done everything in the emulator. It's time we rebuilt the application for ARMI and installed it onto a real Smartphone based on the UIQ user interface.

Building the application for the target Smartphone is simple. As before, you can build it from the CodeWarrior IDE, after selecting the ARMI UREL target, or you can build it from the command line by typing:

```
abld build armi urel
```

This builds a releasable `HelloGUI` application, which is then ready to be copied onto the phone itself. But what exactly do we copy? By now, our application consists of:

- a `.app`, in `\epoc32\release\armi\rel`
- a `.rsc`, a `.mbm` and a `.aif` in `\epoc32\release\winscw\udeb\z\system\apps\HelloGui`.

To install the application onto the smartphone, we *could* transfer all four files into `C:\System\Apps\HelloGui` – as a programmer you should be able to do this easily enough by using suitable PC connectivity software, but this would be unappealing for end users. If that wasn't reason enough to want an installer, large programs such as Battleships make the process more complicated as they need to be installed to more than one target directory. A direct copy may not even be feasible if the PC connectivity software does not support a view of the filing system on the Smartphone.

So for these reasons, Symbian OS provides a powerful yet simple installation system that offers a simple and consistent user interface for installing applications, data, or configuration information onto smartphones.

The basic idea is that end users install components packaged in Symbian OS installation (`.sis`) files. These can be transferred to the phone from a PC using connectivity software, Bluetooth, infrared or even e-mail and Internet connection, as illustrated in Figure 14.9.

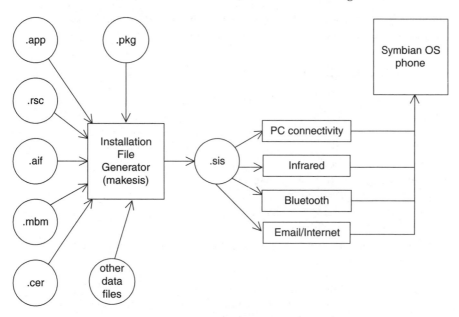

Figure 14.9

The SIS files are generally very small in size (at most a few hundred kilobytes) and are therefore very quick to transfer to the target phone.

Two command line tools are provided to enable you to create these SIS files:

- The Installation File Generator, `makesis.exe`, creates the installation files from information specified in the package file.

- The Certificate Generator, `makekeys.exe`, creates private/public key pairs and a certificate file, used by the Installation File Generator to digitally 'secure' the installation files.

A GUI application called `SISAR` is also available, which provides a simple way to create `.sis` files, but in this section we'll only discuss how to create package files by hand and then use `makesis.exe` to produce the `.sis` file. More information on using `SISAR` can be found in the Tools & Utilities section of the UIQ SDK.

There are three steps involved in generating the `.sis` file:

- If the application is to be secure and protected, you need to produce a certificate file to be included in the `.sis` file.
- Then you need to create a package file that specifies all the files that go into the `.sis` file and where they are to be installed on the target phone.
- Finally, after all the input files are correct, you can run the `makesis` command line tool that generates the `.sis` file.

14.3.1 Securing Your Installation File

Part of producing an installable file is also to make sure that the application is secure, protected against viruses and has not been tampered with before transferring it to the phone itself. For these reasons, Symbian's secure software installation system provides a Certificate Generator – a command line tool called `makekeys` – that allows you to produce a secure digital certificate to be included in the `.sis` file.

The aim of this certificate is to:

- identify the software vendor,
- verify that the installation file has not been tampered with since it was created.

The way the Certificate Generator works is to create a public/private key pair, which are then used by the Installation File Generator (`makesis`) to create a digitally signed and secure installation file.

The private key must be kept secret and is generated straight away, along with a self-signed certificate containing information for its corresponding public key. The self-signed certificate is then used to create a certificate request, which you submit to a Certification Authority in order to get the public key – an authenticated digital certificate that is made generally available. The private key and the authenticated certificate should then be referenced in the package file used for creating the `.sis` file.

You cannot decrypt one key without the other, which means that nobody apart from you (who has the private key) can change the `.sis` file without this being noticed by the installer software, which will alert the user to this fact.

To create a private key for our `HelloGUI` application, type the following from the command line:

```
makekeys -cert -dname "CN=Symbian OR=Symbian Ltd CO=GB"
   Hellogui.key  Hellogui.cer
```

The `-cert` flag specifies that a certificate is to be created, and the `-dname` flag introduces the following distinguished name string that is passed to the Certificate Generator. Within that string:

CN	supplies the common name for the vendor
OR	supplies the organization name of the supplier
CO	supplies the country of origin.
Hellogui.key	is the name for the private key, which will be generated in the current directory.
Hellogui.cer	is the name for the self-signed public certificate, which is also generated in the current directory.

One thing to note is that the Certification Generator uses a mouse input mechanism to generate the private key. The MS-DOS property QuickEdit must be turned off in order for sampling of random data from the mouse to work.

The next section describes how to obtain the public key, but if you want to continue with the next step in the process straight away, go to *Producing the package file.*

Obtaining the public key

After creating the self-signed certificate, you are able to produce a certificate request using `makekeys` again:

```
makekeys -req -dname "CN=Symbian OR=Symbian Ltd CO=GB"
   Hellogui.key Hellogui.cer Helloguireq.p10
```

The only changes are the use of the `-req` option, which specifies that a certificate request is to be created, and the additional name for the request file, which is again generated in the current directory.

This certificate request file should then be sent to a Certification Authority, who will verify your identity and return an authenticated digital certificate that replaces the self-signed certificate previously created.

Certification Authorities are trusted third party organizations that issue digital certificates after checking that they belong to the owner specified by the requester. These organizations may or may not be specific to licensees or operators. A well-known Certification Authority is VeriSign.

14.3.2 Producing the package file

Before you create the `.sis` file you need to specify information about the application in a package file. This is a plain text file but because of its flexibility, the format is one of the most obscure of any Symbian OS tool control file.

For our example, we have created two package files in `\scmp\Hello-GUIfull\Group`. The first one is called `HelloGui.pkg`, which is used to produce a nonsecure `.sis` file that should be installable on any Symbian OS smartphone. The second one, `HelloGuiSecure.pkg`, can be used to produce a secure `.sis` file, which will not be installable on any Smartphone, as that would require a matching public key.

Here is the content of `HelloGuiSecure.pkg`:

```
;
; HelloGui Installation script.
;
;
; Specify the supported languages; items must appear in this
    order subsequently
&EN
;
; The installation name and header data
#{"HelloGui"},(0x101f74aa),1,0,0
;
*"Hellogui.key", "Hellogui.cer"
;
; The files to install
"\epoc32\release\armi\urel\HelloGui.app"-
      "!:\System\Apps\HelloGui\HelloGui.app"
"\epoc32\release\winscw\udeb\z\system\apps\HelloGui\HelloGui.rsc"-
      "!:\System\Apps\HelloGui\HelloGui.rsc"
"\epoc32\release\winscw\udeb\z\system\apps\HelloGui\HelloGui.aif"-
      "!:\System\Apps\HelloGui\HelloGui.aif"
"\epoc32\release\winscw\udeb\z\system\apps\HelloGui\HelloGui.mbm"-
      "!:\System\Apps\Hellogui\HelloGui.mbm"
```

The body of this file should be obvious enough: all source files on your PC that will be packed into the .sis file are listed, along with information about where they should be unpacked upon installation. The "!" drive specifier means 'the chosen installation drive'. If you require a file to reside on the C: drive of the smartphone for instance, you could have specified that instead – but this is rare, and should only be used if absolutely necessary.

The header is more interesting:

- The \& line specifies the languages supported by this installation file: in this case, English only.

- The # line specifies the application's caption to be used at install time, its final UID, and its three-part version number – 1,0,0 is specified here, which will be displayed as 1.0(000), indicating major, minor, and build.

- The * line specifies the private key and the certificate to be used for secure installation. This line is omitted in the non-secure package file. For our example, we have included the self-signed certificate, instead of an authenticated certificate, as an aid to testing the installation, as described later. Obviously, you should include an authenticated certificate in any real product.

This is the general form of a secure .pkg file for a basic, single-language application. In fact, the .pkg file format supports more options than this, including nested packages, multilanguage installation files, and required dependencies. All these are documented in the UIQ SDK.

Even though we specified several languages in the AIF file, we are only using English for the .pkg file, as the .rss file is available only in English. The AIF languages are different and unrelated to this, so if you install this as is on a French language machine, the caption will be in French, but everything else will be in English.

One thing to note is that, because Symbian OS is so flexible and allows many different UIs to be developed, you should always ensure that your SIS file can only be installed on the UI for which it was designed. If you don't, you – or, more importantly, your end users – may experience problems after accidentally installing a SIS file produced for another UI.

To enable you to do this, each product is assigned a unique ID that is used by the application installation mechanism to ensure that only compatible applications can be installed on the phone. Also, each platform version is assigned a unique ID. It is assumed that a platform version is compatible with all earlier platform versions, but not with any later versions. You can specify the platform information by including in the package file a line of the following form:

```
# (0x101F617B), 2, 0, 0, {"UIQ20ProductID"}
```

where:

(0x101F617B)	Represents the product/platform version UID
2, 0, 0,	Represent the major version, minor version and build numbers of the platform (not your application)
{"UIQ20ProductID"}	Represents a feature identification string

Not all products support this mechanism: for those that do, you can find exact details of the UIDs, feature identification strings, and version numbers to use in the appropriate product-specific SDK. On products that support this mechanism, the installation software will refuse to install any SIS file that does not contain the correct platform information.

14.3.3　Generating the final SIS file

When all the source files and application information have been specified in the .pkg file, you can finally run makesis by issuing the command:

```
makesis HelloGui.pkg HelloGui.sis
```

This creates an installable HelloGui.sis file in your current directory. To generate a secure installation file, you need to use the secure version of the .pkg file:

```
makesis HelloGuiSecure.pkg HelloGuiSecure.sis
```

14.3.4　Installing SIS Files

Once you have a working .sis file, it's then incredibly easy to install this onto your Smartphone: all you need to do is transfer it to your Smartphone and launch the .sis file!

There are several transfer methods available – choose what's best for you depending on what is supported on your PC or phone and what is most convenient for you at the time:

- *Serial or USB connection*: this means that you need to install the PC Connectivity software suite accompanying your phone and then send files across through either a serial or USB* cable.

- *Infrared*: this means that you need to have infrared enabled on both your PC (either internally or externally in the form of an infrared pod) and the phone, and also a software program that allows you to send files between two infrared ports. This is normally part of the PC Connectivity software accompanying your phone.

- *Bluetooth*: this means that you need Bluetooth enabled on both your PC and phone.

- *E-mail*: this means that you can send the .sis file to your phone as an e-mail attachment. It requires that you have the e-mail account on your phone activated and, as you will use the mobile phone network to send the file, you may incur costs.

- *Internet*: this means that you can download a .sis file straight from a website. As with e-mail connection, you may incur costs when using the mobile phone network to connect to the relevant website.

*USB is not supported under Microsoft Windows NT.

Once you have transferred the .sis file, it should appear on the phone as a 'received' file – simply tap the icon to launch the file and begin the installation process. The program is then visible from the Application Launcher.

Once the. sis file is available on your UIQ phone, you are also able to send it to other smartphones, using either the infrared, Bluetooth or e-mail facility.

Testing on the emulator

By including the self-signed certificate, Hellogui.cer, in HelloGuiSecure.sis we enable the secure installation to be tested on the emulator. In order for this to work, however, you first need to insert a data file called cacerts.dat into \epoc32\winscw\c\system\data. A suitable file is supplied with the example software, and can be found in \scmp\HelloGUIfull\Group. With this file in place, copy the .sis file into \epoc32\winscw\C. To install it, select 'Install' from the Application Launcher's Launcher menu. Select the relevant application to install and press the Install button.

Don't expect the installed application, built for an ARM target, to run on the emulator! If you want to check that the application runs correctly after being installed, modify the package file to use an application built for the winscw target and rebuild the SIS file.

14.3.5   Checking and uninstalling SIS files

A more authoritative way to ensure that your program has installed successfully is to check that it is listed in the Storage Manager, available through the Control Panel of a UIQ smartphone illustrated in Figure 14.10.

Figure 14.10

Select the Control Panel icon on the Application Launcher, launch Storage Manager from the list of options available, and you'll see something: Figure 14.10.

One of the benefits of installing applications using the .sis file format rather than by copying files directly is that you get an entry in this listing. This means that you can use it to remove HelloGUI if you like, just tap on Uninstall and select your application from the list of uninstallable programs. Then press Uninstall again to confirm. Go back to the Application Launcher and you'll see that it's gone.

Even though the Storage Manager is available on the emulator, there is no point installing .sis files onto it, as these files are built to run only on ARM-based Symbian OS phones. The Storage Manager on the emulator is only useful for installing data or pure Java programs, or if required, checking that a secure .sis file is working.

14.3.6   Delivering Applications to End Users

We've now described how you can use the Installation File Generator (makesis) to produce a .sis file and how to transfer and install that file from a PC onto your phone. In reality, the main customer of a .sis file is an end user, so it's important to think carefully about delivering applications:

- Make sure you replace test UIDs with releasable UIDs allocated by Symbian, as explained in Chapter 4.

- Provide enough instructions to make it easy for end users to install and uninstall your software.

14.4   Designing Applications for UIQ – Some Guidelines

By now, you may feel that you're beginning to get used to the style of applications built using the UIQ GUI. Style here is not in the way you program, but in the way that users see your programs. Good application style (like good programming) is partly something provided and enabled by the system, and partly something you have to do yourself.

The application style guidelines for UIQ are designed to take some fundamental things into account:

- End users may not be very knowledgeable about computers. They might be frightened by many kinds of technology, but we want them to feel comfortable with Symbian OS and its applications. If you've read this book this far, you're probably not one of them – so you have to think *hard* if you're going to deliver friendly, easy to use applications to average end users.

- The physical parameters of the hardware -208×320 pixels or 240×320 pixels VGA portrait screen, pen-operation, no keyboard, hardware keys for up, down and confirm, built-in phone with speaker and microphone.

- Basic usability criteria such as task-orientation, browse-mostly, single-tap to open, hiding the file system and task list, use of views, menus, folders and dialogs.

The different physical parameters of phones dictate that Symbian and its licensees support several different user interfaces and the changing application suite requirements will affect the things we consider important in a GUI. For the moment, however, we'll concentrate on UIQ as it is.

Apart from following the UIQ style guide, it is also worth looking at the built-in applications and taking inspiration from them. The more time you spend planning and designing your application, the easier you will find it is to create and the more rewarding the end result will be.

The following sections list some of the most important GUI style rules for UIQ. More detailed information on designing for UIQ smartphones can be found in the UIQ Style Guide, and Designing for UIQ within the Technical Papers section of the Developer Library in the UIQ SDK.

14.4.1 Planning the GUI

Before you start creating the GUI of your application, the following stylistic aspects are worth considering:

- Every application should fill the screen. The screen is small enough as it is, and there's no point in making your user interface smaller than that.

  There are exceptions to this rule – in fact, there are good exception cases for most style guide rules. For instance, an application designed to add handwriting recognition to a device that doesn't build it in would need to use a window that floats above the current application. A full-screen window would be entirely the wrong approach in this case. Arguably, the word 'application' is also wrong for such a program.

- Each application should have a menu bar, usually containing an Application menu, an Edit menu and an optional Folders menu. A button bar is optional, but the status bar should always be visible.

- Each application should take into account the different methods of inputting text, whether it's through handwriting recognition or the virtual keyboard – or both.

- Good color contrast (a color that stands out from the background) and a nonserif font of reasonable size and type should be used for all non-dimmed items in the user interface.

- Make your application's main features easy to find out about (and easy to use) through the menus, button bar, different views, and folders.

- Try to fit most of the application features within the designated application space – only use scroll bars and scroll arrows where absolutely necessary.

- Create icons that have specific meanings within the user interface (not just for decoration), and apply them where most appropriate – within the application space, menus or on the button bar. Make sure they are dimmed out when not in use.

- Design the user interface so that it renders well for the different zoom levels supported by the system.

14.4.2   Designing List and Detail Views

In UIQ, users do not have access to a file system, so applications must provide their own means of selecting data to view and edit. This is done through List views and Detail views.

List views display multiple entries vertically, allowing users to browse to, navigate to and open a specific entry:

- Make sure that they are scrollable where necessary.
- Don't include menu options that act on individual entries in the list.

Detail views are reachable from list views and focus on data to be edited, through tabs:

- Provide multiple tabs on the button bar to show different parts of the editable data.
- Reuse system icons or provide text for the tabs.

14.4.3   Designing Menus and Folders

The menu bar should always be visible. As mentioned earlier, it will normally include an Application menu, the standard Edit menu and, optionally, a Folders menu:

- Use the name of the application for the Application menu.
- Keep the menus consistent as far as possible between different views.
- Avoid using cascading menus unless absolutely necessary.
- Make the text for menu commands as short as possible and use the ampersand instead of the word 'and'.
- Don't repeat the wording on the menu bar or cascading menu item (if applicable).
- Dim unavailable menu items in most cases.,
- Use radio buttons for 'Sort by' commands.
- Avoid ellipses to indicate commands that lead to dialogs.
- Use dividers to group similar commands together.
- Incorporate folders to divide data into user definable sets.

14.4.4   Standard Menu Items

UIQ defines a set of standard items that appear on the Edit menu such as **Cut**, **Copy**, **Paste**, **Zoom** and **Preferences**. If a **Folders** menu is used, the standard items are **Business**, **Personal** and **Unfiled**, with slight variations depending on whether you are in a List view or a Detail view.

Clearly, not all applications need to feature all these items on their menu lists. Equally clearly, most applications have their own menus too. The UIQ Style Guide recommends you provide your application-specific menu items within the Application menu.

14.4.5 Creating Dialogs

Dialogs are messages that prompt the user to respond. They always appear as wide as the screen and can be categorized into information dialogs, setting dialogs, query dialogs, notification dialogs, and process dialogs:

- Make sure you use the correct type of dialog for the purpose.
- Apply a dialog title with clear context.
- Use buttons such as Done, No/Yes, and Continue instead of OK.
- Keep the dialog layout as uncluttered as possible.
- Align control labels to the right and controls to the left.
- Make sure buttons are right justified and at the bottom of the dialog.
- Use multipage dialogs only when there's no other alternative and avoid adding more tabs than will fit on the screen.

14.4.6 Considering Text Input

Handwriting recognition, in which pen gestures are interpreted as letters, numbers, or other characters and sent to the application as text, is UIQ's main method for text input. Using a virtual keyboard is the other alternative.

- Handwriting should only be activated when the focus is on text control.
- Bear in mind that the virtual keyboard takes up screen space when activated and might hide important features.

14.4.7 Providing Text and Messages

Providing concise wording and clear instructions on menus, dialogs, notifications, infoprints, and other information messages is very important for the usability of your application.

- Use the standard vocabulary set by UIQ, its applications, and the style guide. Avoid ambiguity and programmer-centric vocabulary ('show the hidden window'); instead use user-centric vocabulary ('hide the game'). Take particular care over text that has to be short – on buttons, for example. Make sure that yes/no questions can only have those answers. Reassure the user: make it plain what the result of any action will be

- Remember that the pointer is a pen, not a mouse. Use pen-centric metaphors: use single-tap rather than double-click, and avoid drag-and-drop

- Avoid unnecessary jargon and use only standard computer vocabulary

- Use complete sentences where possible, rather than truncating them

- Be consistent with terms – use the same terms for the same things, regardless of whether you use standard UIQ terms or your own

- Don't end error messages with an exclamation mark or infoprints with a full stop

- Use bold style for active text and for text entered in input fields, menus, dialogs and so on

- Use plain style for descriptive texts and pop-out lists.

14.4.8 Using Scroll Arrows and Scroll Bars

Scroll bars and scroll arrows are used in List views and notes fields of Detail views for scrolling up and down:

- Make sure scroll arrows always appear on top of the content.

- Only include scroll arrows or scroll bars when scrolling is needed.

14.4.9 Designing for Various Zoom Levels

UIQ recommends different zoom levels – Small, Medium, and Large – to ensure that all users can easily read what is on the screen. It's important that your application is designed to take these zoom levels into account and that the result is visually pleasing:

- Make sure that most views with text, including list views, are zoomable and render correctly at each level.

- Provide two icon sizes if using icons next to text within your application; one for Small and Medium levels, and one for the Large level. The large icon should correspond to the font height specified for the Large level.

14.4.10 Linking between Applications

In UIQ, it is possible to provide direct navigation links (DNLs) to other applications from within your application. Where these are implemented:

- make sure they are easily identifiable, through bold highlighting or use of icons as it should be obvious to the user that it is a link
- make sure that the function of the link is clear, and that the user will know what happens when you tap on it
- be consistent – when using the same type of link at several places, always apply the same icon and text in the same way.

14.5   Handling Data

UIQ is designed specifically with handheld mobile devices in mind, and in keeping with the task-based approach, UIQ does not present a file system to the user. This has the following consequences:

- The user should not explicitly have to save data or close applications.
- The data should be stored automatically upon switching between applications.

UIQ automatically detects when the phone is running low in memory and can close down any background applications, but not the system applications or busy applications. In these cases, it is important that the application's user interface helps by:

- informing the user, who then has the option of freeing up storage space
- allowing the user to delete old, unused information and messages.

14.6   Summary

In this chapter, we have included plenty of examples of the application build commands that you need to use at the very end of the Symbian OS build tool chain. In practice, it is a good idea to familiarize yourself with all these build tools, processes and commands available so that when you are ready to produce your application, most of it becomes as automatic as possible.

We have introduced four important command line tools:

- `bmconv`
  Bitmaps are built from Windows `.bmp`s into Symbian OS `.mbm`s and (optionally) a `.mbg` header file containing symbolic IDs. The `bmconv` tool handles this conversion as part of the application build process (`abld`), but it can also be used in its own right to convert `.mbm`s back into `.bmp`s.
- `aiftool`
  Application information files (AIFs) contain icons, natural-language captions, and some other information about applications. They are

constructed from a resource file and an MBM and are created by `aiftool`. Aiftool is called as part of the main application build process (via `abld`).

- `makekeys`
  This tool allows you to produce a digital certificate for securing the installation file. The aim of this certificate is to identify the software vendor and to make sure that the installation file has not been tampered with since it was created. Makekeys is a stand-alone tool.

- `makesis`
  Installation uses a `.sis` file that is built by `makesis` from files specified in a file. The `.sis` file is then distributed to end users and it can be transferred from a PC to a Smartphone using connectivity software, infrared, Bluetooth, e-mail or over an Internet connection. Makesis is a stand-alone tool.

By including the relevant information in the project specification file, bitmap conversion and the generation of AIFs can be incorporated in the overall build process, which can be run either from the command line, or from the CodeWarrior IDE. However, `makekeys` and `makesis` have to be run from the command line.

Get to know all these tools and how to get the most out of them – they're an essential part of programming for all Symbian OS GUIs. All the build tools available for Symbian OS are described in greater detail within the Tools & Utilities section of the Developer Library supplied on the UIQ SDK.

If you're writing Symbian OS applications based on UIQ that you want to be widely used, style and finesse are also essential.

It's worth taking time to plan and design your application thoroughly before even thinking about implementing it. This includes studying the behavior of the built-in applications and the UIQ style guide. There are several usability aspects of the GUI that are fundamental to the way your application should be laid out – these include the use of List and Detail views, menus, dialogs and the button bar.

You shouldn't follow every suggestion or every precedent without thinking about it; most applications depart from the style guide at one point or another. But if you can stay with the style guide where possible, it will help users to understand – and become comfortable using your applications very quickly.

15

Device- and Size-independent Graphics

This chapter discusses writing graphics code that is independent of the size of the graphic and the device to be drawn to. Firstly, we study the writing of graphical applications – a topic that involves both size- and device-independent drawing code. Later, we cover the subjects of blitting, fonts and color, as these subjects are heavily influenced by the need for size- and device-independence. The last section of the chapter looks at the evolution and adaptation of GUI systems for Symbian OS phones.

The chapter covers three areas of code independence:

- *Size-independent drawing*: necessary when drawing to different targets, such as different sized screens. It allows zooming of a graphic, as we have previously seen used by the `FleetView` application in Chapter 9.

- *Target-independent drawing*: an application may need to draw to more than one kind of device, such as a screen and a printer.

- *Device-independent graphical user interfaces*: a GUI requires not just on-screen drawing, but interaction as well. CONE's principles of pointer and key distribution, focus, and dimming, which I described in Chapter 12, can be used for any GUI. But the way that Uikon (and, particularly, a customized layer such as UIQ) builds on CONE is optimized for a particular device and its targeted end users. It turns out that the design of the GUI, as a whole, is heavily dependent on device characteristics.

Printing is a capability that featured in earlier Symbian OS devices. Although it is not used in the current range of Symbian OS smartphones, it continues to be supported and is likely to be reintroduced in future products. Symbian OS has supported target-independent drawing from

Symbian OS C++ for Mobile Phones.   Edited by Richard Harrison
© 2003 John Wiley & Sons, Ltd   ISBN: 0-470-85611-4

the start, so that all components can be written to print or draw to a screen. This is in contrast to its SIBO predecessor, which added printing late in the product cycle, necessitating substantial changes to components that needed to print.

15.1   Size- and Target-independent Drawing for Applications

Graphical applications may need to support zooming and drawing to different targets. Possible targets are the screen or a printer. Different targets may also mean different types of smartphones, with different screen sizes. Later in the section we describe how to draw to printers. GUIs for different smartphones are discussed later in the chapter.

The classes that support size independence are the same for each target – they are CGraphicsContext, MGraphicsDeviceMap, Mgraphics-Device, and TZoomFactor. These have to be implemented differently for the different targets. They are a part of the Graphics Device Interface (GDI).

Application drawing code uses the functions and settings of the GDI classes.

The example 'drawing' application in this chapter demonstrates how to implement zooming on a screen device. For screen devices, controls are used (derived from CCoeControl) to allow user interaction with the graphic, and a GUI is used to display and provide an interface to the application. After going through the example code, we see how the size-independent code may be reused when drawing to a printer.

Figure 15.1 shows a screen shot of the running example. You can find the example code in \scmp\drawing\.

The top-left and bottom-right rectangles illustrate the need to do a 'full redraw' rather than drawing over the existing graphic, which we'll come back to.

The application structure is shown in Figure 15.2.

In order to set the size-independent drawing code in context, an overview of how the example application classes work together is as follows. For greater detail see the example code.

Class	Description
CExampleApplication	On application startup, used by the UI framework to create a CexampleDocument.
CExampleDocument	As soon as it has been created, used by the UI framework to create a CExampleAppUi.

Figure 15.1

CExampleAppUi	On construction, creates a CExampleAppView and passes it a rectangle to work with. The rectangle is the client area: the area of the screen available to the application for drawing, which does not include the space that is available for the menu and application bands on the screen. The CExampleAppUi class also handles all user interface commands, notably 'zoom out' and 'zoom in'. (The Drawing.rss file defines the menu options for implementing these commands.)
CExampleAppView	On construction, splits its rectangle into four quarters and creates a control, CExampleHelloControl, for each; in each case, the control is passed a rectangle to work with (and for two of the controls the 'full redraw' parameter is set to false). Passes user-interface commands to each of these controls.

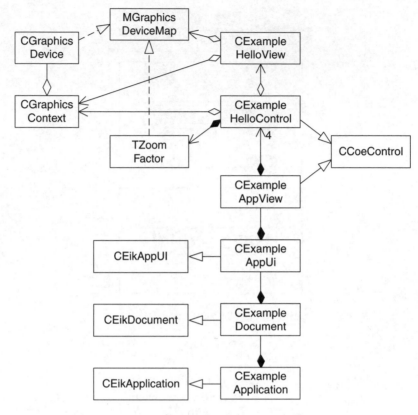

Figure 15.2

CExampleHelloControl	On construction, creates a view, CExampleHelloView, and sets the text for the view. The class is also used for drawing on application startup and whenever the view changes. It draws a border, then calls its view's draw function.
CExampleHelloView	Draws 'Hello World' into the center of the rectangle, then draws a box around the text.

`CExampleAppView` and `CExampleHelloControl` are both controls – they are derived from `CCoeControl`. Such controls are specific to the screen devices and so cannot draw to a printer.

The example separates out the target-independent drawing code into a separate class, which is not a control–`CExampleHelloView`. CExampleHelloView has a drawing function that uses graphics context (CGraphicsContext), drawing functions and a graphics device map

(MGraphicsDeviceMap) to access a graphics device's scaling functions. This drawing code is examined in the following pages.

We use a CExampleHelloView class to separate out size- and target-independent code into an independent unit. This allows the size-independent code to be reused to draw to a printer, as we will see later.

Nevertheless, CExampleHelloControl contains part of the size- and target-independent code: it allocates the size and device-dependent font with which to do the drawing, using an independent font specification. The reason for this is explained later.

The role of each of the GDI classes is summarized below:

Class	Description
CGraphicsContext	The abstract base class created by a graphics device, CGraphicsDevice. Contains the main drawing functions.
	Provides the 'context' in which you are drawing to the associated device in the sense that it holds the pen and brush settings (e.g. color, line styles) and font settings (e.g. bold, underline, italic) for drawing, and also the clipping region (the visible drawing area). These can all be updated while drawing.
	Deals with pixels of device-dependent size and uses fonts with device-dependent size and representation. The sizes and fonts to be passed to CGraphicsContext functions therefore need to be converted from size-independent units to size-dependent units beforehand. This is done by an MGraphicsDeviceMap derived class. This may be a TZoomFactor or the CGraphicsDevice.
	Note that the CGraphicsContext class was described in more detail in Chapter 11 and its handling of colors and bitmaps is described later in this chapter.
MGraphicsDeviceMap	The abstract base class for both graphic devices and zoom factors. Defines the size-dependent functions in a graphics device. These functions convert

Class	Description
	between pixels and twips and perform font allocation and release. Font allocation involves finding the font supported by the device that is the closest to a device-independent font specification.
CGraphicsDevice	The abstract base class for all graphics devices that represents the medium being drawn to.
	Manufactures a graphics context suitable for drawing to itself (using CreateContext()), which takes into account the attributes of the device, such as the size and display mode.
	Allocates (and releases) fonts suitable for drawing to itself and converts between twips and pixels.
	Important graphic devices are CScreenDevice, CBitmapDevice and CPrinterDevice. (Bitmap applications are discussed later in the chapter.)
TZoomFactor	Defines a zoom factor and implements the MGraphicsDeviceMap interface.
	Allocates and releases device-dependent fonts and converts between twips and pixels.
	Facilitates zooming, because it allows the size of the graphic to become independent of the target size.
	This class is recursive, because a TZoomFactor object can use an MGraphicsDeviceMap, which could be a TZoomFactor itself, as illustrated below. This allows a zoom factor object to contain another zoom factor object, multiplying the effect of the child zoom factor.

The top-level zoom factor, however, uses a `CGraphicsDevice` (not another `TZoomFactor`).

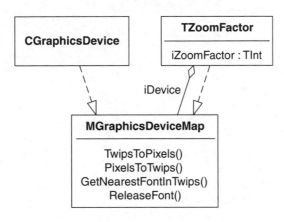

Figure 15.3

In the example, the zoom factor uses the screen device. This is set up as follows in the hello control constructor:

```
iZoomFactor.SetGraphicsDeviceMap(iCoeEnv->ScreenDevice());
```

iZoomFactor is used for getting the appropriate size and device-dependent font and size for drawing.

The following section studies the device-independent drawing code in the CExampleHelloView class and the small amount of device-independent code for allocating a font in the CExampleHelloControl class. In the next section, we'll see how the drawing code is used by the CExampleHelloControl class.

15.1.1 Device-independent Drawing

Device-independent drawing is largely conducted by the `CExample-HelloView` class. Here is its declaration:

```
class CExampleHelloView : public CBase
    {
public:
    // Construct/destruct
    static CExampleHelloView* NewL();
```

```
    ~CExampleHelloView();

    // Settings
    void SetTextL(const TDesC& aText);
    void SetFullRedraw(TBool aFullRedraw);

    // Draw
    void DrawInRect(const MGraphicsDeviceMap& aMap, CGraphicsContext& aGc,
            const TRect& aDeviceRect, CFont* aFont) const;
private:
    void ConstructL();
private:
    HBufC* iText;
    TBool iFullRedraw;
    };
```

Firstly, note that it's derived from `CBase`, not `CCoeControl`. No `CCoeControl`-derived class can be device-independent, because controls are heavily tied to the screen. The drawing code is located in the `DrawInRect()` function, which takes a device map, a graphics context, and a rectangle within which to draw. We'll now look at the drawing code in detail.

`DrawInRect()` can be divided into two sections:

- drawing the text,

- drawing a box around the text.

Drawing a box

It's easiest to start by looking at the part that draws the box:

```
void CExampleHelloView::DrawInRect(const MGraphicsDeviceMap&
aMap,CGraphicsContext& aGc, const TRect& aDeviceRect, CFont* aFont) const
    {
    //Draw text
    ...
    //Draw a box:
    //Allocates a device-independent size for the box to be drawn around
        the text.
    TSize boxInTwips(1440,288); // 1" x 1/5" surrounding box
    //Converts twips to pixels, using iZoomFactor, to get a box of
        1" x 1/5"
    TSize boxInPixels;
    boxInPixels.iWidth=aMap.HorizontalTwipsToPixels(boxInTwips.iWidth);
    boxInPixels.iHeight=aMap.VerticalTwipsToPixels(boxInTwips.iHeight);
    // draws the box
    TRect box(         //this creates a TRect using boxInPixels
            TPoint(
                    aDeviceRect.Center().iX - boxInPixels.iWidth/2,
```

```
                              aDeviceRect.Center().iY - boxInPixels.iHeight/2
                              ),
                  boxInPixels);
    aGc.SetBrushStyle(CGraphicsContext::ENullBrush);
    aGc.SetClippingRect(aDeviceRect);
    aGc.SetPenColor(KRgbDarkGray);
    aGc.DrawRect(box);
    box.Grow(1,1);
    aGc.SetPenColor(KRgbBlack);
    aGc.DrawRect(box);          //this makes the box 2 pixels thick
    }
```

The purpose of this code is to draw a box of a device-independent size. The box is slightly fancy – it is two pixels wide, dark gray on the 'inside', and black on the 'outside'.

All graphics context-drawing functions are specified in pixels. So, before I can draw the rectangle, I have to convert the size I wanted, specified in twips, to a size in pixels. I do this using the twips-to-pixels functions of the graphics device map. Remember that a Map ultimately uses the screen device, which is how it can get the information about the number of pixels to a twip.

Then I have a problem: I can't be sure that this `surround` rectangle is actually contained entirely within `aDeviceRect`, and I should not draw outside `aDeviceRect`. There is no guarantee, either on screen or another device, that drawing will be clipped to `aDeviceRect`. But because I need that guarantee here, I set up a clipping rectangle explicitly.

Even device-independent drawing code must take device realities into account.

In Chapter 9, we saw that we had to take rounding errors into account when sizing the grid for the Solo Ship's fleet view. Here, I am taking device realities into account in a different way: although I calculate the size of the surrounding rectangle beginning with twips units, I do the expansion explicitly in pixels. Whatever display I draw this rectangle to, I want it to consist of a two-pixel border – two lines spaced apart by a certain number of *twips* would overlap at small zoom states, and be spaced apart at large zoom states.

Getting a font

I want to draw the message in 12-point Swiss bold text.
'12-point Swiss bold' is a device-independent way of specifying a font. What I need to do is get a device-dependent font that meets this specification, taking into account the zoom state.

Symbian OS supports fonts using the following classes:

Class	Description
`TFontSpec`	A device-independent font specification, supporting a name, height and style. The style includes italic/normal, bold/normal, and superscript/subscript/normal attributes. Other font-related attributes, such as color, underline, and strikethrough, are implemented algorithmically by drawing functions.
`CFont`	A device-dependent font. Always accessed by `CFont*`. Most `CFont` functions, such as `AscentInPixels()`, provide fast access to pixel sizes either for any character or a particular string.

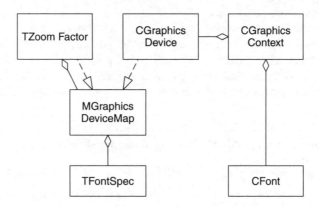

Figure 15.4

We *could* allocate (and release) the font along with the rest of the size- and target-independent code in **DrawInRect( )**. However, **DrawInRect( )** is not a leave function, and should not be, as it is called from the hello control **Draw( )** function, yet font allocation could fail and might leave.

It is possible to get around the above problem using a trap harness, and thereby allocate the font in **DrawInRect( )**. We don't do this as it is not a good practice and it is also bad practice to allocate resources while drawing, so it would make the code unsuitable for large-scale use. Therefore, I have found another place to allocate and release fonts. This is in **CExampleHelloControl::SetZoomAndDeviceDependentFontL( )**. This is an appropriate location for the font allocation as it is called on class construction and as a result of zooming in and out, and the size-dependent font is expected to change when the user zooms in or out (and at no other time).

Here's the code:

```
void CExampleHelloControl::SetZoomAndDeviceDependentFontL(TInt
   aZoomFactor)

{
    //Set zoom factor
    ...
    //Allocate a device dependent font for drawing
    iZoomFactor.ReleaseFont(iFont);
    iFont=NULL;
    _LIT(fontName, "SwissA");
    TFontSpec fontSpec(fontName, 240); // font size is 12 point
    fontSpec.iFontStyle=TFontStyle(EPostureUpright, EStrokeWeightBold,
        EPrintPosNormal);
    User::LeaveIfError(iZoomFactor.GetNearestFontInTwips(iFont,
        fontSpec));
}
```

The key function here is GetNearestFontInTwips(), a member function of MGraphicsDeviceMap. You pass a TFontSpec to this function, and you get back a pointer to a device-dependent **font** (a CFont*).

The mapping from TFontSpec to CFont* is ultimately handled by a graphics device (though the font specification may be zoomed by a zoom factor). Once you have a CFont*, you can only use it on the device that allocated it – more precisely, you can only use it for drawing through a graphics context to the device that allocated it.

This function usually finds a match, but in the unlikely case that it doesn't, it returns an error code. I propagate any error by calling User::LeaveIfError()

Notice the need to release the font after use; when you no longer need a CFont*, you *must* ask the device to release it.

If you forget to release a font, the effect will be the same as a memory leak: your program will get panicked on exit from emulator debug builds.

There is no error if ReleaseFont() is called when the font has not been allocated yet, the function just returns. Therefore, it is safe to release the font at the start of the function. The last font to be allocated before the application is closed is released in the hello control class destructor.

The code iFont=NULL exists because of the possibility of the function leaving before the font is reallocated. In this case, the destructor would attempt to release the font again if iFont wasn't set to NULL, causing the application to crash.

The font specification uses the `TFontSpec` class, defined by the GDI, in `gdi.h`. Here's its declaration:

```
class TFontSpec
    {
public:
    IMPORT_C TFontSpec();
    IMPORT_C TFontSpec(const TDesC& aTypefaceName, TInt aHeight);
    IMPORT_C TBool operator==(const TFontSpec& aFontSpec) const;
    IMPORT_C void InternalizeL(RReadStream& aStream);
    IMPORT_C void ExternalizeL(RWriteStream& aStream) const;
public:
    TTypeface iTypeface;
    TInt iHeight;
    TFontStyle iFontStyle;
    };
```

A font specification consists of a typeface, a height, and a font style. The `TTypeface` class is also defined by the GDI. It has several attributes, of which the most important is the name.

When you use `GetNearestFontInTwips()`, the height is expected to be in twips, though some devices support a `GetNearestFontInPixels()` function that allows the height to be specified in pixels.

The font style covers **posture** (upright or italic), **stroke weight** (bold or normal), and **print position** (normal, subscript, or superscript).

Other font attributes, such as underline and strikethrough, aren't font attributes at all – they're drawing effects, and you can apply them using the `CGraphicsContext` functions.

`TFontSpec` has a handy constructor that I used above for specifying a font in a single statement. It also has a default constructor and assignment operator. Finally, `TFontSpec` has `ExternalizeL()` and `InternalizeL()` functions. These are important: a rich text object, for instance, must be able to externalize `TFontSpec`s when storing, and reinternalize them when restoring. `TFontSpec` objects can be stored with a rich text object (i.e. a document of some sort). They provide necessary information about how to display the document on a screen or printer. The `TFontSpec` class comes into use whenever a rich text object is stored in or restored from a file. `TFontSpec` is actually useful to any application that can display text in more than one size or to more than one target, including our example.

Zooming and fonts

Before we move on we will look at some problems associated with zooming text at very low zoom factors. The first is a fairly straightforward issue, but the second is slightly more subtle.

Figure 15.5

The first case occurs when the zoom factor is set too low for the smallest available size of a font, as is illustrated in Figure 15.5 for the lowest zoom state of our example application, running on a P800 phone. The only sensible solution is to restrict the range of the zoom factor, dependent on the range of font sizes that are available on a particular phone.

> **The fonts on most target phones will be supplied in a smaller number of heights than you will find on the emulator. This can lead to different behaviors on the emulator and the real hardware, as is illustrated in our example in its lowest zoom state. This is another example of the need to take device realities into account – and it emphasizes the importance of thorough testing of your application on the target phone.**

The second case will cause problems even when the phone supplies fonts with a wide range of sizes. If this situation is not dealt with properly, then the width of text at low zoom factors does not scale in proportion to the height. As can be seen in the following diagram, this can cause the text to be too wide for the surrounding box:

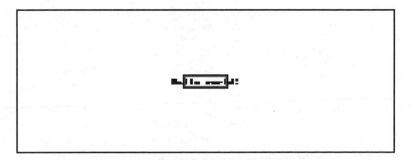

Figure 15.6

When getting a CFont from a TFontSpec you are converting from a twips size to a pixel size and, as you would expect, the height and width of a font will scale proportionately. But a character cannot be displayed in a width smaller than that of a pixel so, if the width scales to less than this, the excess will accumulate, character by character, along the length of the text.

One sensible solution at such low zoom factors is to calculate the width of the whole string in twips, convert the result to pixels and simply draw a horizontal line of the appropriate length.

You might want to support these low zoom factors in order to display the shape of words, sentences and paragraphs even when the font is too small to read, so eventually you can just make out where the capital letters are and, with even lower zoom states, just where the paragraphs are. This requirement applies to a print preview, for example. The Symbian OS rich text view, unlike our example application, can properly handle and display fonts even smaller than a pixel, using techniques such as that described above. Not all smartphone views will have to support this, so for many phones it is not actually an issue.

Drawing the text

We are now back to the hello view class's DrawInRect() function, in which we will see how our allocated font is used.

```
void CExampleHelloView:: DrawInRect(const MGraphicsDeviceMap& aMap,
CGraphicsContext& aGc, const  TRect& aDeviceRect, CFont* aFont)
{
    //Draw some text
    if (iFullRedraw)
        {
        aGc.SetBrushStyle(CGraphicsContext::ESolidBrush);
        aGc.SetBrushColor(KRgbWhite);
        }
    else aGc.SetBrushStyle(CGraphicsContext::ENullBrush);
    aGc.SetPenStyle(CGraphicsContext::ESolidPen);
    aGc.SetPenColor(KRgbBlack);
    aGc.UseFont(aFont);
    TInt baseline=aDeviceRect.Height()/2 + font->AscentInPixels()/2;
    aGc.DrawText(*iText, aDeviceRect, baseline,
        CGraphicsContext::ECenter);
    aGc.DiscardFont();
    //Draw a surrounding box
}
```

This uses the version of DrawText() that specifies a rectangle and guarantees that drawing will be clipped within that rectangle.

However, the DrawText() function will only white out the background before drawing, if iFullRedraw is set. When it is set, the brush setting is solid and white, which will cause the background to be whited;

when it is not set, there is effectively no brush and the text is drawn straight over whatever was there before. This is the reason for the messy display in the upper rectangle in the screen shot of our application. In a real program, the whiting out of the background would not be conditional.

15.1.2   Using the View

`CExampleHelloControl` *has a* `CExampleHelloView`, which it uses for drawing its model. Here's `CExampleHelloControl::Draw()`:

```
void CExampleHelloControl::Draw(const TRect& /*aRect*/) const
    {
    CWindowGc& gc = SystemGc();//CWindowGc is derived from
        CGraphicsContext
    TRect rect = Rect();
    rect.Shrink(10,10);
    gc.SetPenStyle(CGraphicsContext::ENullPen);
    gc.SetBrushStyle(CGraphicsContext::ESolidBrush);
    gc.SetBrushColor(KRgbWhite);
    DrawUtils::DrawBetweenRects(gc, Rect(), rect);//whitens the border
    gc.SetPenStyle(CGraphicsContext::ESolidPen);
    gc.SetPenColor(KRgbBlack);
    gc.SetBrushStyle(CGraphicsContext::ENullBrush);
    gc.DrawRect(rect);
    rect.Shrink(1,1);
    iView->DrawInRect(iZoomFactor, gc, rect, iFont);
    }
```

First, `CExampleHelloControl` draws a surrounding 10-pixel border. This code is guaranteed to be on the screen, and the 10-pixel border is chosen independently of the zoom state (as it would look silly if the border were to scale to honor the zoom state).

I'm careful to draw every pixel: I use `DrawUtils::DrawBetween-Rects()` to whiten the region between the rectangle to be drawn and the outside of the control. I then use the pen to draw a rectangle on the inside of the border. Then the rectangle is shrunk by a pixel and passed to the hello view's draw function.

The fundamental point that this example does show is that you can implement drawing code that's completely independent of a control and completely independent of the screen device.

15.1.3   Managing the Zoom Factor

We have already claimed that a device map ultimately uses a real device. `CExampleHelloControl` shows how the zoom factor relates to the device. Here's the declaration of `TZoomFactor` in `gdi.h`:

```
class TZoomFactor : public MGraphicsDeviceMap
    {
public:
    IMPORT_C TZoomFactor();
    IMPORT_C ~TZoomFactor();
    inline TZoomFactor(const MGraphicsDeviceMap* aDevice);
    IMPORT_C TInt ZoomFactor() const;
    IMPORT_C void SetZoomFactor(TInt aZoomFactor);
    inline void SetGraphicsDeviceMap(const MGraphicsDeviceMap* aDevice);
    inline const MGraphicsDeviceMap* GraphicsDeviceMap() const;
    IMPORT_C void SetTwipToPixelMapping(const TSize& aSizeInPixels,
                                        const TSize& aSizeInTwips);
    IMPORT_C TInt HorizontalTwipsToPixels(TInt aTwipWidth) const;
    IMPORT_C TInt VerticalTwipsToPixels(TInt aTwipHeight) const;
    IMPORT_C TInt HorizontalPixelsToTwips(TInt aPixelWidth) const;
    IMPORT_C TInt VerticalPixelsToTwips(TInt aPixelHeight) const;
    IMPORT_C TInt GetNearestFontInTwips(CFont*& aFont,
                                        const TFontSpec& aFontSpec);
    IMPORT_C void ReleaseFont(CFont* aFont);
public:
    enum {EZoomOneToOne = 1000};
private:
    TInt iZoomFactor;
    const MGraphicsDeviceMap* iDevice;
    };
```

`TZoomFactor` both implements `MGraphicsDeviceMap`'s interface and uses a `MGraphicsDeviceMap`. `TZoomFactor` contains an integer, `iZoomFactor`, which is set to 1000 to indicate a one-to-one zoom, and proportionately for any other zoom factor.

In order to implement a function such as `VerticalTwipsToPixels()`, `TZoomFactor` uses code such as this:

```
EXPORT_C TInt TZoomFactor::VerticalTwipsToPixels(TInt aTwipHeight) const
    {
    return iDevice->VerticalTwipsToPixels((aTwipHeight * iZoomFactor) /
      1000);
    }
```

`TZoomFactor` scales the arguments before passing the function call on to its `MGraphicsDeviceMap`. Other functions combine the zoom and conversion between pixels and twips:

- A pixels-to-twips function scales *after* calling pixels-to-twips on the device map.

- A get-nearest-font function scales the font's point size before calling get-nearest-font on the device map.

The function names in `TZoomFactor` indicate several ways to set the zoom factor. The method I use in `CExampleHelloControl` is the most obvious one,

```
void CExampleHelloControl::SetZoomAndDeviceDependentFontL(TInt
  aZoomFactor)
    {
    iZoomFactor.SetZoomFactor(aZoomFactor);...
     ...
    }
```

and GetZoom() is equally simple:

```
TInt CExampleHelloControl::GetZoom() const
    {
    return iZoomFactor.ZoomFactor();
    }
```

The SetZoomInL() function works like CFleetView::SetZoomL() that we saw in Chapter 9:

```
void CExampleHelloControl::SetZoomInL()
    {
    TInt zoom = GetZoom();
    zoom =
        zoom < 500 ? 500 :
        zoom < 1000 ? 1000 :
        zoom < 1500 ? 1500 :
        zoom < 2000 ? 2000 :
        500;
    SetZoomAndDeviceDependentFontL(zoom);
    }
```

SetZoomOutL() works the other way round. The rest of CExample-HelloControl is the usual kind of housekeeping that, by now, should hold few surprises. See the source code for the full details.

15.1.4 Views and Reuse

CExampleHelloView is a device-independent view: it contains no dependencies at all on any screen device. That means it can be reused in some interesting contexts – especially printing.

Symbian OS contains other views that are designed in a manner similar to CExampleHelloView. The best example of this is rich text views, delivered by the CTextView class. We come back to this later in this section.

Printing

CExampleHelloView can be used for printing and print preview, where this is supported by the UI, without any changes. The GDI specifies a conventional banded printing model that contains, at its heart, an interface class with a single virtual function:

```
class MPageRegionPrinter
    {
public:
    virtual void PrintBandL(CGraphicsDevice* aDevice,
                            TInt aPageNo
                            const TBandAttributes& aBandInPixels) = 0;
    };
```

TBandAttributes is defined as:

```
class TBandAttributes
    {
public:
    TRect iRect;
    TBool iTextIsIgnored;
    TBool iGraphicsIsIgnored;
    TBool iFirstBandOnPage;
    };
```

If you want your application to support printing, use GUI dialogs to set up a print job and start printing. You need to write a class that implements MPageRegionPrinter, and then pass a pointer (an MPageRegionPrinter*) to the print job. The printer driver then calls your PrintBandL() as often as necessary to do the job.

The way in which a driver calls PrintBandL() depends on the characteristics of the printer. Clearly, the driver calls PrintBandL() at least once per page. Drivers may

- call PrintBandL() exactly once per page, specifying the page number and, in band attributes, a rectangle covering the whole page;
- save memory by covering the page in more than one band;
- work more efficiently by treating text and graphics separately so that, for instance, all text is covered in one band, while graphics are covered in several small bands.

As the implementer of PrintBandL(), you clearly have to honor the page number, so that you print the relevant text on every page.

Whether you take any notice of the band attributes is rather like whether you take any notice of the bounding rectangle passed to CCoeControl::Draw(): if you ignore these parameters, your code may work fine. But you may be able to substantially speed printing up by printing only what's necessary for each band.

Anyway, you could reuse the view code in CExampleHelloView to implement PrintBandL() for a single-page print:

```
void CMyApplication::PrintBandL(CGraphicsDevice* aDevice,
                                TInt /* aPageNo */
                                const TBandAttributes& /* aBandInPixels */)
    {
    //Allocate the font
    TFontSpec fontSpec(_L("Arial"), 6*10);
    fontSpec.iFontStyle=TFontStyle(EPostureUpright, EStrokeWeightBold,
EPrintPosNormal);
    CFont* font;
    User::LeaveIfError(aDevice.GetNearestFontInTwips(font, fontSpec));
    //Draw the view
    CGraphicsContext* gc;
    User::LeaveIfError(aDevice->CreateContext(gc));
    TRect rectInTwips(2880, 2880, 2880, 1440):
    TRect rect(aDevice->HorizontalTwipsToPixels(rectInTwips.iTl.iX),
               aDevice->VerticalTwipsToPixels(rectInTwips.iTl.iY),
               aDevice->HorizontalTwipsToPixels(rectInTwips.iBr.iX),
               aDevice->VerticalTwipsToPixels(rectInTwips.iBr.iY));
    iView->DrawInRect(*aDevice, *gc, rect, font);
    delete gc;
    //Release the font
    aDevice.ReleaseFont(font);


    }
```

Because **PrintBandL( )** is a leave function, we are able to allocate the font within it.

This prints the text in a rectangle 2 × 1 inches big, whose top-left corner is 2 inches down, 2 inches right, from the edge of the paper – regardless of paper size and margins. More realistic print code would take account of the paper size and margins, which are set up by the GUI dialogs.

The `CGraphicsDevice::CreateContext()` function creates a graphics context suitable for drawing to the device. It is *the* way to create a graphics context. Calls such as `CCoeControl::SystemGc()` simply get a graphics context that has been created earlier by the CONE environment, using `iScreenDevice->CreateContext()`.

Rich text and other views

The `CTextView` rich text view class provides a powerful view, which hides extremely complex functionality underneath a moderately complex API. This class supports formatting and display of rich text (model objects derived from `CEditableText`), printing, editing, and very fast on-screen update – enough to allow high-speed typing in documents scores of pages long, even on an 18 MHz ARM processor. UIs can provide controls derived from `CEikEdwin`, such as UIQ's `CEikRichTextEditor`, which applications can use for rich text editing.

15.1.5   Summary of Device-independent Drawing

In this chapter, we have looked at size and device-independent drawing, for the purposes of zooming and printing.

`CExampleHelloView` is a rather trivial example of the art. In particular, interaction is only via the toolbar. A more complex example might use a cursor. The Boss Puzzle in the Symbian OS C++ SDK contains a view that supports graphical interaction. See source code in `UIQExamples\papers\boss\view\v3`.

We've also seen that the GDI provides a vital toolkit for device-independent drawing, including the `MGraphicsDeviceMap` class, graphics devices and zoom factors, and font specs and fonts.

15.2   More on the GDI

Device-independent drawing is supported by the Symbian OS GDI (graphics device interface). All graphics components and all components that require a graphics object, such as text content, depend ultimately on the GDI.

The GDI defines

- basic units of measurement – pixels and twips – that are used by all drawing code,
- basic definitions for color,
- graphics devices and graphics contexts,
- fonts,
- bitmapped graphics,
- device mapping and zooming,
- printing.

The GDI is reasonably documented and well illustrated by examples in the Symbian OS C++ SDK. The catchall example program is `\Examples\Graphics\GraphicsShell.mmp`, which as its name suggests is a shell with several examples inside it, including the basic drawing functions supported by `CGraphicsContext`, bitmapped graphics, the `CPicture` class, zooming, offscreen bitmap manipulation, and the built-in fonts.

We've already seen most of the GDI. In this brief section, I'll review and develop more themes:

- introducing bitmap handling
- more on font management

- more on printing
- color and display modes
- web browsing.

15.2.1 Blitting and Bitmaps

These days, displays are fundamentally bitmap-oriented. Graphics primitives such as `DrawLine()` have to **rasterize** or **render** the line – that is, they must determine which pixels should be drawn in order to create a bitmap image that looks like the desired line. Rasterizing in the Symbian OS is the responsibility of the BITGDI, which implements all drawing functions specified by `CGraphicsContext` for drawing to on- and offscreen bitmaps.

Another approach to updating a bitmapped display is simply to **blit** to it: to copy a bitmap whose format is compatible with the format on the display. Blitting is extremely efficient if the source and destination bitmaps are of identical format.

Any GUI worth its salt takes advantage of the efficiency of blitting to optimize certain operations:

- On-screen icons are not rendered using drawing primitives, but pre-constructed in a paint program and blitted to screen when required.

- Flicker-free screen redraws are performed by rendering to an offscreen bitmap (potentially slow), and then blitting to the screen when needed (usually quick). The offscreen bitmap is maintained by `RBackedUp-Window`, a class of the window server. This provides an aid to getting flicker-free fast redraws as whenever a redraw is required and the window contents haven't changed. The image can simply be blitted from the screen and not redrawn from scratch. Redrawing is done entirely from the backup bitmap.

- Animation is a special case of flicker-free update that can be implemented using a sequence of blits, one for each image frame.

- Screen fonts are blitted from the font bitmaps onto the screen (or offscreen bitmap).

Blitting is great, but it isn't always the best thing to use.

- Bitmaps use a lot of memory, so if you can construct a picture from a short sequence of drawing primitives, it's often more compact than storing the picture as a bitmap.

- Bitmaps can't be scaled effectively: you lose information if you scale them down and they look chunky if you scale them up. Also, scaling is generally slow, which eliminates one of the major advantages of using bitmaps.

- Bitmaps are fixed. You can only use them to store predrawn or precalculated pictures, or to cache calculated images for reuse over a short period of time.

- Bitmaps are highly efficient for screens, but highly inefficient for printers, because they involve large data transfers over relatively slow links. They also involve scaling, but usually that's acceptable on printers because the scaling is to a size similar to that which would have been used for the bitmap on screen anyway.

Here's a UML diagram of bitmap support classes in Symbian OS:

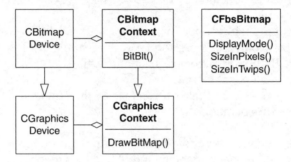

Figure 15.7

The base class for bitmaps is `CFbsBitmap`, which is defined in `fbs.h`. Key properties of a bitmap include the following:

- Its display mode – the number of bits per pixel, and color/gray encoding scheme. See the `TDisplayMode` enumeration in `gdi.h`, and the list below.
- Its size in pixels.
- Its size in twips.
- Its bitmap data, which you can get using `GetScanLine()` and similar functions.

Note that the `CFbsBitmap` constructor sets a pixel size only, but the size in twips can set subsequently. (The size in twips is optional and defaults to 0.0). There are two functions for setting the size in twips: you can pass in a size in twips directly or pass in a Graphics Device Map, which can be used for scaling, to convert the internally stored size in pixels to a size in twips. These functions are defined as follows:

```
void SetSizeInTwips(const MGraphicsDeviceMap* aMap);
void SetSizeInTwips(const TSize& aSizeInTwips);
```

Functions are provided to set and access all the bitmap properties, and also to internalize and externalize bitmaps using streams. CGraphicsContext requires that any graphics device (and hence any graphics context) can do four basic operations with any CFbsBitmap object. Here are CGraphicsContext's bitmap drawing functions:

- DrawBitmap() from the source bitmap to a region of the device, identified by its top-left corner. The bitmap draw size depends on its size in twips, which must be specified. The target device converts from twips to pixels. This means that the bitmap can be drawn according to an actual size, regardless of the twips-to-pixels mappings of the source and target.

- DrawBitmap() from the source bitmap to a region of the device, identified by its bounding rectangle. The bitmap is scaled to fit the target rectangle, which is specified in pixels.

- DrawBitmap() from a rectangular region of the source bitmap to a rectangular region of the device. The bitmap region is scaled to fit the target rectangle. Both rectangles are specified in pixels.

- Use of a bitmap in UseBrushPattern(), for background painting.

The GDI defines a bitmapped graphics device, CBitmapDevice and a bitmapped graphics context, CBitmapContext. You can read pixels and scan lines from a CBitmapDevice and create a CBitmapContext for drawing. You can perform actions such as clear, copy rectangles, blit and 'blit under mask' to a CBitmapContext. (The blit-under-mask functions are used for drawing icons with transparent backgrounds.) CBitmapContext::BitBlt() will always do one-for-one pixel blitting, regardless of pixel size. Compare this with CGraphicsContext::DrawBitmap(), which always scales if it needs to, even when copying from a bitmap to a bitmapped device.

Bitmaps are managed by the font and bitmap server. Prebuilt bitmaps are built into .mbm files, from Windows .bmps, using bmconv – usually one per application or component. .mbms can be built into ROM in a format corresponding to bitmap layout of the Symbian OS device's normal screen mode – this makes blitting from them particularly efficient. Bitmaps delivered with non-ROM components can be built into a compressed .mbm file from which bitmaps are loaded into the FBS's shared heap as needed, before being blitted elsewhere. Offscreen bitmaps may be allocated by applications: they reside in the FBS's shared heap.

15.2.2 More on Fonts

Symbian OS can use both bitmap fonts, for which character bitmaps at various sizes are stored and scalable fonts, for which algorithms to draw characters are stored. Bitmap fonts are stored in a preset range of sizes; for other sizes they can be algorithmically scaled, but the quality is unlikely to be good. Scalable fonts, however, as the name implies, can produce any size to be produced with equal quality.

For Western locales, scalable fonts are clearly useful, but for Far Eastern they are the difference between night and day, since the font information for even a single point size is enormous. By using scalable font technology, information is only *needed* for one size. Other sizes and rasterization for printers, can be handled by the scalable font system.

A number of systems for scalable fonts have been invented, well-known instances being Apple's TrueType, and the open-source FreeType. Symbian OS has a framework called the Open Font System, that allows rasterizer plug-ins DLLs to be supplied that support particular systems. Such a plug-in recognizes and reads font files stored in a particular format, and generates character bitmaps, which are then handled exactly as bitmap fonts.

However, scalable fonts are not always required. For example, the P800 uses only a small and fixed number of bitmap fonts. This is because it has no application requirement for a large set of fonts, such as a word processor. All applications will generally know which font/size they require for any specific widget. As such fonts 'could be' requested by UID rather than a device-independent TFontSpec specification. This is an optimization issue.

If using TFontSpecs, whether a font originated as a bitmap, or from a scalable font rasterizer, is transparent to clients. The same method that we have already seen is always used:

- You use a `TFontSpec` (and its supporting classes) to specify a font in a device-independent way.

- You use an `MGraphicsDeviceMap`, which ultimately leads to a `CGraphicsDevice`, to get a device-dependent font, using `GetNearestFontInTwips()` and the `TFontSpec`.

You can find out what fonts are available on a device through its **typeface store**, implemented by `CTypefaceStore`. You can ask how many typefaces there are, iterate through them all, and get their properties. The `FontsShell` example in the SDK (`\examples\graphics\fonts\FontsShell.mmp`) does exactly this.

Fonts for the screen (and offscreen bitmaps) are managed by the font and bitmap server (FBS). When you allocate a font, using `GetNearest Font...()` or similar functions, it creates a small client-side `CFont*` for the device, and also ensures that the bitmaps for the font are available for blitting to the screen (or offscreen bitmap). For built-in bitmap fonts, the font bitmaps can be in ROM, in which case the `CFont*` acts as a handle to the memory address. Getting a font is a low-cost operation for such fonts. Alternatively, fonts that are installed or generated are loaded into RAM and made accessible so that all programs can blit them efficiently from a shared heap. The CFont* acts as a handle to an address in this heap.

Releasing a font releases the client-side CFont* and in the case of an installable font, decrements a usage count, which will cause the font to be released when the usage count reaches zero.

Installable fonts are a main reason why GetNearestFont...() calls may fail (because of a potential out-of-memory error). Also note that it is worth releasing them as soon as possible, to free up the memory.

Sometimes, you want a device-dependent font. For instance, you may want a font of a particular pixel size, without going through the trouble of mapping from pixels to twips and then back to pixels again. For this, you can use GetNearestFontInPixels() on most graphics devices: this uses a font spec but interprets its iHeight in pixels rather than twips. Or, you may want one to use a special character from a particular symbol font. For this, you can use GetFontById(), which requires you to specify a UID rather than a font spec.

A word processor-like application will potentially utilize a large number of fonts, because there would generally be a number of fonts to select from and each would have a large range of sizes.

Sometimes the device-independence implied by a TFontSpec isn't device-independent enough. You can rely on Arial, Times New Roman, and Courier fonts or similar being present on any Western-locale device but not, for example, in Far Eastern locales. In response, Symbian applications usually contain font specification information in resource files, so that this aspect of an application can be localized: there are FONT and NAMED_FONT resource structures for this.

But don't overgeneralize: text *layout* conventions are different for Far Eastern applications too, so you may have to change other things if you want to support Far Eastern locales. You don't need to make any special font-related changes to support Unicode-based Western-locale machines.

15.2.3 More on Printing

A comprehensive print model is built into the GDI and implemented by higher-level components of Symbian OS. Note, however, that printing is usually only relevant to larger, communicator-type phones. More compact devices usually find it unnecessary, and cut the support.

To provide system support for particular printers, printer drivers are written as plug-ins that implement the CPrinterDevice interface, and stored in the /system/printers directory. On the client-side, the usual approach is to use the provided GUI dialog (if present, it's usually called `CEikPrinterSetupDialog`) to allow the user to set up and start printing. Print set up options can include:

- *page setup*: supports paper size, margins, rich text for header and footer, and options for page numbering, header on first page, footer on first page;
- *print setup*: the number of copies you want to print, and the driver you want to use;
- *print preview*: a preview showing the layout.

In your document model, you should externalize print settings.

Of course, such options can also be set up directly: `CPrintSetup` is the key class here. On-screen print preview support is provided through `CPrintPreviewImage`. And as we've already seen, at the heart of any print-enabled application, you have to implement `MPrintProcessOb-server`'s `PrintBandL()`.

Programming printer support to this extent is not very difficult. However, if you do need to print from your application, then the requirements to write code that is independent of a particular UI increase hugely, to where it becomes a key consideration for many elements of the application.

For example, an application toolbar can be highly device-dependent, as can the menu and the dialogs (though some size independence in the latter case would be useful). On the other hand, a text view intended for a word processor should be highly device-independent.

The requirements for printing text efficiently and for fast interactive editing of potentially enormous documents, however, are substantially different. So the Symbian OS text view component contains much shared code, but also a lot of quite distinct code, for these two purposes. Less demanding applications will have a greater proportion of shared code.

To take another example, without considering printing, a spreadsheet could be considered to be an abstract grid of cells, each containing text, numbers, or a formula. If you take that view, your drawing code can be pixel-oriented, and it won't be too difficult if you decide to support zooming. But this changes if people need to print the spreadsheet, including sensible page breaks and embedded charts. If you write a spreadsheet that supports all this, you need to design for printing from the beginning and to optimize your on-screen views as representations of the printed page.

15.2.4 Color

The basic class for color is the `TRgb`: a red-green-blue color specification. A `TRgb` object is a 32-bit quantity, in which eight bits are each available for red (R), green (G), and blue (B). Eight bits are wasted.

What? Wasted memory in such a fundamental class in Symbian OS? Actually, the waste is small. First, it's hard to process 24-bit quantities efficiently in any processor architecture. More importantly, TRgbs don't exist in large numbers – unlike, say, the pixels on a screen or bitmap. Bitmaps are stored using only the minimum necessary number of bits.

Constants are defined for the set of 16 EGA colors (so named after the IBM PC's 'Enhanced Graphics Adapter', which supported them and also introduced them into character set attributes). Here are the definitions in `gdi.h` that also show how the R, G, and B values are combined in a TRgb:

```
#define KRgbBlack        TRgb(0x000000)
#define KRgbDarkGray     TRgb(0x555555)
#define KRgbDarkRed      TRgb(0x000080)
#define KRgbDarkGreen    TRgb(0x008000)
#define KRgbDarkYellow   TRgb(0x008080)
#define KRgbDarkBlue     TRgb(0x800000)
#define KRgbDarkMagenta  TRgb(0x800080)
#define KRgbDarkCyan     TRgb(0x808000)
#define KRgbRed          TRgb(0x0000ff)
#define KRgbGreen        TRgb(0x00ff00)
#define KRgbYellow       TRgb(0x00ffff)
#define KRgbBlue         TRgb(0xff0000)
#define KRgbMagenta      TRgb(0xff00ff)
#define KRgbCyan         TRgb(0xffff00)
#define KRgbGray         TRgb(0xaaaaaa)
#define KRgbWhite        TRgb(0xffffff)
```

We use `#define` rather than (say) `const TRgb KRgbWhite = TRgb(0xffffff)` because GCC 2.7.2 didn't support build-time initialization of class constants. Plus initialization of any TRgb from one of these 'constants' is no more expensive than if 'proper' `const` TRgbs were used.

> **All `CGraphicsContext` color specifications for pens and brushes use `TRgb` values. The graphics device then converts these into device-dependent color values internally.**

With measurements and fonts, you have to convert to device-dependent units (pixels and CFonts) before calling CGraphicsContext functions. The same approach could have been taken with colors but it wasn't, because the meaning of a color is less device-dependent than the size of a pixel or the bitmap for a font.

Concrete color values such as `KRgbBlack` are useful in many situations. There are also some logical color values, defined in `TLogicalColor` that refer to colors used in the UI scheme, such as the colors used in menus, menu highlights, toolbar buttons, or window shadows. The choice of which physical colors these logical values correspond to belongs to the device OEM. The mapping is stored in a `CColorList` object. The color list

- supports logical-to-RGB color mappings loaded from resource files or specified programmatically;
- supports independent sections for the system and applications: a section is identified by a UID and a logical color by an enumerated constant;
- supports mappings for both four-gray and 256-color schemes: the 256-color scheme will be used and will look good, if the screen mode supports 16 or more colors. Otherwise, the four-gray scheme will be used.

An application can get the color list through `CEikonEnv::ColorList()`.

A key thing you have to know about a device is how many colors it supports. Actually, the number of supported colors depends not only on the device, but also on the current display mode of the device. Most devices have a preferred display mode, and some support multiple display modes: you can check the display modes supported by a window server screen device and set your window to use a required display mode, if it's supported. Some display modes consume more power than others, so the window server will change the display mode in use, to the one with the minimum power requirement for any visible window.

You can create bitmaps with any display mode. When you blit them onto another bitmap, or display them in a particular mode, the bitmap data is contracted or expanded as necessary, to match the display mode of the target bitmap.

The display modes supported by Symbian OS are defined in the `TDisplayMode` enumeration in `gdi.h`. They are

Mode	Bits	Type	Comment
ENone			A null value that shouldn't be present in any initialized `TDisplayMode` object.
EGray2	1	Grayscale	Black and white: displays `KRgbBlack` and `KRgbWhite`.

EGray4	2	Grayscale	Minimal grayscale: displays KRgbBlack, KRgbDarkGray, KRgbGray, and KRgbWhite exactly.
EGray16	4	Grayscale	16 shades of gray.
EGray256	8	Grayscale	256 shades of gray.
EColor16	4	Color	Full EGA color set (named after the IBM PC's 'Enhanced Graphics Adapter'): displays all standard KRgbXxx values exactly.
EColor256	8	Color	Netscape color cube: exactly represents all 216 combinations of R, G, B in multiples of 0x33, plus all remaining 40 combinations of pure R in multiples of 0x11, pure G, pure B, and pure RGB gray likewise.
EColor64K	16	Color	High color: represents 5 bits of R, 6 bits of G, and 5 bits of B, so that 0xrrrrrrrr, 0xgggggggg, 0xbbbbbbbb will convert to TRgb(0xrrrrr000, 0xgggggg00, 0xbbbbb000), with the least significant bits of each color being dropped.
EColor16M	32	Color	8 bits each for R, G, and B. 8 bits wasted.
ERgb	32	Color	Like EColor16M
EColor4K	16	Color	Uses 4 bits each for R, G, and B. 4 bits are wasted.

The window server sets the screen's display mode to the most capable mode required by any currently visible window and supported by the hardware. So the 'preferred *screen* display mode' is actually implemented as a 'default *window* display mode'. There may be a trade-off here between higher display modes and higher power consumption.

ROM bitmaps are generated in the preferred display mode, so that typically no bitmap transformations are required when blitting to screen.

If a color is passed to a CGraphicsContext function that is not supported exactly on the device, then the nearest supported color is used instead. We are quite used to having real-world colors mapped down onto black and white or low-fidelity color. The best approach for real-world colors is simply to allow the `TDisplayMode` mappings to do their thing.

> *This nearest-color transformation is done before any other operation uses the color,* including *logical operations such as XOR. This can produce unexpected effects, but logical operations are in any case of dubious value on windowing systems – with the exception of XORing with KRgbWhite, which has its uses and will always work as expected.*
>
> *You'll see mentions of 'palettes' in some of the GDI definitions; these were added to the design before support for any form of color display was implemented (in Symbian OS v5). When color display mode support was added, not to use palettes to optimize (say) the shades available in a 256-color display mode. Instead, we use the fixed Netscape color cube set, and that's it. This reduces the complexity of the API, loses no worthwhile features for devices in this class, and avoids the funnies you occasionally see (or remember seeing!) on Windows PCs when the palette was optimized for a foreground window while other visible windows' palettes went wild.*

Finally, you want the user to select a color, for example, in a drawing program. Some GUI's supply controls for this – in UIQ its `CQikColorSelector`.

15.2.5   Web Browsing

Web browsing technology is often confused on the subjects of target and size independence. It has evolved without clear distinctions between print and on-screen graphics and without a consistent way of resizing. Text is specified in html – which is device-independent – while pictures (gifs, jpgs, or bmps) are specified in pixels. The physical size of the graphics will then vary depending on the resolution of the monitor/screen – they will be smaller for higher resolution monitor/screen displays. There is no clear relation between text sizes and graphics, which means they don't scale together on most browsers. However, there are ways around these problems and text and pictures scale together on Symbian OS browsers.

15.3   The Developer's Quest for Device-independent Code

It is not always worthwhile making your code target-independent. So, when should you make your drawing code device-independent and when don't you need to? There are two extreme cases in which the answer is quite clear:

- If your code is designed exclusively for a screen-based UI, then it should not be target-independent (though you may wish to build in size independence).

- If your code is designed primarily for printing, with a screen-based UI for editing, then it should be device-independent.

An application toolbar can be highly device-dependent, as can the menu and the dialogs (though some size independence in the latter case may be useful). On the other hand, a text view intended for a word processor should be highly device-independent; so too should a mapping program that might well be printer-oriented.

15.3.1   Real Devices Intrude

This answer is only a starting point, however, and there are many awkward, intermediate cases. You have to take the realities of the device into account, even when writing the most device-independent code.

The influence of target devices on your code is even greater when the devices are relatively limited in CPU power and display resolution. With high-resolution displays and near-infinite CPU power, you can render everything with no thought for rounding errors, scale and so on. With small displays, slower CPUs, and no floating-point processor, you have to take much more care in both graphical design and programming.

In a map application, zooming introduces considerations not only of scaling, but also of visibility. In a high-level view of the map, you want to see any coastline, a few major cities, big rivers, and any borders. In a zoomed-in view of a city, you want to see district names, underground train stations, public buildings, and so on.

In the high-level view, you wouldn't try to draw these details at small scale: you would omit them altogether. And this omission is device-dependent: you can include more minor features in a printed view than in an on-screen view at the same zoom level, because most printers have higher resolution than a screen.

There are plenty of other complications with maps, such as aligning labels with features and transverse scaling of linear elements such as roads – and many other applications share these considerations. Fortunately, you don't usually edit maps on a handheld device, so there isn't the need for very quick reformatting code that there is with word processors. As a result, there may actually be better code sharing between printer and screen views.

15.4   GUI Systems

Up to this point in the chapter, we have been looking at size and device-independence of graphical applications. Now we will look at the smartphone independence of the GUI systems supporting these graphical applications on mobile phones.

The GUI determines the appearance and the interactive mechanisms of a smartphone: the 'look-and-feel'. These are clearly going to vary between smartphones: the GUI is, like the applications that use it, highly device-dependent.

This section is divided into a history of the Symbian OS GUI approach and the current situation.

15.4.1   The Evolution of the Symbian OS GUI System

In 1987, Psion conceived SIBO as a multiplatform operating system that would use the same system and application engines, but a different GUI (and different application GUIs) to support different devices. Within narrow parameters of variation, the concept worked for a range of quite distinct devices released between 1989 and 1997.

From the beginning, Symbian OS was designed with the same system structure, most of which would be independent of the GUI. Application code would be carefully structured into engines (or models) that were also GUI independent. A replaceable GUI framework would define the system's look and feel and applications' GUIs would be separate from their engines. If the GUI were replaced, only a small proportion of the total code in the system would need to be replaced along with it:

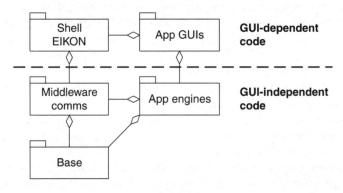

Figure 15.8

The GUI used in the Psion PDAs was called Eikon (hence `CEikAppUi` etc.). During early development of Symbian OS, Eikon was rewritten twice, but neither event had an impact on application engines – testimony to the effectiveness of the original architectural design. The first rewrite factored the GUI (originally titled HCIL) into CONE (a key item of GUI infrastructure and yet completely independent of any specific GUI) and Eikon. CONE survives to this day, but Eikon has been reworked, as different kinds of phones must be supported.

As Psion and others produced new PDAs, ports of Eikon and the applications were made. In 1997, Geofox released its Geofox One, with a 640 × 320 display and a track pad. Psion produced the netBook/Series 7 with a 640 × 480 color screen, and the Psion Revo with a 480 × 160 display, with three-button toolbars.

These ports showed that Eikon could scale to a different screen with only relatively minor modifications to itself and its applications. But they also show that even a minor change in screen size can make some software look odd, or prevent it from running altogether unless modified. The Battleships game's application view would look odd on a Psion netBook (lots of spare space above and below) and may not fit on a Revo at all.

To get Java certification in 1998, Symbian OS needed (among other things), to be able to run GUI applications and applets written in Java – often ones originally targeted at a much larger device. Symbian did that by adding some more tweaks to Eikon and some support in the AWT implementation that links Java to Eikon. For instance, Symbian added scrolling menu panes, metaphors for emulating right-click, and scrolling app views. Although that meant that Java applications and applets could run without modification under Eikon, it did not guarantee a satisfactory experience with applications whose menus and views were too big to see all at once, and required contortions to produce frequently needed pointer gestures.

So by the time Symbian split from Psion to become a mobile phone software vendor, it had became clear that different mobile phone manufacturers would have different requirements for which Eikon, however modified, would not be suitable.

Replacing Eikon

For some time, Symbian thought that each new device would have its own GUI that would replace Eikon altogether. The first opportunity to do this was during 1997 and early 1998, with the Ilium Accent produced by Philips Consumer Communications (never generally released).

With a 640 × 200 display and no keyboard, the device parameters were radically different from those that drove Eikon, as shown below:

The experience of the Ilium Accent showed clearly that Eikon *could* be replaced, but it also demonstrated the costs and the risks involved.

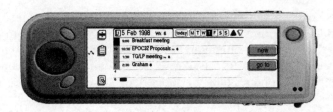

Figure 15.9

In 1999 a new strategy evolved, in which Symbian would support only a limited number of GUI *reference designs*, targeted at different types of hardware, from smartphones to higher-end communicators.

The first ideas for reference designs had a range of device parameters for each design, for example,

- display in the range 320+ x 120+ pixels,
- optional pointer,
- keypad.

But it soon became clear that this kind of flexibility leads to significant nonoptimality in a design.

Let's say that you design a view for exactly 320 x 120 pixels: you go to all kinds of trouble to cram in text and abbreviate words; you use icons instead of text, popup displays, very simple views that sacrifice functionality, perfect alignment of various columns and icons, and so on. A display such as this will break if you make it even a single pixel smaller, but *it will look equally odd if you make it larger*: Users will wonder why you didn't use the extra space.

To give another example, imagine that when you design, you take the view that there may or may not be a pointer. Then you have pointer-driven ways of doing everything and alternative joystick/keypad-driven ways of doing everything. These may not be very compatible: a screen with a pointer can have many more buttons, for example, than one using a joystick or arrow keys. If we deliver a design that is obviously optimized for pointers rather than a keypad or joystick, users with only the keypad/joystick will consider it even more awkward. Worse still, we might forget to optimize the design sufficiently for the needs of keypad/joystick-only users.

It's clear that a GUI – especially at the smartphone-end of the market – has to be quite *in*flexible in its design parameters. It must specify screen size and input devices exactly, and allow for no variation.

Architectural consequences

The reference design strategy required that the GUI system (previously Eikon) be split into three elements:

- A core GUI framework called Uikon (Universal or Unified Eikon). This includes the Uikon Core (eikcore.dll) that contains framework classes such as `CEikAppUi` and `CEikonEnv` and the Uikon Core Controls (eikcoctl.dll) that contained a number of concrete controls, such as menus and list boxes, that are expected to be present on all devices. You could write an application using Uikon alone that would be good enough to test some engine APIs, but not much more.

- the reference design itself that would add more concrete controls,

- a look-and-feel (LAF) module, which could differ for each device. The LAF allows the manufacturer to specify such things as system fonts and bitmaps.

In terms of applications, the application engines would continue to be part of the core OS, while reference designs would supply appropriate application user interfaces.

Symbian OS reference designs

There are several broad categories of smartphone:

- a design for keyboard-based mobile phones
- a design for a pen-based mobile phone with input via a touch screen
- a phone design for one-handed use, with interaction via a keypad.

A GUI for keyboard-based mobile phones was developed closely with Nokia for a top-of-the-range communicator device that was launched as the Nokia 9210. Key aspects of the design were keyboard input, a full-color landscape display, a device status pane area, and soft hardware keys that made for quick access to common commands.

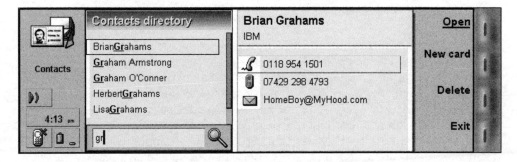

Figure 15.10

A reference design for pen-based mobile phones is developed by a dedicated team (and Symbian subsidiary) called UIQ Technology, based in Ronneby, Sweden. This design has a tabletlike screen with stylus operation and handwriting recognition. The initial design was completed for the launch of Symbian OS v6.0. For Symbian OS v7.0, it was released under the commercial name UIQ. We've been using this GUI throughout this book.

Open platform development

Reference designs for one-handed use smartphones are seen by mobile phone manufacturers as an area in which product differentiation is most vital. Here, Symbian offers just the core of Symbian OS, allowing mobile phone manufacturers complete freedom to innovate with Symbian OS as a common starting point. To make sure that GUI innovation is encouraged, Symbian OS source code is open to key parts of the development community to engage and compete in platform development.

A platform can include not just a GUI framework, but also new applications, and tools and source code to enable customization and application development. The outstanding initial success here has been the Nokia Series 60 Platform, a design for a color-screen smartphone, optimized for one-handed use. In Europe, Nokia has already released a Series 60 phone, the Nokia 7650, and announced two more – the Nokia 3650 and N-Gage. Series 60 Platform and Symbian OS have also been licensed to manufacturers including Matsushita (Panasonic), Samsung, Sendo and Siemens. In Japan, NTT DoCoMo has launched the F2051 FOMA phone, built by Symbian OS licensee, Fujitsu.

15.4.2   Major GUI Components

The major GUI components of Symbian OS are

- *CONE*: the UI control framework, which provides a general-purpose application user interface and controls for this interface. This component is essential to the GUI but independent of any specific GUI. Comprises the classes `CCoeAppUi`, `CCoeEnv` and `CCoeControl`.
- *Uikon*: this ties together CONE and the Application Architecture to give a framework for applications and a set of core controls that are present on all UI variants. Application developers can use the Uikon APIs directly, as we have in the 'drawing' example, or use a more appropriate device-specific API available in the UI variant's own libraries. We did not need to, but we could have used `CQikApplication`, `CQikDocument`, and `CQikAppUi` in our example. Uikon and the application architecture are documented in the Symbian Developer Library.

- *A phone UI framework*: this provides a specific set of custom UI components used by a licensee for their device, including items such as widget libraries, fonts, status bars, and indicators. Mobile phone manufacturers can either build their own, which they can then license so that it is available to other manufacturers, or customize one of the existing frameworks.

  - *the phone UI*: the collection of application UIs provided on the phone, which defines the 'UI' as the user understands it. In most cases, the phone UI constituents will be constructed out of components provided by the phone UI framework.

The current phone UI frameworks are as follows:

- UIQ which supports a touch screen with stylus operation and handwriting recognition and is developed by UIQ Technology (Sony Ericsson P800);
- Nokia Series 60 Platform, a design for a color-screen smartphone, optimized for one-handed operation (Nokia 7650, Nokia 3650, Nokia N-Gage);
- Nokia Series 80 Platform, a design for keyboard-based mobile phones (Nokia 9210I, Nokia 9290);
- Techview – a Symbian-developed framework for testing purposes only;
- other – largely specialized frameworks for Japanese phones.

An application developed for a given UI Framework will be compatible with all phones using that framework. It is possible to dramatically change the look and feel of the phone whilst maintaining compatibility. Mobile phone manufacturers may substitute new application UIs without affecting the operation of other applications. So, phones can be enhanced and branded through: modifying the existing application UIs, changing application skins (colors, appearance, icons), introducing new applications, or adding dynamic web content.

Further, the fact that the UI frameworks available for Symbian OS are based on Symbian's GUI foundation means that porting an application from one UI platform to another is a relatively straightforward task.

15.5 Summary

The theme of size and device-independent graphics has taken us through several topics related to UI applications, namely zooming, printing, fonts, rich text, colors, blitting, and web browsing. It has also taken us through a discussion of whole GUI systems.

After reading this chapter, you will hopefully be able to write applications that support the above-mentioned capabilities and are portable between smartphones using the same UI Framework.

Resources, in the shape of CONE and Uikon, are available to reduce the cost involved for writing new UI frameworks. The writing of these frameworks has increasingly become a licensee responsibility, because of the too high costs of maintaining a family of reference designs within Symbian. The expense of writing a new framework may be avoidable by using UIQ or Nokia Series which can be customized for a new look and feel.

16

A Multiuser Application

So far in this book, we've been learning mainly about how to program the GUI aspects of Symbian OS applications. Now, we're going to move on to a new set of topics: communications and system programming. The attractiveness of Symbian OS as an application platform arises both from the portability of Symbian OS phones, and from their ability to use wireless communications at short range (via infrared and Bluetooth) and long range, using various communications technologies over mobile networks (SMS, MMS and TCP/IP amongst others).

I'll start this chapter with a brief review of the communications facilities in Symbian OS. Then, I'll say a little more about the two-player version (tp-ships, that we first met near the end of Chapter 9), which was written without any communications support. Finally, I'll introduce the communications issues faced by Battleships.

I'll be describing the full Battleships application and incorporating the transaction-oriented games stack (TOGS) that is specified in Appendix 3. I'll also give a brief account of the TOGS components, which are described in more detail in Chapters 19 and 20. Together with the active object and client-server system programming frameworks (the subjects of Chapters 17 and 18 respectively) they demonstrate the abilities of Symbian OS with short- and long-range communication.

If you're not already familiar with communications programming, there are two ways you can approach the following chapters:

- *avoid communications*: just read Chapters 17 and 18, so you can learn about the Symbian OS active object and client-server frameworks, which are useful for all kinds of system programming tasks, whether or not they are communications-related,

- take this as an opportunity to learn a few things about communications: you'll find parts of the last couple of chapters a heavy read, and

Symbian OS C++ for Mobile Phones.   Edited by Richard Harrison
© 2003 John Wiley & Sons, Ltd   ISBN: 0-470-85611-4

you may need to refer to other communications literature for better explanations of some of the ideas.

16.1   Communications in Symbian OS

Symbian OS offers a wide-ranging communications infrastructure, including serial communications over RS232 and infrared, Bluetooth, TCP/IP, IrDA, USB, fax, and communication via GSM and GPRS. Future releases of Symbian OS will offer an even wider range of possibilities, with further support for 3G communications.

The main communications facilities that are currently available are seen in Figure 16.1

The servers are:

- C32 – the serial communications server, which drives the communications ports and also runs serial-like protocols that use other communications services.
- ESOCK – the sockets server, which provides a sockets-based API that's used for dial-up TCP/IP, IrDA via the infrared port and Bluetooth.
- ETEL – the telephony server, which is used to control phone-like devices including landline and mobile-phone modems, enabling them to make fax, data, and voice calls.
- Messaging – provides SMS and MMS send/receive, and Internet e-mail using dial-up TCP/IP for SMTP send, POP3/IMAP4 receive.

The first three of these servers work closely together to offer communications facilities to higher-level programs, including the communications applications included in Symbian OS:

Battleships uses two communications protocols:

- Bluetooth sockets, for short-range links
- SMS (short message service) text messages, for long-range links.

I'll give descriptions for the implementations for drivers for both these protocols in Chapters 19 and 20. By reading these chapters, you'll gain specific insight into how the main communications components of Symbian OS fit together.

You can also find good overview technical information on the communications facilities in technical papers on Symbian's website, and detailed API references in the product-specific SDKs available from the product manufacturers.

16.2   Battleships without Communications

The main problem with communications is that it's *complicated*. The final Battleships application has over 70 classes. There is simply no way

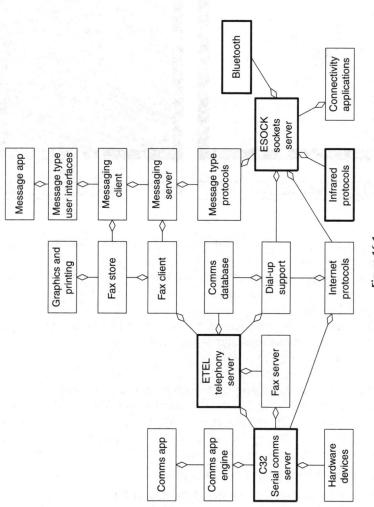

Figure 16.1

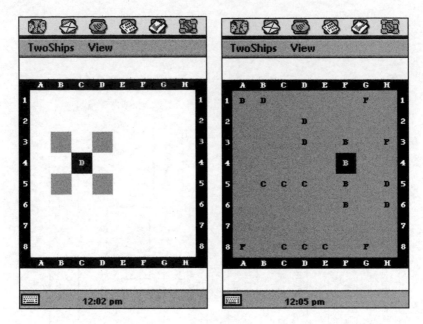

Figure 16.2

to produce an application like that without planning carefully how to get from here to there.

The starting point for Battleships was the two-player game, `tp-ships` for short. The purpose of `tp-ships` was to prove that the Battleships idea would work.

Here are some screenshots of a game that's just begun, showing my view of the opponents fleet and my own:

You play this game as follows:

- Player 1 starts the game by launching it from the menu. The initial view is player 1's view of the opponent's fleet. The player can use the **View** menu to change the view to **My fleet**, as shown in the second picture. Player 1 requests a hit on the 'opponent's fleet', and then chooses **Hide** from the **View** menu.

- Player 1 passes the game over to player 2.

- Player 2 can then select either **Target fleet** for selecting a location to hit or **My fleet** to view any hits on their fleet before selecting **Hide** and handing back over.

And so play continues. The hide system is good enough to make a playable game between friends, though it isn't really coded to stop determined cheats.

The game design in UML is given in Figure 16.3.

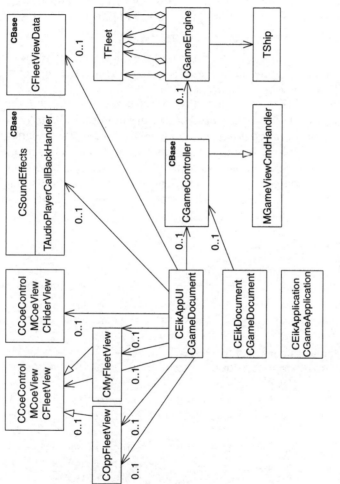

Figure 16.3

If you compare this with the Solo Ships design described in Chapter 9, you'll notice that

- instead of the document owning the controller, and the app UI owning the view, the app UI owns the controller, and the controller owns the view,
- the engine and view classes perform the same function as in Solo Ships,
- there are actually five views in this application: each player's fleet view and their view of the opponent's fleet plus a hider view.

There are two interesting aspects to this program:

- *It does some interesting Uikon tricks*: it has five app views, and it manipulates its menu options. It's useful to know this kind of trick, so we'll have a closer look at the code.
- *It doesn't use communications to get information about the other player's fleet*: instead, it uses object-oriented plumbing. We'll have a look at the code involved, and we'll explain why changing the ten or so lines of plumbing to support communications is going to take several hard chapters to explain a suitable replacement for them. Or, put another way, we'll understand why I wanted to take the `tp-ships` shortcut before doing all that communications work.

Let's look at these two things in turn.

16.2.1   View and Menu Tricks

As we saw in Chapter 9, a GUI application can have more than a single view. Many of the built-in Symbian OS applications have more than one view:

- Agenda can have day, week, month, year, to-do, and anniversary views
- Contacts has card and list views
- Messaging supports various types of message types and has appropriate views for each of them.

In a Symbian OS GUI application, it's very easy to change views without changing the app UI, because the view is quite a distinct entity from the app UI.

The app UI has seven members, five of which are views:

```
CGameController* iController; //Controller for both players
COppFleetView* iP1OppFleetView; //Fleet views for P1 fleet and
   target fleet
CMyFleetView* iP1MyFleetView;
COppFleetView* iP2OppFleetView; //Fleet views for P2 fleet and
   target fleet
CMyFleetView* iP2MyFleetView;
CHiderView* iHiderView; //Hider view
CFleetView* iActiveView;
```

There are also three associated flags that control whether the views are visible. The **Hide** menu option calls `CmdHideL()` which is coded as:

```
void CGameAppUi::CmdHideL()
    {
    ActivateViewL(TVwsViewId(KUidTpShips,KHiderViewUID));
    iActiveView = NULL;
    // toggle player turn
    if (iController->Engine().IsMyTurn())
        iController->Engine().SetSecondPlayer();
    else
        iController->Engine().SetFirstPlayer();
    }
```

This function calls `ActivateViewL` with the UIDs for the application and a view, which causes the framework to change the view, and then sets the flags in the engine that control the visibility of the menu options:

```
void CGameAppUi::DynInitMenuPaneL(TInt aResourceId, CEikMenuPane*
    aMenuPane)
    {
    if (aResourceId != R_GAME_VIEW_MENU) return;
    // toggle menu item dimming on view displayed
    TBool view = ((iActiveView == iP1MyFleetView)||(iActiveView ==
        iP2MyFleetView));
    aMenuPane->SetItemDimmed(EGameCmdMyFleet,view);
    view = ((iActiveView == iP1OppFleetView)||(iActiveView ==
        iP2OppFleetView));
    aMenuPane->SetItemDimmed(EGameCmdOppFleet,view);
    // if hider view is active, can't zoom or hide
    if (iActiveView == NULL)
        {
        aMenuPane->SetItemDimmed(EEikCmdZoomOut,ETrue);
        aMenuPane->SetItemDimmed(EEikCmdZoomIn,ETrue);
        aMenuPane->SetItemDimmed(EGameCmdHider,ETrue);
        }
    }
```

Controlling the visibility of menu options is more interesting. In the code above, you can see that I check the `iActiveView` member against each type of view and then dim the appropriate menu options depending on the value of the corresponding flag. This is reflected in the application when the framework calls `ViewActivatedL()`:

```
void CFleetView::ViewActivatedL(const TVwsViewId& /*aPrevViewId*/, TUid
/*aCustomMessageId*/,const TDesC8& /*aCustomMessage*/)
        {
Window().SetOrdinalPosition(0);
(static_cast<CGameAppUi*>(iEikonEnv->EikAppUi()))->SetActiveView(*this);
        }
```

16.2.2 Object-oriented Plumbing

Here's part of the controller code that runs when the player using that controller requests a hit to a square on their opponent's fleet:

```
void CGameController::ViewCmdHitFleet(TInt aX, TInt aY)
    {
    __ASSERT_ALWAYS(!(iEngine->iOppFleet.IsKnown(aX, aY)), Panic
        (EHitFleetAlreadyKnown));
    // hit fleet
    iEngine->iOppFleet.SetShipType(aX, aY, iEngine->iMyFleet.ShipType
        (aX, aY));
    TFleet::THitResult result = (TFleet::THitResult)iEngine->
        iOppFleet.SetHit(aX,aY);
    iEngine->iMyFleet.SetHit(aX,aY);
    // update view and play sounds
    -
    }
```

The purpose of this code is to

- find out what was on the opponent's fleet on that square,
- mark the information as found on my view of the opponent's fleet,
- mark the opponent's fleet as hit on that square.

The code simply follows pointers to find the opponent's real fleet:

```
TFleet& oppFleet = iOtherController->iGameEngine->iMyFleet;
```

Then, the code calls functions on both my view of the opponent's fleet, and on the opponent's fleet itself.

From a design point of view, this is all very elegant. I was able to reuse the same engine and view in Battleships, and testing them here just involved the plumbing above.

16.2.3 Communications is Different

But this implementation of CGameController::ViewCmd HitFleet() simply won't do in a real communications world. It uses a synchronous, reliable function call, with parameters in internal format, to find out the state of my opponent's fleet, and to inform my opponent that I have hit a square.

With real communications, I have to tackle some fundamental issues:

- Communications is asynchronous. I might have to wait an arbitrary time between sending the hit request, and receiving the response. While I'm waiting, I might change my mind and decide to abandon the game. So, while I am waiting for the response from the other player, the game must be able to handle other kinds of input from me.

- Communications is unreliable. I need to use protocols that ensure that either my message gets to the other player, or that I am reliably informed that it has not got there for a reason that I can understand. I may need to be able to retry sending my message.

- Communications uses external formats. I will have to decide on the binary format that I use to communicate between the two games. I will have to change to external format when sending, and change back into internal format when receiving data. Furthermore, if I wish to play this game against an implementation running on a non-Symbian OS device, I will have to publish these formats in sufficient detail that someone else will be able to reproduce them using a non-Symbian OS format.

- Communications is basically awkward for end users. It involves lots of setup, obscure error messages, and endless frustrations. Sometimes, there is no choice but to expose these things in the user interface of a communications-related application. But if at all possible, it's best to hide them.

- There are several forms of communications that would be attractive to the end user. Two that instantly spring to mind are playing the game between two machines in the same vicinity, using a Symbian OS phone's Bluetooth capabilities, and playing between players in different locations, using text messages between the players' mobile phones. I have to design the communications strategy to be able to work effectively with both.

- Finally, I will have to be able to get there from here. You can't just write thousands of lines of communications code and expect it to work. You have to test at each step of the way. That means you have to broadly envisage a reasonable scheme for each step along the way, before you set out along it.

16.3 TOGS

Later in this chapter I'll be describing how we turned `tp-ships` into a real, playable, Battleships game with Bluetooth and SMS text messaging communications.

The basic communications infrastructure for Battleships is provided by the **transaction-oriented games stack**, TOGS. TOGS was designed with the following objectives in mind:

- it should support Battleships over Bluetooth and SMS,
- it should support other games, and other protocols,

- it should take into account the realities of both SMS and Bluetooth as communications media, including the fact that text messaging is slow, expensive, and unreliable,

- it should allow an individual to play many games simultaneously with the same or different players,

- it should support games played over a very long period (weeks, for instance),

- it should allow the machine to be backed up during the course of a game: since backup closes all applications, this means it should save the state of a game in a file and allow it to be restored again.

It also had some objectives for the book:

- TOGS should be a good demonstration of Symbian OS capabilities as a communications platform

- it should provide excellent example code for the Symbian OS client-server architecture

- it should be fun.

TOGS and Battleships together meet all these objectives. But even so, TOGS is a work in progress. There are design issues that I'll highlight along the way and there's plenty of scope for future development of the stack and of its applications.

16.3.1 The Shape of TOGS

Just as in conventional programming, communications designers solve complex problems by breaking them into parts. Each part is called a **protocol**: a protocol is a set of conventions by which two sides can communicate.

It's customary to design protocols at various levels, and to layer one protocol on top of another. The result is a **stack** of protocols, in which data typically gets sent from an application down the stack to some physical transport medium, and then at the other end comes back up the stack to an application again.

Each layer in the stack is there for a definite purpose, and the layer above should assume that the layer below performs its purpose to its stated specification.

While you're developing the stack, you test each layer in turn, so that you can develop the whole stack gradually rather than writing everything and then trying to debug it all at once.

The stack I designed for Battleships has five layers. Here it is:

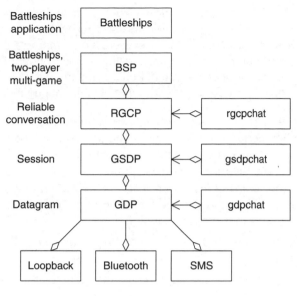

Figure 16.4

These layers are explained in more detail in Appendix 3. Meanwhile, here's a very brief description of each layer.

16.3.2   Starting Points – Datagram and Conversation

It's best to start with GDP and RGCP.

- GDP, **game datagram protocol**, addresses the underlying realities of communications. With just about any real communications protocol, it's easy to send a datagram (a short packet of binary data). With just about any real communications protocol, the datagram may not arrive: datagrams may arrive out of sequence, may not arrive at all, or may even arrive more than once. The GDP specification allows all these things to happen, and is therefore very easy to implement for anyone with knowledge about specific real communications protocols.

- RGCP, **reliable game conversation protocol**, addresses the requirements of application programs. Firstly, two programs communicate to each other using a **session**: once the session is set up, there is simply an 'other' partner, and there's no need to choose who it is. Secondly, each party in the session takes it in turns to send a request, and receive a response – or, looked at the other way, to receive a request, and send a response. This is a **conversation**, and it's clearly a natural paradigm for a turn-based game. Finally programs want **reliable** behavior rather than the underlying unreliability of real communications systems.

'Reliability', in the communications sense, is a technical term with a precise definition. Clearly, 'reliable' doesn't mean that any packet you send is 100 percent certain to be received. You simply can't guarantee that: there could be network failures, or the recipient might have turned their machine off, lost it or even died. Reliable means that system behavior, as you perceive it, is consistent with what actually happened. So:

- If your request didn't get through, you can never receive a response, or another request from your opponent, and you cannot issue another request yourself – you get panicked by RGCP if you try, because that would violate the rules of conversation.
- If you receive a response, then that's an indication that your request did get through and was handled according to conversation rules by the other party. You can rely on that.
- If the packet you send arrives, then it's guaranteed to be the packet you sent.
- *and so on*: we could spell out many other particular cases.

In summary, GDP addresses the realities of communications, and RGCP meets the basic requirements of application programs.

16.3.3   GSDP – Game Session Datagram Protocol

We could have written RGCP as a layer directly over GDP. Then, loopback would be useful simply for testing GDP test code, and real communication with Bluetooth would be used for playing a single game between two Symbian OS phones:

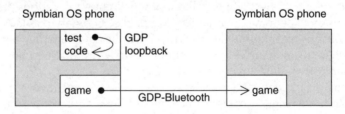

Figure 16.5

Admittedly you could play another game using another GDP protocol, say SMS. But having only one game on each Symbian OS phone per GDP protocol is too restrictive.

I wanted to be able to run more than one game on each phone, each using whatever protocol was really appropriate for it. And it would be quite nice to have a local facility, to play a game against another program on the same machine for test or demo purposes.

So I decided to run *all* GDP implementations inside a server. Contrast the previous diagram in which you can have only a single *application* per GDP protocol on each Symbian OS phone (which is a restriction), with the diagram below. There you can have only a single *server* on each Symbian OS phone, which shares GDP between all client applications (which is exactly the kind of thing that servers are there for).

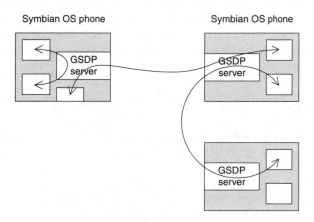

Figure 16.6

Given that a server can share GDP implementations, the question is, *how* should it share? It's easy enough to ensure that a client *sending* a datagram can get it to the right destination, but what happens when it reaches that device? Which client should *receive* it?

The answer is that a session – a connection between two specific games – must be built in at this level in TOGS, instead of the RGCP level (which moves up the stack). Hence the name of this level: **game session datagram protocol** or GSDP. Each time a client program sends data using GSDP, the GSDP server adds session information into the GDP packet that is actually sent. When a GSDP server receives a packet, it interprets that session information to ensure that the packet is routed to the correct recipient. GSDP also includes the support for setting up a session.

GSDP is still unreliable: a datagram that is sent may not be received. GSDP also makes no assumptions about the order in which packets are sent. RGCP, the next layer up in the stack, adds reliability and conversation enforcement, which is useful for games like Battleships. GSDP could also be used as the basis of other protocols that add reliability in different ways for use by different types of application.

16.3.4   BSP – Battleships Protocol

Given that RGCP delivers reliable conversation facilities for use by games, the final task was to specify the protocols to be used by the Battleships game itself.

I called this protocol Battleships Protocol (BSP). BSP allows you to set up an RGCP session, and play multiple Battleships games. It supports first-move arbitration, turns as hit requests, abandoning a game midway through, starting a new game after the previous one was won, lost, or abandoned, and terminating a game including the RGCP session.

In fact, many of these requirements are generic, and BSP could be split into two layers – one for two-player multigame sessions, and another for the Battleships game itself.

16.3.5   Test Programs

One of the reasons for splitting up a communications stack into layers is simply to be able to understand each layer, and to understand the stack as a whole.

So, at each stage, we could rely on the tested aspects of our code being good, and concentrate debugging activity on the new aspects.

Of course, in real life, the development isn't a one-way cascade. You don't catch *all* bugs in your test code, and you can never expect to. Even if you do, you make feature changes at previously tested levels of the stack. You have to go up and down the stack a few times, and maintain your test code along with the production code.

It's a fact of life that test code tends to take shortcuts, as you concentrate your efforts on the production code. The TOGS test code can be found on the supporting website at ***www.symbian.com/books***.

16.3.6   Pattern Reuse

Reuse is good for software systems: it keeps them small and enhances their reliability. In Chapter 3, it was pointed out that sometimes you can't reuse *code*: you have to reuse *patterns* instead.

Communications is full of this kind of thing. TOGS includes patterns that are standard in the communications industry:

- an unreliable datagram protocol – like IP in TCP/IP,

- the use of port IDs, in GSDP, to identify sessions – like TCP/IP and IrDA,

- the use of a higher-level protocol, RGCP, to add reliability – like TCP in TCP/IP.

Pattern reuse feels like reinventing the wheel – a bad thing, if there was already a wheel around which you could have used. It turns out that, for all the cases above, there was no wheel already available.

- I couldn't use IP, because IP assumes bigger packets than would be supported, say, by SMS, and IP's addressing scheme isn't appropriate either for GDP-SMS (which uses phone numbers for an address), or GDP-Bluetooth, as a Bluetooth device is not guaranteed to have an IP address.
- I couldn't even use sockets as provided by the ESOCK API, because although sockets are superficially attractive, the whole protocol is too heavyweight for the requirements of GDP, and I would still have to write my own reliability layer for the specialist requirements of SMS. So I would have written more code to fit in with ESOCK than I would save by using it.
- I had to write the GSDP server instead of using, say, the address books associated with IrDA, because although IrDA can multiplex client sessions on one physical layer, other GDP protocols such as SMS cannot.
- I couldn't use TCP or even the same kind of approach that TCP uses to deliver reliability, because it's far too heavyweight for the requirements of conversational games. Also, TCP makes assumptions about the cost of sending packets that aren't appropriate for SMS.

Although it sounds as if SMS is the problem here, causing all these rewrites, the same is true of the loopback protocol, and the same would be true of other possible GDP protocols such as IR, or WAP Wireless Datagram Protocol (WDP) datagrams.

If pattern reuse is justified, then the main issue isn't code bloat. What you really need to watch out for is that you reuse the right patterns. The patterns used in communications have been established over five decades of computer communication, two centuries of electronic communication, and three millennia of civilization. Good communications books don't date quickly, and are worth their weight in gold. Tanenbaum's *Computer Networks* (Prentice Hall, 3rd ed, 1996, ISBN 0 133 499 456) is a particularly good read.

16.3.7   Building on TOGS

Unlike the other source code, there are some special licensing restrictions on TOGS, which you need to observe if you're modifying the TOGS code.

The restrictions are intended to prevent incompatible communications protocols appearing in the marketplace, so that players find they can't interoperate their Battleships game, or the GDP-SMS implementation.

So, for instance, *any* implementation of BSP, identified by BSP's GSDP game protocol UID, must conform to the RGCP and BSP specifications in Appendix 3. And *any* implementation of GDP-SMS must conform to the specification in Chapter 20. You can write different games, different GDP drivers, or different ways of putting GDP onto SMS – just so long as you change the identifying features, so that existing implementations don't get confused.

The restrictions are *not* intended to prevent lots of activity, enterprise, and fun based on the TOGS source code. You can write new games, new servers – even servers not related to communications purposes – different reliability layers, and anything else you like, using the existing TOGS specification as a starting point. You could even write non-Symbian OS implementations of TOGS.

We've now seen an overview of Symbian OS communications facilities, the test version of Battleships without any communications at all, and the TOGS stack that is going to make programming Battleships and other applications easier. The next step is taking these separate components and producing the full Battleships game.

16.4 Using the Game

I explained the rules for Battleships back in Chapter 9, and you will by now, probably have got used to them by playing Solo Ships.

The main issues with using Battleships are with connecting to the other player, and making sure that the communications works OK. A reminder, of what happens when you install Battleships and launch it from the UIQ menu, and select Start game (*Ctrl+N*) can be seen in Figure 6.7.

Your options here are related to the communications possibilities, and the way things work through the GSDP server are shown in Figure 16.8.

As this diagram shows, you can play for real using Bluetooth or SMS, or for test purposes using loopback.

16.4.1 Playing for Real

If you are playing for real against someone else, with a real Symbian OS phone, then your easiest option is to

- specify a protocol – Bluetooth or SMS,
- choose who will initiate, and who will listen; if you're using SMS, then the initiating partner will have to enter the listening partner's phone number,
- select your first move preferences (if you can't agree, then the game will choose in favor of the initiating partner's preference).

Figure 16.7

The game will connect and you can start playing it. Throughout the connection sequence and turn-based play, each player can see the state changing by choosing the status option from the Battleships menu.

16.4.2 Reliability from RGCP

You can use the loopback function to demonstrate RGCP resend, and packet dropping based on sequence numbers.

Say it's **Initiate's** turn to play: go into the **Listen** game and close it. Go into **Initiate**, and make a move. You know that the move is sent to the GSDP server, which can't deliver it immediately, but because of GSDP's present rules, it keeps the datagram in its receive queue. So, close the **Initiate** game (and any other GSDP clients), wait for two seconds for the GSDP server to close itself down, and then launch both games – **Initiate** and **Listen** – again.

The net effect of these gymnastics is that you have sent a hit request from one game to the other, but it hasn't got through. You'll notice that both games think it's the other player's turn – **Listen**, because it sent a message that did get through, but it hasn't received the acknowledgement yet, and **Initiate**, because it sent a message that didn't get through.

You can resend (*Ctrl+Y*, the standard shortcut key for **Redo**) from either application: a resend from **Listen** will be ignored by **Initiate** (RGCP checks the sequence number and notes that the packet has already been

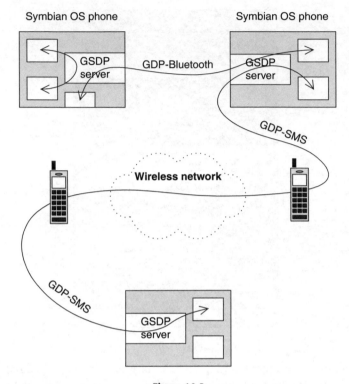

Figure 16.8

received). A resend from **Initiate** will be received by **Listen** and will cause **Listen** to change state to **My turn**.

If you like, you can run this code under the debugger: set breakpoints in RGCP code and you can see the packet dropping in action.

16.4.3   SMS

If you can arrange a game with SMS as a communications medium, you'll need to take care of sending and receiving.

The initiating player must specify the SMS protocol, and the other player's phone number. The listening player just specifies the SMS protocol.

Depending on the phone, there may not be any visible indication that your message has been sent successfully. Also, GDP doesn't have to indicate success. Finally, the performance of SMS text messaging in the UK networks that I've used is quite variable. Most networks are pretty good most of the time, but occasionally messages can get held up for a couple of hours, and this is more likely to happen when messages cross from one network to another.

All this can lead to an amount of uncertainty as to whether a message sent over SMS has ever reached its destination. If you think a message may not have got through, you can resend the last message by pressing **Resend** (*Ctrl+Y*).

Receiving messages is managed by the GDP-SMS implementation; this watches for messages of the appropriate type and ignores any other SMS's that may have been received. GSDP channels any messages received by GDP-SMS to the right game – that's not necessarily the game from which you issued **Receive All**.

The ability of GSDP to get the message to the right game is handy for a final test scenario as follows. Just as in the loopback case, you set up two games on the same Symbian OS phone. You then initiate one game, specifying your own phone number as the other player. Then, you can play a game against yourself! – remember you're paying for the privilege of each move.

This is certainly useful for testing whether you've got the basic setup right, and I found it useful during development to test software functionality.

Much of the SMS based phone to phone communication can be quite time consuming, with varying wait times between sending and receiving messages. Note that this tends to be much more apparent on GSM networks with much of this time spent making the connection. GPRS promises to be much faster with instant connection times.

It's pretty easy to set up connections for sending and receiving messages, but there are just enough things that can go wrong to make life potentially uncomfortable for some average users. Even so, the SMS implementation shows what can be done. I have some more detailed analysis of the issues here, and some suggestions for future work, at the end of the chapter.

16.5 From the Inside

Let's have a look at the design of Battleships, and some features of its implementation in C++. As a reminder, the class design for Solo Ships is given in Figure 16.9.

The controller contains the persistent document data required by the Symbian OS application architecture. The app view uses a single component control, the fleet view of the opponent's fleet. The fleet view uses an `MGameViewCmdHandler`, an interface implemented by the controller. The trick with Solo Ships was to use object-oriented plumbing to connect the opponent's fleet with the engine's `iMyFleet`.

The design for Battleships is given in Figure 16.10.

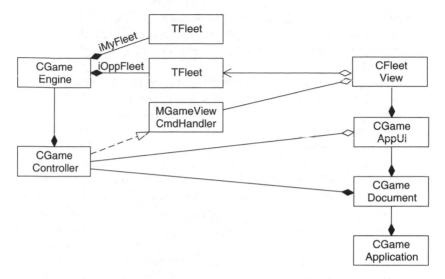

Figure 16.9

Compared with Solo Ships, the most important difference is that the controller now owns an RGCP stack, shown here by the CRgcpSession class. This implies three things:

- the controller uses CRgcpSession to send BSP requests and responses to the other player
- the controller implements MRgcpHandler to handle BSP requests and responses from the other player
- RGCP's persistent state is now included, along with the engine's persistent state, in the persistent state of the game as a whole.

The Battleships controller is somewhat more complicated than the Solo Ships controller, with nearly 30 public and 23 private functions. It implements the full specification of BSP. The controller has been designed so that all function calls (except construction, destruction, persistence-related functions, and some const queries) correspond directly with an event in BSP. That includes events from the app UI, the view, and incoming data from the RGCP session. To assure system integrity, the controller asserts that all events occur in an allowed BSP state, and panics the program if it finds otherwise. It is the responsibility of the event source to precheck validity – for instance, the opponent's fleet view should not issue a hit request unless the BSP state is **my-turn**: this particular condition is assured by dimming the view unless the BSP state is **my-turn**.

Finally the app UI uses two dialogs, to initiate a game (and RGCP communications session), and to start a new game (in the context of the same RGCP communications session).

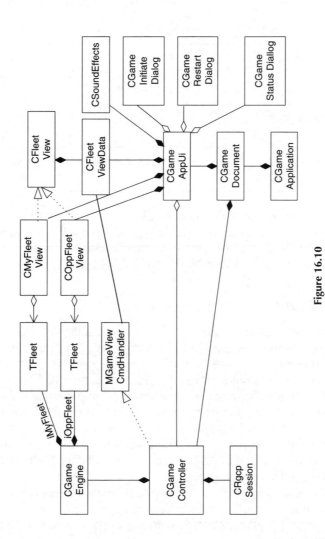

Figure 16.10

You can find the code for Battleships on the website:

Source Files	Purpose
`engine.cpp`, `engine.h`	The engine, exactly as in Solo Ships
`view.cpp`, `view.h`	The fleet view, exactly as in Solo Ships
`appui.cpp`, `appui.h`	The application, document, and app UI classes, along with the initiate and new game dialogs
`controller.cpp`, `controller.h`	The controller
`battleships.rss`, `battleships.hrh`	Resource file and shared constants

If you've read the walkthrough of Solo Ships and the description of BSP, then the implementation of Battleships will contain few surprises.

I'll pause to look at a few representative functions, so you can get a flavor of how things hang together in a real TOGS game.

16.5.1   The Status View

Before we get involved in the communications code, let's look at the status view. There's not much to learn from its code, but there are a couple of reminders of the realities of graphic design here.

My choice for the design of the status view was primarily determined by the capabilities of the device it was designed for, in particular the portrait screen format used by UIQ based devices. In the Battleships application the fleet view occupies the majority of the screen, leaving very little space for a permanent status view. The simplest, but not necessarily the best option is to use a dialog to display the game status available and make this available from the menu options.

This has the advantage of leaving the maximum available area for playing the game and also offers a more flexible approach to the presentation of the status information rather than trying to squeeze the information into too small a space.

16.5.2   Handling Hit Requests

Remember that, in Solo Ships, the controller handled hit requests from the view using

```
void CGameController::ViewCmdHitFleet(TInt aX, TInt aY)
    {
    __ASSERT_ALWAYS(!(iEngine->iOppFleet.IsKnown(aX, aY)),
        Panic(EHitFleetAlreadyKnown));
    // hit fleet
    iEngine->iOppFleet.SetShipType(aX, aY, iEngine->iMyFleet.
        ShipType(aX, aY));
    TFleet::THitResult result = (TFleet::THitResult)iEngine->
        iOppFleet.SetHit(aX,aY);
    iEngine->iMyFleet.SetHit(aX,aY);
    // update view and play sounds
    switch (result)
            {
            case TFleet::EMiss:
                    ActiveView().MissSound();
                    break;
            case TFleet::EShip:
                    ActiveView().ExplSound();
                    break;
            case TFleet::ESunk:
                    ActiveView().SunkSound();
                    break;
            };
    ActiveView().DrawTilesNow();
    // if game is won, transition to finished
    if (iEngine->IsWon())
            {
            iState = EFinished;
            iEnv->InfoMsg(R_GAME_CONGRATULATIONS);
            }
    }
```

This code clearly won't work in a communications environment, because its most important two lines,

```
// Hit fleet
iEngine->iOppFleet.SetShipType(aX, aY, iEngine->iMyFleet.ShipType
    (aX, aY));
iEngine->iOppFleet.SetHit(aX, aY);
```

use object-oriented plumbing, rather than communications, to find the state of the 'opponent's' fleet. In Battleships, the code reads simply:

```
void CGameController::ViewCmdHitFleet(TInt aX, TInt aY)
    {
    __ASSERT_ALWAYS(IsMyTurn(),
Panic(EHitFleetNotMyTurn));
    SendHitRequest(aX, aY);
    SetState(EOppTurn);
    }
```

This code:

- Asserts that it is called only under the correct conditions – namely, that it's my turn.

- Sends a hit request, using another controller function, which as we'll see shortly, calls an RGCP send

- Sets the state to the opponent's turn, which as we'll see shortly, has some side effects on the view.

16.5.3   Checking Conditions

Before any controller function is called, the caller should ensure the correct conditions. The controller itself asserts the correct conditions, and panics the application if they do not apply.

These assertions – which, because I use __ASSERT_ALWAYS, are built into release programs too – are vital aids to ensuring the quality of the released program.

There is a fine judgment call about whether to choose __ASSERT_DEBUG or __ASSERT_ALWAYS: asserts have a code space and runtime performance impact, and you should test code properly before releasing it. So, if you can, it's better to use __ASSERT_DEBUG than __ASSERT_ALWAYS. I felt this code was complicated enough to warrant __ASSERT_ALWAYS.

If the choice between always and debug-only is finely balanced, the choice between using asserts and not using them at all is a no-brainer. You simply can't expect to debug state machine code like this effectively unless you use asserts extensively. Their role in ensuring quality is second only to taking care over the design – the kind of thinking that went into the specifications in Appendix 3.

So, the caller is responsible for making sure controller functions get called in the right conditions. There are three interesting cases here:

- *the caller is the app UI*: the user can issue commands at any time, so the app UI has to check whether the circumstances are valid, and if not, tell the user it can't execute the command, and preferably also indicate why not,

- *the caller is the fleet view*: the fleet view is prevented from issuing commands to the controller, by being dimmed when the controller is not in my-turn state,

- the caller is RGCP, using one of its handler functions, in response to a datagram received from the other player. The rules of BSP should ensure that only good data gets sent from the other player, and the rules of GDP, GSDP and RGCP should ensure that only the data that was sent, over the intended session, is received. But for one reason or another this might not be the case, so the controller asserts the correct conditions on its RGCP handler functions.

We'll take a brief look at the code involved in the app UI and app view cases. We'll comment further on the RGCP case when we encounter it later.

App UI checking

Here's an example of the app UI case. When it's my turn, I can choose, instead of playing a turn, to abandon the game. But I can't do that when it's not my turn. The app UI makes the check:

```
void CGameAppUi::CmdAbandonL()
    {
    // User-friendly check
    if(!iController->IsMyTurn())
        iEikonEnv->LeaveWithInfoMsg(R_GAME_NOT_YOUR_TURN);
    // Confirm with the user
    if(!iEikonEnv->QueryWinL(R_GAME_QUERY_ABANDON))
        return;
    // Do it
    iController->Abandon();
    }
```

The app UI also checks whether the user really wants to abandon the game, before calling the controller. So the controller's `Abandon()` function should only be called in the right circumstances, and asserts that this is the case:

```
void CGameController::Abandon()
    {
    __ASSERT_ALWAYS(IsMyTurn(), Panic(EAbandonNotMyTurn));
    SendAbandonRequest();
    SetState(EFinished);
    iEnv->InfoMsg(R_GAME_ABANDONED);
    ActiveView().DrawNow();
    }
```

Again, the MVC-style implementation of this function is very simple.

Dimming the view

The view is prevented from issuing commands at the wrong time by dimming it unless the state is my-turn. The controller includes support for this, in `SetState()`:

```
void CGameController::SetState(TState aState)
    {
    // update app view, if needed, for change in "my turn" status
```

```
        if (iAppView)
            {
            if (Engine().IsMyTurn())
                    Engine().SetSecondPlayer();
            else
                    Engine().SetFirstPlayer();
            }
        // set state as requested
        iState=aState;
        }
```

Key processing is the responsibility of the opponent fleet view:

```
TKeyResponse COppFleetView::OfferKeyEventL(const TKeyEvent& aKeyEvent,
    TEventCode aType)
    {
    if (aType!=EEventKey)
        return EKeyWasNotConsumed;
    if (aKeyEvent.iCode==EQuartzKeyFourWayLeft ||
        aKeyEvent.iCode==EQuartzKeyFourWayRight ||
        aKeyEvent.iCode==EQuartzKeyFourWayUp ||
        aKeyEvent.iCode==EQuartzKeyFourWayDown)
        {
        // move cursor
        if (aKeyEvent.iCode==EQuartzKeyFourWayLeft)
            MoveCursor(-1,0);
        else if (aKeyEvent.iCode==EQuartzKeyFourWayRight)
            MoveCursor(1,0);
        else if (aKeyEvent.iCode==EQuartzKeyFourWayUp)
            MoveCursor(0,-1);
        else if (aKeyEvent.iCode==EQuartzKeyFourWayDown)
            MoveCursor(0,1);
        // redraw board
        DrawTilesNow();
        return EKeyWasConsumed;
        }
    else if (aKeyEvent.iCode==EQuartzKeyConfirm)
        {
        if (iFleet.IsKnown(iCursorX, iCursorY))
            iEikonEnv->InfoMsg(R_GAME_ALREADY_KNOWN);
        else
            iData.iCmdHandler.ViewCmdHitFleet(iCursorX,
                iCursorY);
        return EKeyWasConsumed;
        }
    return EKeyWasNotConsumed;
    }
```

16.5.4   Hit Processing: the Full Story

We saw how the view command handler initiated hit request processing, in CGameController::ViewCmdHitFleet(). The request now proceeds in four stages:

- I send a request using RGCP in my game
- the opponent receives the request using RGCP in their game
- the opponent sends a response using RGCP in their game
- I receive the response using RGCP in my game.

Of course, the opponent's response doesn't actually get sent until the opponent makes the next request. But the handling of my hit request doesn't depend on the opponent's next request, so we can look at these four operations in isolation.

Sending the request

Here's `SendHitRequest()`:

```
void CGameController::SendHitRequest(TInt aX, TInt aY)
    {
    TBuf8<10> buffer;
    RDesWriteStream writer(buffer);
    writer.WriteUint8L(aX);
    writer.WriteUint8L(aY);
    writer.CommitL();
    iRgcp->SendRequest(KGameOpcodeHit, buffer);
    }
```

I use a descriptor write stream to formulate a request, encoding the (x,y) coordinates I want to hit. I then call RGCP `SendRequest()` specifying the hit-request opcode. The descriptor write stream is a convenient API, and it enables me to guarantee the external format of the BSP hit request.

RGCP then sends the hit request (along with my previous response) using GSDP. The datagram arrives at the other player's machine and is routed by GSDP to the correct game instance. RGCP cracks the datagram into response and request, gets the previous response handled, and then calls code to handle the current request.

If the GSDP datagram *didn't* get through, then I could resend it. If I resend but the datagram *did* get through, then RGCP at the other player's end would drop the duplicate packet.

Either way, there is a guarantee that my request is handled only once.

Handling the request

Here's how the request is handled. By implementing `MRgcpHandler`, the game controller in the other player's game codes a first-level handler that calls a specific handler function depending on the request opcode:

```
void CGameController::RgcpHandleRequest(TInt aOpcode, const TDesC8& aData)
    {
```

```
switch(aOpcode)
    {
case KGameOpcodeStart:
    HandleStartRequest(aData);
    break;
case KGameOpcodeRestart:
    HandleRestartRequest();
    break;
case KGameOpcodeNop:
    break;
case KGameOpcodeAbandon:
    HandleAbandonRequest();
    break;
case KGameOpcodeHit:
    HandleHitRequest(aData);
    break;
default:
    Panic(EHandleRequestBadOpcode);
    }
}
```

The hit request will be handled using the `HandleHitRequest()` function, which starts as follows:

```
void CGameController::HandleHitRequest(const TDesC8& aData)
    {
    __ASSERT_ALWAYS(IsOppTurn(), Panic(EHandleHitReqNotOppTurn));
    // crack parameters
    RDesReadStream reader(aData);
    TInt x=reader.ReadUint8L();
    TInt y=reader.ReadUint8L();
```

This code begins by asserting that the hit request is being handled in BSP opp-turn state, and then uses a readstream to get the function parameters.

This code will panic if the hit-request opcode is received at the wrong time, and will leave if the hit-request opcode comes at the right time but lacks the two further bytes in the request data. Downstream, the engine will panic in debug builds only if the (x, y) coordinates are outside the range 0 to7.

This exemplifies a difficulty in communications programming: checking incoming data, and responding to problems in it:

- Should you panic (because there's nothing you can do, and there's no point in carrying on the game)?

- Should you drop the packet (say, leaving with `KErrCorrupt`) and issue a meaningful error dialog to the user? If so, how do you get really *meaningful* error information to the user?

- Should you drop the packet silently, so that the user, who is more interested in Battleships, a nice easy game, than the complexities of BSP and RGCP, doesn't get confused by communications-type error messages?

- Or should you trust that nothing will ever go wrong, and not implement any error checking at all?

Programmers are too fallible for it to be safe to assume that nothing will ever go wrong. We are technically fallible; our programs have bugs. We're morally fallible too – at least someone must be, because malicious programs keep on getting written. So, it's a good idea to check incoming data, and so minimize exposure to all the things that could potentially be wrong with it. And it may not always be right to panic the user's program because of bad incoming data from another place.

Checks on incoming data should be made with the following goals in mind:

- Programming errors must be caught early – ideally, during debug phase, and in debug builds only.

- Neither programming errors nor malicious acts should cause a program to lose truly personal data.

- In cases where the *program* can't tell whether incoming data could cause personal data loss, the program should either reject the data out of hand, or give the *user* an opportunity to decide what to do, or should come with a general warning to the user to take precautions.

Battleships, along with the specifications of TOGS and GDP implementations, meets these criteria. If the program panics, the only thing that's lost is the state of the shared game between the person whose program caused the crash and me – not truly personal data. Admittedly Battleships meets the third criterion trivially, because the circumstances it caters for don't arise. In more complex situations (say, launching an executable e-mail attachment, which could be a virus) a user check *is* required.

Whole books could be written about this, but we'll move on to see what happens to the hit request. `HandleHitRequest()` continues:

```
// hit the square
    TFleet& fleet=iEngine->iMyFleet;
    fleet.SetHit(x,y);
    SendHitResponse(x,y,fleet.ShipType(x,y));
```

This is the code that ties together my hit request with the real state of the opponent's fleet, and sends the response back. We'll consider `SendHitResponse()` below.

BSP is designed so that the hit response contains a hit response – no more. So the code above was unconditional, and uncomplicated. However, after receiving a hit, the opponent has to analyze the consequences – whether it means the opponent has lost the game, for instance – and deal with them. `HandleHitRequest()` continues:

```
// update view
   iAppView->SetCursor(x,y);
   iAppView->DrawTilesNow();
   // if game is lost, issue a nop-request and transition to finished
   if (iEngine->IsLost())
       {
       SendNopRequest();
       SetState(EFinished);
       }
   else // transition to my-turn
       SetState(EMyTurn);
   // update view
   ActiveView().DrawNow();
   }
```

Firstly, there's an MVC view update: the other player updates 'my fleet' by setting the cursor to the hit location and redrawing its tiles.

If the game is lost, the other player's controller immediately sends a NOP request back to me. That flushes the buffer and causes the hit response to be sent immediately. That means I get to see the hit response, and to find out that I've won the game, without waiting for the other player to do anything. The state then changes, and the view is redrawn.

Sending the response

Here's the code used to send the hit response:

```
void CGameController::SendHitResponse(TInt aX, TInt aY, TShip::TShipType
   aShipType)
   {
   TBuf8<10> buffer;
   RDesWriteStream writer(buffer);
   writer.WriteUint8L(aX);
   writer.WriteUint8L(aY);
   writer.WriteUint8L(aShipType);
   writer.CommitL();
   iRgcp->SendResponse(KGameOpcodeHit, buffer);
   }
```

The code is pretty similar to the code for sending the hit request.

Handling the response

When the response gets back to me, my controller handles it initially in `RgcpHandleResponse()`, which contains a switch statement identical in form to that in `RgcpHandleRequest()`. All being well, control then transfers to `HandleHitResponse()`:

```
void CGameController::HandleHitResponse(const TDesC8& aData)
    {
    __ASSERT_ALWAYS(IsOppTurn(), Panic(EHandleHitRespNotOppTurn));
    // crack parameters
    RDesReadStream reader(aData);
    TInt x=reader.ReadUint8L();
    TInt y=reader.ReadUint8L();
    TShip::TShipType type=(TShip::TShipType) reader.ReadUint8L();
    // update engine
    TFleet& fleet=iEngine->iOppFleet;
    if (type==TShip::ESea)
        fleet.SetSea(x,y);
    else
        fleet.SetShipType(x,y,type);
    fleet.SetHit(x,y);
    // update view
    iAppView->SetCursor(x,y);
    iAppView->DrawTilesNow();
    // if you won the game, transition to finished.
    if (iEngine->IsWon())
        SetState(EFinished);
    // update view
    iAppView->DrawNow();
    }
```

The code starts off similar to the request handler. I assert the right condition, and crack the response data using a descriptor read stream.

Then I update my internal representation of the opponent's fleet, and my view of it. I check whether this means I've won and, if so, change state to finished. Finally, I update the status display – which will show either that it's my turn, or that I've won the game.

That completes the story of a hit request. What we could do with two lines of code in Solo Ships, and only a few lines in tp-ships, requires the entire TOGS stack, and all the processing you've seen in the Battleships controller, to play a game against a different player.

16.6 Taking Battleships Further

We had a lot of fun developing Battleships as the flagship example program for the book. We aimed to deliver an example that would show the capabilities of Symbian OS as a communications platform and provide lots of excellent code for you to use as a starting point for your own programs. With Battleships as it is, I think we've achieved that. But there's plenty more to do, so I'll conclude this chapter with some thoughts on taking Battleships and TOGS further.

The main categories of improvement are:

- adding finesse to the Battleships game
- other games
- single-player games
- improvements in Symbian OS and TOGS.

16.6.1   Better Battleships

It's not hard to see how the Battleships game itself could be improved to increase its end-user appeal. Here are a few ideas.

Integration with address book

Instead of requiring the user to type in the phone number manually, you could integrate with the Contacts Model API to allow you to select from people in your Contacts database who have a mobile phone number. The Contacts Model documentation is available in the UIQ SDK documentation.

User-controlled layout

As it's presently implemented, Battleships is really two single-player games. The computer sets up both users' games, and the moves I make are largely independent of the moves of my opponent. Result: we're really playing two games against the clock, and the one who completes the game first is not really so much the winner, as the best (or luckiest) player. In fact if we took this view to an extreme, we could considerably save on the messaging involved in a game of Battleships.

In a real Battleships game, fleet layout is what really enables each player to test the psychology of the other. Adding an option for user-specified layout would make a good addition to the interest in playing the game. The code required to support this, in engine, app UI and view, would be nontrivial.

Random first-move preferences

The first-move preference specification is awkward, and gives the users one more thing to think about. There's a strong argument for removing it from the UI and replacing it with a randomly set first-move preference. This could be done without changing BSP: the first-move calculation is still done from within the controller when the initiating player receives the listening player's initiate-response.

This is a good example of how to make a UI simpler, by removing choices that generate more confusion than flexibility.

To remove the choice about initiate/listen from the UI isn't possible with SMS, or even desirable. But the idea of making a call or sending a message is well enough understood. Perhaps the terminology could be better – Send First Message and Receive First Message – but this terminology doesn't suit protocols like Bluetooth.

Little things maybe, but these are among the first UI features our end users will see, and if they appear hard to use, some would-be users won't take a second look.

Chat channel

BSP is sufficient to implement the rules of Battleships and to make the game playable. But a game of Battleships played between two players in a room, using old-style paper-and-pencil methods, wouldn't be simply a set of hit requests and responses. There would be comments, taunts, discussions about when to play the next game, and so on. Chat adds to the psychology and interest of a real Battleships game, just as much as user-controlled fleet layout.

It would be nice to build in a chat channel to Battleships to support this kind of chat.

However, BSP is built on RGCP – a *conversational* protocol in which players take it in turn to send and receive messages. That's basically unsuitable for chat.

My favorite scheme to implement a chat channel would be to implement it at the GSDP level, so that the Battleships game has another connection to the GSDP server. That's easy – the Symbian OS client-server architecture supports multiple independent connections from any thread.

At the UI level, there would be no need to alter the start-new-game dialog (chat initiate/listen details are the same as for the game). The view would have to be altered to include chat data – or, as in tp-ships, a multiview approach could be used. The persistent form of the game would have to be altered to include a log of chat – perhaps on a different stream from the main game data, so that chat could be added optionally without breaking the format of existing games.

With this scheme, each chat message would be a different GDP datagram from the datagram containing each BSP response/request pair. That's unfortunate, because of the real cost of a text message. There are many options for addressing this:

- *Do nothing*: the user pays the cost of each message, but no higher-level protocol needs to be altered.
- *Save datagrams at connection time*: you don't need to send two initiate packets. In the BSP initiate packet, send the initiating partner's

GSDP port ID allocated for chat, and in the BSP initiate-response, send the listening partner's GSDP port ID allocated for chat. This relies on GSDP's support for allocating a port ID without participating strictly in the initiate/listen protocol.

- *Save datagrams at hit-request time*: include an optional message with each hit request, and also with the initiate request.
- *Low-level protocol piggy-backing*: implement a completely different piggy-backing and reliability scheme than that supported by RGCP, which takes the chat requirement into account, and optimizes accordingly.

In my view, the first option isn't going far enough, while the final option is probably going too far. The middle two options require alterations to BSP, but not to the underlying TOGS protocols. If BSP is changed, then a different game protocol ID needs to be assigned, otherwise the changed BSP won't successfully interoperate with the existing BSP.

Incidentally the chat requirement – whether verbal or protocol-assisted – is a good reason for not removing the legend on the borders of the fleet view. At first I thought those legends were redundant. But having played the game with some humans, it quickly became apparent that without them it would be hard to talk about the game.

Midgame protocol change

It would be nice if the GDP protocol could be changed midgame, so that a game could be played using long-range messaging (SMS) when the players are far apart, and short-range, free, messaging (Bluetooth) when they meet.

If the GDP protocol were changed, this would imply a change in the addressing information used by the players. However, the GSDP port IDs must not be changed: they are what define the game session.

This could probably be kludged without changing TOGS, but I suspect the optimal solution would involve changes at the GSDP level, if only for proper encapsulation.

Better capability and state support in TOGS

A characteristic of protocol stacks is that they don't represent complete encapsulation. For instance, it doesn't make much sense to use RGCP without being aware of the underlying realities of GSDP and GDP, even perhaps the realities of a particular GDP protocol. RGCP adds value to these layers, but doesn't entirely encapsulate them.

As a consequence, an RGCP application such as Battleships needs access to the underlying specifics. For instance, is the current GDP protocol networked, and does it require receive-all? What GDP protocols does the GSDP server support? How can I set, store, and restore a particular GDP and GSDP configuration?

At the moment, TOGS handles these issues with ad hoc exposure and pass-through of getter and setter functions up and down the stack. A better system would be to use the capabilities pattern at each level:

- define a struct that includes all capability and setting information for a given level,

- define getter and setter functions for this struct, which change its members,

- define getter and setter functions for a protocol level, which get a struct from the level's current settings, or set the settings from the values in a struct.

Clearly, a different struct would be required for each level in the protocol stack.

The capabilities pattern is particularly worthwhile for a protocol layer with many capabilities, especially if that layer often resides towards the bottom of a stack.

16.6.2 Other Games

TOGS isn't designed for Battleships alone. Any turn-based game could be implemented on a TOGS stack, using GDP, GSDP, and RGCP unchanged, but replacing BSP and the Battleships game.

Suitable games would include Chess, Checkers (Draughts), Backgammon, Tic-Tac-Toe (Noughts and Crosses), Connect Four, Scrabble, and many other games, including two-player card games.

The developer investment required to produce a good game on top of TOGS is not high, since you don't have to build in the intelligence required for a human-versus-computer version of these games. You only have to specify and implement the protocols required to communicate moves, the rules needed to referee attempted moves, and a GUI.

Other games will probably use a variant of BSP. However, the details will be subtly different. Board games are based on a single shared game state (the board and pieces on it) that is public to both players. Battleships is a game of two halves, in which each player's initial fleet disposition is initially unshared, but as the game progresses it is selectively revealed to the other player.

Many games include some combination of shared state and unshared state. In Scrabble, for instance, the knowledge of your letters is private to you, the unused letter pool is unknown to anyone, and the board state is completely public. Many card games rely on guessing unshared state for much of the interest in the game. The whole intrigue in multiplayer Diplomacy (though not the two-player version) arises from selective, and not always truthful, sharing of state between players.

The degree of shared state will have some effects on your game protocol. A key rule in gaming is to minimize disclosure – that is, don't share things at the protocol level and hide them at the UI level. For instance, in Battleships, it would be possible to exchange the entire state of each player's board at the beginning of the game, so that the response to hit requests could be instant. But then a knowledgeable player could find out the state of the opponent's game, and win in 20 moves.

16.6.3   Single-player Games

Solo Ships was an attractive and useful single-player version of Battleships – possible because Battleships is really two half-games.

Another single-player form of Battleships would involve the computer as a genuine opponent. The basic game design would use two `CGameEngines`, and two `CFleetViews` (one for my fleet and one for the opponent fleet). Instead of having an app view, the other engine would have a `CGameComputerPlayer`, which would make hit requests on behalf of the computer.

For a minimum level of skill, the computer player could make random hits on unknown squares. The engine includes some trivial logic to mark squares as sea, if they are diagonally adjacent to partial known ships, or directly adjacent to complete known ships, so random hits on unknown squares would not be a bad start for the computer player.

For a greater level of skill, the computer player could treat Battleships as a **constraint satisfaction problem** (CSP), a standard form of problem for which artificial intelligence research has developed many general methods. A little data from a few hits will generate enough constraints that a CSP approach can considerably improve on a purely random hit request. And a carefully chosen initial hit strategy will do better than a purely random initial hit strategy.

Many of the two-player games mentioned above could be improved by the addition of the option for a computer opponent. In the case of the more complex games, such as Chess, the barriers to entry are high, as engines suitable for deployment on Symbian OS already exist, and have in some cases been ported from bodies of code that have evolved over two or three decades.

16.6.4 Infrastructure Improvements

Besides improving the game – or using a different game – there are many things we could do to improve the facilities for games in general, as provided by both Symbian OS and the TOGS stack.

Symbian OS v7.0 and TOGS as they stand provide an attractive enough platform for turn-based games played at both short and long distance.

TOGS is designed specifically for two-player turn-based games and can also support two-player chat. For multiplayer games, or for real-time games, GDP and GSDP will still be useful, but alternatives to RGCP, tailored for multiplayer or real-time requirements, will be needed. Some GDP implementations are clearly more suitable than others for real-time: any kind of pull protocol (such as receive-all for GDP-SMS) is clearly ruled out.

Symbian OS also includes a media framework with better sound facilities and support for an increasing range of graphics and video formats. This has obvious application to games. In the longer term, it's possible that many aspects of the Symbian OS architecture may evolve to support real-time games; this requires sound, graphics, and communications improvements, some of which will be delivered in software, some on silicon.

Here are some more ideas for taking TOGS forward with Symbian OS.

Better long-distance messaging

GDP-SMS provides a messaging transport that shows the viability and usefulness of long-range wireless messaging. It is the cheapest and fastest way to send a message to another Symbian OS phone, using only a mobile phone to access a communications network.

Ultimately, the UI behind this kind of messaging will be very simple, because the phone metaphors (initiate = send first message, listen = wait to receive first message, address = phone number) are well understood by end users.

More tightly integrated telephony will address push, setup, and the distinction between GDP and other messages.

In the SMS world, Nokia has defined Smart Messaging protocols that could be exploited to deliver a much more satisfactory method of distinguishing between GDP messages, end-user text messages, and other types of Smart Message.

Smart Messaging is a solution for today. Another option available is WAP; however, WDP is a better and more general solution. WDP, like all aspects of WAP, is bearer-independent, and so will support other bearers such as North America's TDMA.

Symbian OS will also deliver its own bearer-independent messaging infrastructure designed to ensure that an even wider generality of incoming messages are routed to the application that is intended to handle them.

Better short-distance messaging

The Bluetooth implementation of GDP works very well, and enables two players in the same room to have a good game of Battleships.

That makes sense in the case of SMS, where you're not in direct touch with the player and can't distinguish between time being used to send a message, and the time the other player needs in order to think about the next move.

With Bluetooth, the decision will be more difficult. The whole point of Bluetooth is to enable piconets to be setup up between multiple devices in sub-10-meter proximity but without line-of-sight communication. During the lifetime of a piconet, Bluetooth nodes enter and exit in arbitrary sequence. During the lifetime of a particular set of nodes in a piconet, each node has a fixed address – but that address has no lasting meaning, and is probably even more user-hostile than the addresses used in IrDA.

GDP-BT will probably need to be a networked GDP protocol, and will need to use a friendly form of Symbian OS device address rather than the transient Bluetooth node address.

16.7   Summary

In the book so far, we've concentrated on the basic Symbian OS C++ APIs, concentrating mainly on graphics and the GUI. In this chapter, we've used an evolution of the Battleships application to introduce the topics of communications and system programming, which are described in more detail in the following chapters.

We've also seen an overview of the Symbian OS communications facilities – and of the TOGS stack, which helps to ease the task of writing a communicating application.

17

Active Objects

Back in Chapter 2, we saw that the system design of Symbian OS is optimized for event handling and that all events are handled by active-object `RunL()` functions. We noted that the major frameworks in Symbian OS – CONE's application framework, and E32's server framework – are built as event handlers, so that typical applications and typical servers are just a single thread, using active objects to implement multitasking in response to events.

An event-handling thread has a single **active scheduler** that is responsible for deciding the order in which events are handled. It also has one or more **active objects**, derived from `CActive`, which are responsible both for issuing requests (that will later result in an event happening) and for handling the event when the request completes. The active scheduler calls `RunL()` on the active object associated with the completed event.

CONE maintains an outstanding request to the window server for user input and other system events. When this request completes, it calls the right app UI and control functions to ensure that Uikon and, eventually your application, handles the event properly. So, as we saw in Chapter 4, all application code is ultimately handled under the control of an active-object `RunL()`.

It's because CONE provides this framework for you that you can get started with Symbian OS application programming without knowing exactly how active objects work. But for more advanced GUI programming and for anything to do with servers, we do need to know how they work in detail.

I'll tackle that in three stages:

- Firstly, I'll use the `active` example to show how a simple active object can request an event and then handle it.
- Then, I'll describe the active object framework.

Symbian OS C++ for Mobile Phones.   Edited by Richard Harrison
© 2003 John Wiley & Sons, Ltd   ISBN: 0-470-85611-4

- Finally, I'll conclude with an overview of some well-established patterns for using active objects.

As we cover servers in the next two chapters, there'll be many more opportunities to see how active objects work in practice.

17.1   A Simple Active Object

The \scmp\active\ example demonstrates two active objects in use. It's derived from our GUI Hello World! example from Chapter 4. Here's what it looks like when you launch it and select the 'Other' menu item in Figure 17.1:

Figure 17.1

The **Set Hello** menu item triggers a Symbian OS timer. That timer completes 3 s later, causing an event. This event is handled by an active object, which puts an info-message saying **Hello world!** on the screen. During that 3-s period, you can select the **Cancel** menu item to cancel the timer, so that the info-message never appears.

The **Start flashing** menu item starts the **Hello world!** text in the center of the display flashing and **Stop flashing** stops it. The flashing is implemented by an active object that creates regular timer events and then handles them. Its event handling changes the visibility of the **Hello World!** text and redraws the view.

We'll start by looking at CDelayedHello, a derived active object class that implements **Set Hello** and its associated Cancel. Then I'll explain some of the underlying fundamentals, before taking a closer look at CFlashingHello (which implements Start flashing and Stop flashing) along with some other active object patterns.

```
class CDelayedHello : public CActive
    {
public:
    // Construct/destruct
    static CDelayedHello* NewL();
    ~CDelayedHello();

    // Request
    void SetHello(TTimeIntervalMicroSeconds32 aDelay);

private:
    // Construct/destruct
    CDelayedHello();
    void ConstructL(CEikonEnv* aEnv);

    // From CActive
    void RunL();
    void DoCancel();

private:
    RTimer iTimer;      // Has
    CEikonEnv* iEnv;    // Uses
    };
```

From the class declaration, you can see that CDelayedHello:

- *Is-a* active object, derived from CActive
- *Has-a* event generator – an RTimer, whose API we'll see below
- Includes a **request function**, SetHello() that requests an event from the RTimer
- Implements RunL(), to handle the event generated when the request completes
- Implements DoCancel(), to cancel any outstanding request.

All active object classes share this pattern. They are derived from CActive and implement its RunL() and DoCancel() functions. They include an event generator and at least one request function (Figure 17.2).

It is also worth mentioning here a change from Symbian OS v5. The CActive class now provides an error handling function called RunError(). This is a virtual function for which CActive provides a default implementation. The active scheduler calls the RunError()

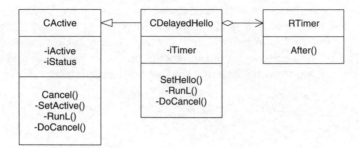

Figure 17.2

function of an active object if a leave occurs in that object's `RunL()`. This will be explained in more detail later in the chapter.

17.1.1 Construction and Destruction

Let's walk through the implementations of all the functions declared above. First, the easy bits: here are two of the members involved in construction and destruction.

```
CDelayedHello* CDelayedHello::NewL()
    {
    CDelayedHello* self = new(ELeave) CDelayedHello;
    CleanupStack::PushL(self);
    self->ConstructL();
    CleanupStack::Pop(self);
    return self;
    }

CDelayedHello::CDelayedHello()
: CActive(0)
    {
    CActiveScheduler::Add(this);
    }
```

`NewL()` is a static function that follows the standard constructor-encapsulation pattern that we saw in Chapter 6.

The C++ constructor is required for any derived active object class; inside it, you call `CActive`'s constructor to specify the active object's **priority**, which is used for tiebreaking when more than one event occurs while another is being handled. You should specify zero here unless there are good reasons to specify something lower or something higher. I'll cover those reasons below. The new object adds itself to the active scheduler, so that the active scheduler can include it in event handling.

```
void CDelayedHello::ConstructL()
    {
```

```
    iEnv = CEikonEnv::Static();
    User::LeaveIfError(iTimer.CreateLocal());
    }

CDelayedHello::~CDelayedHello()
    {
    Cancel();
    iTimer.Close();
    }
```

The second-phase constructor gets a pointer to the Uikon environment and then uses `iTimer.CreateLocal()` to request that the kernel creates a kernel-side timer object, which we access through the `RTimer` handle. If there is any problem here, we leave.

The destructor starts by canceling any events requested by the active object. `Cancel()` is a standard `CActive` function that checks to see whether a request for an event is outstanding and if so calls `DoCancel()` to handle it.

> **Any active object class that implements a `DoCancel()` function must also call `Cancel()` in its destructor.**

The destructor closes the `RTimer` object, which destroys the corresponding kernel-side object. After this, the base `CActive` destructor will remove the active object from the active scheduler.

17.1.2 Requesting and Handling Events

`SetHello()` requests a timer event after a given delay:

```
void CDelayedHello::SetHello(TTimeIntervalMicroSeconds32 aDelay)
    {
    _LIT(KDelayedHelloPanic, "CDelayedHello");
    __ASSERT_ALWAYS(!IsActive(), User::Panic(KDelayedHelloPanic, 1));

    iTimer.After(iStatus, aDelay);
    SetActive();
    }
```

Every line in this function is important:

- First, we assert that no request is already outstanding (that is, that `IsActive()` is false). The client program must ensure that this is the case, either by refusing to issue another request when one is already outstanding or by canceling the previous request.
- Then, we request the timer to generate an event after `aDelay` microseconds. The first parameter to `iTimer.After()` is a `TRequestStatus&` that refers to the `iStatus` member that we inherit

from `CActive`. As I'll explain below, `TRequestStatus` plays a key role in event handling.

- Finally, we indicate that a request is outstanding by calling `Set-Active()`.

This is the invariable pattern for active object request functions. Assert that no request is already active (or, in rare cases, cancel it). Then issue a request, passing your `iStatus` to some function that will later generate an event. Then call `SetActive()` to indicate that the request has been issued.

You can deduce from this that an active object can be responsible for only one outstanding request at a time. You can also deduce that all request functions take a `TRequestStatus&` parameter – or, put the other way round, any function you see with a `TRequestStatus&` parameter is a request function, which will complete asynchronously and generate an event.

Our GUI program calls this function from `HandleCommandL()` using

```
iDelayedHello->Cancel();              // Just in case
iDelayedHello->SetHello(3000000);     // 3-second delay
```

In other words, it cancels any request so that the assertion in `SetHello()` is guaranteed to succeed and then requests a delayed info-message to appear after 3 s.

When the timer event occurs, it is handled by the active object framework, as we'll describe below, and results in `RunL()` being called:

```
void CDelayedHello::RunL()
    {
    iEnv->InfoMsg(R_ACTIVEHELLO_TEXT_HELLO);
    }
```

Clearly, this code is very simple: it's a one-line function that produces an info-message with the usual greeting text.

The degree of sophistication in an active object's `RunL()` function can vary enormously from one active object to another. CONE's `CCoeEnv::RunL()` function initiates an extremely sophisticated chain of processing; we have plenty of evidence for that from the debug session in Chapter 4. In contrast, the function above was a simple one-liner.

If an active object's `RunL()` calls a leaving function, then that active object should provide an override of the `RunError()` function.

17.1.3 Canceling a Request

If your active object can issue requests, it *must* also be able to cancel them. CActive provides a Cancel() function that checks whether a request is active and if so calls DoCancel() to cancel it. As the implementer of the active object, you have to implement DoCancel():

```
void CDelayedHello::DoCancel()
    {
    iTimer.Cancel();
    }
```

There is no need for any checking here. Because CActive has already checked that a request is active, there is no need for you to check this or to reset the active flag.

> **There is an obligation on any class with request functions, to provide corresponding cancel functions also.**

17.2 How it Works

It's time we looked beneath the surface to see how an event-handling thread, with an active scheduler and active objects, works. The general structure of such a thread can be seen in Figure 17.3.

The thread may have many active objects. Each active object is associated with just one object that has request functions – functions taking a TRequestStatus& parameter. These request functions complete their requests asynchronously, resulting in an event. Because of this, objects with request functions are often referred to as **asynchronous service providers**.

Actually, we'll see below that this need not be a precisely one-to-one relationship. But it's easier to explain as if it were.

When a program calls an active-object request function, the active object passes on the request to the asynchronous service provider. It passes its own iStatus as a TRequestStatus& parameter to the request function and, having called the request function, it immediately calls SetActive().

The TRequestStatus is a 32-bit object intended to take a completion code. Before starting to execute the request function, the asynchronous service provider sets the value of the TRequestStatus to KRequest-Pending, which is defined as 0×80000001.

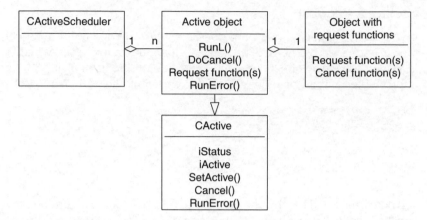

Figure 17.3

When the asynchronous service provider finishes processing the request, it generates an event. This means it signals the requesting thread's **request semaphore** and *also* posts a completion code (such as `KErrNone` or any other standard error code – anything except `KRequestPending` is permissible) into the `TRequestStatus`.

The active scheduler is responsible for detecting the occurrence of an event so as to associate it with the active object that requested it, and call `RunL()` on that active object.

The active scheduler calls `User::WaitForAnyRequest()` to detect an event. This function suspends the thread until one or more requests have completed. The active scheduler then scans through all its active objects, searching for one that has issued a request (`iActive` is set) and for which the request has completed (`iStatus` is some value *other* than `KRequestPending`). It clears that object's `iActive` and calls its `RunL()`. When the `RunL()` has completed, the scheduler issues `User::WaitForAnyRequest()` again.

So the scheduler handles precisely one event per `User::WaitForAnyRequest()`. If more than one event is outstanding, there's no problem: the next `User::WaitForAnyRequest()` will complete immediately without suspending the thread and the scheduler will find the active object associated with the completed event.

If the scheduler can't find the active object associated with an event, this indicates a programming error known as **stray signal**. The active scheduler panics the thread.

Given the delicacy of this description, you might expect writing an active object to be difficult. In fact, as we've already seen with `CDelayedHello`, it's not. You simply have to:

- issue request functions to an asynchronous service provider, remembering to call `SetActive()` after you have done so,

- handle completed requests with `RunL()`,
- be able to cancel requests with `DoCancel()`,
- set an appropriate priority,
- handle leaving from the `RunL()` with `RunError()`.

17.2.1 More on Canceling Requests

All asynchronous service providers must implement cancel functions corresponding to their request functions. All active objects that issue request functions must also provide a `DoCancel()` to cancel an outstanding request.

A cancel is actually a request for early completion. *Every* asynchronous request issued must complete precisely once – whether normally or by a cancel. The cancel must return synchronously and quickly and when it has returned, the original asynchronous request must have completed.

When a request is issued, it can complete in roughly four ways:

- The request can't even begin to execute, perhaps because there is no memory for the relevant resources or there is a bad parameter. If this happens, the requesting function should not leave or return a nonzero error code. Instead, it should post its completion code into the request status, so that the request completes just once (Figure 17.4).

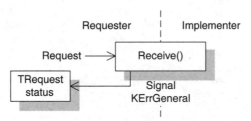

Figure 17.4

- The request is issued successfully and completes successfully some time later. This is the normal case, as in Figure 17.5.
- The request is cancelled before it completes. As part of its cancel processing, the service provider posts the request complete with `KErrCancel`. This can be seen in Figure 17.6.
- The client issued a cancel, but the request completed normally before the service provider got to process the cancel. In this case, the service provider should ignore the cancel. In its turn, the client, through the `CActive::Cancel()` protocol, will ignore the normal completion. The client should be careful; however, normal completion might involve writing data to some buffers whose address was passed as part of the initial request. The client should be sure to issue

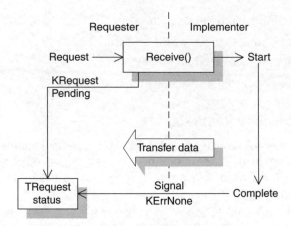

Figure 17.5

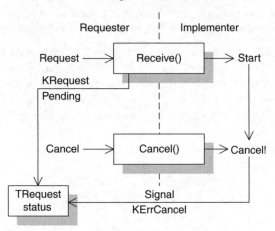

Figure 17.6

`Cancel()` *before* destroying the buffers just in case the request completes normally, which can be seen in Figure 17.7.

The GSDP server in Chapter 19 provides asynchronous service. We'll see in that chapter how cancel looks from the service provider's side.

`CActive` implements the `Cancel()` function as follows:

- Check `iActive` to see whether there is an outstanding request. If not, nothing needs to be done.

- Call `DoCancel()` to cause the request to complete (if it hasn't completed already).

- Issue `User::WaitForRequest()`, specifying `iStatus`. This is guaranteed to complete immediately, but it also decrements the thread

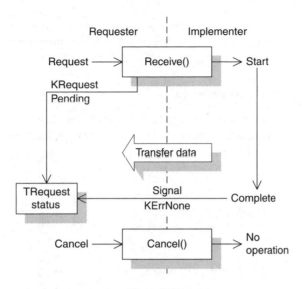

Figure 17.7

semaphore's value so that a subsequent call to User::WaitFor AnyRequest() by the active scheduler will not falsely complete because of this cancelled request.

- Reset iActive to indicate that there is no longer an outstanding request associated with this active object.

This logic is implemented in CActive::Cancel(); all you have to do when you're writing a derived active object class is to implement DoCancel(). You don't have to – in fact, you must not – do any of the other things that CActive::Cancel() does.

How can a request still be outstanding, if it has already completed? Easy: the request has completed (in another thread), but it iActive *indicates whether the request has been handled by this hasn't been handled by this thread.*

17.2.2 Error Handling

The active object framework provides error handling support through the CActive::RunError() function. This function is called by the active scheduler when the RunL() of the current active object leaves. It takes the leave code as its only argument and returns an error code indicating whether the leave has been handled.

A default implementation is provided by CActive, which just returns the leave code. This indicates that the leave has not been handled. Only a return value of KErrNone informs the active scheduler that the leave has been handled. If the leave has not been handled, then the active scheduler calls the central error handler Error().

17.2.3 Non-preemption and Priority

Active objects in the same thread handle events non-preemptively. Only when one RunL() has completed is the active scheduler able to detect and handle another event.

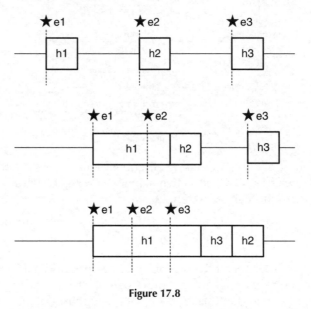

Figure 17.8

Figure 17.8 shows some interesting scenarios: three events, e1, e2, and e3, occur in sequence. They are handled by h1, h2, and h3. Each handler takes a finite amount of time to execute.

If events are widely spaced, as in the top line of the figure, then they are handled as soon as they occur.

If one event happens while a first event is being handled, as in the second line, then it is not handled until after the first handler has completed. This is non-preemptive event handling.

If *two* events happen while a first event is being handled, as in the third line, then when the first handler has completed, the thread can choose which of the two outstanding events to handle first. There is no obligation to handle events in the sequence in which they occurred. The active scheduler checks active objects in order of priority – highest priority first, lowest priority last, and calls RunL() on the first one it finds with a completed request. So if h3 has higher priority than h2, and if both e2 and e3 have happened by the time the active scheduler processes the completion of `User::WaitForAnyRequest()`, then h3 will be called first.

The priority of active objects is governed by a constructor parameter, which I set to zero in `CDelayedHello`. Awkwardly, there are two rival enumerations for active-object priorities. You would expect `CActive::TPriority` to be definitive – but it isn't. The CONE GUI framework defines `TActivePriority` in `coemain.h`, which is used to set CONE's active object priorities. Since, as an application program, you'll be jostling with CONE active objects, there's a strong argument for this being the definitive definition, for application code at least.

In any case, zero is zero, and that's the priority you should use unless you have a good reason not to. I'll cover such good reasons later. In order to avoid the conflict generated above, I code zero explicitly in all my active objects, unless I have a good reason not to.

17.2.4 Starting and Stopping the Scheduler

Application programmers never have to call `CActiveScheduler::Start()` and `Stop()`. The CONE framework does that for you.

The active scheduler's wait loop is started by issuing `CActiveScheduler::Start()`. Clearly, before this happens, at least one request function should be issued so that the first `User::WaitForAnyRequest()` will actually complete. From that point on, any completed events will cause one of your active object's `RunL()` functions to be called.

A `RunL()` function can stop the active scheduler by issuing `CActiveScheduler::Stop()`. When that `RunL()` returns, the function call to `CActiveScheduler::Start()` will complete.

Stopping the active scheduler will bring down the thread's event-handling framework, which is not something you should do lightly. Only do it if you are the main active object that controls the thread.

As a server programmer, you have to provide server bootstrap code that includes creating and starting the active scheduler for your server thread. I'll show you how to do that in Chapter 19.

The active scheduler offers a nesting facility, whereby you can issue `CActiveScheduler::Start()` from within a `RunL()` function. This is used to keep the active scheduler going while ostensibly handling a synchronous function. The end of the 'synchronous' function is indicated by a matching `CActiveScheduler::Stop()`. The net effect is like `Yield()` in some systems. This method is used by modal Uikon dialogs.

You shouldn't nest the active scheduler however, unless you have thought carefully through the implications of doing so. In particular, you must ensure strict nesting of all `CActiveScheduler::Start()` and `CActiveScheduler::Stop()` functions – which is of course a natural property of modal dialogs.

17.2.5 Adding Functionality to the Active Scheduler

CActiveScheduler is a concrete class that can be used as is. It also provides two virtual functions that can be used if you need them for additional purposes:

- If a RunL() called by the active scheduler leaves and that leave is not handled by the active object's RunError(), then Error(TInt) is called with the leave code. By default, this function does nothing.

- WaitForAnyRequest() may be used to perform some standard processing before issuing User::WaitForAnyRequest(). By default, this function simply issues User::WaitForAnyRequest() and any override must also ensure that it calls User::WaitForAny Request().

The Uikon environment sets up an active scheduler whose Error() function displays a natural-language version of the KErrXxx error code with which RunL() left.

The CONE environment overrides CActiveScheduler::WaitFor AnyRequest() to ensure that the window server's client-side buffer is flushed before issuing User::WaitForAnyRequest(). That means that any drawing done by the client during a RunL() is sent to the window server for execution so that while waiting for the next user input, there is no outstanding drawing.

If you're implementing a server, it's useful to implement a scheduler with its own Error() function. We'll show how to do this in the GDSP server implementation, in Chapter 19.

Don't write over-elaborate overrides for CActiveScheduler:: Error() and CActiveScheduler::WaitForAnyRequest(). If they're too elaborate, you create dependencies between your active objects and your active scheduler, which means that your active objects won't run in any other environment than (say) the server for which you designed them. That might be fine – just make sure you've thought about it.

17.2.6 Framework Summary

We can now understand all the functions in CActive and CActiveScheduler, as in Figure 17.9.

The CActive class

Here's CActive's declaration, from e32base.h:

```
class CActive : public CBase
    {
```

```
public:
enum TPriority
    {
    EPriorityIdle=-100,
    EPriorityLow=-20,
    EPriorityStandard=0,
    EPriorityUserInput=10,
    EPriorityHigh=20,
    };
public:
    IMPORT_C ~CActive();
    IMPORT_C void Cancel();
    IMPORT_C void Deque();
    IMPORT_C void SetPriority(TInt aPriority);
    inline TBool IsActive() const;
    inline TBool IsAdded() const;
    inline TInt Priority() const;
protected:
    IMPORT_C CActive(TInt aPriority);
    IMPORT_C void SetActive();
// Pure virtual
    virtual void DoCancel() =0;
    virtual void RunL() =0;
    IMPORT_C virtual TInt RunError(TInt aError);
public:
    TRequestStatus iStatus;
private:
    TBool iActive;
    TPriQueLink iLink;
    friend class CActiveScheduler;
    friend class CServer;
    };
```

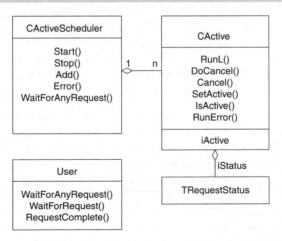

Figure 17.9

Member functions here are as follows:

Function	Description
~CActive()	Virtual destructor. Calls Deque() to de-queue the object from the active scheduler.
Cancel()	Cancels a request. If a request is active, calls DoCancel(), waits synchronously for completion on iStatus and sets the request as no longer active.
SetPriority()	Changes the priority after construction.
Priority()	Returns the active object's priority.
IsActive()	Indicates whether a request is outstanding.
Deque()	Removes this object from the active scheduler.
IsAdded()	Indicates whether the object has been added to the active scheduler.
Cactive()	C++ constructor: you must specify the active object's priority as a constructor parameter.
SetActive()	Call this function after you have issued a request function to indicate that a request is outstanding.
DoCancel()	Implement this in a derived class to cancel a request issued to an asynchronous service provider. This function can only be called if a request is active. You must provide a way to cancel requests.
RunL()	Implement this in a derived class to handle the completion of a request that was issued.
RunError()	This should be overridden if the RunL() implementation calls a leaving function. The TInt parameter is the leave code from RunL().

I've not yet met a case in which Deque() or SetPriority() are necessary or couldn't be handled by a different design approach. Bizarrely, although Deque() is a member of CActive, which enables you to remove an active object from the scheduler, you can only *add* an active object to the scheduler using a member function of CActiveScheduler (which is a friend class).

The *CActiveScheduler* class

Here's CActiveScheduler's definition, from e32base.h:

```
class CActiveScheduler : public CBase
    {
public:
    IMPORT_C CActiveScheduler();
    IMPORT_C ~CActiveScheduler();
    IMPORT_C static void Install(CActiveScheduler* aScheduler);
    IMPORT_C static CActiveScheduler* Current();
    IMPORT_C static void Add(CActive* anActive);
    IMPORT_C static void Start();
    IMPORT_C static void Stop();
    IMPORT_C virtual void WaitForAnyRequest();
    IMPORT_C virtual void Error(TInt anError) const;
protected:
    inline TInt Level() const;
private:
    TInt iLevel;
    TPriQue<CActive> iActiveQ;
    };
```

Member functions here are as follows:

Function	Description
CActiveScheduler()	Default C++ constructor, invoked when you create an active scheduler with new.
~CActiveScheduler()	C++ destructor.
Install()	Installs a pointer to the active scheduler specified in a privileged TLS location that is very fast to access – nearly as fast as a pointer. You can then access the active scheduler with the function CActiveScheduler::Current().
Current()	Returns a pointer to the currently installed active scheduler (or zero, if there isn't one).
Add()	Adds an active object to the scheduler. You should call this as part of the construction of all active objects, using CActiveScheduler::Add(this).

`Start()`	Starts the scheduler or increases the scheduler nesting level. There should be at least one outstanding request on an active object, otherwise this will cause a thread to hang. This function includes the active scheduler's central wait loop handler. This function does not return until a corresponding `Stop()` has been issued.
`Stop()`	Decreases the nesting level. When the current `RunL()` or `Error()` has completed, it will cause the currently active `Start()` function to return.
`WaitForAnyRequest()`	Override this function if you want to perform special processing before calling `user::WaitForAnyRequest()`.
`Error()`	Override this function to handle leaves from `RunL()` of any active object scheduled by the active scheduler. The `TInt` parameter is the leave code from `RunL()`.

The declaration of CActive and CActiveScheduler are rather bizarre C++ and reflect, perhaps, the fact that these were among the earliest classes to be implemented in Symbian OS. However, in practice, this bizarreness doesn't get in the way of working with active objects or of building well-engineered, large-scale, object-oriented systems with the active object framework.

The *TRequestStatus* class

Here's the declaration of TRequestStatus, from e32std.h:

```
class TRequestStatus
    {
public:
    inline TRequestStatus();
    inline TRequestStatus(TInt aVal);
    inline TInt operator=(TInt aVal);
    inline TInt operator==(TInt aVal) const;
    inline TInt operator!=(TInt aVal) const;
    inline TInt operator>=(TInt aVal) const;
```

```
    inline TInt operator<=(TInt aVal) const;
    inline TInt operator>(TInt aVal) const;
    inline TInt operator<(TInt aVal) const;
    inline TInt Int() const;
private:
    TInt iStatus;
    };
```

A `TRequestStatus` is simply a well-encapsulated integer, which you can't do anything with except compare and assign.

Priority enumerations

Here, again, is `CActive`'s `TPriority` enumeration,

```
class CActive : public CBase
    {
public:
    enum TPriority
        {
        EPriorityIdle=-100,
        EPriorityLow=-20,
        EPriorityStandard=0,
        EPriorityUserInput=10,
        EPriorityHigh=20,
        };
...
```

And here's `TActivePriority` from `coemain.h`:

```
enum TActivePriority
    { // an alternative set to the TPriority in E32BASE.H
    EActivePriorityClockTimer=300,
    EActivePriorityIpcEventsHigh=200,
    EActivePriorityFepLoader=150,
    EActivePriorityWsEvents=100,
    EActivePriorityRedrawEvents=50,
    EActivePriorityDefault=0,
    EActivePriorityLogonA=-10
    };
```

In my opinion, the really important ones are as follows:

Symbolic Name	Value	Description
EActivePriority WsEvents	100	User input events from the window server are handled at this priority.

EActivePriority RedrawEvents	50	Redraw events from the window server are handled at this priority.
EActivePriority Default	0	Active objects should have this priority unless there is very good reason.
EPriorityIdle	−100	Priority for background task active objects, which run in the idle time of all other active objects on the same thread.

If you need to code nonzero priorities for your own active objects, you need to know the real (not just symbolic) values of the priorities above so that you can ensure you fit in with them. EPriorityHigh and EPriorityLow would be good values to use if you want to be a little higher than zero, or a little lower. I'll cover priorities again in the sections below.

17.3   Active Object Patterns

CDelayedHello demonstrates the one-shot active object pattern. A single request is made through the Application Programming Interface (API), and it's handled through the RunL() function. There are many other ways you can use active objects.

17.3.1   Maintaining an Outstanding Request

The active example shows the outstanding-request active object pattern, in which RunL() handles the completion of a previous request and then issues a new request.

Here's the declaration of CFlashingHello:

```
class CFlashingHello : public CActive
    {
public:
    // Construct/destruct
    static CFlashingHello* NewL(CActiveHelloAppView* aAppView);
    ~CFlashingHello();

    // Request
    void Start(TTimeIntervalMicroSeconds32 aHalfPeriod);

private:
    // Construct/destruct
```

```
    CFlashingHello();
    void ConstructL(CActiveHelloAppView* aAppView);

    // from CActive
    void RunL();
    void DoCancel();

    // Utility
    void ShowText(TBool eShowText);

private:
    // Member variables
    RTimer iTimer;
    TTimeIntervalMicroSeconds32 iHalfPeriod;

    // Pointers elsewhere
    CActiveHelloAppView* iAppView;
    };
```

Figure 17.10 shows how it fits into the application program:

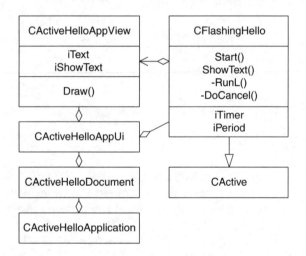

Figure 17.10

The `Start()`, `RunL()`, and `DoCancel()` functions of `CFlashing Hello` show how to maintain an outstanding request. Here's `Start()`:

```
void CFlashingHello::Start(TTimeIntervalMicroSeconds32 aHalfPeriod)
    {
    _LIT(KFlashingHelloPeriodPanic, "CFlashingHello");
    __ASSERT_ALWAYS(!IsActive(), User::Panic
      (KFlashingHelloPeriodPanic, 1));
    // Remember half-period
```

```
    iHalfPeriod=aHalfPeriod;

    // Hide the text, to begin with
    ShowText(EFalse);

    // Issue request
    iTimer.After(iStatus, iHalfPeriod);
    SetActive();
    }
```

Start() begins by asserting that a request is not already active and ends by issuing a request – just as before. Because a whole series of requests will be issued, Start() doesn't merely pass the half-period parameter to the iTimer.After(), but stores it as a member variable for later use.

Start() also starts off the visible aspect of the flashing process by immediately hiding the text (which is visible until Start() is called).

When the timer completes, RunL() is called:

```
void CFlashingHello::RunL()
    {
    // Change visibility of app view text
    ShowText(!iAppView->iShowText);

    // Reissue request
    iTimer.After(iStatus, iHalfPeriod);
    SetActive();
    }
```

RunL() changes the visibility of the text to implement the flashing effect. Then, it simply renews the request to the timer with the same iHalfPeriod parameter as before. As always, the renewed request is followed by SetActive().

The only way to stop the flashing is to issue Cancel() which, as usual, checks whether a request is outstanding and, if so, calls our DoCancel() implementation:

```
void CFlashingHello::DoCancel()
    {
    // Ensure text is showing
    ShowText(ETrue);

    // Cancel timer
    iTimer.Cancel();
    }
```

We make sure the text is showing and then cancel the timer.

`ShowText()` is the utility function that sets the visibility of the text; it simply changes the `iShowText` in the app view and then redraws the app view.

```
void CFlashingHello::ShowText(TBool aShowText)
    {
    iAppView->iShowText = aShowText;
    iAppView->DrawNow();
    }
```

In summary, a continuously running active object is little harder to implement than a one-shot object.

Although this example looks simple enough, I experienced an unexpected half-hour of frustration with the debugger while testing it. Here's what I learned:

> **Because the `DoCancel()` contains drawing code, it must be executed when there is still an environment in which drawing is possible. This means you must cancel or destroy the flashing hello *before* calling `CEikAppUi::Exit()`.**

Here's my command handler for `EEikCmdExit`:

```
case EEikCmdExit:
    iFlashingHello->Cancel();
    Exit();
    break;
```

I just cancel the active object from here. I *destroy* it from its owning class's destructor – this is the right place to destroy it, since the destructor gets called in cleanup situations, while the command handler does not. Prior to coding this `Cancel()` explicitly in the exit command handling code, my `DoCancel()` was being called from the active object's destructor and was trying to draw to an environment that by then had been destroyed.

You should always be careful about doing anything fancy from an active object's `DoCancel()`. Nothing in a `DoCancel()` should leave or allocate resources, and `DoCancel()` should complete very quickly. I got myself into trouble because I don't simply *stop* flashing when I cancel; instead, I restore the visibility state, which involves drawing. In fact, this is a good rule for any kind of cleanup or destructor; just cleanup and destroy – don't do anything else.

17.3.2 State Machines

Active objects can be used to implement state machines. As an example, say, we wish to amalgamate the `CDelayedHello` and

CFlashingHello functionality and so produce the effect of the Hello world! message flashing on/off once followed by the info-message with the same greeting. This cycle repeats until cancelled. This behavior can be represented by the state diagram shown in Figure 17.11.

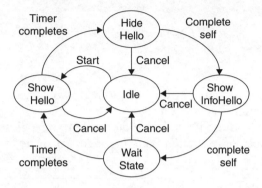

Figure 17.11

The state machine is initially in the Idle state and starting the machine moves it into the Show Hello state. The greeting is shown and a timer is started. When the timer completes, it notifies the state machine and it moves to the Hide Hello state – the greeting is removed. The state machine then moves itself into the Show Info Hello state. Here, an info-message is produced showing the greeting text. Again, the state machine moves itself into the Wait State in which it starts a timer. When the timer completes, it notifies the state machine and it moves to Show Hello state and the cycle repeats. Cancelling the state machine can be done in any state and simply moves the state machine into the Idle state.

This is a very simple state machine but demonstrates the idea. The CMultiPartHello object implements the state machine. Essentially, it is an active object that maintains an outstanding request once it has been started until it is cancelled.

```
class CMultiPartHello : public CActive
    {
public:
    static CMultiPartHello* NewL(CActiveHelloAppView* aAppView);
    virtual ~CMultiPartHello();

    void Start(TTimeIntervalMicroSeconds32 aDelay);

private:   // From CActive
    virtual void RunL();
    virtual void DoCancel();
```

```
private:
    CMultiPartHello();
    void ConstructL(CActiveHelloAppView* aAppView);

    void CompleteSelf();
    void ShowText(TBool aShowText);

private:    // Enums

    enum THelloState
        {
        EIdle                   = 0,
        EShowHello,
        EHideHello,
        EShowInfoHello,
        EWaitState
        };

private:
    RTimer iTimer;
    TTimeIntervalMicroSeconds32 iDelay;
    THelloState iState;

    CEikonEnv* iEnv;
    CActiveHelloAppView*    iAppView;
    };
```

We will now discuss some of the new functionality introduced by this class. The `CompleteSelf()` function makes the active object eligible for `RunL()` next time control is returned to the active scheduler.

```
void CMultiPartHello::CompleteSelf()
    {
    TRequestStatus* pStat = &iStatus;
    User::RequestComplete(pStat, KErrNone);
    SetActive();
    }
```

The `User::RequestComplete()` is applied to the active object's own `iStatus` and active object sets itself active. This has the effect of appearing as if the active object issued a request and that request has been completed. Therefore, the active scheduler can call the `RunL()`.

The `RunL()` is where the behavior for the different states is implemented.

```
void CMultiPartHello::RunL()
    {
    THelloState nextState = iState;
    switch( iState )
```

```
        {
case EShowHello:
        {
        ShowText(ETrue);

        // issue request
        iTimer.After(iStatus, iDelay);
        SetActive();

        nextState = EHideHello;
        } break;
case EHideHello:
        {
        ShowText(EFalse);

        CompleteSelf();
        nextState = EShowInfoHello;
        } break;
case EShowInfoHello:
        {
        iEnv->InfoMsg(R_ACTIVEHELLO_TEXT_HELLO);

        CompleteSelf();
        nextState = EWaitState;
        } break;
case EWaitState:
        {
        // issue request
        iTimer.After(iStatus, iDelay);
        SetActive();

        nextState = EShowHello;
        } break;
default:
        break;
        }
iState = nextState;
}
```

When the RunL() is entered, a switch is done on the current state. In each state, the desired behavior is performed and the next state is set. The next state is performed when the RunL() is next called. This can be due to either an asynchronous request being completed (the timer in this example) or by the state machine completing itself.

You could argue that there is no need for the complete-self functionality – the states can be amalgamated. Applying this to our example would mean that the Hide Hello, Show Info Hello and Wait State could be amalgamated into a single state. So why not do this? One reason would be to allow control to be returned to the active scheduler and thereby maintain responsiveness, that is, allow higher priority active objects like CONE's user input handler to be given some processor time. Also, state machines will not be as simple as this and the processing path can vary depending on some conditions. For example, in a given state A, the next state can be state B (which uses a `CompleteSelf()` call to get there) or state C (which is entered once an asynchronous request completes) depending on some changing condition.

Checking the state need not only be done in the `RunL()`. For example, the `DoCancel()` function in our example checks to see if the timer needs to be cancelled.

```
void CMultiPartHello::DoCancel()
    {
    switch( iState )
        {
    case EHideHello:
    case EShowHello:
        {
        iTimer.Cancel();
        } break;
    default:
        break;
        }
    ShowText(ETrue);
    iState = EIdle;
    }
```

Looking back at the `RunL()` we can see that the timer was started in the `Show Hello` state and then the state changed to `Hide Hello`. Similarly, in `Wait State`, the state was changed to `Show Hello` once the timer had been started. Therefore, the timer is only cancelled if the state machine is in the `Hide Hello` or `Show Hello` states, when `Cancel()` is called.

17.3.3 Interfaces for Handling Completion

The `CDelayedHello`, `CFlashingHello`, and `CMultiPartHello` active objects are concrete classes that both define the requests and handle their completion. Often, though, active objects are used for implementing abstract interfaces. For instance, in a communications stack such as TOGS, you issue a request for some received data. When the data comes, you want a function to be called.

You know by now that in Symbian OS, a requirement like this is going to be implemented with an active object that requests the received data using an asynchronous function and then handles the completed receive with a `RunL()`. It's tempting to provide an interface that exposes this by

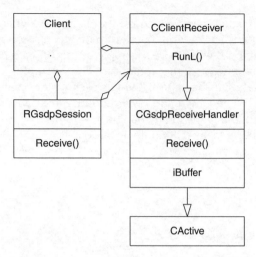

Figure 17.12

including `CActive`-derived objects in your API and inviting the client to implement a `RunL()` to handle the received data. For instance, you could use the design in Figure 17.12.

With this design, you write the following specification for `RunL()`:

- 'When implementing `RunL()`, the data requested by the client is in `iBuffer`. Handle this buffer according to the requirements of your protocol.'

This works, but I don't like it. The client has to implement a derived active object and has to use that object from the main client class. The API includes active objects, which cloud the real issues that the API is answering. And the design isn't portable to a non-Symbian OS system.

I prefer to hide active objects from APIs like this. Instead, I define an interface such as `MGsdpHandler` and use a hidden active object whose `RunL()` function calls `MGsdpHandler::GsdpHandle()` with the buffer reference passed as a parameter. In your API description, you can now say,

- 'Your client class should implement the `MGsdpHandler` interface. The function `MGsdpHandler::GsdpHandle(const TDesC8& aData)` will be called to handle received data.'

The GSDP client interface can be seen in Figure 17.13. For good measure, the interface uses `Listen()` and `StopListening()` functions to maintain an outstanding request.

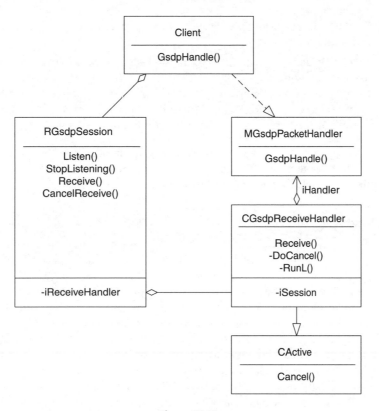

Figure 17.13

17.3.4 Long-running Tasks and Incremental Interfaces

Sometimes, you want to be able to implement a long-running task alongside your application's (or server's) main task. In non-Symbian OS systems, you might implement such a task with a 'background thread'. In Symbian OS, the best way to implement long-running tasks is with a low-priority active object that runs in the idle time of event-handling active objects. It is essentially a low-priority state machine that always completes itself. The paradigm for a long-running task is as follows:

- You design the state machine to do the background task through a series of states.
- In the Start() function, use a self-complete style function to make you eligible for RunL() next time control returns to the active scheduler. When your RunL() is called, you should do some processing for the long-running task – say, about 100 ms worth of processing.
- If your task is not complete, you then need to self-complete to continue processing.

- If your task is complete, you simply stop and don't reissue your artificial request.

- In `DoCancel()`, you may wish to destroy any intermediate data associated with the long-running task. You don't need to issue `User::RequestComplete()` because you already did that from either `Start()` or `RunL()`. Remember, any request you issue should complete *precisely once*, so you don't need to complete it again.

Some Symbian APIs for long-running tasks are designed to be called from active objects in this way. For example, the DBMS provides compaction APIs whose specification is roughly:

```
void Start(parameters) ...;
TInt Step(TInt& aStep);
void Close();
```

Here, `Start()` starts off an operation, while `Step()` performs a single step (or state), returns an error code, and sets its reference parameter to `ETrue` if the operation has finished, or `EFalse` otherwise. `Close()` releases all the resources associated with the incremental operation so that it can be cancelled even if it hasn't completed.

There are two advantages to providing a long-running task with an API like this. Firstly, it hides the active objects from the API and therefore prevents the active object paradigm from clouding your thinking about the real issue – namely, how to provide an incremental version of the function for your long-running task. Secondly, it allows the API to be used by code in which there is no active scheduler and hence no active objects – for instance, from a native method in a Java thread.

In `e32base.h`, the `CIdle` class provides a ready-made (if near-trivial) wrapper for idle-time processing like this. Idle-time active objects should use a priority such as `CActive::EPriorityIdle`, which equates to −100. It's probably not a great idea to have too many idle-time active objects running in a single thread – if you want these objects to run together, you'll have to construct your own scheduling algorithm between the 'idle-time' objects.

It should be noted that a task cannot always be easily broken down into small `Step()` functions or states. Care must be taken in the design of the idle object architecture and code to maintain responsiveness. It can be difficult to retrofit this.

17.3.5 Prioritizing and Maintaining Responsiveness

As we've seen, active objects specify a priority, which is passed as a parameter to their constructor and is used to order the active objects on the active scheduler's queue. In turn, this governs the search order after a `User::WaitForAnyRequest()` completes: the higher the active object's priority, the earlier it will be checked by this scan. Therefore, if two or more active objects' requests complete while another request is being handled, it's the highest-priority object that will be handled first.

Normally, you should code your active objects so that priority doesn't matter. You should give most active objects a priority of zero. There can be good reasons to go higher or lower than this: some events really are more important than others. On a GUI thread, for instance, user responsiveness is critical. CONE's user input-handling active object, therefore, specifies a priority of 100. Keeping the view up-to-date is also important (though not as important as handling the events, which might change the view). So CONE handles window-server redraw events at a priority of 50.

Any long-running tasks implemented by active objects are not really events at all, and should, therefore, be prioritized below anything that is an event. Long-running tasks should always have a negative priority: the recommended value is `CActive::EPriorityIdle`, which is −100.

In active objects, higher priority means simply that you get handled sooner than others, if your event completes along with several others during the handling of the previous `RunL()`. But active object priority does *not* cause preemption: if the previous `RunL()` takes a long time, nothing you can do with active object priority will get you scheduled sooner.

> **Make sure that you understand what active object priority means and be sensible in allocating active object priorities.**

So, if you try to use an ultra-high-priority event to do something that requires a certain response time, it won't always work. An active object used to keep a sound channel going, say, could get held up behind a step in a long-running task doing printing or database compaction. In other words, you can't use high-priority active objects to achieve responses that *must* occur within a certain time of an event. For that, you need to use the Symbian OS preemptive thread system, appropriate buffering, and appropriate thread priorities.

17.4 Summary

Symbian OS has a highly responsive preemptive multithreaded architecture. However, most application and server tasks are by nature event handlers and active objects are a very suitable paradigm with which to handle events.

Because they are non-preemptive, it's very easy to program with active objects, because you don't need to code mutual exclusion with other threads trying to access the same resources.

Active objects also use fewer system resources than full-blown threads: thread overheads start around 4 k kernel-side and 12 k for the user-side stack, whereas active objects need be only a few bytes in size. Additionally, switching between active objects is much cheaper than switching between threads, even in the same process. The time difference can be up to a factor 10.

This combination of ease of use and low resource requirement is a major factor in the overall efficiency of Symbian OS. However, performance can be impaired if there are many active objects since the active scheduler needs to iterate through the list of active objects to find which object's RunL() to call.

For some purposes, threads are necessary. In the next chapter, we'll include them as we discuss the Symbian OS client-server architecture.

18

Client-server Framework

Back in Chapter 2 we showed the process and privilege boundaries that Symbian OS uses to assure a high level of system integrity. We noted that the design of Symbian OS is optimized for event handling and, in the last chapter, covered active objects in detail: active objects assure a high level of system integrity because they don't require the same kind of sharing disciplines as are needed in multithreaded systems.

The final major building block for the system integrity of Symbian OS is the client-server framework in which servers handle system resources on behalf of multiple clients. Examples of servers in Symbian OS include

- the file server, which shares all file-related resources between all clients;

- the window server, which shares UI resources – keyboard, pointer, and screen – between all applications;

- the font and bitmap server, which manages shared, system-wide resources for fonts and bitmaps;

- the database server, which is used to control database sharing where shared access is desired; the database server is optional: if a database is not intended to be shared between applications, you can drive it directly without using the server;

- the serial communications server, which shares the serial port and other virtual serial protocols between all client programs;

- the sockets server, which maintains sockets protocol resources and allows them to be shared between programs;

- many other servers, associated with particular applications or subsystems such as messaging, multimedia handling, and so on.

Symbian OS C++ for Mobile Phones.   Edited by Richard Harrison
© 2003 John Wiley & Sons, Ltd   ISBN: 0-470-85611-4

In this chapter, I'll describe the client-server's framework design in general terms, building on and reinforcing the treatment of active objects from the previous chapter. Specifically, I'll cover

- the general design of the client-server framework;

- optimization for better performance, including examples taken from the standard servers in Symbian OS;

- the relationship between servers and preemptively-scheduled threads, including some recommendations on thread and active object priorities (and a little myth-busting in the bargain);

- some reference information about the classes related to servers and threads.

In the next chapter, we'll put the ideas from Chapter 16, Chapter 17, and this chapter together, as we describe the implementation of the GSDP server in detail.

18.1   Introduction

As we saw in Chapter 2, a server runs in a different thread – and usually a different process – from any of its clients. The kernel supports two ways to cross this thread boundary:

- Message passing – used in all client-server transactions.

- Interthread data transfer – used, if necessary, to transfer more data between client and server.

A client-server message consists of a 32-bit request code and up to four 32-bit parameters (see Figure 18.1). The result passed back from the server is a single 32-bit value. The 32 bits of the return value may be interpreted as a `TInt`, a `TInt32`, a `TUint32`, or any kind of pointer depending on the needs of the particular message.

Some simple server transactions can be handled using message passing alone, but more complex transactions may need more data (see Figure 18.2). In this case, the client should pass a pointer to a descriptor containing the data. Then, the server can use interthread read or write functions to transfer the data.

The **client-server framework** has to manage all this communication. In the remainder of this section, we'll look at several interesting aspects of that framework.

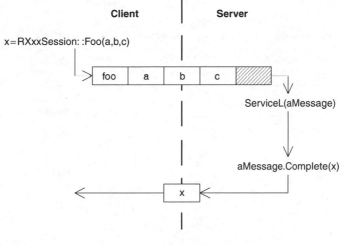

Figure 18.1

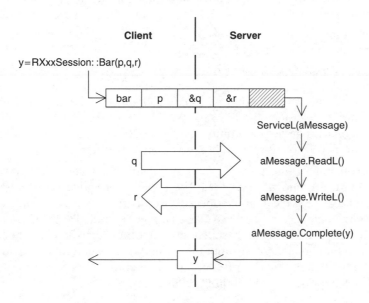

Figure 18.2

18.1.1 Handling Routine Requests

A routine message is passed from client to server over the client-server session as shown in Figure 18.3.

There are three parts to this session:

- A client-side class, derived from `RSessionBase`, which contains functions offered to the client. Each function is implemented by

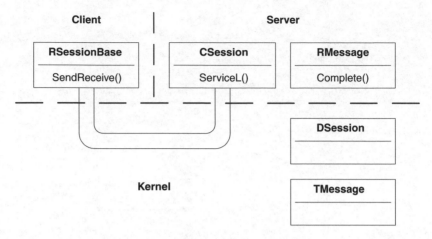

Figure 18.3

converting its parameters into suitable message parameters and issuing a `SendReceive()` or `Send()` call to send the message to the server.

- A kernel-owned class, `DSession`, which links the client and the server and is also used for cleanup.

- A server-side class, derived from `CSharableSession`, which contains a `ServiceL()` function. The derived class uses this function to check the request code and to call a handler function. The handler function interprets the message parameters and performs its service. When the handler has completed, it calls `Complete()` with a 32-bit return code that is returned to the client via the kernel.

You can find all the user-side classes for the client-server architecture in `e32std.h` and `e32base.h`. The services provided by the kernel-side classes are clear and concrete, transparent to both client-server implementation. As such, these classes are not included in the C++ SDK. But as we'll see during the discussions that follow, the kernel's role in the client-server architecture is important, and it's helpful to understand that the kernel has an object representing the session.

The key principles of handling routine requests are

- to wrap up the parameters appropriately,
- to minimize client-server communication (because of the overheads – this is discussed at length later in this chapter),
- to use interthread reads and writes where necessary.

18.1.2   Setting up Sessions

One question you may be asking is, 'How did a session for message passing get there in the first place?'

When a client thread wishes to start using a server, it connects a client-server session to that server. The `Connect()` function (in `RSessionBase`) takes the name of the server. When it's called, the kernel creates a new session object. At that point, the server recognizes the connect message and calls `NewSessionL()` in the derived server class to create a new `CSharableSession`-derived object, representing the server end of the session, to which future routine messages will be routed using `ServiceL()`.

A new session has a handle that is also passed as a parameter in any message. A single client thread may have multiple sessions to any individual server – each session identified by a different handle.

18.1.3 Starting Servers

That begs another question: 'How did the server get there, so that sessions could be set up in the first place?' There are three possibilities:

- The server is a system server, started by Symbian OS startup code: such servers are essentially a part of Symbian OS itself and the system is effectively dead without them. Examples include the file server, the window server, and the font and bitmap server.

- The server is not needed, except when certain applications or servers that require it are active. When these are not active, the server can terminate in order to save resources. Such servers should be started from the client API prior to connecting a session, if they are not already active. They should terminate themselves if their last session is closed – perhaps after a short time-out period. Most servers are of this so-called **transient server** type in which startup and shutdown issues are particularly delicate. The GSDP server code in the next chapter shows how it's done.

- The server is not a shared system service at all, but a convenience for each instance of a running application. Examples include the POSIX and AWT servers in the Symbian OS Java implementation. In this case, the server should be started as part of application startup, and terminated as part of application shutdown.

When a server starts, it declares a name that must be used by all client connect messages. System servers should use a unique name, so that clients (through the client API) can easily find them and connect to them. Private servers should use a name that is private to the instance of the application that started them – for example, by including the application's main thread name in the server name. This leads to the question, as a client how do I work out the name of the server? The answer is, of course, that the name is intimate between client-side API and server, so the client-side implementation is implemented to use this name (i.e. hard-coded or by inclusion of a mutual header). What is most important here is to make clear that you should implement the client side of your server to hide this name from users of the client side. (Classic example is RFs:Connect(). You don't need to know the server name as a user of the client side – it's inside the function implementation.)

After a session is connected, routine client-server communication uses the `RSessionBase`, `DSession`, and `CSession`/`CSharableSession` objects in client, kernel, and server, respectively, and the name isn't needed anymore.

18.1.4 Handling Asynchronous Requests

Servers are often associated with asynchronous request functions. As we saw in the last chapter, when examining the client API to a server, you can tell asynchronous request functions by their `TRequestStatus&` parameter. Asynchronous requests are handled by the client interface by using a form of `RSessionBase::SendReceive()` that also takes a `TRequestStatus&` parameter. The server keeps the message corresponding to the asynchronous request until that request has been completed.

When the server completes the request, the kernel puts the completion code into the `TRequestStatus` passed by the client, and then signals the client's request semaphore to indicate that the request has been completed.

In contrast, synchronous functions are sent as a message using a form of `RSessionBase::SendReceive()` that does *not* take a `TRequest-Status&` parameter. The message for a synchronous function must be completed synchronously by the server's handler function.

We saw in the previous chapter that any asynchronous request must be completed precisely once – whatever the circumstances. The kernel guarantees that even if the server thread dies an asynchronous request by a client is completed as part of the kernel's thread-death cleanup.

We also saw that any API that provides asynchronous request functions should also offer corresponding cancel functions: so cancel functions too, are often present in servers' client APIs.

18.1.5 Ending a Session and Cleanup after Client Death

When a client has finished with a client-server session, it should end it with `RSessionBase::Close()`. This sends a disconnect message to the server, which responds by simply destroying its end of the session: by calling the destructor of the `CSharableSession`-derived object. Then, the kernel destroys its representation of the session. The handle of the client-side `RSessionBase` is set to zero.

It is safe for a client to call `Close()` on the `RSessionBase` again: when the handle is zero, `Close()` does nothing.

If the client thread dies, the kernel will perform thread-death cleanup and send a disconnect message to the server end of all the sessions for which it was a client thread. Thus, the server-side end of the session is destroyed, even if the client thread dies.

Servers must perform effective cleanup. When a `CSession`/`CSharableSession` object is destroyed, any resources associated with it should also be destroyed. Usually, this is just standard C++ destructor processing.

18.1.6   Cleanup after Server Death

Servers should be written with the greatest care, so they do not terminate prematurely. But if a server does die, then the kernel gives the opportunity for clients to recover. Any outstanding messages from asynchronous requests will be completed with the `KErrDied` return code.

After a server has died, the client should clean up all `RSessionBase` objects relating to that server. Any `SendReceive()` issued on an `RSessionBase` to a dead server will result in a completion code of `KErrServerTerminated`. This will be the case even if a new instance of the server is started: old sessions will not be reconnected.

18.1.7   Handling Multiple Objects from One Session

Sometimes it is useful for a single client-side thread to have multiple sessions with a given server. A good example of this is the socket server; each socket created by the client could be a separate session. However, creating a new session consumes resources in the server and in the kernel to support it. Therefore, creating many sessions for one client thread may be inefficient. An alternative approach is provided by the use of subsessions (see Figure 18.4). In general, if your client thread needs multiple sessions, try to use subsessions.

Before any subsessions can be created, an initial session must be created as normal. However, subsequently the client thread can create subsessions – these consume fewer server-side resources and can be created in less time. However, the code to support subsessions is slightly more complex (and larger).

Typically, the server will have to implement a reference-counting scheme to manage the lifetime of the server-side session. It is possible to implement reference counting using `CObject` and associated classes but

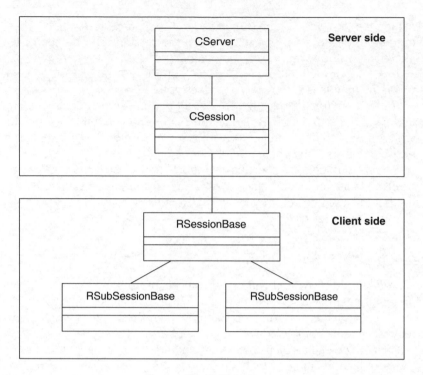

Figure 18.4

these classes are more complicated than is necessary for most require-ments, and it's possible to use RSubSessionBase successfully without using CObjects. We will not consider these classes any further in this chapter but they are described in the SDK. In any case, the idea is to route requests from the CServer::ServiceL() function to the particular server-side object that should handle the request.

A subsession is created by calling RSubSessionBase::CreateSub-Session() with a reference to the original session and a function code. Subsequently, the subsession can be used to make SendReceive() and Send() calls as normal. When the subsession is finished with, RSubSes-sionBase::CloseSubSession() should be called.

The GSDP server doesn't use multiple objects in this way.

18.2　Performance

The fundamental means of communication between client and server is a transaction based on a message send, some optional interthread data transfer, and message completion. Compared with conventional multithreading, shared heaps, blocking I/O, mutex synchronization, and full-blown concurrent programming disciplines, this transaction-based model is less demanding for programmers and system resources. In itself, this is a significant boost to client-server performance and to the overall system performance of Symbian OS.

However, when you implement a server – or even when, as a client, you use one – you should be aware of the performance issues that still exist and what you can do to tackle them.

The issue that matters more than anything else is the frequency of transactions and the cost of the main operations involved in a transaction:

- Context switching between processes is the most expensive operation – that is, sending a message to a server in a different process, or sending the response back from server to client.

- Context switching between *threads*, when the client and server are in the same process, is much more efficient.

- Interthread data transfers between processes are fairly expensive.

- Interthread data transfers between threads in the same process are quite cheap.

Compared with these costs, the difference between a small interthread data transfer and a large one is trivial.

Clearly, the actual cost of a client-server context switch depends on the specific Symbian OS hardware and on the present state of execution. You should think in terms of a few hundred microseconds – that is, around ten thousand cycles – expended largely on MMU manipulation or cache misses. The actual cost of a data transfer depends similarly on hardware specifics.

Bearing these costs in mind, some standard techniques have evolved to improve client-server performance. Some techniques are about server design (including the implementation of the client interface), some are about sensible client programming, and some are down to Symbian OS system configuration. The main techniques are as follows:

- Design the server and client interface to support client-side buffering and high-level transactions.

- As a client programmer, cache server-side data. Some server support may be needed to keep the cache up to date.

- Configure the system so that related servers run in the same process.

- As a last resort, design the server and its client interface to support shared memory.

We will review these techniques in the following sections.

18.2.1 Client-side Buffering

The Symbian OS window server uses client-side buffering in order to minimize the number of transactions between the client and server. In a code sequence such as this:

```
CWindowGc& gc = SystemGc();
...
gc.SetPenStyle(CGraphicsContext::ESolidPen);
gc.SetBrushStyle(CGraphicsContext::ENullBrush);
gc.SetPenColor(KRgbBlack);
gc.DrawRect(rect);
rect.Shrink(1, 1);
gc.SetPenColor(KRgbWhite);
gc.DrawRect(rect);
```

A naïve server implementation would result in six client-server transactions and cripplingly slow graphics. Instead of passing each function call to the server directly, the window server's client interface converts the call into an operating code with parameters that are stored in a client-side buffer (see Figure 18.5). The above sequence requires only a few hundred machine instructions in the client thread, and a few hundred machine instructions in the server thread, plus one transaction that will also be used for other drawing.

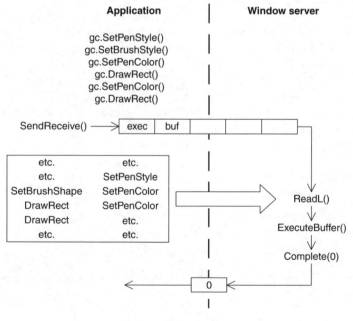

Figure 18.5

When the buffer is full, or when a function such as `DrawPolyLine()` (which requires large-scale data transfer) is invoked, the client interface 'flushes' the buffer – that is, it requests the window server to execute all operations stored in the buffer. The window server reads the buffer using an interthread read, and then executes all the drawing commands stored in it. The buffer is reasonably large and its format has been optimized so that the above commands and more would easily fit inside one buffer.

Symbian has had two opportunities to see what happens if you don't do this kind of client-side buffering. During early Symbian OS development, prior to implementing the window server buffer, graphics were very slow. During the Java implementation project, prior to implementing client-side buffering from Java programs to the AWT server, a similar thing happened. You can see the effects for yourself on the emulator by putting the window server client interface into 'auto-flush' mode, which empties the buffer after each command. Use Ctrl + Alt + Shift + F to turn on auto-flush, and Ctrl + Alt + Shift + G to turn it off again. The difference is noticeable – and bear in mind that your PC's clock is probably many times faster than that of a Symbian OS phone.

Client-side buffering has some consequences in the API. First, the operations that are buffered cannot return a result. Second, you sometimes need to be able to force a buffer flush. With the window server, a `Flush()` function is available for this purpose. `Flush()` is called by UIKON's active scheduler prior to waiting for the next request, so there are relatively few circumstances in which you would need to call it yourself.

The stream store also implements client-side buffering. A write stream uses a buffer that is flushed to the destination file only when it is full, or when you call `RWriteStream::CommitL()`. A read stream uses a buffer, which it prefills from the source file and which it uses until another buffer of source data is needed.

18.2.2 High-level Transactions

If the overheads of a transaction are high, then if possible you should specify the client API so that one transaction does a lot of work.

A good example of this is the capability and setup pattern used by the serial communications and other Symbian OS servers. Rather than providing many getter/setter functions to test and change the state of a communications port – speed, parity, data bits, stop bits, XON/XOFF, and the like – the communications server provides a single `struct` containing all the settings. You use a `Config()` call to get this `struct`;

you can then change values in it, and use `SetConfig()` to send it back to the communications server.

18.2.3   Data Caching

Rather than read a data value from a server every time you need it, it's sometimes a good idea to cache the data value client side. When they need it, your client programs can use this value without a client-server transaction. However, you must ensure that the cached value is updated when necessary.

For example, the Battleships status display shows the GDP protocol in use for the communication session. You can get this protocol from the GSDP server using a simple client API call, but that would require a client-server transaction every time the status view is drawn. It seemed simpler to cache the value in the status view and to update it whenever the client changes the settings.

If the setting is updated by the server outside the client's control, then you need to use another method to update the client-side cache. For example, the system shell displays files in the current folder and highlights any files that are currently open. Files may be deleted and programs opened and closed, outside the shell's control. To keep the display up to date, the shell uses notification APIs in both the file server (which knows about all files) and the window server (which knows about all open GUI applications). The notification API generates an event when a relevant change occurs. The shell then responds to this event by asking the server for updated information. Thus, there is no need for the shell to poll (which would waste power and time), or to require the user to refresh the display (which would be confusing).

On the emulator, you can alter a file visible to the shell by using the Windows shell rather than that of Symbian OS. If you want Symbian OS to pick up this change and notify its shell, press and release F5. This simulates opening and closing the door on a real machine for a removable media device. In turn, this causes the file server to notify the shell of general changes and the shell does a complete rescan.

18.2.4   Related Servers in the Same Process

Two key aspects of a client-server transaction are cheaper if the client and server are on different threads in the same process:

- Context switching is much cheaper, because no changes to the MMU are required.
- Interthread data transfer is cheaper, because there is no need to map the relevant client's data space into the server's address space.

These effects are particularly relevant in the hardware implementations used by most current available Symbian OS phones. Like the Psion Series 5, they use a postcache MMU in order to save power. This means that the entire cache has to be flushed after a context switch, which slows things down.

There are several ways to ensure that a client-server call avoids these overheads.

Related servers can be run in the same process. Symbian OS v7.0, for example, runs the serial communications server, sockets server, and telephony server all in a single process. This process is isolated from clients so that system integrity is maintained. But the many interactions between these servers are of much lighter weight than they would be if they were all in different processes.

Symbian OS v7.0 uses **fixed processes** to minimize the effects of context switching between the kernel, the four important server processes, and a client process (such as an application) that might be using them.

Conventional processes are mapped to different addresses, depending on whether they're running or not. Fixed processes and the kernel server are always mapped to the same address. If context is switching between one conventional process, the kernel, and the fixed processes, then no MMU remapping or associated cache flushing needs to take place. This boosts overall system efficiency significantly.

The choice of which servers are fixed is made at the time a ROM is built, so it can vary between devices. However, the four important server processes are the communications server process (including all the communications-related servers), the file server, the window server, and the font and bitmap server (FBS), and these four are normally set to fixed. Figure 18.6 shows the different types of servers and the processes in which they run.

If you write a server for a specialist application, you may be able to run that server in the same process as the application that uses it. It's not always going to be possible, but where it *is* possible, it's easy enough to do so.

18.2.5 Shared Memory

If the techniques outlined above don't deliver the performance you need, you can as a last resort, use shared memory to avoid the transaction model of client-server communication altogether. You have to replace the transaction model with conventional mutex-type synchronization.

The font and bitmap server (FBS) uses this design, in conjunction with a client program and the window server (see Figure 18.7).

If Symbian OS had no FBS, then a client program that wanted to blit a bitmap to the screen would have two options:

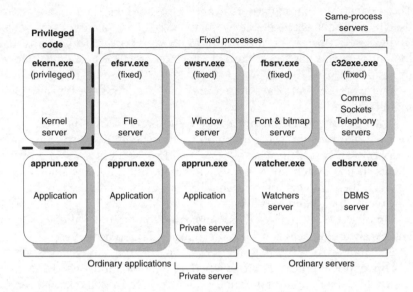

Figure 18.6

- If the bitmap is already in ROM, the client can pass the address of the bitmap to the window server, and the window server can use the BITGDI to blit the bitmap from the ROM to the screen window.

- If the bitmap is created and drawn by the client in its own RAM, the client uses the BITGDI to draw the bitmap and then sends a message to the window server to draw the bitmap. The window server must then copy the bitmap from the client into its own memory and use the BITGDI to blit it to screen. This would involve considerable interthread data transfer from client to window server and also serious space allocation issues inside the window server.

To avoid the waste of both space and time involved in this kind of data transfer, the font and bitmap server mediates access to a shared heap that contains all RAM-based bitmaps. With the FBS, the following take place:

- The usage patterns for ROM-based bitmaps are essentially unchanged.

- To use a RAM-based bitmap, the client requests the FBS create the bitmap in its shared heap. The client uses the BITGDI to draw to the bitmap. The client locks the bitmap before it starts drawing and unlocks it again when it has finished.

- When the client wants the window server to draw the bitmap to the screen, it sends a request specifying the bitmap's FBS-owned handle. The window server then uses this handle to identify the address of the bitmap in its address space (which may be different from the address

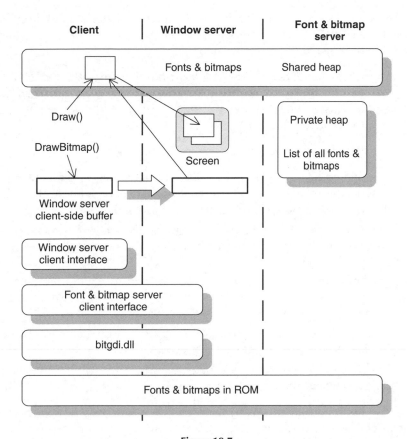

Figure 18.7

in the client's). Finally, the window server uses the BITGDI to blit the bitmap to the screen.

This makes Symbian OS very efficient at handling bitmaps, which in turn contributes significantly to the speed of its graphics.

There is an interesting interaction between the FBS and the window server's client-side buffer. Imagine a client specifies the following sequence of events:

- Create a shared, RAM-loaded bitmap
- Use the BITGDI to initialize the bitmap contents
- Request the window server to draw the bitmap
- Delete the bitmap.

By the time the client's buffer is flushed and executed by the window server (the third step above), the bitmap could have been deleted. To avoid this, the FBS client-side implementation calls the window server client's `Flush()` function (through an interface function) before sending the FBS message to delete the bitmap.

> *As well as handling ROM- and RAM-resident bitmaps, the FBS also handles ROM- and RAM-resident fonts. The precise details are different from those for bitmaps, but the motivations for using a server to share font data are essentially the same.*

The shared memory technique is not without cost:

- The shared heap is mapped to different addresses in different processes. This means that conventional pointers cannot be used within the heap: a handle system has to be used instead.

- The shared heap is not the default heap for either client or server process. This means that the default `operator new()` and `operator delete()` don't work.

- For both the above reasons, objects designed for use in the default user heap cannot be placed onto a shared heap. New classes must be written specially for this purpose.

- You have to use mutex synchronization to control access to objects on the shared heap.

- You have to make sure that things stay in sync when you delete objects that are shared with other servers.

For the font and bitmap server, these costs are low because the shared heap contains large-scale objects (fonts and bitmaps) and because the usage patterns of these objects minimize the difficulties of mutex-based sharing. In addition, the benefits of sharing are very high because of

- the intensity of operations involving bit-blitting,

- the enormous difference the efficient implementation makes to users' perceptions of system efficiency

- the size of the objects that would have to be exchanged using client-server transactions if the shared heap was not available.

Other shared-heap server designs have been implemented on Symbian OS, but in each case the design issues and the cost/benefit analysis have had to be thought through very carefully.

18.3   Servers and Threads

The session between a client and a server is owned by the kernel. The kernel specifies that the session is between the client *thread* or *process* and the server *thread*. This has simple, but profound implications for any program that uses servers:

- Unless sharable sessions are used, client-side resources representing server-side objects may only be used and destroyed by the client thread that created them.
- Server responsiveness to clients is governed by the duration of the longest possible `RunL()` of any active object running on the server thread.

Most of the time, these implications are not onerous. Most Symbian OS clients and servers are single-threaded, and the active object paradigm for event-handling threads is perfectly adequate. In a few cases, though, these implications raise issues for Symbian OS programmers and, as usual, some techniques have been developed to tackle them.

18.3.1 Sharing Client-side Objects between Threads

Many programs written for non-Symbian OS platforms are multithreaded: one thread might be used to handle user input, while another thread handles communication with a sockets protocol. The threads may communicate by means of a file and either thread may wish to update the display. Naturally, the two threads synchronize using mutexes or semaphores. In ER5, this was not possible because servers treated client sessions as being owned by a single client thread. For pure Symbian OS programs, this restriction does not matter. A pure Symbian OS program uses a single thread where other programs would use a process, and active objects where other programs would use threads.

However, if you want to port a program from another operating system, this restriction can cause difficulties. For this reason, versions of Symbian OS v6.x onwards provide sharable sessions that allow a server to provide sessions that can be shared between multiple threads in the client process. These sharable sessions do not allow sharing between processes and servers are not obliged to implement sharable sessions, so consider the functions provided by any server that you use.

If you need to use a server that does not provide sharable sessions, then the one technique that is available to work around this limitation is to implement a private server in the process that will own the session with the external server. Then, each thread in the process can own its own session with the private server. It can be seen that this is more complex to implement than directly using a server that provides sharable sessions.

18.3.2 Multithreading in the Server

A server is implemented as a *single* event-handling thread. As usual, in Symbian OS, the server thread implements event handling by using active objects.

In fact, the server itself is an active object: `CServer::RunL()` is coded to perform first-level handling of incoming requests, which usually results in a `CSharableSession::ServiceL()` call to handle a message. `CServer::NewSessionL()` and `CSession/CSharable-Session` destructors are also called in response to connect and disconnect messages.

The server may contain other active objects for handling input events, timeouts, and so on.

18.3.3   Time-critical Server Performance

For some time-critical applications, we can ask two very specific questions about server performance:

- What is the fastest guaranteed service time required to process a client message?

- What is the fastest guaranteed time needed to respond to an event on some I/O device owned by the server?

By 'fastest guaranteed time', I mean the time that would be required *in the worst possible circumstances*. Often, the service time or response time will be much better than this, but that can't be guaranteed because something else might be happening instead. It only takes a little thought to arrive at the following important conclusion:

> **The fastest guaranteed service time is limited by the duration of the longest-running `RunL()` of any active object in the server's main thread.**

This is because, when a client thread makes a request, the server thread may already be running a `RunL()` for another client request, or for some other activity within the server. The `CServer::RunL()` for the client request cannot preempt a `RunL()` that is already in progress. So the client will have to wait until the current `RunL()` has finished before its request even starts to be handled.

We can easily see that the same applies to responding to external events.

> **The fastest guaranteed response time to an external event is limited by the duration of the longest-running `RunL()` in the thread that drives a device.**

This has important implications for server design. If you are designing a high-performance server, you should not call long-running operations from *any* of your service functions. Furthermore, you should not perform long-running operations from *any* of the RunL()s of any other active object in your server's main thread.

If you need to deliver long-running operations to clients, or to perform long-running operations for internal reasons, you must run them on a different thread. The most obvious way to structure that thread would be as a server – clearly, a low-performance server. You might run the server as a private server in the main server's process. The low-performance server's 'client' API should deliver its long-running functions asynchronously, so that the high-performance server can kick them off quickly and then handle their completion with an active object.

If your server specifies a server-side interface – for instance, one that allows plug-in protocol implementations – and if your server has any critical response time requirements, you should be very clear about the responsibilities of anyone implementing that interface.

Obviously, the kernel thread has the highest priority of any thread in the system. Device driver code – in interrupt service routines, device drivers, or delayed function calls (DFCs) – can block any other code. So device driver code, which is a special case of a plug-in API, should be particularly quick.

18.3.4 Thread Priorities

If you are interested in server performance, then you are probably also interested in thread priorities. The basic rule is simple.

> **A server with a shorter guaranteed response time should have a higher thread priority than a server with a longer guaranteed response time.**

Otherwise, a long-running RunL() in a low-performance server with a mistakenly high thread priority will block what might otherwise have been a short-running RunL() in a high-performance server whose priority has been sensibly chosen, but is lower than that of the misbehaved low-performance server.

> **Don't kid yourself about response times. Awarding a server a high thread priority doesn't necessarily give it a short guaranteed response time.**

To get short guaranteed response time, you must analyze *all* the RunL()s in your server's main thread. If any of them is longer than is justified by the thread priority you have awarded your server, then it's effectively a low-performance server and you're compromising the ability of any lower-priority server to deliver on its response time promise. That's antisocial behavior: don't do it.

Another rule about server thread priorities can be deduced from the rule above:

> **All system servers should have a priority that is higher than all applications.**

This is because an application might include arbitrarily long-running code. If it was allowed to run at higher priority than any server, the application would block the server from servicing any of its clients.

18.4   The Client-server APIs

It's now time for a brief review of the client-server APIs. I'll highlight the main features here, and in the next chapter, I'll demonstrate how they're used in practice. The main classes are all defined in `e32std.h` and `e32base.h`. They are:

Class	Purpose
`RThread`	A thread
`RHandleBase, RSessionBase`	Client-side session classes
`CServer, CSession, CSharableSession, RMessage`	Server-side server, session, and message classes.
`TPckg<T>, TPckgC<T>, TPckgBuf<T>`	Type-safe buffer and pointer descriptors for any kind of data.
`RSubSessionBase`	Client-side subsession.
`CObject, CObjectCon, CObjectIx, CObjectConIx`	Server-side classes related to subsessions.

We will look at the main features of the most important classes (see Figure 18.8).

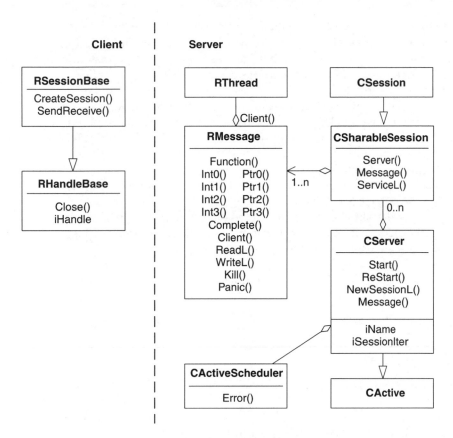

Figure 18.8

18.4.1 Thread Basics

Threads are basic for client-server programming because the client and the server are separate threads and must be able to refer to each other.

The `RThread` class enables one thread to create or refer to another, to manipulate the other thread, and to transfer data between itself and the other thread. `RThread`'s default constructor is set up to create an `RThread` object for your own thread:

```
RThread me;
```

Usually, however, a program uses an `RThread` to refer to another thread. In the context of client-server programming, the server uses `RThreads` to refer to its clients. The server's client interface may use an `RThread` to create an instance of the server. It isn't usually necessary for client code to use `RThreads` directly.

The main things you can do with an `RThread` are as follows:

- `Create()` a new thread. You specify a function in which execution is to begin and parameters to that function. The thread is created in suspended state, and you have to `Resume()` it to start it executing.
- `Open()` a handle to an existing thread.
- `Kill()` and `Panic()` the thread. Killing is the normal way to end another thread; panicking indicates that the thread had a programming error. Servers use this to panic their client when the client passes a bad request.
- Set and query the thread's priority.
- Cause an asynchronous request issued by that thread to complete, using `RequestComplete()`.
- `ReadL()` data from a descriptor in the other thread's address space, or `WriteL()` data to a descriptor in the other thread's address space.

The full `RThread` API is well documented in the C++ SDK.

Many `RThread` functions are mirrored in the `User` class's API. Functions such as `Kill()`, `Panic()`, and `RequestComplete()` affect the currently running thread, so that

```
User::Kill(KErrNone);
```

is equivalent to

```
RThread me;
me.Kill(KErrNone);
```

18.4.2 Interthread Data Transfer and the Package Classes

In the Symbian OS documentation, interthread data transfer is referred to as interthread communication or ITC. I have used 'data transfer' rather than 'communication' here, because in reality communication is about much more than just data transfer.

All transfer of data between threads is based on six member functions of `RThread` (found in `E32std.h`):

```
TInt GetDesLength(const TAny* aPtr) const;
TInt GetDesMaxLength(const TAny* aPtr) const;
void ReadL(const TAny* aPtr, TDes8& aDes, TInt anOffset) const;
void ReadL(const TAny* aPtr, TDes16& aDes, TInt anOffset) const;
void WriteL(const TAny* aPtr, const TDesC8& aDes, TInt anOffset) const;
void WriteL(const TAny* aPtr, const TDesC16& aDes, TInt anOffset) const;
```

Interthread data transfer is performed from data buffers identified by a descriptor in both the currently running thread and the 'other' thread identified by the `RThread` object. The descriptor in the currently running thread is identified by a conventional descriptor reference (such as `const TDesC8&` for an 8-bit descriptor) from which an interthread write will take data. The descriptor in the other thread is identified by an address, passed as a `const TAny*`, which is the address of a descriptor *in the other thread's address space* (the address will probably have been passed from client to server, as one of the four 32-bit message parameters). Bearing this in mind

- `GetDesLength(const TAny*)` returns the `Length()` of the descriptor referred to in the other thread's address space.

- `GetMaxDesLength(const TAny*)` returns the `MaxLength()` of the descriptor referred to in the other thread's address space.

- `ReadL(const TAny*, TDes8&, TInt)` reads data from the other thread into a descriptor in this thread. Data is transferred from the `anOffset`'th byte of the source. The amount of data transferred is the smaller of the number of bytes between `anOffset` and `Get-DesLength()` of the source descriptor in the other thread, and the `MaxLength()` of the destination descriptor in the current thread. There is also a 16-bit version of `ReadL()`.

- `WriteL(const TAny*, const TDesc8&, TInt)` writes data from this thread into a descriptor in the other thread. A 16-bit variant is also provided.

If any of these functions is called with a `TAny*` that is not the address of a valid descriptor in the other thread, then a `KErrBadDescriptor` error results. `GetMaxDesLength()` and `GetDesLength()` return this as their result – you can distinguish it from a true descriptor length, because all Symbian OS error codes are negative. `ReadL()` and `WriteL()` leave with this as their error code.

> **A bad descriptor almost certainly indicates a bad client program. Any server detecting `KErrBadDescriptor` should panic the offending client.**

All interthread data transfer uses descriptors. This is appropriate, because descriptors contain an address and a length. If you wanted to transfer a floating-point number from a client to a server, you could use the following client code:

```
TInt p[4];                    // Message parameter array
TReal x = 3.1415926535;       // A 64-bit quantity
```

```
TPtrC8 xPtrC(&x, sizeof(x)); // Address and length in a descriptor
p[0] = &xPtrC;              // Pass address of descriptor as zeroth message
                               parameter
```

And this code on the server side:

```
TReal x;
TPtr8 xPtr(&x, 0, sizeof(x));               // Address, length, max-length
Client().ReadL(Message.Ptr0(), &xPtr, 0);  // Transfer data
```

But this code isn't very type-safe, and it's not exactly straightforward either. The package classes offer a type-safe alternative. On the client side:

```
TInt p[4];                      // Message parameter array
TReal x = 3.1415926535;         // A 64-bit quantity
TPckgC<TReal> xPackage(x);      // Package into a descriptor
p[0] = &xPackage;               // Address of package
```

And the server side:

```
TReal x;
TPckg<TReal> xPackage(x);                       // Package it up
Client().ReadL(Message.Ptr0(), &xPackage, 0);   // Transfer data
```

The `TPckg<T>` and `TPckgC<T>` classes are simply type-safe, thin template wrappers around `TPtr8` and `TPtrC8`. A third package class, `TPckgBuf<T>`, performs a similar function for `TBuf8<sizeof(T)>`.

We can picture the APIs related to interthread data transfer, and other thread functions, as shown in Figure 18.9.

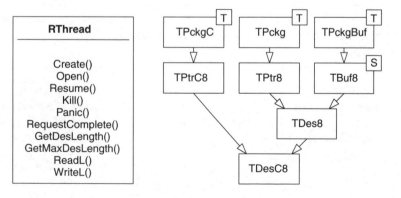

Figure 18.9

18.4.3   Client-side Objects

The main client-side object is `RSessionBase`, derived from `RHandle-Base`. As a server provider, your client interface should include a class derived from `RSessionBase` that handles communications from client to server.

A client-side handle for server-side (and kernel-side) objects

`RHandleBase`, which is defined in `e32std.h`, is the base class for client-side objects that refer to a number of kernel-side objects, and also for `RSessionBase`, which refers to server-side objects.

For our purposes, the only relevant aspects of the `RHandleBase` class declaration are

```
class RHandleBase
    {
public:
    ...
    inline RHandleBase();
    IMPORT_C void Close();
    ...
    inline TInt Handle() const;
    ...
protected:
    TInt iHandle;
    };
```

In brief, an `RHandleBase` has a 32-bit handle that's used by the client to refer to a particular session with the server. From the server's perspective, the client's thread ID, combined with this handle, uniquely identifies a server-side session.

RSessionBase – client-side session

`RSessionBase` is the base class for any client-side session with a server. Here are the relevant parts of its declaration:

```
class RSessionBase : public RHandleBase
    {
public:
    enum TAttachMode {EExplicitAttach, EAutoAttach};
    IMPORT_C TInt Share(TAttachMode aAttachMode=EExplicitAttach);
    IMPORT_C TInt Attach() const;
    ...
protected:
    IMPORT_C TInt CreateSession(
```

```
        const TDesC& aServer,
        const TVersion& aVersion) ;
   IMPORT_C TInt CreateSession(
        const TDesC& aServer,
        const TVersion& aVersion,
        TInt aMessageSlots) ;
   ...
   IMPORT_C void SendReceive(TInt aFunction, TAny* aPtr,
                             TRequestStatus& aStatus) const;
   IMPORT_C TInt SendReceive(TInt aFunction, TAny* aPtr) const;
   IMPORT_C TInt Send(TInt aFunction,TAny* aPtr) const;
   ...
   };
```

You use one of the versions of `CreateSession()` to create a new session with the server – referring to it by name. This allocates a handle in the base `RHandleBase` class. Normally, your derived client class will call `CreateSession()` from a friendlier client function, such as `Open()`, `Connect()`, and so on, as is done with the `RFsSession` class.

Use `Close()`, defined in the `RHandleBase` class, to close the session. After being closed, the handle is set to zero, so that the object can no longer be used. The handle is also set to zero by `RHandleBase`'s inline constructor, prior to connecting the session.

Messages are sent using `SendReceive()` or `Send()`. The synchronous form of `SendReceive()` (the one that returns a `TInt`) is expected to complete immediately by means of a synchronous `ServiceL()` function in the server. The asynchronous form may complete some time later. The `Send()` routine sends a blind message to the server and no reply is expected.

Both forms of `SendReceive()` take a 32-bit `TInt` argument called `aFunction` that specifies the request code for the message. The `TAny* aPtr` argument should point to four 32-bit words containing pointers or 32-bit integers that carry the parameters of the message.

The synchronous form of `SendReceive()` is implemented (in private Symbian OS code) as

```
      TInt SendReceive(TInt aRequest, TAny* aPtr)
        {
        TRequestStatus status;
        SendReceive(aRequest, aPtr, status);
        User::WaitForRequest(status);
        return status.Int();
        }
```

The asynchronous version is the more fundamental. It causes a message to be sent containing

- the request code

- four 32-bit parameters
- the client's thread ID
- the handle from the `RHandleBase`
- the address in the client's process of the `TRequestStatus` to be used to complete the message.

This is all wrapped up into a single message. When the server completes the message, it posts the 32-bit result back to the client's request status.

We can now explain the other parameters to the variants of `Create-Session()`.

The `TInt aMessageSlots` parameter tells the kernel how many messages to reserve for this client-server session. If the session supported only synchronous function calls, only one message slot could ever be used. If the session supports asynchronous requests, then one additional message is needed per asynchronous request that could possibly be outstanding, in most cases, that is, one or two – very rarely are any more needed. If the version of `CreateSession()` that omits the `aMessageSlots` parameter is used, then the session will use message slots from a pool held by the kernel. This allows more efficient use of resources overall. If your session can use a large number of message slots (i.e. it can have a large number of concurrent asynchronous operations), then you should allocate your own message slots to avoid taking too many of the common message slots.

The `TVersion` contains three version numbers:

- Major, as in 7 for Symbian OS v7.0
- Minor, indicating a minor feature release
- Build, indicating the build number – effectively a maintenance level.

The `TVersion` is intended to ensure that the client API and the server implementation, which may be provided in separate DLLs, are at compatible levels.

> Using `TVersion` is probably no more or less effective than using a host of other disciplines to make sure these programs are in sync. In the GSDP sample code, the same DLL is used to implement client interface and server, so we are happy to pass a `TVersion(0, 0, 0)` in our `CreateSession()`.

The `Share()` and `Attach()` functions are used to manage shared sessions. If the server to which you are connecting does not use sharable sessions (i.e. sessions that can be shared between multiple threads in the client process), then these functions are irrelevant. If you are using a server that supports sharable sessions, then the first session is created as normal and then `RSessionBase::Share()` is called on that session to make it sharable. If the `EExplicitAttach` value is used for the `aAttachMode` parameter, then any other thread that wants to share the session will need to call `RSessionBase::Attach()` on the session. If the `EAutoAttach` value is used, then all threads are automatically attached to the session.

RSubSessionBase – *client-side sub-session*

`RSubSessionBase` is the base class for a client-side subsession. Here are the relevant parts of its declaration:

```
class RSubSessionBase
    {
public:
    inline TInt SubSessionHandle() const;
protected:
    inline RSubSessionBase();
    inline RSessionBase& Session();
    ...
    IMPORT_C TInt CreateSubSession(
        RSessionBase& aSession,
        TInt aFunction,
        const TAny* aPtr);
    IMPORT_C void CloseSubSession(TInt aFunction);
    ...
    IMPORT_C TInt Send(TInt aFunction,const TAny* aPtr) const;
    IMPORT_C void SendReceive(
        TInt aFunction,
        const TAny* aPtr,
        TRequestStatus& aStatus) const;
    IMPORT_C TInt SendReceive(TInt aFunction,const TAny* aPtr) const;
    ...
    };
```

An `RSubSession` object is created by calling `CreateSubSession()` with references to an existing `RSessionBase` object and a function to perform. Once the subsession has been created, the normal `SendReceive()` and `Send()` functions can be used to communicate with the server. When the subsession has been finished with, the `CloseSubSession()` function should be called. If access to the `RSessionBase` object is required, then the `Session()` function can be used.

18.4.4 Server-side Objects

The three main server-side objects are `CServer`, the base class for the entire server, `CSharableSession`, the base class for a server-side object, and `RMessage`, which contains the message sent from a client.

There is just one `CServer` object per server, but there are as many `CSession`/`CSharableSession` objects as there are `RSessionBase` objects currently in session with this server. There is one server-side `RMessage` object for each outstanding request: that means at most one `RMessage` for a request being handled synchronously from one client, and any number of `RMessages` for requests being handled asynchronously.

CServer – a server

`CServer` is the active object that fields messages from all potential clients and channels them to the right `CSession`/`CSharableSession` object to be interpreted and executed. Here are the relevant parts of the `CServer` declaration:

```
class CServer : public CActive
    {
protected:
    enum TserverType {EUnsharableSessions,ESharableSessions};
    ...
public:
    IMPORT_C ~CServer() =0;
    IMPORT_C TInt Start(const TDesC& aName);
    IMPORT_C void StartL(const TDesC& aName);
    IMPORT_C void ReStart();
    inline const RMessage& Message() const;
protected:
    IMPORT CServer(Tint aPriority,TServerType aType=EUnsharableSessions);
    IMPORT_C void DoCancel();
    IMPORT_C void RunL();
private:
    virtual CSharableSession* NewSessionL(const TVersion& aVersion)
      const = 0;
    ...
protected:
    TSessionControl iControl;
    HBufC* iName;
private:
    const TServerType iSessionType;
    RServer iServer;
    TDblQue<CSharableSession> iSessionQ;
protected:
    TDblQueIter<CSharableSession> iSessionIter;
    };
```

The server *is-a* active object. It issues a request to the kernel for a message from any client. Its `RunL()` function (which should really be `private`) then handles the message, usually by finding the appropriate session and calling its `ServiceL()` function.

Bootstrap code for a server should create an active scheduler, a cleanup stack, and a server object; create a name for the server (stored in `iName`), and then issue `Start()` (or `StartL()`) to cause the server to issue its first request. The server name must be specified by any client wishing to connect to the server, as the `aServer` parameter to `RSessionBase::CreateSession()`.

The server's `RunL()` function renews the request to the kernel automatically. However, it *doesn't* renew the request if (say) the session's `ServiceL()` function leaves. You have to handle this from the active scheduler's `Error()` function, and issue `ReStart()` to renew the request.

When the server handles a connect message, it invokes `NewSessionL()` to create a new server-side `CSharableSession`-derived object. Oddly this function is `const`: you usually need to cast away `const`-ness when you implement this function in order to be able to increment usage counts and do other housekeeping in your derived server class.

When the server handles a disconnect message, it simply deletes the affected session, which also causes its C++ destructor to be invoked.

Unlike in earlier versions of Symbian OS, the server does not store the message – it is available to the session's `ServiceL()` routine and that must store it if necessary.

You can iterate through the sessions owned by the server, using the protected `iSessionIter` member.

`CSession` and `CSharableSession` – a server-side session

In earlier versions of Symbian OS, the `CSession` class was used on the server side to encapsulate a client-side session. The `CSession` class was thread-specific – that is, it was owned by one thread on the client side. Symbian OS v6.x introduced sharable sessions with the `CSharableSession` class. A `CSharableSession` object can be shared between multiple threads in the same process on the client side. When you design your server, you can choose whether to provide sharable or unsharable sessions. Classes derived directly from `CSharableSession` rather than from `CSession` will not be able to make use of the thread-specific functions in `CSession` to transfer data between the server and the client thread so it may be easier to derive from `CSession` if you do not need to share a session between threads. Note that a sharable session requires the client-side session to call `Share()` and possibly `Attach()` to actually be shared – these functions are described under the `RSessionBase` class details.

The `CSession` class derives from `CSharableSession` and whichever you choose forms the base class for the server side end of a session. You implement its `ServiceL()` function to interpret and handle client requests. These classes also provide many convenience functions for

accessing the client and the current message. The relevant parts of
CSharableSession and CSession are

```
Class CSharableSession : public CBase
    {
    friend class CServer;
public:
    IMPORT_C ~CSharableSession() =0;
    IMPORT_C virtual void CreateL(const Cserver& aServer);
    ...
    inline const CServer* Server() const;
    inline const RMessage& Message() const;
    ...
    virtual void ServiceL(const RMessage& aMessage) =0;
    ...
private:
    TInt iResourceCountMark;
    TDblQueLink iLink;
    const CServer* iServer;
    };
```

```
class CSession : public CSharableSession
    {
public:
    IMPORT_C ~CSession() =0;
    ...
    IMPORT_C void ReadL(const TAny* aPtr,TDes8& aDes) const;
    IMPORT_C void ReadL(const TAny* aPtr,TDes8& aDes,TInt anOffset) const;
    IMPORT_C void ReadL(const TAny* aPtr,TDes16& aDes) const;
    IMPORT_C void ReadL(const TAny* aPtr,TDes16& aDes,TInt anOffset)
      const;
    IMPORT_C void WriteL(const TAny* aPtr,const TDesC8& aDes) const;
    IMPORT_C void WriteL(const TAny* aPtr,const TDesC8& aDes,TInt
      anOffset)const;
    IMPORT_C void WriteL(const TAny* aPtr,const TDesC16& aDes) const;
    IMPORT_C void WriteL(const TAny* aPtr,const TDesC16& aDes,TInt
      anOffset)const;
    ...
protected:
    IMPORT_C CSession(RThread aClient);
    private:
        RThread iClient;
    };
```

Your derived class may include any kind of C++ constructor and second-
phase constructor in order to initialize the session properly, with the
proviso that if it is derived from CSession, the derived class's C++
constructor should pass the client thread's RThread to the protected
CSession constructor.

The CSharableSession class provides utility functions to access the
server and the current message and the CSession class provides utility
functions to read and write data between server and client address spaces.

You should implement ServiceL() to handle a message from the client. Interpret the request code and parameters in the RMessage from the client; when ServiceL() is complete, use aMessage.Complete() to pass a result back to the client.

Handling asynchronous client requests is more complex than handling simple, synchronous, requests. The aMessage parameter to ServiceL() must be stored because the next time ServiceL() is called it will be different. Each asynchronous request supported by the server will require an active object to service it. When the asynchronous operation is completed, the stored RMessage must be retrieved and Complete() called on it. Note that if the active object request function returns an error, the client request should be completed at that point.

The other functions provided by the CSharableSession API are either for specialist use or deprecated, so I haven't described them here.

RMessage – a server-side message

When a client issues a message, it arrives at the server as an RMessage object. You can use the functions of RMessage to access the request code and message parameters. You also get a Client() function that returns an RThread for the client, and convenience functions to read from, write to, panic, terminate, or kill the client thread. You call Complete() to complete the handling of a message.

Here is the complete definition of RMessage:

```
class RMessage
    {
    friend class CServer;
public:
    enum TSessionMessages  EConnect = -1, EDisConnect = -2 ;
public:
    IMPORT_C RMessage();
    IMPORT_C RMessage(const RMessage& aMessage);
    IMPORT_C RMessage& operator=(const RMessage& aMessage);
    IMPORT_C void Complete(TInt aReason) const;
    IMPORT_C void ReadL(const TAny* aPtr, TDes8& aDes) const;
    IMPORT_C void ReadL(const TAny* aPtr, TDes8& aDes, TInt anOffset)
      const;
    IMPORT_C void ReadL(const TAny* aPtr, TDes16& aDes) const;
    IMPORT_C void ReadL(const TAny* aPtr, TDes16& aDes, TInt anOffset)
      const;
    IMPORT_C void WriteL(const TAny* aPtr, const TDesC8& aDes) const;
    IMPORT_C void WriteL(const TAny* aPtr, const TDesC8& aDes,
                         TInt anOffset) const;
    IMPORT_C void WriteL(const TAny* aPtr, const TDesC16& aDes) const;
    IMPORT_C void WriteL(const TAny* aPtr, const TDesC16& aDes,
                         TInt anOffset) const;
```

```
    IMPORT_C void Panic(const TDesC& aCategory, TInt aReason) const;
    IMPORT_C void Kill(TInt aReason) const;
    IMPORT_C void Terminate(TInt aReason) const;
    inline TInt Function() const;
    inline const RThread& Client() const;
    inline TInt Int0() const;
    inline TInt Int1() const;
    inline TInt Int2() const;
    inline TInt Int3() const;
    inline const TAny* Ptr0() const;
    inline const TAny* Ptr1() const;
    inline const TAny* Ptr2() const;
    inline const TAny* Ptr3() const;
    inline const RMessagePtr MessagePtr() const;
protected:
    TInt iFunction;
    TInt iArgs[KMaxMessageArguments];
    RThread iClient;
    const TAny* iSessionPtr;
    const RMessage* iMessagePtr;
    };
```

The following are the most important functions you use when writing a server:

- `Client()` returns a `const RThread&` representing the client thread that sent the message.

- `Function()` returns the 32-bit request code for the function requested by the client.

- `Int0()`, `Int1()`, `Int2()`, and `Int3()` access the four message parameters, interpreting them as `TInt`s. `Ptr0()`, `Ptr1()`, `Ptr2()`, and `Ptr3()` access the four parameters as well, interpreting them as `TAny*` pointers.

- When you have handled the function, you convey the 32-bit result to the client using `Complete()`. This causes the client's request status for the message to be posted with the completion code. The kernel also releases the kernel-side message slot.

- `RMessage`'s `ReadL()` and `WriteL()` functions are convenience wrappers for corresponding `RThread` functions: they read and write from client's memory. Use these to communicate data that is too large to communicate within the operating code, parameters, and completion code. Likewise, `RMessage`'s `Panic()`, `Kill()`, and `Terminate()` functions are convenience wrappers for corresponding `RThread` functions on the client thread.

`RMessage` is not intended for derivation, so its `protected` members should really be `private`.

18.5   Summary

In this chapter, I've

- outlined the basic workings of servers,
- given some hints and tips for optimizing their performance,
- reviewed the API elements associated with servers.

In the next chapter, we'll get practical: with the GSDP server, we can see all these facets of the client-server framework working together.

19

The GSDP Server

In the previous two chapters, I've described the Symbian OS active object and client-server frameworks – the foundations for system programming. I'm now in a position to describe the Game Session Datagram Protocol (GSDP) server we implemented for sharing GDP datagrams among multiple client games on a single Symbian OS phone. Along the way, we will encounter all the most important practical techniques needed to program a Symbian OS server.

We've already seen that the purpose of the GSDP server is to allow GDP drivers to be shared between multiple games on a single Symbian OS phone. To achieve this, the GSDP server

- runs the GDP implementations on behalf of all games on a Symbian OS phone;
- associates an origin address, a destination address, a destination port number, and a game protocol with each client session so that from the client's perspective, GSDP is a session protocol rather than a stateless datagram protocol;
- when sending a packet selects the right GDP implementation and adds the correct port numbers and game protocol ID into the packet's datagram content;
- when receiving a packet uses the port number and protocol ID to select the client that should receive it.

This functionality is reflected in both the server's client interface and in its internal structure.

The GSDP server presented here has its own specific task to perform, but in many ways it's typical, and I'll describe it in sufficient detail here so that you can use it with confidence as a basis for implementing your own servers.

Symbian OS C++ for Mobile Phones. Edited by Richard Harrison
© 2003 John Wiley & Sons, Ltd   ISBN: 0-470-85611-4

19.1   Software Structure

You can find the source code for the GSDP server and its client interface in \scmp\gsdp\. Here's the structure of the server and its interfaces in Figure 19.1:

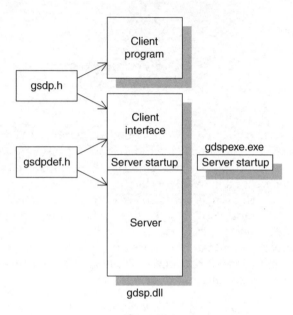

Figure 19.1

The server and its client interface are delivered in a single DLL, gsdp.dll. The client interface consists of a header file, gsdp.h, which we'll look at shortly, and corresponding functions exported from the DLL. Another header file, gsdpdef.h, includes definitions that tie the client interface and the server together – mainly the request codes passed in the various messages. All other header files are private either to the client alone or to the server alone.

The GSDP server is a **transient server**: it's started when a client needs it, and terminates itself when no more clients need it. The startup code launches the server as a new thread under the WINS emulator, or as a new process on target machines. The startup code for the emulator is *also* contained in gsdp.dll; for MARM, it's a separate, very small, .exe.

This design minimizes the differences between the two platforms. It follows the design of the Symbian OS DBMS server, which was written for Symbian OS v5. Among the useful aspects of this single-DLL design are that it eases debugging under the emulator because there's only a single project, and it ensures that client and server code are in sync because they're delivered in the same DLL.

Many Symbian OS servers use a separate DLL for the client interface, and a relatively large program for the server, delivered as a .exe for target machines, and a DLL for the emulator.

One reason that may compel server authors to use a .exe is if they cannot do anything to eliminate writable static data in the server code, perhaps because they are porting code.

19.2 The Client Interface

You can see a UML diagram representing the client interface in Appendix 3. Here is the code from gsdp.h:

```
#ifndef __GSDP_H
#define __GSDP_H


#include <e32std.h>


// game session datagram protocol interface specification


class MGsdpPacketHandler
    {
public:
    virtual void GsdpHandleL(const TDesC8& aData)=0;
    };


const TInt KMaxGdpDisplayName = 0x20;


class TGdpProtocolInfo
    {
public:
    TUid iUid;
    TBuf<KMaxGdpDisplayName> iDisplayName;
    TBool iNetworked;
    };


class CGsdpReceiveHandler;


class RGsdpSession : public RSessionBase
    {
public:
    // construct
    inline RGsdpSession() : iHandler(0) {};
    // open/close
```

```
        IMPORT_C void ConnectL(MGsdpPacketHandler& aHandler);
        IMPORT_C void Close();
        // Query supported protocols
        IMPORT_C TInt CountGdpProtocols() const;
             ///< Returns count of protocols
        /// Retrieve info for particular protocol
        IMPORT_C TInt GetGdpProtocolInfo(TInt aProto,
            TGdpProtocolInfo& aInfo) const;

     // load and get GDP protocol
        IMPORT_C void SetGdpProtocolL(TUid aProtocol);
        IMPORT_C TUid GetGdpProtocol() const;
        IMPORT_C TBool GdpIsNetworked() const;

        // game protocol
        IMPORT_C void SetGameProtocol(TUint32 aProtocol);
        IMPORT_C TUint32 GetGameProtocol() const;
        // set and get my address and port
        IMPORT_C void SetMyPort(TUint32 aPort);
        IMPORT_C TUint32 GetMyPort() const;
        IMPORT_C TUint32 AllocMyNextPort();
        // set and get other address and port
        IMPORT_C void SetOtherAddress(const TDesC& aAddress);
        IMPORT_C void GetOtherAddress(TDes& aAddress) const;
        IMPORT_C void SetOtherPort(TUint32 aPort);
        IMPORT_C TUint32 GetOtherPort() const;
        // main protocol functions
        IMPORT_C void Listen();
        IMPORT_C void StopListening();
        IMPORT_C void Send(const TDesC8& aData);
        // initiate receive-all for "pull" protocols
        IMPORT_C void ReceiveAll() const;
private:
friend class CGsdpReceiveHandler;
        void Receive(TDes8& aBuffer, TRequestStatus& aStatus);
        void CancelReceive();
        CGsdpReceiveHandler* iHandler;
        };

const TInt KMaxGsdpAddress=40;
const TInt KMaxGsdpData=100;

#endif
```

The interface follows the communications stack pattern that we've talked about before; requests are passed down the stack using functions of the RGsdpSession class, while data is passed up the stack by calling MGsdpPacketHandler::GsdpHandleL().

The client interface uses active objects, but hides them from the interface, so that:

- as a client, you do not have to write a derived active object class – you just derive from `MGsdpPacketHandler` and implement `GsdpHandleL()`,

- as a client, you can issue `Listen()` to receive any number of packets and `StopListening()` to stop receiving them.

Internal to the client interface there is a derived active object class – `CGsdpReceiveHandler` that maintains the outstanding receive request, calling `RGsdpSession`'s private, nonexported `Receive()` and `CancelReceive()` functions in order to do so.

19.2.1 Message-passing Functions

The majority of the client interface consists of message-passing functions that compose the parameters of the client interface function into a message, send the message, receive the response, and return the result to the client.

Here is a simple example:

```
EXPORT_C void RGsdpSession::SetGameProtocol(TUint32 aProtocol)
    {
    TInt p[KMaxMessageArguments];
    p[0] = (TInt)aProtocol;
    SendReceive(EGsdpReqSetGameProtocol, p);
    }
```

The basic idea is to `SendReceive()` the request code, `EGsdpReqSetGameProtocol`, along with as many of the four 32-bit parameters as you actually need. In this case, we only need one – the game protocol.

`EGsdpReqSetGameProtocol` is defined symbolically in `gsdpdef.h`. This header file is included in the server so that the same constants are used when interpreting the request.

There is no client-side equivalent of `RMessage`. You have to allocate your own array of integers, store as many of the parameters as necessary, and then pass the array to `SendReceive()`. The number of message slots is heavily built in to the Symbian OS architecture, but it is still better to use a symbolic constant. In this case, the constant is KMaxMessageArguments from e32std.h.

The corresponding getter function is

```
EXPORT_C TUint32 RGsdpSession::GetGameProtocol() const
    {
    return SendReceive(EGsdpReqGetGameProtocol, 0);
    }
```

In this case, there's nothing to send, so we pass 0 to indicate that there's no message parameter array. The kernel will fill out the four parameters with undefined values – but this saves you from having to waste code on them in your client interface.

You, therefore, have two choices about the message parameter array: if your request doesn't need any parameters, you don't need one, and can pass 0 (null) to SendReceive(); if you need any parameters, you must allocate an entire message array with all four slots, even if you don't need them all. This is because the kernel will copy all the values out of the array when creating the message.

This time, we return a result – the game protocol ID. We simply return the 32-bit result that was passed back by the server as the return code of the message to the client.

Incidentally, it's fine to pass any 32-bit quantity back as the result of a synchronous message. The same pattern cannot be used for asynchronous messages because the active scheduler uses the special value KRequestPending to indicate that a request has not completed. So, if an asynchronous request completed with a value that just happened to be KRequestPending (0x80000001), the active scheduler would panic because of a stray signal. An asynchronous version of the same function would have to use an interthread write to pass back any numeric value if there is the remotest possibility that the value might ever be KRequest Pending.

In some cases, the return value from SendReceive() is used as a genuine error code, like this:

```
EXPORT_C TInt RGsdpSession::SetGdpProtocol(TUid aProtocol)
    {
    TInt p[KMaxMessageArguments];
    p[0]= aProtocol.iUid;
    return SendReceive(EGsdpReqSetGdpProtocol, p);
    }
```

For setting and getting the other machine's address, we have to pass the address of a descriptor:

```
EXPORT_C void RGsdpSession::SetOtherAddress(const TDesC8& aAddress)
    {
    TInt p[KMaxMessageArguments];
    p[0] = (TInt)&aAddress;
    SendReceive(EGsdpReqSetOtherAddress, p);
    }
```

```
EXPORT_C void RGsdpSession::GetOtherAddress(TDes8& aAddress) const
    {
    TInt p[KMaxMessageArguments];
    p[0] = (TInt)&aAddress;
    SendReceive(EGsdpReqGetOtherAddress, p);
    }
```

The setter function takes a const TDesC8& parameter and the getter a TDes8& so that, from the client's perspective, the correct type-safe objects are passed. In either case, however, the client interface code simply passes a pointer to the descriptor as a message parameter. These pointers will eventually be used in calls to RThread::ReadL() and RThread::WriteL(), which we saw in the previous chapter. The exact format of the addressing information is protocol specific, for example, a telephone number for SMS and a device address for Bluetooth. That's why I use an 8-bit descriptor to pass this information as a binary blob rather than as text. It's then up to the protocol implementation to interpret this appropriately.

Coding the message-passing functions is easy enough, if a little tedious for large client interface APIs. Inevitably, in practice, it involves a lot of copying and tweaking. Be very careful that you tweak *everything* you're supposed to. It's easy to copy to a new function name, change the parameter names a little, and then forget to change the request code. The results will be puzzling.

The best way to prevent this sort of error is to first write a test for the new function that assumes that it will work. Then run the test before writing the server-side code; the result should be a failure. If not, there is a good chance that the request code is referring to the wrong function. You can then write the server-side code to implement the function, adding new test cases as you go. When all the tests pass, you can be sure that you've finished implementing that feature.

19.2.2 Listening and Receiving

The client interface for receiving data consists of two functions:

```
IMPORT_C void Listen();
IMPORT_C void StopListening();
```

RGsdpSession::Listen() makes the client interface start maintaining an outstanding receive request, using an outstanding-request pattern active object, such as the one we saw in Chapter 17. StopListening() tells the client interface to stop maintaining that request.

The corresponding active object is declared in the private header file gsdpclient.h as

```
class CGsdpReceiveHandler : public CActive
    {
public:
    // Construct/destruct
    CGsdpReceiveHandler(MGsdpPacketHandler& aHandler,
        RGsdpSession& aSession);
    ~CGsdpReceiveHandler();

    // Operation
    void Receive();
private:
    // From CActive
    void RunL();
    void DoCancel();
private:
    RGsdpSession& iSession;
    MGsdpPacketHandler& iHandler;
    TBuf8 <KMaxGsdpData> iBuffer;
    };
```

RGsdpSession::Listen() calls CGsdpReceiveHandler::
Receive(), which is implemented as

```
void CGsdpReceiveHandler::Receive()
    {
    iSession.Receive(iBuffer, iStatus);
    SetActive();
    }
```

In other words, the receive handler issues the initial request to receive
data into its own receive buffer. When this first receive completes, the
receive handler's RunL() is invoked:

```
void CGsdpReceiveHandler::RunL()
    {
    iHandler.GsdpHandleL(iBuffer);
    // Initiate next receive
    Receive();
    }
```

This calls the client's handler to pass the received data up the stack,
and then issues another receive. If the client issues StopListening(),
the client interface Cancel()s the receive handler active object, which
in turn causes DoCancel() to be called:

```
void CGsdpReceiveHandler::DoCancel()
    {
    iSession.CancelReceive();
    }
```

This simply passes on the cancel to the server.

The receive message is the *only* asynchronous message supported by the client interface. It is implemented using the asynchronous version of `SendReceive()`:

```
void RGsdpSession::Receive(TDes8& aBuffer, TRequestStatus& aStatus)
    {
    TInt p[KMaxMessageArguments];
    p[0] = (TInt)&aBuffer;
    SendReceive(EGsdpReqReceive, p, aStatus);
    }
```

Every asynchronous request should have a corresponding cancel function. Here it is:

```
void RGsdpSession::CancelReceive()
    {
    SendReceive(EGsdpReqCancelReceive, 0);
    }
```

One important thing to ensure when writing (and calling) asynchronous functions that the parameters passed must not be local stack variables. If they are, the function may have returned, destroying the stack frame containing the variables, before the server has accessed the memory locations. This results in subtle, hard to reproduce bugs.

These functions are very simple on the client side. On the server side, however, their implementation is more interesting. Let's now begin to see how things look from that perspective.

19.2.3 Connecting and Disconnecting

The client interface and receive handler functions assume the server is already there, and that the client has already connected to it.

The client connects to the server using `ConnectL()`, which sets a handle value to associate with the session. The client disconnects using `Close()`, which zeroes the handle. As a precaution, the `RGsdpSession`'s C++ constructor sets the handle value to zero so that the session is clearly closed before `ConnectL()` is called. If you try to invoke any `SendReceive()` function on a zero-handle object, you'll get a panic.

If the GSDP server was a system server, guaranteed to be alive all the time (otherwise the system as a whole has effectively died), then all these functions would be very simple. This would do the trick for opening the session:

```
EXPORT_C void RGsdpSession::ConnectL(MGsdpPacketHandler* aHandler)
    {
    // Connect to server
    User::LeaveIfError(CreateSession(KGsdpServerName, TVersion(0,0,0)));

    // Create active object receive handler and add it to scheduler
    iHandler = new CGsdpReceiveHandler(aHandler, *this);
    if(!iHandler)
        {
        RSessionBase::Close();
        User::Leave(KErrNoMemory);
        }
    CActiveScheduler::Add(iHandler);
    }
```

And this would close it:

```
EXPORT_C void RGsdpSession::Close()
    {
    // Destroy receiver-handler
    delete iHandler;
    iHandler = 0;

    // Destroy server session
    RSessionBase::Close();
    }
```

Finally, the inline constructor,

```
inline RGsdpSession() : iHandler(0) {};
```

would set the handle to zero before the session was first connected.

You can see that this logic is very simple. If I wasn't using the hidden iHandler active object, this would all boil down to

```
EXPORT_C void RGsdpSession::ConnectL(MGsdpPacketHandler*
    aHandler)
    {
    // Connect to server
    User::LeaveIfError(CreateSession(KGsdpServerName,
        TVersion(0,0,0)));
    }
```

To connect, I simply issue CreateSession(), specifying the name of the server (defined in gsdpdef.h):

```
_LIT(KGsdpServerName,"GSDP server");
```

The other parameter is a zeroed-out version struct (which I don't check in the server because I know the server code must be the same as the client's).

Up to (but not including) Symbian OS v6.0, each client/server session had dedicated message slots. Because the slots were preallocated, this ensured that asynchronous calls could not fail for lack of slots. However, it also meant that memory was wasted since it is extremely rare for a server to be using all its message slots. From Symbian OS v6.0, a global pool of 255 message slots was introduced that can be shared between all client/server sessions. Now clients can choose whether to rely on the global pool, which in extreme situations may be exhausted, or to allocate exclusive slots.

The GSDP server is not an essential system server, therefore, I *can* afford to take the risk that very occasionally the message slot pool will be exhausted, so the CreateSession() call specifies the use of the global pool.

To avoid wasting memory terminates itself when no longer needed. That creates special difficulties for `ConnectL()`. The *purpose* of `ConnectL()` is to connect reliably in the communications sense of 'reliable', namely, that

- either it succeeds in connecting to the GSDP server in such a way that there is exactly one GSDP server in the system, and the client is connected to it;
- or it fails to connect and leaves so that the client understands that connection has failed.

It is *not* acceptable to

- accidentally launch a second instance of the GSDP server and connect to it;
- silently fail to connect so that the client believes there is a connection, but there isn't one.

The code that ensures reliable server connection is in `RGsdpSession::ConnectL()`. It relies on code that ensures reliable server launch, running in the server itself. We'll look at that code below.

Here's the beginning of `RGsdpSession::ConnectL()` as implemented in `gsdpclient.cpp`:

```
EXPORT_C void RGsdpSession::ConnectL(MGsdpPacketHandler* aHandler)
    {
```

```
// Connect to server
TInt err = KErrNone;
for(TInt tries = 0; tries  < 2; tries++)
    {
    err = CreateSession(KGsdpServerName, TVersion(0,0,0));

    if(!err) break; // Connected to existing server - OK

    if(err != KErrNotFound && err != KErrServerTerminated)
        break; // Problems other than server not here
                    - propagate error

    err = CGsdpScheduler::LaunchFromClient();

    if(!err) continue; // If server launched OK, try again to connect

    // If someone else got there first, try again to connect
    if(err == KErrAlreadyExists) continue;
    break; // Server not launched: don't cycle round again
    }
User::LeaveIfError(err);
...
```

We try to connect the server, hoping it is already alive. If our original session returns with KErrNone, all's well.

We can deal with two possible error conditions – server not found, and server terminated – as we'll see shortly. If the attempt to connect to the server produced any other error, then we leave with the error code.

We can get KErrNotFound simply because the server hasn't been launched.

It may seem very unreasonable to get KErrServerTerminated when you're trying to connect – if the server has terminated, shouldn't we get KErrNotFound? The answer lies in the two-phase process for connection:

- Firstly, the kernel checks whether a server with the given name exists and creates a DSession for it.
- Then, the server sends a connect message to the server, which is handled by the CServer class and results in a NewSessionL() function call to create the server-side session.

If, in the first step above, the kernel can't find the server, this process will return KErrNotFound. But if in the second step, the server happened to be in the process of terminating, it won't handle the message and you'll get KErrServerTerminated.

Whether the server wasn't found, or whether it was terminated, the response is the same: try to launch a new instance of the server, with the static CGsdpScheduler::LaunchFromClient(), which we'll look at below.

There are three potential outcomes to launching the server:

- We launched it successfully and so received `KErrNone`.

- Some other client tried to launch it so we received `KErrAlreadyExists`.

- Some other problem occurred.

In the case of `KErrNone` and `KErrAlreadyExists`, we consider it a successful launch, and loop a second time to connect to the now-launched server.

In the case of another problem, we give up and leave with the error code.

Assuming that `CGsdpScheduler::LaunchFromClient()` does its job, then `RGsdpSession::ConnectL()` connects reliably to the server – it either connects to precisely one server or it fails to connect and lets the client know.

19.2.4 The Client API as a DLL

The GSDP server's client API is the first code we've looked at closely that is delivered as a static interface DLL with exported functions. It's worth a quick break from the server theme to look at the DLL specifics here. Firstly, the `.mmp` file is as follows:

```
// GSDP.MMP
TARGET          GSDP.DLL
TARGETTYPE      DLL
UID             0x1000008d 0x101F8B57
SOURCEPATH      .
SOURCE          gsdpclient.cpp gsdpserver.cpp
SOURCE          gsdpsession.cpp gsdpport.cpp
SOURCE          gsdprxq.cpp gsdpgdpadapter.cpp


SYSTEMINCLUDE   \epoc32\include \epoc32\include\kernel
LIBRARY         euser.lib efsrv.lib estor.lib ecom.lib gdp.lib
```

The DLL specifics here are

- `TARGETTYPE dll`, along with the `.dll` extension on the filename, tell `makmake` we want a static library DLL,

- a second UID of `0x1 000 008d`, which should be used for all static library DLLs (yes, the second UID: remember that the first one is specified implicitly by `makmake`). I also specify a third UID of `0x101F8B57` that is specific to GSDP.DLL.

As we saw with GUI applications, every DLL has to have an E32Dll()
function, even though this does nothing. Ours is in gsdpclient.cpp,
right at the bottom:

```
EXPORT_C TInt E32Dll(TDllReason)
    {
    return 0;
    }
```

Functions that we wish to make available to clients of the DLL have to
be exported from it. You mark them with IMPORT_C in header files, and
EXPORT_C in source files.

Functions not exported from the DLL are effectively private, and so
cannot be accessed by code outside the DLL. Thus, no client can call
RGsdpSession::Receive() because it is private in the C++ sense
and is, in any case, not exported from the DLL.

There is no need to export the server-side functions to the client – in
fact, it would be quite wrong to do so. Even if a client could get the
header files, which have plenty of C++ public functions in them, the
client shouldn't be able to call these functions directly.

19.3   The Server Implementation

The server is *much* more complicated than the client, as you'll soon see
by taking a look at its class definitions in gsdpserver.h. For a quick
overview, they are seen in Figure 19.2 in UML.

We're going to have to cover this one step at a time. The majority
of the work for clients – including all the explicit message handling – is
done in CGsdpSession, so we'll cover that class first.

GSDP datagrams are sent by simply wrapping them up and then using
one of the GDP protocol implementations to send them. On receipt,
the process is reversed. The server owns a list of CGsdpGdpAdapters,
which handle both datagram wrapping and unwrapping, and own the
specific GDP protocol implementations so that there is one adapter for
each GDP protocol. Adapters are shared between all sessions that use
them. The server also uses a port number allocator to allocate unique
port IDs to GSDP-initiating clients.

The GSDP client API implements a synchronous Send(), but the
underlying GDP SendL is asynchronous. To bridge between the two,
the CGsdpGdpAdapter uses a queue to store outgoing datagrams if the
protocol is busy sending one.

Receiving datagrams is more complicated than sending them because
an incoming datagram must be associated with the correct session.

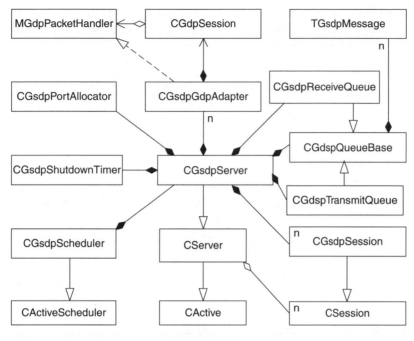

Figure 19.2

Incoming datagrams go onto a queue, and items are only taken off that queue when an appropriate client issues a receive request. Queue management can become complex and, when we look at this, we will be able to understand how asynchronous messages are handled server side. Unsurprisingly, the transmit and receive queues have a lot of code in common, and this is reflected in the common base class, CGsdpQueue.

As you can see from the diagram, it's the server class CGsdpServer that holds everything together. Startup and shutdown are handled in conjunction with a derived active scheduler class, a shutdown timer, and some support from the CGsdpSession destructor. We'll end the description of the server with detailed coverage of startup and shutdown.

The patterns involved in this server are surprisingly common in other Symbian OS servers, such as the window server and the serial communications server. Some of the important ones are as follows:

- Client requests are handled by a CSharableSession-derived class. Symbian OS v6.0 and above supports sharing of client/server sessions between threads in a process. CShareableSession is the base class for such shareable sessions.
- A variety of protocol implementations are supported: the server has to maintain an adapter for each implementation and associate each client with the correct one.

- Incoming events must be associated with a client and must fulfill a receive request when one is outstanding. This requires that incoming events be managed by a queue.

Although the standard Symbian OS servers are often more sophisticated in their handling of these issues than the GSDP server, the latter provides a very solid foundation for building on, and for adding extra sophistication where it is required.

19.3.1 Message Handling

When a client sends a message, the kernel gets it to the right server and the server gets it to the right session. The session then handles the message using its ServiceL() function, which has to analyze the request code and parameters, decide which function to perform, and complete the message either synchronously or asynchronously.

ServiceL() is, therefore, a huge switch statement. Here is CGsdpSession::ServiceL(), edited to remove some repetition from the case clauses:

```
void CGsdpSession::ServiceL(const RMessage& aMessage)
    {
    switch (aMessage.Function())
        {
    case EGsdpReqCountProtocols: // return TInt
        aMessage.Complete(Server()->CountProtocols());
        break;
    case EGsdpReqProtocolInfo: // retrieves protocol info
        GetProtocolInfoL(aMessage.Int0(), aMessage.Ptr1());
        aMessage.Complete(KErrNone);
        break;
    case EGsdpReqSetGameProtocol: // TUint32 aProtocol
        SetGameProtocol(aMessage.Int0());
        aMessage.Complete(KErrNone);
        break;
    case EGsdpReqGetGameProtocol: // returns TUint32
        aMessage.Complete(GetGameProtocol());
        break;
    case EGsdpReqSetGdpProtocol: // RGsdpSession::GdpProtocol aProtocol
        aMessage.Complete(SetGdpProtocol(TUid::Uid(aMessage.Int0())));
        break;
    case EGsdpReqGetGdpProtocol: // returns RGsdpSession::Protocol
        aMessage.Complete(GetGdpProtocol().iUid);
        break;
    case EGsdpReqGdpIsNetworked: // returns TBool
        aMessage.Complete(GdpIsNetworked());
        break;
        . . . .
    case EGsdpReqSend: // const TAny& aData
        aMessage.Complete(Send(aMessage.Ptr0()));
        break;
```

```
    case EGsdpReqReceiveAll:
         ReceiveAll();
         aMessage.Complete(0);
         break;
    case EGsdpReqReceive: // TAny& aBuffer - async
         Receive(aMessage.Ptr0());
         break;
    case EGsdpReqCancelReceive:
         CancelReceive();
         aMessage.Complete(0);
         break;
    default:
         Server()->PanicClient(EBadRequest);
         };
    }
```

I use four key functions of the RMessage class in this Servi-ceL() code:

- Function() gives me the request code;

- Int0(), Int1(), Int2(), and Int3() return the message parameters as integers;

- Ptr0(), Ptr1(), Ptr2(), and Ptr3() return the message parameters as TAny* pointers;

- Complete() completes the message, returning the TInt completion code.

My philosophy here is to get beyond the shaky, type-unsafe stage as soon as possible. So I contain all knowledge of RMessage parameter numbers and types within ServiceL(), and call second-level handler functions, specified with type-safe parameters, to do the real work. Most second-level functions don't then need to use any RMessage-related functions at all. Those that do can get the current RMessage by using the server's Message() function.

Error handling

If the client request is malformed, we panic the client thread – just as we would panic our own thread if a conventional function call were made in error. The default case is handled by

```
Server()->PanicClient(EBadRequest);
```

This calls my `CGsdpServer::PanicClient()` function:

```
void CGsdpServer::PanicClient(TInt aPanic) const
    {
    // Let's have a look before we panic the client
    __DEBUGGER()

    // OK, go for it
    const_cast <RThread&>(Message().Client()).
        Panic(_L("GSDP-Server"),aPanic);
    }
```

The heart of this function is the `RThread::Panic()` call to panic the client thread, that is, the client for the message currently being processed.

Until I put the `__DEBUGGER()` line in, it was impossible to analyze the cause of the panic under the emulator. The thread being panicked was in the middle of a send–receive function, somewhere in the kernel executive, and the corresponding debug information was not supplied in the SDK. However, the real cause of the panic is hidden inside the state of the *server*, so invoking the debugger from the server thread gives a much better clue about why the client thread was about to be panicked. The `__DEBUGGER()` macro only affects emulator debug builds, so I don't need to put any conditional compilation constructs around it.

If the function servicing the client request leaves, then the `CGsdpServer::RunError()` function is called:

```
TInt CGsdpServer::RunError(TInt aErr)
    {
    // if it's a bad descriptor, panic the client
    if (aErr==KErrBadDescriptor)     // client had a bad descriptor
        {
        PanicClient(EBadDescriptor);
        }
    // anyway, complete the outstanding message
    Message().Complete(aErr);
    ReStart(); // really means just continue reading client requests
    return KErrNone;
    }
```

Normally the `RunError()` function responds to the error by completing the outstanding message with the error code. However, if the error is KErrBadDescriptor, then the client is panicked, since this is a programming error.

The `CGsdpServer::PanicClient()` function should only be called when a client message is being processed. You should never need or want to panic a client thread at any other stage than when processing a message from it. I guarantee this as the RunError() function can only be called as a result of the CGsdpSession::ServiceL() function leaving, and this function is only called to process client requests.

The RunError() *function was introduced in Symbian OS v6.0. Previously, any leave from a* RunL() *resulted in the active scheduler's* Error() *function being called. Unfortunately, there was rarely enough context as to what had gone wrong to do much useful error recovery. This led to the rule of thumb that* RunL() *functions should not leave.*

By tying the error handling into the active object rather than the active scheduler, the RunError() function makes the context available, allowing effective error handling to be introduced. As a result, new code will increasingly be designed with RunL()s that do leave, and RunError() functions to trap the leaves.

Synchronous message handling

Once we're out of the type-unsafe territory of the ServiceL() switch statement, the message handler functions are usually quite straightforward. Here are the ones for setting and getting the game protocol:

```
void CGsdpSession::SetGameProtocol(TUint32 aProtocol)
    {
    // Set protocol
    iGameProtocol = aProtocol;

    // Check whether we can now receive anything
    if(iReceiveActive)
        (Server)->iReceiveQueue->CheckPackets(this);
    }

TUint32 CGsdpSession::GetGameProtocol()
    {
    return iGameProtocol;
    }
```

Basically, the setter stores its argument in a member variable, while the getter returns the member variable as a result. The setter, though, includes some important code. If the game protocol supported by a session changes, then it's possible that a session in listening mode might be able to receive packets intended for the newly specified game protocol, so a function is called to check this. We'll return to that later.

The handlers for setting and getting the partner's address involve interthread data transfer:

```
void CGsdpSession::SetOtherAddress(const TAny* aAddress)
    {
```

```
    Message().ReadL(aAddress, iOtherAddress);
    }

void CGsdpSession::GetOtherAddress(TAny* aAddress)
    {
    Message().WriteL(aAddress, iOtherAddress);
    }
```

It's that simple, thanks to the effectiveness of the ReadL() and WriteL() functions, and thanks to the error handling performed by CGsdpServer::RunError().

Note here that we're using a TAny* as the address of a descriptor in *the other thread's* address space. I passed this address from the other thread as a message parameter. The ReadL() and WriteL() functions will check that the contents of the address look like a descriptor, and will leave with KErrBadDescriptor if they aren't convinced. The RunError() function we saw above will then panic the client.

The send handler does an interthread read to get the data to send, and then sends it using the correct GDP implementation:

```
TInt CGsdpSession::SendL(const TAny* aData)
    {
    __ASSERT_ALWAYS(iMyPort != 0, iServer->PanicClient(ESendFromZeroPort));
    TBuf8 <KMaxGsdpData> buffer;
    Client().ReadL(aData, buffer, 0);
    iGdpProtocol->SendL(iGameProtocol, iMyPort,
                        iOtherAddress, iOtherPort, buffer);
    return 0;
    }
```

As with all communications systems, a send may take a long time and the success or otherwise of the send is only detected asynchronously. So, in most communications-related servers, send would be asynchronous.

The GSDP specification makes Send synchronous, while the GDP send is asynchronous. This is implemented via a queue of datagrams that is drained by the protocol asynchronously. If the queue is full, the GSDP specification allows us to silently drop the packet.

> **The design would be cleaner if the GSDP send was also a synchronous and with hindsight this change should have been made.**

Asynchronous message handling

The only asynchronous function available from the client-side session is Receive(). On the client side, this uses the asynchronous form of `SendReceive()`. On the server side, the case handler for this function omits the `aMessage.Complete()` that is present in all other case handlers because the message will only complete when the receive completes. Here's some of that code from `CGsdpSession::ServiceL()` again:

```
case EGsdpReqSend: // const TAny& aData
    aMessage.Complete(Send(aMessage.Ptr0()));
    break;
    ...
case EGsdpReqReceive: // TAny& aBuffer - async
    Receive(aMessage.Ptr0());
    break;
case EGsdpReqCancelReceive:
    CancelReceive();
    aMessage.Complete(0);
    break;
```

The receive function is implemented as follows:

```
void CGsdpSession::Receive(const TAny* aBuffer)
    {
    __ASSERT_DEBUG(!iReceiveActive,
                   PanicServer(EReceiveReceiveAlreadyActive));

    // Remember receive request
    iReceiveMessage = Message();
    iReceiveBuffer = aBuffer;
    iReceiveActive = ETrue;

    // Check for immediate fulfillment
    iServer->iReceiveQueue->CheckPackets(this);
    }
```

The session indicates that a receive request has been issued and it stores the `RMessage` associated with the receive. It also stores the pointer to the client-side buffer, which is the destination for the received data.

Sometime later, the receive will complete. This will cause the data to be transferred to the client's buffer using an interthread write and the receive message will then complete. Then there are two possibilities:

- There is already data for this session to receive, waiting on the server's receive queue: in this case, the receive completes essentially synchronously from the `CheckPackets()` function shown above.

- Some data arrives later and is handled by the `RunL()` of an active object other than the server. This completes the receive asynchronously.

As we saw in Chapter 17, any API providing an asynchronous request should also offer a corresponding cancel. Our server-side cancel code is

```
void CGsdpSession::CancelReceive()
    {
    if(!iReceiveActive)
        return;
    iReceiveMessage.Complete(KErrCancel);
    iReceiveActive = EFalse;
    }
```

This is astonishingly simple, and yet fulfils the contract required by cancel:

- If no request has been issued, no action needs to be taken. This could be the case if, say, the client thread issued a cancel just as an active object in the server thread was handling an incoming packet that satisfied the receive request. When the server gets to handle the cancel, it is executing the server's own RunL() and, by this time, the receive request is no longer active.

  This could also happen in the much simpler case that no Receive() had been issued at all (although, thanks to the active object pattern I've used in the client API, that can't happen with this particular server: CancelReceive() is private in the client API, and is only issued if there was actually a request to cancel).

- Otherwise, the message is completed with KErrCancel, and the server-side flag is cleared to indicate that a receive is no longer active.
- In either case, the client's cancel message completes synchronously.
- In either case, by the time the client's cancel message completes, the original receive message is guaranteed to have completed as well.

The details of receive-queue handling are specific to receive-type requests: we'll look at that in more detail later in this chapter. However, all asynchronous messages, and their cancel requests, should be handled server side by a pattern based on the one shown here.

19.3.2   Sending Datagrams

The purpose of the GSDP server is to send datagrams using a given GDP protocol, but wrapping them to include port IDs and game protocol ID. The diagram opposite shows the processing that happens to a GSDP datagram as it goes from client to the GDP medium:

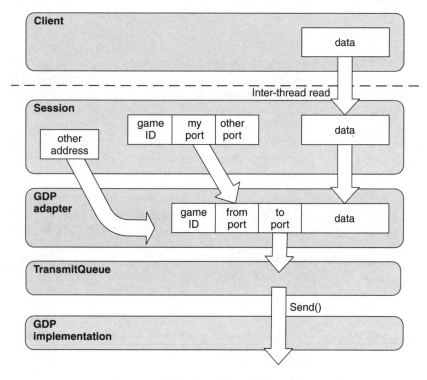

Figure 19.3

Before the processing in this figure takes place, the client has already done some setup using GSDP functions:

- The client has started a GSDP session.
- The client has specified a GDP protocol, which causes the GSDP server to attach that protocol to the server-side session.
- The client has specified a game protocol ID, its own port ID, and the other partner's address and port ID. The server simply stores these values in its CGsdpSession member variables.

The client is then ready to send a GSDP datagram:

- The client calls Send(), which causes an EGsdpReqSend message to be sent to the server.
- The server-side session uses an interthread read to get the data to send.
- The data is then wrapped up to include the game protocol ID, my-port ID as the from-port ID, and the other port ID as the to-port ID.
- The datagram to send is then placed on the transmit queue.

- When the GDP implementation is able to, a datagram is taken off the transmit queue and sent.

The GDP implementation may then do any processing required to send the datagram. A loopback implementation will make a simple function call. A real implementation, such as the Bluetooth implementation, will use a system of active objects to open the communications resources needed, send the datagram, and then close the communications resources again. GDP then reports the success or failure of the send operation via the `MGdpHandler::SendComplete()` function.

Two parts of this sending process are worth discussing in more detail:

- Setting up the GDP protocol implementation required by the session: many servers implement some kind of polymorphism and, although we take some shortcuts in the GSDP server, we can highlight some of the standard issues.
- Manipulating the packet on its way from the session through the GDP adapter to the GDP `Send()` function.

Polymorphism

Many Symbian OS servers use polymorphism to provide a consistent interface to multiple implementations of a service. File systems, communications protocols, media drivers, messaging transports – all have a standard interface with a range of implementations that the client can select between.

In Symbian OS, polymorphism is usually (though not always) implemented via servers that load plug-in DLLs. These are DLLs that contain a factory function that returns an instance of a class that implements an interface the server expects. Typically, the interface that a DLL supports is identified by UID. This allows servers to dynamically load implementations and provide the different implementations to the server's clients. Many of the servers in Symbian OS implement this pattern, including the file server, socket server, and messaging server.

Previously, each of these servers has implemented their own code to manage the plug-in scheme. Recognizing the scope for simplification, Symbian OS v7.0 introduced ECOM, a plug-in framework. This handles all the complexity of finding, loading and unloading plug-in DLLs, managing implementations and providing the correct implementation to satisfy a client request. For the GSDP server, I shall be using the ECOM framework to provide access to the different GDP implementations.

- The client can enumerate the GDP implementations using CountProtocols() and GetProtocolInfo(). Each protocol is identified by a UID and includes a text description suitable for display to the user.
- The client uses SetProtocol() to select the protocol.

- The server uses REcomSession::ListImplementations() to discover which protocols exist and to answer client queries.
- The server then uses CGdpSession::NewL() to create the desired protocols.
- Each session holds a pointer to the protocol that the client selects.

19.3.3   Using the ECOM Framework

There are three roles in the ECOM framework:

- *Interface clients*: these are the users of the services provided by the interface.
- *Interface definer*: the definition of the service provided.
- *Implementation provider*: provides implementation(s) of the service.

The GSDP server acts as the interface client. The interface is defined by GDP, and the details of the implementations are covered in the next chapter. Here, we'll concentrate on the definition of the service and how the server makes use of it. We'll also touch briefly on writing ECOM plug-ins, using the loopback GDP as an example.

We'll start with the interface definition. This is provided in \scmp\gdp\ src\gdp.h:

```
class MGdpPacketHandler
    {
public:
    virtual void GdpHandleL(const TDesC8& aFromAddress,
        const TDesC8& aData)=0;
    virtual void SendComplete(TInt aErr) = 0;
    };

class CGdpSession : public CBase
    {
public:
    IMPORT_C static CGdpSession* NewL(TUid aUid);
    virtual void OpenL(MGdpPacketHandler* aHandler)=0;
        // start up, and set handler for packets received
    virtual void SendL(const TDesC8& aToAddress, const TDesC8& aData)=0;
        // send packet
    virtual void ReceiveAllL()=0; // do a pull if necessary
    virtual TInt GetMaxPacketLength() const =0; // max packet length
    virtual TBool IsNetworked() const =0;
        // requires addresses, if networked
    inline TUid Uid() const {return iDtor_ID_Key;};
protected:
    TUid iDtor_ID_Key;
    };
```

The interface is defined in two classes, CgdpSession, and MGdpHandler. CGdpSession is the pure virtual base class from which I will derive the different implementations. It has methods for the basic functions supported by GDP. Two features make this class a little different from normal virtual base classes:

- *A static* NewL() *function*. Clients call this function to acquire an implementation, specified using a UID.

- *The* iDtor_ID *member*. This is used by the ECOM framework to identify implementations when creating and destroying them.

Let's look at the NewL() function:

```
EXPORT_C CGdpSession* CGdpSession::NewL(TUid aUid)
    {
    return static_cast<CGdpSession*>(REComSession::CreateImplementationL
       (aUid, _FOFF(CGdpSession, iDtor_ID_Key)));
    }
```

This is incredibly simple: the UID parameter is passed to the REcom-Session::CreateImplementation() function, which then returns the right instance. The offset of the iDtor_ID member is also passed to ECOM so that it can track instantiated classes. This is all that is required to create the right implementation – ECOM does all the hard work of finding and loading the right DLL for us.

19.3.4 Using the Protocols

Next, we'll look at how the server makes use of ECOM to find protocols and make them available to its clients. During server startup, ConstructL() is called:

```
void CGsdpServer::ConstructL()
    {
    // construct receive queue
    iReceiveQueue=CGsdpReceiveQueue::NewL(*this);
    // construct shutdown timer
    iShutdown=new(ELeave) CGsdpDelayedShutdown(this);
    iShutdown->ConstructL();
    // construct port allocator
    iPortAllocator=new(ELeave) CGsdpPortAllocator;
    iPortAllocator->ConstructL();

    // Initialize the protocol info
    InitProtocolsL();
    iProtocolUpdater=CGsdpProtocolUpdater::NewL(*this);

    // identify ourselves and open for service
    StartL(KGsdpServerName);
    // initiate shut down unless we get client connections
    iShutdown->Start();
    }
```

`ConstructL()` creates all the objects that the server will need. This includes the receive queue, a port allocator, and a shutdown timer. It then initializes the protocols via `InitProtocolsL()`:

```
void CGsdpServer::InitProtocolsL()
    {
    REComSession::ListImplementationsL(KGdpProtocolImpl, iProtocolInfo);
    CGdpSession* loop = CGdpSession::NewL(KGdpLoopbackUid);
    CleanupStack::PushL(loop);
    CGsdpGdpAdapter* adapter = CGsdpGdpAdapter::NewL(loop, *this);
    CleanupStack::Pop(loop);
    CleanupStack::PushL(adapter);
    User::LeaveIfError(iAdapters.Append(adapter));
    CleanupStack::Pop(adapter);
    }
```

This uses `REcomSession::ListImplementations()` to fill the `iProtocolInfo` array with data on each of the GDP implementations. Since ECOM manages many different plug-in systems, the specific interface required is identified by UID. In the case of the `CGdpSession` interface, this is `KGdpProtocolImpl (0x101F8B52)`.

One of the confusing aspects of the ECOM framework is the number of UIDs involved. There are at least three separate UIDs used:

- *a UID to identify the interface – 0x101F8B52 in the case of GDP*
- *a UID for each DLL that provides implementations – 0x101F8B54 for GDPLOOP.DLL*
- *a UID for each individual implementation – 0x101F8B53 for the GDP loopback protocol.*

Once all the protocols available have been recorded, `InitProtocolsL()` tries to load the loopback protocol. If successful, an adapter is created for it and the adapter is stored in the server. If any of these steps fail, the function leaves, and server startup will fail. This ensures that once the server is created, we can be certain that there is always a loopback implementation available.

It is possible for a user to install new protocol implementations while the GSDP server is running. We need some code to reload the `iProtocolInfo` member when this happens. Fortunately, ECOM provides a `NotifyOnChange()` asynchronous function that we can use to detect changes. The `CGsdpProtocolUpdater` class wraps this up:

```
class CGsdpProtocolUpdater : public CActive
    {
public:
    static CGsdpProtocolUpdater* NewL(CGsdpServer& aServer);
```

```
    ~CGsdpProtocolUpdater();
    void Start();

private:
    CGsdpProtocolUpdater(CGsdpServer& aServer);
    ConstructL();
    void RunL();
    TInt RunError(TInt aErr);
    void DoCancel();
private:
    CGsdpServer& iServer;
    REComSession iEcomSession;
    };
```

This is a standard active object class that owns a session to ECOM. The aim of the class is to keep a notification request outstanding to ECOM. A common pattern with active objects is to have a method that makes the asynchronous request. In this case, I've called it Start(). This is then called from ConstructL() and also from the RunL(), ensuring that the object keeps the notification request outstanding:

```
CGsdpProtocolUpdater::ConstructL()
    {
    iEcomSession = REComSession::OpenL();
    CActiveScheduler::Add(this);
    Start();
    }

void CGsdpProtocolUpdater::Start()
    {
    iEcomSession.NotifyOnChange(iStatus);
    SetActive();
    }

void CGsdpProtocolUpdater::RunL()
    {
    if(iStatus == KErrNone)
            iServer.UpdateProtocolInfo();
    Start();
    }
```

When the asynchronous request completes, we notify the server via the UpdateProtocolInfo() function. This is a function that reloads the protocol list.

One of the most difficult parts of programming is handling error conditions properly. Updating the protocol list is a good example of this. What we want is that when the ECOM registry information changes, we update the server's list of protocols. Since the list is of variable length, this involves a heap allocation, and this can, of course, fail.

When I first wrote the code, `UpdateProtocolInfo()` looked like this:

```
void CGsdpServer::UpdateProtocolInfoL()
    {
    iProtocolInfo.ResetAndDestroy();
    REComSession::ListImplementationsL(KGdpProtocolImpl, iProtocolInfo);
    }
```

In this version, if the `ListImplementationsL()` call leaves, `iProtocolInfo` will be an empty array. I then coded a `RunError()` in the `CGsdpProtocolUpdater` class to catch the leave and ignore it.

But this approach felt wrong. There was no point asking the `CGsdpProtocolUpdater` object to handle the error because it couldn't do anything about it. And would a client really want to get a null list of implementations? I foresaw that this could cause problems with UIs that might assume that at least one protocol would be returned, especially since the loopback protocol is guaranteed to be available.

The second version of `UpdateProtocolInfo()` was as follows:

```
void CGsdpServer::UpdateProtocolInfo()
    {
    RImplInfoPtrArray imps;
    TRAPD(err,REComSession::ListImplementationsL(KGdpProtocolImpl,
        imps));
    if(err == KErrNone)
        {
        iProtocolInfo.ResetAndDestroy();
        iProtocolInfo = imps;
        }
    else
        imps.ResetAndDestroy();
    }
```

In this version, if `ListImplementationsL()` succeeds, we update the `iProtocolInfo` list with the new information, but if it fails then the old `iProtocolInfo` is retained.

This means that protocol info returned to the client may not be up-to-date, and this in turn has some subtle implications:

- It may not be possible to use a protocol that appears in the list because it is no longer installed.

- There may be other protocols that exist but that don't appear in the list.

Neither of these two problems seem as bad as returning an empty protocol list, so this is the version that I chose. You can see here how the error handling of a relatively simple update function ended up changing the guarantees that the server was able to make to its clients. Tempting though it may be to ignore errors, handling them is much, much better than hoping they will never come up. It's my experience that in production code, every possible error condition will occur, and unhandled ones will eventually sneak up and bite you.

Thanks to the combination of initializing `iProtocolInfo` at startup and the updates provided by `UpdateProtocolInfo()`, `CountProtocols()` and `GetGdpProtocolInfo()` are both simple:

```
TInt CGsdpServer::CountProtocols()
    {
    return iProtocolInfo.Count();
    }

void CGsdpServer::GetProtocolInfoL(TInt aProto, TGdpProtocolInfo& aInfo)
    {
    if(aProto < 0 || aProto >= iProtocolInfo.Count())
        User::Leave(KErrArgument);

    aInfo.iUid = iProtocolInfo[aProto]->ImplementationUid();
    aInfo.iDisplayName = iProtocolInfo[aProto]->DisplayName();
    TLex8 lexer(iProtocolInfo[aProto]->OpaqueData());
    User::LeaveIfError(lexer.Val(aInfo.iNetworked));
    }
```

`CountProtocols()` just return the length of the `iProtocolInfo` array. `GetGdpProtocolInfo()` returns the information on a protocol. The protocol to inquire about is selected by an integer parameter, and the function simply copies the information out of the `iProtocolInfo` array and copies it back to the client. Any ECOM implementation can supply some additional information, called *opaque data*, in its definition. Here, I use this to indicate whether the protocol is networked or not. Since it is a blob of binary data, I use `TLex8` to parse it into an integer. If the format isn't right, we leave.

The way I've written the interface requires a client/server call to retrieve the information for each protocol. I've selected this for ease of programming, but it is inefficient and it would be better to return all the information in one go. One way to do this would be to use the `MPublicRegistry` ECOM interface.

GSDP clients can select the protocol that they want to use via the `SetGdpProtocol()` function, which ends up calling the `CGdpSession::SetGdpProtocolL()`:

```
void CGsdpSession::SetGdpProtocolL(TUid aProtocol)
    {
    iGdpProtocol = Server()->GetProtocolL(aProtocol);
    }
```

As the protocols are shared between sessions, they are owned by the `CGdpServer` class. So the `SetGdpProtocolL()` function asks the `CGdpServer` for the right protocol adapter:

```
CGsdpGdpAdapter* CGsdpServer::GetProtocolL(TUid aProtocol)
    {
    // Check if we already have an adaptor
    // TODO: Use Find()?
    TInt i;
    for(i=0; i < iAdapters.Count(); i++)
        {
        if(iAdapters[i]->ProtocolUid() == aProtocol)
            {
            return iAdapters[i];
            }
        }

    // if not, then create one
    CGsdpGdpAdapter* adapter = CGsdpGdpAdapter::NewL(*this);
    CleanupStack::PushL(adapter);
    CGdpSession* protocol = CGdpSession::NewL(aProtocol);
    adapter->SetProtocolL(protocol);
    User::LeaveIfError(iAdapters.Append(adapter));
    CleanupStack::Pop(adapter);
    return adapter;
    }
```

First, `GetProtocolL` checks whether there is already a matching protocol loaded. If not, the GDP session is created, and then wrapped in a `CGsdpGdpAdapter` instance. This is stored in the server in case other sessions want to use the same protocol. Then the new adapter is returned to the session.

This implementation ensures that protocols are only loaded when needed, but doesn't unload them when sessions are closed. A full server should implement a reference-counting scheme to unload the protocols when they are no longer needed.

Creating a GDP implementation

We've looked at using GDP implementations via the ECOM framework. Now we come on to creating an implementation. The simplest protocol, and one that is very useful in testing, is loopback. This does exactly what you might expect – anything sent via SendL() is immediately delivered to GdpHandleL(), looping the data back.

ECOM interface implementations are delivered as DLLs, as you might expect. A single DLL can have both implement multiple interfaces, and provide multiple implementations of each interface. However, we're only interested in one interface (the GDP one) and one implementation (the loopback one).

Here is the MMP file for the GDPLOOP.DLL:

```
// gdploop.mmp

TARGET gdploop.dll
TARGETTYPE ECOMIIC
UID 0x10009D8D 0x101F8B54

SOURCE gdploop.cpp
USERINCLUDE .
SYSTEMINCLUDE \epoc32\include
SYSTEMINCLUDE \epoc32\include\ecom

RESOURCE 101F8B54.rss

LIBRARY gdp.lib euser.lib ecom.lib
```

As this is an ECOM plug-in, the TARGETTYPE is ECOMIIC. All ECOM DLLs must have a second UID of 0x10009D8D, which I duly specify. The third UID identifies the DLL, and ECOM requires this to be present. Although this DLL is not a UI component, it still specifies a resource file (**101F8B54.rss**) via the RESOURCE keyword. This file is central to making ECOM work:

```
#include "RegistryInfo.rh"

RESOURCE REGISTRY_INFO theInfo
{
dll_uid = 0x101F8B54;
interfaces =
    {
    INTERFACE_INFO
        {
        interface_uid = 0x101F8B52;
        implementations =
            {
            IMPLEMENTATION_INFO
                {
                implementation_uid = 0x101F8B53;
                version_no = 1;
                display_name = "Loopback";
                default_data = "";
                opaque_data = "0";
                }
            };
        }
    };
}
```

Let's go through this step by step:

- The resource file is named after the DLL 3rd UID – **101F8B54.rss**. This is a requirement of the ECOM framework. If the two don't match, then my DLL will not be loaded.

- I include **RegistryInfo.rh** as this defines the resource structures ECOM depends on.

- I declare a `REGISTRY_INFO` structure to describe the DLL.

- Its first entry is `dll_uid`, which I set to the GDPLOOP.DLL 3rd UID.

- Its second entry is `interfaces`, which I set to an array of `INTER-FACE_INFO` resources.

- Each `INTERFACE_INFO` resource defines the registration information for the interface. I set the `interface_uid` to the GDP interface UID, `0x101F8B52`.

- The `implementations` member is set to an array of `IMPLEMENTA-TION_INFO` structs.

- I set the `implementation_uid` to the UID of the loopback implementation, and set the `version_no`, `display_name`, `default_data` and `opaque_data` fields appropriately.

The result is a resource file that describes this DLL as having one implementation of the GDP interface.

When ECOM loads a DLL in response to a request, it needs to be able to create the right implementation. The key to this is the `Implementa-tionGroupProxy()` function in **gdploop.cpp**:

```
const TImplementationProxy ImplementationTable[] =
    {
    {{0x101f8b53}, CGdpLoopback::NewL},
    };

EXPORT_C const TImplementationProxy* ImplementationGroupProxy
    (TInt& aTableCount)
    {
    aTableCount = sizeof(ImplementationTable) /
                    sizeof(TImplementationProxy);

    return ImplementationTable;
    }
```

This function which the build tools arrange to be at ordinal 1 in the DLL, returns an array of `TImplementationProxy` elements. Each entry in the array links an implementation UID to a function that returns an instance of that implementation. This allows ECOM to create the desired implementation when required.

There's one more place where ECOM imposes requirements, and that's in the destructor:

```
CGdpLoopback::~CGdpLoopback()
    {
    REComSession::DestroyedImplementation(iDtor_ID_Key);
    }
```

Every ECOM implementation instance must inform ECOM when it is destroyed and this is done by the `DestroyedImplementation()` call.

Once the items above are taken care of, there's no more ECOM scaffolding needed and the code can focus on providing the services required. The rest of the loopback implementation can be found in `\scmp\gdp\src\gdploop.cpp`.

Sending a datagram

When a packet is sent, the GDP adapter is used to send the client's datagram along with the relevant game protocol and port information to the GDP destination address:

```
TInt CGsdpSession::Send(const TAny* aData)
    {
    __ASSERT_ALWAYS(iMyPort != 0, iServer->PanicClient(ESendFromZeroPort));
    TBuf8 <KMaxGsdpData> buffer;
    Client().ReadL(aData, buffer, 0);
    iGdpProtocol->SendL(iGameProtocol, iMyPort,
                        iOtherAddress, iOtherPort, buffer);
    return 0;
    }
```

The GDP adapter's `SendL()` function is implemented as follows:

```
void CGsdpGdpAdapter::SendL(TUint32 aGameProtocol,
                            TUint32 aFromPort,
                            const TDesC& aToAddress,
                            TUint32 aToPort,
                            const TDesC8& aData)
    {
    if(iLastError)
        {
        iLastError = KErrNone;
        User::Leave(iLastError);
        }
    iTransmitQueue->Transmit(aGameProtocol, aToPort, aToAddress,
        aFromPort, aData);
    }
```

If `iLastError` is `KErrNone`, then this simply passes the data and addressing information to the transmit queue. If `iLastError` is not `KErrNone`, then we leave with this error. This enables us to pass back any errors from GDP that occurred on the previous send although they will only be received by the client on the next `Send()`. In the transmit queue, the datagram packet is created and placed on the queue. Then, if the adapter is able to send data at this point (i.e. there is no send outstanding) the datagram is sent:

```
void CGsdpTransmitQueue::Transmit(TUint32 aGameProtocol, TUint32 aToPort,
    const TDesC8& aFromAddress, TUint32 aFromPort, const TDesC8& aData)
    {
    // get first free packet slot - drop packet if there isn't one
    TGsdpPacket* packet=AddPacket(aGameProtocol, aToPort, aFromAddress,
        aFromPort, aData);
    if (!packet)
        return;
    // see if it can be sent
    if(iAdapter.CanSendPacket(*packet))
        iAdapter.SendPacket(*packet);
    }
```

`SendPacket()` is triggered from two places in the code:

- when a datagram is sent, and there is no send outstanding,
- when the GDP protocol calls the `SendComplete()` callback, signalling that it is ready to send the next datagram.

This ensures that if there are packets on the queue, there will be asynchronous events that gradually drain the queue. Finally, in `SendPacket()` the GSDP packet is passed down to the GDP protocol:

```
void CGsdpGdpAdapter::SendPacket(TGsdpPacket& aPacket)
    {
    if(!iSendActive) // otherwise drop
        {
        RDesWriteStream writer(iSendBuffer);
        writer < < aPacket.iGameProtocol;
        writer < < aPacket.iFromPort;
        writer < < aPacket.iToPort;
        writer < < aPacket.iData;
        writer.CommitL();
        iSendActive = ETrue; // do this first in
            case SendComplete is sync
        iGdpSession->SendL(aPacket.iAddress, iSendBuffer);
        }
    }
```

This code works by forming a GDP datagram in the GDP adapter's `iSendBuffer`, containing the game protocol ID, from-port, to-port, and GSDP datagram data.

The `RDesWriteStream` class makes it convenient to write data to a buffer described by a descriptor. We simply write the data we need and commit the stream. Because the stream store documents the external format of items written in this way, we can be confident about the external data format. For instance, the stream store guarantees that integers are written in little-endian byte order. We document this as the external format of a GSDP datagram.

When we look at what happens when a datagram is received, we will see this code in reverse.

19.3.5 Receiving Datagrams

There are big differences between sending and receiving. Sending is initiated by the client, but receiving is initiated by something coming in from elsewhere. The client has to maintain an outstanding request to listen for incoming packets, and the server has to maintain a queue so that incoming packets don't get lost, just in case they arrive between the server-side completion of one receive message, and the client renewing the receive request.

How receiving a packet is managed in the GSDP server is shown in Figure 19.4.

The steps in receiving a datagram are as follows:

- The datagram arrives via a GDP implementation.
- It is handled by the GDP adapter's `GdpHandleL()` function.
- The GDP adapter parses the GSDP header information – game protocol ID, from-port, and to-port – and passes the datagram to the receive queue.
- The receive queue allocates a slot for the incoming datagram and adds it to the queue.
- The receive queue then checks to see whether any session can receive the packet: this means the session must have an outstanding receive request, and either a matching port ID, or a zero-port ID, and a matching game protocol ID.
- If a session can receive it, the message is sent to the client using interthread write, the receive message is completed, and the datagram slot is returned to the receive queue's free list.
- On the client side, this will eventually cause the `CGsdpReceiveHandler` active object's `RunL()` to be scheduled, which calls the client's `GsdpHandleL()` function and then renews the receive request to the server.

This is a classic 'receive' pattern that's used, with minor variations, by all servers that implement receive-type semantics. For instance, the window

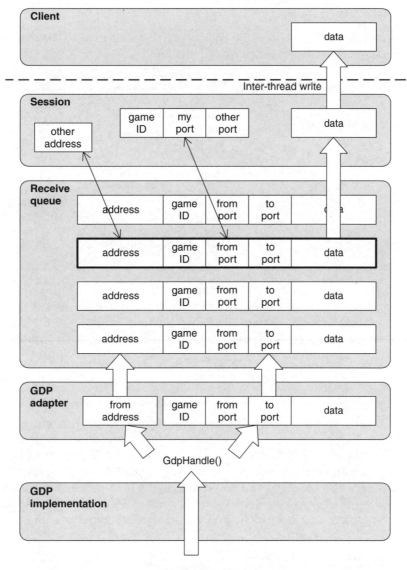

Figure 19.4

server's handling of input events uses a similar system: a raw event is received from the kernel, processed, and added to a queue. The client that should receive the event is identified and the event is sent to the client as soon as possible. The most important differences between one server and another include the type of event that is handled, the means by which the target client is identified, and the queue management algorithms.

Unwrapping the datagram

The datagram is passed from the GDP implementation to the GDP adapter and handled by its GdpHandleL() function:

```
void CGsdpGdpAdapter::GdpHandleL(const TDesC& aFromAddress,
                                 const TDesC8& aData)
    {
    TUint32 gameProtocol;
    TUint32 toPort;
    TUint32 fromPort;
    RDesReadStream reader(aData);
    gameProtocol = reader.ReadUint32L();
    reader >> fromPort;
    reader >> toPort;
    reader >> iReceiveBuffer;
    iServer->iReceiveQueue->Receive(gameProtocol, toPort,
                                    aFromAddress, fromPort,
                                    iReceiveBuffer);

    }
```

This uses a descriptor-read stream to perform a series of read operations that exactly mirror those performed by the descriptor-write stream in CGsdpGdpAdapter::SendL(). When the field values from the datagram have been identified, the datagram is sent to the receive queue's Receive() function.

The receive queue

The receive queue is a list of datagrams (referred to as 'packets' in its API) waiting to be received by clients. Packets are added to the list by the Receive() function. A packet just added to the list is offered to all sessions to see whether it can be received. When a session changes its settings or issues a receive, it checks the receive queue using CheckPackets() to see if there are any packets waiting for that session. By this means, a packet is delivered to a session as soon as possible.

Here is the definition:

```
class CGsdpReceiveQueue : public CGsdpQueueBase
    {
public:
    enum TReceiveCheck { EAllowZero, EDontAllowZero };
public:
    // construct
    static CGsdpReceiveQueue* NewL(CGsdpServer& aServer);
    // functions
    void Receive(TUint32 aGameProtocol, TUint32 aToPort, const TDesC8&
        aFromAddress, TUint32 aFromPort, const TDesC8& aData);
    void CheckPackets(CGsdpSession* aSession);
```

```
private:
    CGsdpReceiveQueue(CGsdpServer& aServer);
private:
    CGsdpServer& iServer;
    };
```

On construction, `CGsdpQueueBase` creates 10 blank packets and adds them to the `iSlots` queue. When a packet is received, the following processing takes place:

```
void CGsdpReceiveQueue::Receive(TUint32 aGameProtocol, TUint32 aToPort,
    const TDesC8& aFromAddress, TUint32 aFromPort, const TDesC8& aData)
    {
    TGsdpPacket* packet=AddPacket(aGameProtocol, aToPort, aFromAddress,
        aFromPort, aData);
    if (!packet)
        return;
    CGsdpSession* session=iServer.SessionForPacket(*packet);
    if (!session)
        return;
    session->ReceivePacket(*packet);
    }
```

`AddPacket()` is a method provided by `CGsdpQueueBase`. It adds a packet onto the queue, by taking a blank packet from the `iSlots` queue, filling it in, and placing it on the `iPackets` queue.

The server is then asked to find a session that can receive the packet using `SessionForPacket()`. This function is a member of the server class because the server class knows about all the sessions.

If the server can identify a session to receive the packet, the packet is received by calling the relevant session's `ReceivePacket()` function. Otherwise, the packet remains on the queue and will hopefully be received later when a session changes its parameters or issues another receive request.

Here's how the server finds a session for the incoming packet:

```
CGsdpSession* CGsdpServer::SessionForPacket(const TGsdpPacket& aPacket)
    {
    CSession* session;
    // Iterate through sessions with nonzero port id
    iSessionIter.SetToFirst();
    for(session = iSessionIter++; session; session = iSessionIter++)
        {
        if(STATIC_CAST(CGsdpSession*, session)->GetMyPort() == 0)
            continue;
        if(STATIC_CAST(CGsdpSession*, session)->CanReceivePacket(aPacket))
            break;
        }
```

```
    if(session)
        return STATIC_CAST(CGsdpSession*, session);

    // Iterate through sessions with zero port id
    iSessionIter.SetToFirst();
    for(session = iSessionIter++; session; session = iSessionIter++)
        {
        if(STATIC_CAST(CGsdpSession*, session)->GetMyPort() != 0)
            continue;
        if(STATIC_CAST(CGsdpSession*, session)->CanReceivePacket(aPacket))
            break;
        }
    return STATIC_CAST(CGsdpSession*, session);
    }
```

The code is simple – it consists of two scans through all sessions. On the first scan, we *ignore* all sessions whose my-port ID is zero. On the second scan, we select *only* sessions whose my-port ID is zero. This means that packets with a nonzero to-port ID are matched against a specific session before being tried on a new listening session.

`CanReceivePacket()` tests whether a receive is active and, if so, whether the packet can be received according to other criteria:

```
TBool CGsdpSession::CanReceivePacket(const TGsdpPacket& aPacket) const
    {
    return iReceiveActive && (
        iMyPort == aPacket.iToPort && (
            iMyPort!=0 || // In session
            iMyPort==0 && iGameProtocol == aPacket.iGameProtocol //
                Listening
            )
        );
    }
```

If the packet can be received, the session receives it with `ReceivePacket()`, as follows:

```
void CGsdpSession::ReceivePacket(TGsdpPacket& aPacket)
    {
    // Decide whether to drop or to receive
    TBool drop = EFalse;
    if(aPacket.iGameProtocol != iGameProtocol)
        drop = ETrue;
    if(iOtherPort != 0 && (aPacket.iFromAddress != iOtherAddress ||
                            aPacket.iFromPort != iOtherPort))
        drop = ETrue;

    // Get remote's port and address information if we haven't already
        got it
    if(iOtherPort == 0)
        {
```

```
    iOtherPort = aPacket.iFromPort;
    iOtherAddress = aPacket.iFromAddress;
    }

// Receive packet if we should
if(!drop)
    {
    iReceiveMessage.WriteL(iReceiveBuffer, aPacket.iData);
    iReceiveMessage.Complete(KErrNone);
    iReceiveActive = EFalse;
    }

// In any case, tell the receive queue to free the packet for
    future use
(Server)->iReceiveQueue->FreePacket(aPacket);
}
```

This divides into three parts:

- We decide whether to drop the packet or not: this is just an interpre- tation of the rules laid down in the GSDP protocol specification.

- If we are going to receive the packet, we write the data to the client, complete the session's outstanding receive request, and note that we no longer have a receive request active.

- In any case, we tell the receive queue to free the packet slot so it can be used again for a new incoming packet.

It's in this function that the client session's outstanding receive request is completed.

Most of the GDP packets received from a GDP implementation will be formed into a GSDP packet, placed on the receive queue, matched against a session with an outstanding receive request, and immediately received using `ReceivePacket()`.

If this was the process for *all* incoming packets, then there would be no need for a queue. But there are two reasons for maintaining a queue rather than simply offering the packet directly to the client.

First, the client could already be busy processing the previously received packet. In this case, the client will shortly renew its receive request, and it will then be possible to receive the packet. For this reason, the session's receive message handler function includes a check to see whether there are already any packets that this client can receive:

```
void CGsdpSession::Receive(const TAny* aBuffer)
    {
    __ASSERT_DEBUG(!iReceiveActive, PanicServer
        (EReceiveReceiveAlreadyActive));
```

```
// Remember receive request
iReceiveMessage = Message();
iReceiveBuffer = aBuffer;
iReceiveActive = ETrue;

// Check for immediate fulfillment
(Server)->iReceiveQueue->CheckPackets(this);
}
```

If the CheckPackets() function finds a packet that can be received, then the receive message handled by this function is immediately completed.

There is also a consideration unique to the design of the GSDP server. It's possible for the server to receive an incoming datagram for a session that has not yet been created, or whose settings have not yet been properly initialized. This allows the GSDP server to hold packets for

- a new game that hasn't been started,

- an existing game that has been closed temporarily and should be reopened.

This applies until the game has been started or reopened.

Queue management

The receive queue is a scarce resource – or, more precisely, the free slots on the receive queue are a scarce resource. They are preallocated so that a receive operation cannot fail because of an out-of-memory error. Precisely because they are preallocated there are not many of them.

> **The receive queue should manage itself carefully to ensure that message slots are not allocated and then never freed.**

This means that

- messages that cannot match any existing client should be dropped before they are added to the queue;

- when a client is terminated, any messages for that client should be deleted from the queue;

- if a client fails to clear a message after a reasonable period – say, 20 s or more – then the client is 'not responding' in the sense that users of Windows applications are familiar with. The server should drop packets for it and should arguably panic the client.

The receive-queue management in the GSDP server follows the specification of GSDP in Appendix 3 closely. But on reflection, that specification is incomplete and may need to be improved to support quicker deletion of nonreceived packets. The fact that the current specification is to keep received packets on the queue, before a game has started that can accept them is the key item of contention here.

19.3.6   Startup and Shutdown

Earlier, we made the point that connection to a transient server was a delicate affair. We need to connect reliably in the sense that either we succeed in connecting to the one GSDP server, or we fail and report it to the client program. We mustn't fail silently, and we mustn't get into any situation in which there are two GSDP servers. Part of the responsibility for this lies with the client interface, in `RGsdpSession::ConnectL()` – the remainder lies with the server startup code, which we'll now describe.

Additionally, server startup has to set up the server environment that will be used to deliver the GSDP server's services. That means, at a minimum

- a new thread, with a cleanup stack, active scheduler, and `CServer` object ready to receive connection requests from clients via the kernel;
- anything specific that the server needs to do its job – in the case of the GSDP server, that means the receive queue, GDP adapters, port number allocator, shutdown timer, and so on.

If the server can't construct all this, then launch is considered to have failed. If it *can* construct all this, then it's ready to receive client connect requests. Client connect requests may fail too – but that's a different matter from server launch.

Server startup is a kind of bootstrap process involving the following general steps:

- The client program launches a new thread (on the emulator) or process (on a real Symbian OS phone) running the server code.
- The client then waits until the server has initialized.

The server, for its part

- allocates a cleanup stack;
- allocates and constructs an active scheduler;
- allocates and constructs the server and all the objects the server owns;

- tells the client it has started or, if any of the above operations fails, tells the client the return code indicating failure;
- starts the active scheduler, which enables the server to start handling requests.

There is a great deal of complexity in server startup, with many possible failure scenarios. This is further complicated by the differences between the emulator and target hardware. The WINS emulator does not support processes, so the server runs as a thread. On hardware, the server will run in its own process.

Thanks to this complexity, the whole startup sequence is quite delicate. The following sections give a blow-by-blow commentary on the whole process. The code presented here is also suitable for copy-and-paste application to your own servers.

You'll remember that the whole launch sequence is initiated by the following line in the client interface (`RGsdpSession::ConnectL()`, in fact) as part of session connect processing:

```
err = CGsdpScheduler::LaunchFromClient();
```

We have included many of the server launch functions as static members of `CGsdpScheduler`, which is defined as

```
class CGsdpScheduler : public CActiveScheduler
    {
public:
    class TServerStart
        {
    public:
        TServerStart(TRequestStatus& aStatus);
        TPtrC AsCommand() const;
        inline TServerStart() {};
        TInt GetCommand();
        void SignalL();
    private:
        TThreadId iId;
        TRequestStatus* iStatus;
        };
    // launch

    static TInt LaunchFromClient();
#ifdef __WINS__
    static TInt ThreadFunction(TAny* aThreadParms);
#endif
    IMPORT_C static TInt ThreadStart(TServerStart& aSignal);
    static void ConstructL(TServerStart& aStart);
```

```
    ~CGsdpScheduler();
    void Error(TInt aError) const; // from CActiveScheduler
private:
    CGsdpServer* iServer;
    };
```

From client interface to server bootstrap

The LaunchFromClient() function is executed in the context of the client thread:

```
TInt CGsdpScheduler::LaunchFromClient()
    {
    TRequestStatus started;
    TServerStart start(started);

    const TUidType serverUid(KNullUid,KNullUid,KNullUid);

#ifdef __WINS__
    TName name(KGsdpServerName);
    name.AppendNum(Math::Random(),EHex);
    RThread server;
    TInt r=server.Create(name,ThreadFunction,
                            KDefaultStackSize*2, KMinHeapSize, 100000,
                            &start, EOwnerProcess);
#else
    RProcess server;
    TInt r=server.Create(KGsdpServerExe,start.AsCommand(),serverUid);
#endif
    if (r!=KErrNone)
        return r;
    TRequestStatus died;
    server.Logon(died);
    if (died!=KRequestPending)
        {
        // logon failed - server is not yet running, so cannot have
            terminated
        User::WaitForRequest(died); // eat signal
        server.Kill(0);                        // abort startup
        server.Close();
        return died.Int();
        }

    User::WaitForRequest(started,died);        // wait for start or death
if (started==KRequestPending)
    {
    // server has died, never made it to the startup signal
    server.Close();
    return died.Int();
    }
server.LogonCancel(died);
server.Close();
```

```
User::WaitForRequest(died);    // eat the signal (from the cancel)
return KErrNone;
}
```

Its main purpose is to launch a new thread (on the emulator) or process (on a real Symbian OS phone) in which the new server code will run. If the new thread or process cannot even be created, then `LaunchFromClient()` returns the error code. If the new thread or process was successfully created, then `LaunchFromClient()` waits for the server to initialize, and then passes the initialization result code (hopefully `KErrNone!`) back to the caller.

The synchronous wait at the end of this function means that the client process is unresponsive during server launch. In turn, that means server launch had better be quick in order to preserve application responsiveness. If you have a server whose launch takes a long time, you have three options:

- Settle for unresponsive applications during server launch.

- Make the launch as quick as possible – for example, loading only the server framework – and then load any specific protocols you need from within the server, asynchronously, after launch.

- Make the launch itself asynchronous, as far as the client is concerned: present this through an asynchronous `Connect()` function.

I would prefer the second option in order to isolate the client from the asynchronicity.

Thread launch on the emulator

On the emulator, the new thread is given a name, stack size, minimum and maximum heap size, an entry point address, and an initial parameter, by the launching code above.

The `EOwnerProcess` indicates that the server thread is owned by the whole emulator process – not the launching client thread. The launching client thread may die while other clients are active and would take the server thread down with it if we specified `EOwnerThread`.

We use a random name so that if we are restarting the server that has just exited, the name of the thread will be unique. Otherwise, we would receive a `KErrAlreadyExists` error from the kernel. This may mean that if two clients start to connect at the same time, two server threads will be launched. We'll see how this is handled later on.

It is possible (although very unlikely) that a thread can be started successfully, but fail before running the thread function. To protect against this, the client thread logs onto the server thread. This allows the client to detect the server thread from dying before the server startup `TRequestStatus` is signaled.

If any of these steps fail, we clean up the server thread and return the error code to the client.

The thread function `CGsdpScheduler::ThreadFunction()` is a wrapper around `CGsdpScheduler::ThreadStart()`:

```
#ifdef __WINS__
TInt CGsdpScheduler::ThreadFunction(TAny* aThreadParms)
    {
    // get a handle to our code to prevent yank on client death
    RLibrary lib;
    lib.Load(_L("gsdp.dll")); // this ought to work, so no error handling
    // go with the thread
    return ThreadStart(*static_cast<TServerStart*>(aThreadParms));
    }
#endif
```

The main purpose of the code above is to act as a type-safe wrapper to `ThreadStart()`, which we'll see below.

The code that loads the `gsdp.dll` is there to ensure that this thread registers an interest in its own DLL code. We launched the thread by simply specifying a function address that happened to be around in RAM. But that function is part of `gsdp.dll`, which was loaded by the client application. If the client dies, it will decrement the usage count on `gsdp.dll` and (if it's the last client) Win32 will, quite correctly, unload the DLL. If the server's lifetime persists beyond that of the client (which it does, by 2 s) the server code will disappear while the server is still trying to use it, resulting in a horrible death for the server and indeed for the whole emulator. So, the server thread registers its long-term interest in `gsdp.dll` by loading it explicitly here. The `RLibrary` is deliberately not closed: it will be cleaned up, and the library unloaded when the thread terminates after the server has finished and returned from `ThreadStart()`.

Process launch on Symbian OS

On real Symbian OS platforms, a process is launched from `gsd-pexe.exe`, whose `E32Main()` function is also designed to pass the parameter quickly to `CGsdpScheduler::ThreadStart()`:

```
#include "gsdpserver.h"

GLDEF_C TInt E32Main()
    {
```

```
CGsdpScheduler::TServerStart start;
TInt r=start.GetCommand();
if (r==KErrNone)
     r=CGsdpScheduler::ThreadStart(start);
return r;
}
```

The `CGsdpScheduler::TServerStart` class is designed solely to allow the client thread to pass a `TRequestStatus` in such a way that it survives either the thread launch or the process launch above. This allows the server to signal success or failure back to the client. `CGsdpScheduler::TServerStart` wraps up a pointer to a `TRequestStatus` and a thread ID. This can then be passed on the command line to the new process using the `AsCommand()` method. Once in the new process, the `TServerStart` object can be reconstituted from the command line using the `GetCommand()` method. Here's the code for this:

```
inline CGsdpScheduler::TServerStart::TServerStart() {};
inline CGsdpScheduler::TServerStart::TServerStart(TRequestStatus& aStatus)
    :iId(RThread().Id()),iStatus(&aStatus)
    {aStatus=KRequestPending;}
inline TPtrC CGsdpScheduler::TServerStart::AsCommand() const
    {return TPtrC(reinterpret_cast <const TText*>(this),
        sizeof(TServerStart)/sizeof(TText));}

TInt CGsdpScheduler::TServerStart::GetCommand()
    {
    RProcess p;
    if (p.CommandLineLength()!=sizeof(TServerStart)/sizeof(TText))
        return KErrGeneral;
    TPtr ptr(reinterpret_cast<TText*>(this),0,sizeof(TServerStart)/
        sizeof(TText));
    p.CommandLine(ptr);
    return KErrNone;
    }
```

It's unusual for Symbian OS code to differ between the emulator and real Symbian OS phones. Server launch is one of the areas where the differences cut in. The design of the GSDP server minimizes those differences down to a couple of carefully controlled hotspots, which we've now dealt with.

Server bootstrap

Whether on the emulator or a real Symbian OS machine, execution in the newly launched server thread now begins in earnest with `CGsdp-Scheduler::ThreadStart()`:

```
EXPORT_C TInt CGsdpScheduler::ThreadStart(TServerStart& aStart)
    {
    // get cleanup stack
    CTrapCleanup* cleanup=CTrapCleanup::New();
#ifdef _DEBUG

        TRAPD(terr,

  for(TInt i=0; i< 20; i++)
            CleanupStack::PushL((TAny*)NULL);
        CleanupStack::Pop(20);

        );

#endif
    __UHEAP_MARK;
    // initialize all up to and including starting scheduler
    TInt err = KErrNoMemory;
    if (cleanup)
        {
        TRAP(err, ConstructL(aStart));
        delete cleanup;
        }
    __UHEAP_MARKEND;
    return err;
    }
```

First, the function constructs a cleanup stack. Next, the function pushes 20 items on to the cleanup stack and then pops them back off again. This strange looking code is to ensure that the cleanup stack will not grow during the code between __UHEAP_MARK and __UHEAP_MARKEND. This is a useful standard trick that's widely used. As we saw in Chapter 6, the bootstrap implements heap marking. If any server code is ever written that causes memory leaks, then the __UHEAP_MARKEND on server shutdown will cause a server panic. This means that sources of memory leaks can be identified and fixed at the earliest point that they occur during development – a significant assurance of robustness in Symbian OS programs.

Once the cleanupstack is set up, the function calls ConstructL() to complete the startup of the server, using a TRAP to catch any errors.

```
void CGsdpScheduler::ConstructL(TServerStart& aStart)
    {
    // construct active scheduler
    CGsdpScheduler* self=new(ELeave) CGsdpScheduler;
    CleanupStack::PushL(self);
    CActiveScheduler::Install(self);
    // construct server
    self->iServer=new(ELeave) CGsdpServer;
    self->iServer->ConstructL();
    // Let the client know we've started OK
    aStart.SignalL();
    CActiveScheduler::Start();
    // Destroy the scheduler
```

```
CleanupStack::PopAndDestroy(self);
}
```

First `ConstructL()` creates a real active scheduler and installs it as the thread's current scheduler. This essentially stores a pointer to the scheduler in a special location reserved by the kernel. You can get at this pointer using `CActiveScheduler::Current()` if you need to. Next the server object is constructed and stored in the `iServer` member. Finally, the client is signaled to indicate that the startup has succeeded. Then all that remains to do is to start the active scheduler, which starts the server processing requests. `CActiveScheduler::Start()` does not return until `CActiveScheduler::Stop()` is called – we'll see how this happens below.

Server construction

By the time we get to the `CGsdpServer` construction, we are in normal C++ territory: we have a cleanup stack and active scheduler, we don't have to think about the thread that's launched us – in short, the code has lost the flavor of a bootstrap. `CGsdpServer`'s C++ constructor passes two parameters down to the base `CServer` constructor: a zero active object priority, and flag specifying that sessions can be shared:

```
CGsdpServer::CGsdpServer() : CServer(0, EShareableSessions)
    {
    }
```

This indicates that the server does not have a higher or lower priority than any other active object in the thread, and that the server sessions can be shared between all threads in a process. By default, a session can only be used by the thread that creates it. Sharing sessions between threads is a useful facility that servers should support if possible, so we enable this here. The main work of construction is handled by `CGsdpServer::ConstructL()` which we looked at earlier.

Server shutdown

The server shuts down 2 s after its last client session is closed. The server keeps a count of client sessions using `IncrementSessions()` and `DecrementSessions()`, which are called from `NewSessionL()` and from the session destructor:

```
void CGsdpServer::IncrementSessions()
    {
    iSessionCount++;
    iShutdown->Cancel();
    }

void CGsdpServer::DecrementSessions()
    {
    iSessionCount--;
    if(iSessionCount > 0)
        return;
    iShutdown->Start();
    }
```

So, the shutdown timer is started whenever there are no sessions (both when the server is constructed, and when the final session has ended), and is cancelled when a new session is created.

The shutdown timer class is a simple active object:

```
class CGsdpDelayedShutdown : public CActive
    {
public:
    CGsdpDelayedShutdown;
    void ConstructL();
    ~CGsdpDelayedShutdown();
    void Start();
private:
    void DoCancel();
    void RunL();
private:
    RTimer iTimer;
    };
```

The Start() function sets the timer:

```
void CGsdpDelayedShutdown::Start()
    {
    iTimer.After(iStatus, KGsdpShutdownInterval);
    SetActive();
    }
```

KGsdpShutdownInterval is defined, in gsdpdef.h, as

```
const TInt KGsdpShutdownInterval=2000000;
```

or two seconds.

If the timer expires, `RunL()` initiates the shutdown:

```
void CGsdpDelayedShutdown::RunL()
    {
     CActiveScheduler::Stop();
    }
```

It simply stops the active scheduler, which causes the scheduler `Start()` function called from `CGsdpScheduler::ConstructL()` to exit when the `RunL()` has completed. In turn, this causes the scheduler to be popped from the cleanup stack and destroyed. The `CGsdpScheduler` class cleans up any active objects it owns, including the server. The stack then unwinds and the thread function exits, terminating the thread.

19.4   Summary

In this chapter, we've walked through the most critical code in the GSDP server, and seen how a server hangs together in practice, making good use of the APIs and paradigms we discussed in the previous chapter. You can see the entire source code for the GSDP server on the Symbian website.

Servers also demonstrate the power of active objects in routine server programming. Active objects are used for everything from the server, to the shutdown timer, to the GDP implementations themselves. Outside the server launch, we never had to think about concurrency issues, even for receive, send, and cancel processing.

There's a lot more work to be done with the design of GSDP server as it stands, including improving queue management and reference counting of GDP implementations. However, the GSDP server demonstrates the patterns used by many Symbian OS servers and is a useful example to copy from.

The next chapter will take us through the development of the GDP Bluetooth and SMS protocols, which implement the communications functionality used by the real Battleships game.

20

GDP Implementations

In this book, we deal with three different GDP implementations. First, there's the loopback implementation, which has been mentioned on a number of occasions in the course of the last few chapters. This simply delivers packets back to the device sending them by passing them straight back up through the GDP handle interface in its `SendL()` function. This is something that's very useful for testing client code and provides the starting point for the two nontrivial implementations that we'll be looking at here:

- GDP-SMS – communicates over SMS, a telephony network protocol,

- GDP-BT – communicates over Bluetooth, a personal area network protocol.

Concrete GDP implementations provide the GDP services through the M class interfaces that are defined in Appendix 3. They take the disparate services offered by the underlying communications protocols and allow the client (that is, the application using the protocol) to access them with minimal knowledge of that protocol.

I will present these implementations in the order they were developed (SMS, then Bluetooth), but first we need a bit of background in order to understand fully the framework used throughout the rest of this chapter.

I've tried my best to minimize the amount of communications-specific terminology used, and to give some clues to what's going on to readers from noncommunication backgrounds when such terminology *is* necessary. If you wish to learn more about communications programming, there are many good textbooks (particularly those by Andrew Tanenbaum and William Stallings) that cover this area excellently.

Symbian OS C++ for Mobile Phones.   Edited by Richard Harrison
© 2003 John Wiley & Sons, Ltd   ISBN: 0-470-85611-4

20.1   Tasks, States and State Machines

Network communications are asynchronous by nature. In Symbian OS, communications services are provided by servers that present application programming interfaces (APIs) through R (resource) objects, which provide asynchronous functions taking a TRequestStatus& parameter. As is explained in Chapter 17, the natural way to handle such functions is to encapsulate them in an active object framework.

A naive GDP implementation would use a single active object to encapsulate every possible asynchronous function. This would result in a large and potentially uncontrollable system of active objects. Instead, we use the **state pattern**, which was also touched upon in Chapter 17. The state-machine framework I present here (Figure 20.1) provides a very robust method for developing state-machine-based communications protocols. For one-time use, it could be considered rather heavyweight, but as I was able to reuse it through all the concrete implementations, particularly the more complex short message service (SMS) and Bluetooth cases, it proved an invaluable benefit.

Typically, a **communications layer** provides a number of operations to clients using that layer. Internally, each operation (or **task**) may map to multiple suboperations, each of which may be asynchronous. However, these operations must be carried out sequentially, which is where the state pattern comes in; it allows one active object effectively to change its class as it threads its way through the suboperations (or **states**) required to carry out the task. Multiple tasks can (potentially) run in parallel, but within a task the states must be executed sequentially. So you can see how these tasks might map to threads in a traditional multithreaded system, but to active objects within Symbian OS.

From the GDP class, CGdpSession, it is clear that SMS and Bluetooth have two principal tasks – sending and receiving – each of which maps to the various asynchronous operations that are required to achieve this

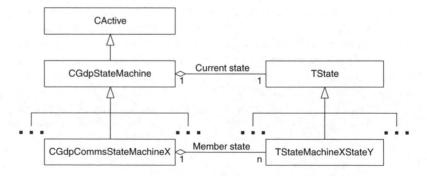

Figure 20.1

communication through the relevant Symbian OS APIs. In addition, SMS and Bluetooth have a further task concerned with resource management.

In order to simplify the development of the specific task implementations, I wrote a generalized GDP state machine class that provides a simple implementation of the state pattern appropriate to our needs. A specific task can be implemented by deriving from the CGdpStateMachine class, which in turn derives from CActive.

Each **state** of such a task is represented by a class derived from the TState base class, which is defined in gdpstatemc.h. Furthermore, each state has four virtual functions:

- EnterL(), which issues a request
- CompleteL(), which handles its completion (called from RunL() when the completion code is KErrNone)
- ErrorL(), which handles an error (called from RunL() when the completion code indicates an error)
- Cancel(), which is called if the task is cancelled by the user.

There are also functions in CGdpStateMachine for transitioning between states.

To design a system using the state machine, you have to draw a state diagram, explain what the states are for, and determine the transitions between them.

20.1.1 GDP State Machines

Here's CGdpStateMachine, the abstract state machine class I used:

```
class CGdpStateMachine : public CActive
    {
public:
    ~CGdpStateMachine();

    class TState
        {
    public:
        virtual void EnterL() = 0;
        virtual TState* CompleteL() = 0;
        virtual TState* ErrorL(TInt aCode) = 0;
        virtual void Cancel() = 0;
        };

protected:
    CGdpStateMachine(TPriority aPriority);
    void ChangeState(TState* aNextState);
```

```
    inline TState* CurrentState();
    void SetNextState(TState* aNextState);
    void ReEnterCurrentState();

    // Methods to be implemented by concrete state machines
    virtual TState* ErrorOnStateEntry(TInt aError) = 0;
    virtual TState* ErrorOnStateExit(TInt aError) = 0;

private:
    // Overrides of CActive functions
    void RunL();
    void DoCancel();

private:
    TState* iState;       // Current state
    };
```

CGdpStateMachine derives from CActive and implements the pure virtual functions declared in that class – namely, RunL() and DoCancel(). This means that any concrete state machine can use the features provided by CActive without actually having to redefine these functions.

Derived concrete classes do, however, have to implement further abstract functions from CGdpStateMachine – namely, ErrorOnStateEntry() and ErrorOnStateExit(), which I will come to in a moment. In addition, the concrete state machine must provide appropriate means to interface with external objects.

The state base class, TState, is defined as a nested class within CGdpStateMachine. Each specific concrete state within a task must provide an instance of a class derived from this base. The iState member in CGdpStateMachine points to the current state – it is private, so derived classes must access it through the various getters and setters provided. This encapsulation is important, as it ensures we cannot inadvertently change state at an inappropriate time.

An initial state change occurs every time the task represented by the state machine is invoked. This may be through an external request (a 'send packet' request, for example) or just by opening the state machine. This can then cause a whole sequence of successive asynchronous requests, each with a corresponding state change. While these asynchronous operations are taking place, the task is 'active', meaning that no further external requests can be serviced until we're finished with this one. Finally, when no more asynchronous requests are required, the task returns to 'inactive' and the state machine is able to accept further client requests.

For every state change, there are two phases, as shown in Figure 20.2. First, we issue a request to enter the new state – which in turn will issue an asynchronous request – and second, we complete the entry to the state,

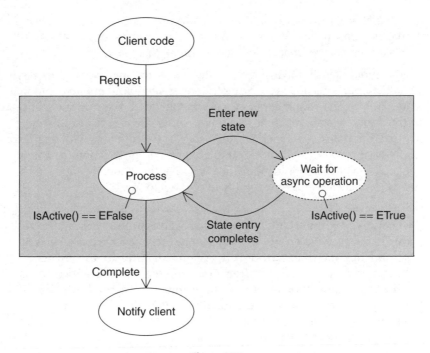

Figure 20.2

when the asynchronous request completes. At this point, there may be *another* state change – and further asynchronous request(s) – or the task may just complete. The latter case signifies the completion of the external request, and the machine moves back to its inactive mode waiting for another request.

The actual code to change state is as follows:

```
void CGdpStateMachine::ChangeState(CGdpStateMachine::TState* aNextState)
    {
    // Enter aNextState, and make it our current state.
    __ASSERT_ALWAYS(!IsActive(),
        GdpUtil::Panic(GdpUtil::EStateMachineStateError));
    TInt err;
    while(aNextState)
        {
        // State change required.
        iState = aNextState;
        TRAP(err, iState->EnterL());
        aNextState = err ? ErrorOnStateEntry(err) : NULL;
        }
    }
```

None of the state-changing functions (ChangeState(), Set-NextState() and ReEnterCurrentState()) should be called while there are any outstanding asynchronous requests (that is, when we are halfway through a state change). The ___ASSERT_ALWAYS() *statement ensures that we don't ever break this invariant.*

This function calls the EnterL() function of the state we wish to move to, which will issue any asynchronous requests required to move into that state. This code is run in a trap harness so that I can pass any leave errors to the concrete state machine's trap handler, ErrorOnStateEntry(). This trap handler can specify a new state to try to enter, allowing it to implement request retries, for example. I keep doing this until something succeeds, which I should point out because the concrete trap handler needs to be aware of the possibility of endless loops.

> **Generally, it is important that communications protocols never leave! This is because we do not own the active scheduler we are running in, and our internal state errors will mean little to the owner of the scheduler. Instead, we trap any potential leaves, and pass an error indication back to our client through an agreed interface if it's appropriate to do so. Additionally, within the client, any Errors returned from function calls are converted into int Leaves with User::LeaveIfError().**

Trap harnesses do have a slight performance cost, so using a single trap harness, which passes any leaves to the implementation-specific error handler, removes the need for every individual state to have its own layer of harness and handler.

The state machine's RunL(), which handles the completion of state entry, is slightly different:

```
void CGdpStateMachine::RunL()
    {
    // Make sure we're in a state, if we're not, something is awry
       so panic
    __ASSERT_ALWAYS(iState != NULL,
    GdpUtil::Panic(GdpUtil::EStateMachineStateError));

    // See how the last SetActive ended up
    TInt err = iStatus.Int();

    // Depending on how that went, decide where to go next
    TState* nextState = NULL;
    if(err == KErrNone)
        {
        // Everything was OK, leave this state
        nextState = iState->CompleteL();
        }
    else
```

```
        {
        // Something went wrong, ERROR
        nextState = iState->ErrorL(err);
        }

    // Just move on to the next state (could be NULL, in which case
       we're done)
    ChangeState(nextState);
    }
```

Here, I selectively call either the `CompleteL()` or `ErrorL()` members of the current state, depending on the outcome of the asynchronous request as returned by `iStatus.Int()`, inherited from `CActive`.

Both these member functions indicate state entry asynchronous request completion. I've just separated them for implementation convenience within the concrete state classes.

Either one of the state completion functions may return a new state to try to enter next. If `NULL` is returned, we drop out of `RunL()` and revert to inactive mode.

Note that unlike `ChangeState()` this is not run within a trap harness. `CActive` provides `RunError()` to handle and leaves occurring within `RunL()`.

```
TInt CGdpStateMachine::RunError(TInt aError)
/**
    This is called if RunL leaves.
    This will be through CompleteL or ErrorL leaving.
    Handle the error in the state machine.
**/
    {
    TState* nextState = ErrorOnStateExit(aError);
    ChangeState(nextState);
    return KErrNone;
    }
```

Any errors are passed to the concrete state machine's `ErrorOnStateExit()` error handler.

Note that `SetActive()` is *never* called by the state machine. This is entirely the responsibility of the concrete state classes, whenever they make an asynchronous request.

The remaining state machine functions are fairly straightforward. The most interesting are `DoCancel()`, which simply calls the `Cancel()` member of the current state, and `SetNextState()`, which sets `iState` to the passed state, without actually going through the process of entering it. This is specifically so that any future calls to `ReEnterCurrentState()` will actually move to this new state, rather than the last one moved into. A more generic implementation would probably use a separate `iNextState` member to achieve this, but for our purpose I got away with using the `iState` variable to represent both current and next state, depending on the context in which it is used.

20.2   SMS Implementation

SMS (short message service) is part of the GSM Specification. The Symbian OS SMS implementation is based upon the ETSI GSM 03.40 v7.4.0 Technical realization of the SMS specification. This specification is available for download from ***http://www.etsi.org***.

One reason SMS is a great protocol is because with it you can play GDP games at a distance.

Here's an overview of how the SMS implementation is intended to work. First, the scenario for sending a message, which is as follows:

- The sender makes a move that is immediately converted to an SMS message

- The phone sends the SMS over the GSM network to the receiver phone.

And then, for receiving a message, which is as follows:

- The receiving player tells their GDP game to receive incoming messages.

- The GSDP server will channel the message to the game session for which it is intended. It will collect the GDP message from the phone. Non-GDP game SMS messages will be unaffected. Note: If the GSDP server is not running, the Messaging App will collect the message in the mail store and this will not work.

GDP-SMS is a networked implementation: the 'address' used to send a message is the other player's telephone number. However, because SMS behaves rather differently with different networks and handsets, an extra complication is added at this point. Some phones require you to specify a **Service Center Address** (SCA) to which messages are initially sent before being routed to their destination. The way SCAs are set depends on the phone and on the network. GDP-SMS uses knowledge about some phones and networks to get the SCA from the phone, but in some cases this can't be done, so the user is required to enter a long-form destination address of the form `destination-number@sca-number`.

The aim of my GDP-SMS implementation is to demonstrate GDP working over a large distance, and to provide you with a working example of how to use some of the sockets-based features within Symbian OS.

20.2.1 ESOCK and Symbian OS Support for SMS

Prior to Symbian OS v6.0, SMS functionality was provided through ETEL, the telephony server. As well as providing the facility for Transmission Control Protocol/Internet Protocol (TCP/IP) and dial-up networking to establish a point-to-point PPP data connection, ETEL also provided access to the phone books stored on GSM phones. This method requires a detailed knowledge of the structure and format of an SMS message.

From Symbian OS v6.0 on, an alternative approach is provided through ESOCK, the Sockets server. ESOCK provides a generic interface to communication end points through plug-in communication protocols, providing the ability to establish connections, send and receive data. A good example of a communication protocol is `tcpip.prt`, which contains a suite of protocols: UDP, TCP, ICMP (Internet Control Message Protocol), IP, and DNS, of which UDP and TCP are accessible via sockets to transfer data over IP.

Access to ESOCK plug-in protocols is via `RSocket`. The Sockets Client API `RSocket` provides functions for socket creation, reading, writing, passive connection, active connection, setting addresses, and querying addresses. Asynchronous commands are also available with `RSocket::Ioctl()`.

From Symbian OS v6.0, the GSM SMS Protocol Module `smsprot.prt` enables clients to send, receive, enumerate, and delete GSM SMS messages.

Our GDP-SMS implementation will be based on ESOCK and the `smsprot.prt` protocol.

Figure 20.3 introduces a few of the Symbian OS classes we will be using during our implementation; it also shows the relationship between some of the components we have been discussing.

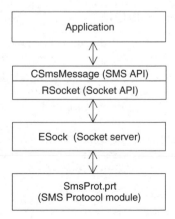

Figure 20.3

20.2.2   The GDP-SMS Message Format

The single requirement for the format of our GDP-SMS message is that it provides a means for our receiver to recognize the GDP-SMS message on the receiving end.

This means that we must define our own internal message format that supports these features (Figure 20.4).

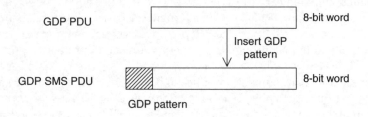

Figure 20.4

The data formats are as follows:

- GDP PDU. This is the chunk of data that's passed down to our GDP implementation through the CGdpSession interface, along with protocol-specific addressing information.

- GDP-SMS PDU. Here, we have put a unique pattern on the front so that we can spot our messages at the far end. In this case, "//BATTLESHIP". Note, alternatively, SMS Port numbers can be used.

The destination address passed down through the GDP interface is encoded directly into the SMS PDU, along with the SCA if present.

20.2.3   The GDP-SMS Implementation

The structure of the SMS code is illustrated below. The main component from the client's point of view is the CGdpSmsComms class shown in the center of the diagram. This implements CGdpSession, through which all client operations are performed (save for initially creating the object). This center class actually provides a simple, thin interface, or **facade**, into the various components used for SMS communications.

The three main subcomponents, CGdpSmsSender, CGdpSmsReceiver, and CGdpSmsResourceManager actually carry out the three tasks required for the GDP session (Figure 20.5).

It is interesting to note that the session class, CGdpSmsComms, has no state itself. This is because GDP-SMS, along with all GDP implementations, is actually stateless. This may come as a surprise after all the talk about state machines, but it does make sense when you consider that GDP is a connectionless, best-efforts datagram protocol. All the state is concerned with is attempting to get a specific packet in or out of the machine – no significant state is held over between packet transmissions.

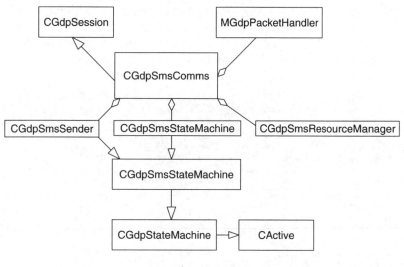

Figure 20.5

Through the following sections, you'll see an in depth example of how to use the Symbian OS sockets API to communicate over SMS.

Sender

The `CGdpSmsSender` class handles the task of creating an SMS message and sending it over via the socket server. It comes to life whenever the client code calls the `SendL()` member of the GDP-SMS session object, and continues asynchronous activity until the packet is successfully sent or has completed a bound number of unsuccessful retries.

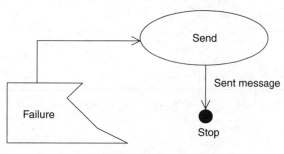

Figure 20.6

As Figure 20.6 shows, it is a very simple process to send GDP packets via SMS:

- *Send*: Open a socket, stream the SMS message to the socket, and call an `RSocket::Ioctl()` to start the send in action.

It is possible that the send operation may fail. I deal with this by using the handler mechanism of the generic state machine to do a limited number of retries before giving up.

The `Ioctl()` call on the socket is asynchronous in nature, so the `CGdpSmsSender` task class is derived from `CGdpSmsStateMachine` as follows (from `gdpsms.h`):

```
class CGdpSmsSender : public CGdpSmsStateMachine
//------------------
        {
public:
    class TSenderState : public CGdpSmsStateMachine::TState
            {
    public:
            TSenderState(CGdpSmsSender& aSmsSender);
    protected:
            CGdpSmsSender& iSender;
            };

    class TSendMsgState : public TSenderState
            {
    public:
            TSendMsgState(CGdpSmsSender& aSmsSender);
            void EnterL();
            TState* CompleteL();
            TState* ErrorL(TInt aCode);
            void Cancel();
    private:
            };

    friend class TSendMsgState;

public:
    CGdpSmsSender(CGdpSmsResourceManager& aResMan);
    ~CGdpSmsSender();
    void OpenL(MGdpPacketHandler& aHandler);
    void Close();
    void SendL(const TDesC8& aAddress, const TDesC8& aData);

protected:
    void Reset();
    TState* ErrorOnStateEntry(TInt aError);
    TState* ErrorOnStateExit(TInt aError);

private:

    RSocket                 iSocket;
    CSmsMessage*            iSmsMsg;
```

```
TPckgBuf<TUint>          iOctlResult;
TBool                    iGotScAddress;
MGdpPacketHandler*       iHandler;
TInt                     iRetries; // Remaining...

// States
TSendMsgState            iSendMsgState;

};
```

As you can see, this class actually inherits from `CGdpSmsStateMachine`, which is a very slightly specialized version of the abstract `CGdpStateMachine` class. I will introduce the facilities it adds when we discuss the resource manager later on.

The sender state machine defines a concrete state class – `TSendMsgState` along with one instance variable. This is the state object that the current state pointer in the abstract state machine will point to. The state class has a reference back to the sender object that it belongs to – this information is not automatically provided by the nesting construct, because it only affects the *class* relationship, not any specific *object* (or instance) relationship. I provide this reference variable in a generalized `TSenderState` class that all the concrete states derive from.

We'll now trace through the operation of sending a GDP-SMS message. The `OpenL()` function is called when the client code calls the corresponding function in `CGdpSmsComms`. This puts it into an initial state, by calling `Reset()`, ready to accept requests to send packets. It also stores a pointer to the GDP packet handler. This will be used to inform the handler of a completed send operation.

The state machine gets kicked into life every time the `CGdpSmsComms::SendL()` is called; this is the only point of contact with the GDP client in the whole process of sending.

```
void CGdpSmsComms::SendL(const TDesC8& aAddress, const TDesC8& aData)
    {
    RDebug::Print(_L("CGdpSmsComms::SendL() Called."));
    __ASSERT_ALWAYS(iSender != NULL, GdpUtil::Fault
      (GdpUtil::EProtocolNotOpen));
    iSender->SendL(aAddress, aData);
    }
```

This invokes the `SendL()` function within the sender class, which looks like this:

```
void CGdpSmsSender::SendL(const TDesC8& aAddress, const TDesC8& aData)
    {
    if (IsActive())
```

```
            return;       // Don't leave -- just quietly drop the
                          overflow packet

// Create the SMS message
CSmsBuffer* smsBuffer = CSmsBuffer::NewL();
iSmsMsg = CSmsMessage::NewL(iResMan.iFs, CSmsPDU::ESmsSubmit,
  smsBuffer);

TSmsUserDataSettings smsSettings;
smsSettings.SetAlphabet(TSmsDataCodingScheme::ESmsAlphabet7Bit);
smsSettings.SetTextCompressed(EFalse);

// Convert address to unicode string required by CSmsMessage
__ASSERT_ALWAYS(aAddress.Length() <= KGdpMaxAddressLen,
GdpUtil::Fault(GdpUtil::EBadSendAddress));

TBuf<KGdpMaxAddressLen> bufAddress;
bufAddress.Copy(aAddress);

iSmsMsg->SetUserDataSettingsL(smsSettings);
iSmsMsg->SetToFromAddressL(bufAddress);
iSmsMsg->SmsPDU().SetServiceCenterAddressL(KServiceCenterNumber);

// Convert data into unicode as CSmsBuffer will convert to
   appropriate
// format for us.
TBuf<KGdpSmsSduMaxSize> buf;
buf.Copy(aData);

// Insert our SMS pattern header so that our receiver is able
   to detect
// the incoming message, and then append the data.
smsBuffer->InsertL(0, KGdpSmsHeaderTag);
smsBuffer->InsertL(KGdpSmsPatternLen, buf);

iRetries = KGdpSmsSendRetries;
Reset();
ReEnterCurrentState();  // Kick us off again.
}
```

If we're already busy sending an SMS message, we simply give up on this new request and return – it is up to the client to ensure we're not given more packets than we can cope with.

We create a CSmsBuffer object to contain the text contents of our message. CSmsBuffer maintains the text content of the message in an array of TText objects, and provides functions to modify the content.

We then create a CSmsMessage object. This object helps to hide the underlying complexity of ensuring that the message is structured in the correct network format of an SMS message. TSmsUserDataSettings allows us to specify that the message is stored in 7-bit format. SMS supports both 7-bit and 8-bit data transport, but only 7-bit is universally implemented, so that's what I've chosen to use here. SMS supports a maximum message length of 160 7-bit characters.

aAddress contains the telephone number of the receiver. The only test we do with this number is to see that it does not exceed a certain size.

Interestingly, CSmsMessage accepts a Unicode string for the address to which the SMS will be sent, so we must convert the address from narrow text. We can do this by copying it into a Unicode descriptor. We then set the service center number.

Before inserting aData into the CSmsBuffer object, we have to ensure that we first insert the pattern that will enable the receiver to recognize the message as a GDP SMS message.

We then set up how many times we're going to retry sending if the first attempt fails and then set the state machine going.

The first time a packet is sent, the current state (iState) will already be set to TSendMsgState, so ReEnterCurrentState() will cause us to enter that state:

```
void CGdpSmsSender::TSendMsgState::EnterL()
{
// Close the socket as it may already be open.
iSender.iSocket.Close();

// Open a socket
TInt ret = iSender.iSocket.Open(iSender.iResMan.iSocketServer,
KSMSAddrFamily, KSockDatagram, KSMSDatagramProtocol);
User::LeaveIfError(ret);

// Bind to SMS port
TSmsAddr smsAddr;
smsAddr.SetSmsAddrFamily(ESmsAddrSendOnly);
iSender.iSocket.Bind(smsAddr);

// Open a write stream on the socket and stream our message.
RSmsSocketWriteStream writeStream(iSender.iSocket);
TRAP(ret, writeStream << *(iSender.iSmsMsg));
User::LeaveIfError(ret);

// message has not been sent at this point
TRAP(ret, writeStream.CommitL());
User::LeaveIfError(ret);

// Send the message
iSender.iSocket.Ioctl(KIoctlSendSmsMessage, iSender.iStatus,
&iSender.iPkgBuf, KSolSmsProv);
iSender.SetActive();
}
```

To initiate a send, I first close various resources that might already be open, reopen them, and then open a socket using the SMS protocol. The socket server is already opened by the resource manager. We specify that we are only sending an SMS when binding to the socket.

Next we open an `RSmsSocketWriteStream` object on the open socket, output the `CSmsMessage`, and commit the write.

At this point however the message has not been sent. We need to make an asynchronous call on the socket to complete the send.

If this fails and leaves, the appropriate sender state machine error handler is invoked:

```
CGdpStateMachine::TState* CGdpSmsSender::ErrorOnStateEntry
  (TInt /*aError*/)
    {
    return NULL;       // Give up!
    }
```

If the state entry completes successfully, however, the following code will be invoked by the `RunL()`:

```
CGdpStateMachine::TState* CGdpSmsSender::TSendMsgState::CompleteL()
    {
    RDebug::Print(_L("Message sent!!"));
    delete iSender.iSmsMsg;
    iSender.iSocket.Close();

    // Let handler know message sent successfully
    iSender.iHandler->SendComplete(KErrNone);
    return NULL;        // Last state so return NULL
    }
```

This simply releases the resources we have used to send the message and notifies the handler that we have successfully sent the message. Notice how we specify the next state to enter as NULL as we have completed the process.

If the send fails, the `ErrorL()` function is invoked:

```
CGdpStateMachine::TState* CGdpSmsSender::TSendMsgState::ErrorL(TInt aCode)
    {
    // Let handler know the reason for sent failure
    iSender.iHandler->SendComplete(aCode);
    User::Leave(aCode);
    return NULL;      // Never actually hit
    }
```

The sender state machine's error handler performs the following:

```
CGdpStateMachine::TState* CGdpSmsSender::ErrorOnStateExit(TInt /*aError*/)
    {
    Reset();
    if (--iRetries < 0)
        return NULL;
```

```
    return CurrentState(); // Force re-entry to initial state
    }
```

As asynchronous errors often indicate a communications error, we attempt to reset the state machine and do a limited number of retries.

If, for some reason, the operation is cancelled while the send is taking place (generally because the user chose to quit, and hence `Close()` has been called), the following will be called:

```
void CGdpSmsSender::TSendMsgState::Cancel()
    {
    if (iSender.iSocket.SubSessionHandle())
        iSender.iSocket.CancelIoctl();
    }
```

Receiver

Receiving an SMS message is a bit more complex than sending an SMS message. Once the client issues a `ReceiveAllL()`, we need to enter a wait state, and wait for SMS messages containing our GDP SMS specific pattern. In fact, this is all that we can do as all other SMS messages will be consumed by the messaging application. Once received, we can read the SMS message and then go back into the wait state and wait for the next GDP specific SMS message (Figure 20.7).

Note: If the SMS message is received before the `CGdpSmsReceiver` is Active, the message will be collected by the messaging app and will no longer be available to the GDP application.

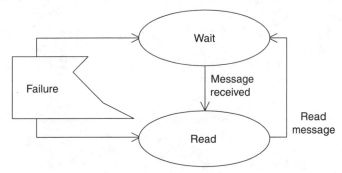

Figure 20.7

This requires the following asynchronous requests:

- Issue an `RSocket::Ioctl()` call on an SMS socket to find a matching pattern on incoming SMS messages.
- Issue an `RSocket::Ioctl()` call on an SMS socket to read an SMS message.

This directly tells us the states that are required, as shown in the state diagram above.

Issuing the wait for a GDP SMS message works like this:

```
void CGdpSmsReceiver::TWaitState::EnterL()
    {
    // Close the socket, as it may already be open.
    iReceiver.iSocket.Close();

    // Open a socket
    User::LeaveIfError(iReceiver.iSocket.Open(
    iReceiver.iResMan.iSocketServer, KSMSAddrFamily, KSockDatagram,
    KSMSDatagramProtocol));

    // Set the pattern to search for in incoming SMS messages. Messages,
       which do
    // not have our signature will be consumed by the Messaging
       Application.

    TSmsAddr smsAddr;
    smsAddr.SetSmsAddrFamily(ESmsAddrMatchText);
    smsAddr.SetTextMatch(KGdpSmsHeaderTag8());

    User::LeaveIfError(iReceiver.iSocket.Bind(smsAddr));

    // Wait for incoming messages
    iReceiver.iOctlResult()= KSockSelectRead;
    iReceiver.iSocket.Ioctl(KIOctlSelect, iReceiver.iStatus,
    &(iReceiver.iOctlResult), KSOLSocket);
    iReceiver.SetActive();
    }
```

First of all, as with `TSendMsgState::EnterL()` I close various resources that might already be open, reopen them, and then open a socket using the SMS. We then set the pattern match so that we receive only those messages matching the pattern and call an asynchronous control command on the socket to accept incoming SMS messages.

Successful completion invokes the following:

```
CGdpStateMachine::TState* CGdpSmsReceiver::TWaitState::CompleteL()
    {
    // Received a message so move to read state.
    return static_cast<TState*> (&iReceiver.iReadMsgState);
    }
```

As we have a message matching our pattern, we simply move on to the `TReadMsgState`.

```
void CGdpSmsReceiver::TReadMsgState::EnterL()
    {
```

```
// Create an empty message and buffer for our incoming message.
CSmsBuffer* buffer=NULL;
buffer=CSmsBuffer::NewL();
iReceiver.iSmsMsg = CSmsMessage::NewL(iReceiver.iResMan.iFs,
CSmsPDU::ESmsSubmit, buffer);

// Read the message.
RSmsSocketReadStream readstream(iReceiver.iSocket);
TRAPD(ret, readstream >> *(iReceiver.iSmsMsg));
User::LeaveIfError(ret);

// Let the socket know that we have read the message and it
   can be removed
// from the message store.
iReceiver.iSocket.Ioctl(KIoctlReadMessageSucceeded,
iReceiver.iStatus, NULL, KSolSmsProv);
iReceiver.SetActive();
}
```

Here I create a `CSmsBuffer` and `CSmsMessage` to store our extracted SMS message; then I open an `RSmsSocketReadStream` object on the open socket and stream the SMS from the socket into the `CSmsMessage`. As we have collected and dealt with the message, we send an Acknowledge to the SMS Service Center so it doesn't attempt to resend it.

On successful completion, we need to extract the contents of the message and pass this to the handler as follows:

```
CGdpStateMachine::TState* CGdpSmsReceiver::TReadMsgState::CompleteL()
    {
    // Extract the message contents
    CSmsBuffer& smsbuffer = (CSmsBuffer&)iReceiver.iSmsMsg->Buffer();
    TInt len = smsbuffer.Length();
    HBufC* hbuf = HBufC::NewLC(len);
    TPtr ptr = hbuf->Des();

    // We are only interested in the message contents, not our pattern.
    smsbuffer.Extract(ptr, KGdpSmsPatternLen, len-KGdpSmsPatternLen);

    // Convert from unicode data and send to handler
    TBuf8<KGdpSmsSduMaxSize> buf;
    buf.Copy(*hbuf);
    iReceiver.iHandler->GdpHandleL(KNullDesC8, buf);

    // Cleanup
    CleanupStack::PopAndDestroy(); // hbuf;
    delete iReceiver.iSmsMsg;
    iReceiver.iSocket.Close();

    // Return to wait state for next message.
    return static_cast<TState*> (&iReceiver.iWaitState);
    }
```

We remove the GDP SMS pattern from the message and convert the data into narrow format before passing the data to the handler. Finally, we set the next state to `TWaitState` to receive further GDP SMS messages.

As with `CGdpSmsSender`, if we discover an error, we attempt to reset the state machine and wait for the message to arrive again.

Resource manager

So far, I've just been using the resource manager without actually explaining what it is. Basically, it's a convenient place to hold onto any resources that are shared between the sender and the receiver. Specifically, these are connections to the socket server and the file server. It also implements a bit of functionality in the form of `OpenL()`, `Close()` and `ResetL()` functions to set up and maintain these resources. Keeping this stuff out of the central `CGdpSmsComms` class keeps the façade interface clean from any implementation detail, minimizing the dependency of the GDP client on the implementation's internals.

`CGdpSmsResourceManager` is a simple class, as it does not have any of the active management functionality and does not carry out any asynchronous task by itself, but simply acts as a single point of contact (or proxy) for resource access.

```
class CGdpSmsResourceManager : public CBase
//--------------------------
    {
public:
    ~CGdpSmsResourceManager();
    void OpenL();
    void Close();
public:
    RSocketServ iSocketServer;
    RFs iFs;
    };
```

As you can see, the resource manager is not `CActive` derived – it just holds the shared resources used by both the sender and the receiver.

20.3   Bluetooth Implementation

Bluetooth is a short-range wireless communications technology, standardized by the Bluetooth Special Interest Group (SIG) in 1998. With an operating range of approximately 10 m, it is an ideal technology for sharing information between devices. It also does not suffer the shortcomings of infrared, which requires direct line of sight.

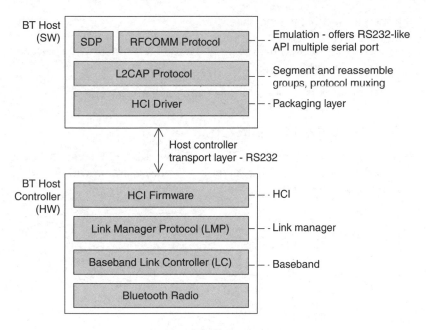

Figure 20.8

Bluetooth is composed of a hierarchy of components, referred to as the Bluetooth stack. This consists of both physical hardware components and software components. Figure 20.8 shows this relationship.

The hardware components that implement the lower level of the stack are typically not directly accessible by software applications. The interface to the hardware is provided by the Host Controller Interface (HCI) driver.

Applications typically interact with Bluetooth using either of the following:

- RFCOMM: which provides a serial like interface.

- L2CAP: which provides finer control over the bluetoothBT connection.

For our GDP-BT implementation, we will be using the second approach (L2CAP) which is the typical approach used by applications. RFCOMM is used more often for legacy code requiring a serial interface.

Here's an overview of how the BT implementation is intended to work. First, the scenario for sending a message:

- sender makes a move,

- sender selects a BT device in range to send move to,

- a connection is made to the receiver and the packet is transferred.

And then for receiving a message:

- receiving player tells their GDP game to receive incoming messages,
- receiver listens for incoming GDP connections,
- receiver accepts GDP connection and packet is accepted.

20.3.1   Symbian OS Support for Bluetooth

Access to either RFCOMM or L2CAP from Symbian OS is via the socket server client APIs. The socket server has a plug-in framework for which BT.PRT implements the generic interface specified by ESOCK.

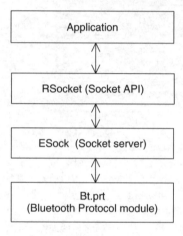

Figure 20.9

Figure 20.9 shows that the Socket Server can provide access to both RFCOMM and L2CAP. We can also see that there is no direct access to the HCI or Link Layer components. However, it is possible to indirectly access the HCI and Link Layer components through Ioctl() commands via RFCOMM and L2CAP sockets.

However, we will not need access to the Link Layer or HCI for our application and will concentrate on the L2CAP layer.

20.3.2   The GDP-BT Protocol Mapping

Unlike GDP-SMS, we do not need to modify the GDP-PDU that is passed down to our GDP-BT implementation from the CGdpSession interface.

We will simply store the data as an 8-bit descriptor array. In this format, the data can be sent to the receiver using standard Socket APIs.

The destination address passed down through the GDP interface is used to create a connection to a specific BT device and on a specific port.

20.3.3 The GDP-BT Implementation

The structure of GDP-BT is intentionally very similar to that of GDP-SMS, so there is no need to explain a whole new class structure. In particular, the BT communications component has two `CGdpStateMachine`-derived classes called `CGdpBTSender` and `CGdpBTReceiver` in much the same way that the SMS communications component did for carrying out the basic sending and receiving tasks.

Sender

The `CGdpBTSender` class handles sending a data packet over Bluetooth. The most difficult part of the operation is the initial connection.

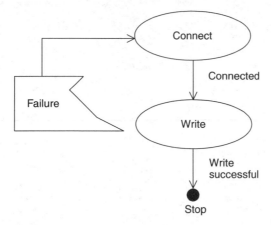

Figure 20.10

Figure 20.10 shows two sequential operations are required to send GDP packets via Bluetooth:

- *Connect*: make an L2CAP connection.
- *Write*: send the packet to the connected receiving device.

At any point in the sequence, an operation may fail. As with SMS, I deal with this – the generic state machine handler mechanism. Both the operations are asynchronous, so `CGdpBTSender` is derived from `CGdpBtStateMachine` as follows (from gdpbt.h):

```
class CGdpBTSender : public CGdpBTStateMachine
    {
    // State value declarations
```

```
    class TSenderState : public CGdpStateMachine::TState
        {
    public:
        TSenderState(CGdpBTSender& aSender);
        virtual TState* ErrorL(TInt aCode);
    protected:
        CGdpBTSender& iSender;
        };

    class TConnectState : public TSenderState
        {
    public:
        TConnectState(CGdpBTSender& aSender);
        void EnterL();
        TState* CompleteL();
        void Cancel();
        };

    class TWriteState : public TSenderState
        {
    public:
        TWriteState(CGdpBTSender& aSender);
        void EnterL();
        TState* CompleteL();
        void Cancel();
        };

    friend class TConnectState;
    friend class TWriteState;

public:
    CGdpBTSender(CGdpBTResourceManager& aResMan);
    void SendL(const TDesC8& aAddress, const TDesC8& aData);
    ~CGdpBTSender();
    void OpenL();
    void Close();

protected:
    void Reset();

    TState* ErrorOnStateEntry(TInt aError); ///< Override from
        CGdpStateMachine
    TState* ErrorOnStateExit(TInt aError); ///< Override from
        CGdpStateMachine

private:
    RSocket         iWriteSocket; ///< Socket to use for sending data out
    TBuf8<KGdpBTMaxPacketLength> iPacket; ///< Place to store the data
        to go out
    TBTSockAddr     iAddr; ///< BT address of device to send packet to
    TInt            iRetries; ///< No. of (potential) remaining retires.

    // States
    TConnectState   iConnectState;
    TWriteState     iWriteState;
};
```

The sender state machine defines two concrete state classes – `TConnectState` and `TWriteState` along with two instance variables.

As with GDP-SMS, the `SendL()` function is called when the client code calls the corresponding function in `CGdpBTComms`.

```
void CGdpBTSender::SendL(const TDesC8& aAddress, const TDesC8& aData)
    {
    TLex8 lex(aAddress);
    lex.Mark();
    if (!(lex.Get() == "0" && lex.Get() == "x"))
        lex.UnGetToMark();

    TInt64 addr;
    User::LeaveIfError(lex.Val(addr, EHex));

    iAddr.SetBTAddr(TBTDevAddr(addr));

    // Check the size of the data we're trying to send:
    //   - if it's bigger than the maximum we can store in our member
    //     variable then call GdpUtil::Fault()
    __ASSERT_ALWAYS(aData.Size() <= iPacket.MaxSize(),
    GdpUtil::Fault(GdpUtil::EBadSendDescriptor));
    // If the current state is NULL then panic as we're not ready
    __ASSERT_DEBUG(CurrentState() != NULL,
    GdpUtil::Panic(GdpUtil::ESenderNotReady));
    // If we're already waiting for another packet to be sent, we can't
    // send another one but we don't leave, we just quietly drop the
    // overflow packet and quit this send
    if (IsActive())
        return;
    // We're still here so it's OK to send the packet.
    iPacket.Copy(aData); // Buffer packet until needed
    // Set the number of retries to the maximum allowed
    iRetries = KGdpBTSendRetries;
    ReEnterCurrentState(); // Kick us off again but remain connected...
    }
```

As well as checking the content and size of the `aData`, `SendL()` also checks for a valid BT device address, and stores this for when a connection is required. The data we wish to send is also stored at this point. We then enter the `TConnectState`.

```
void CGdpBTSender::TConnectState::EnterL()
    {
    DEBUG_PRINT(_L("Entering CGdpBTSender::TConnectState::EnterL"));

    // What are we going to attempt to connect to?
    TProtocolDesc& desc = iSender.iResMan.ProtocolDesc();

    // Close the socket, just in case it's already open
    iSender.iWriteSocket.Close();

    // Now open the socket according to what the Resource Manager told us
```

```
User::LeaveIfError(
    iSender.iWriteSocket.Open (iSender.iResMan.SocketServer(),
                               desc.iAddrFamily,
                               desc.iSockType,
                               desc.iProtocol)
                    );

// Use this hard-coded port num for GDP
// GDP will be port 11, 1 = SDP, 3 = RFCOMM, etc
iSender.iAddr.SetPort(0x000b);

// Send the connect request out of the socket!
iSender.iWriteSocket.Connect(iSender.iAddr, iSender.iStatus);
// and wait until we're connected...
iSender.SetActive();
DEBUG_PRINT(_L("Leaving CGdpBTSender::TConnectState::EnterL"));
}
```

The protocol we will be connecting to is the Bluetooth L2CAP protocol. Our resource manager is responsible for ensuring that such a protocol exists. We specify an unused port that we will send data through, in this case port 11, and of course we must ensure that the receiver is listening to port 11.

Using the BT device address we received in `CGdpBTSender::SendL()` we try to connect to that device, and if successful we move to the write state.

```
CGdpStateMachine::TState* CGdpBTSender::TConnectState::CompleteL()
    {
    // The connect worked, so now we want to write same data
    //  - go into the write state
    return &(iSender.iWriteState);
    }
```

Once connected, the sockets interface makes sending the packet straightforward and we can simply call `RSocket::Write()`.

```
void CGdpBTSender::TWriteState::EnterL()
    {
    // Initiate a write...
    iSender.iWriteSocket.Write(iSender.iPacket, iSender.iStatus);
    // and wait...
    iSender.SetActive();
    }
```

If the write is successful, we can let the handler know. We have also finished the send operation, have no further states to go onto, and can return NULL.

```
CGdpStateMachine::TState* CGdpBTSender::TWriteState::CompleteL()
    {
    // Hey! We're done! Stop going through the state machine
    iSender.iHandler->SendComplete(KErrNone);
    return NULL;
    }
```

Receiver

Receiving a GDP packet over Bluetooth differs very little from how you would receive data from any other socket. We must **listen** for a connection, **accept** the connection, and then read our data from the socket.

Figure 20.11 shows, two sequential operations are required to receive GDP packets via Bluetooth:

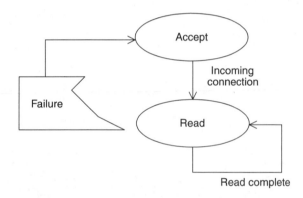

Figure 20.11

- Accept: the incoming connection from sender.
- Read: read the packet from the connected device and while we have an active connection we can wait for additional packets.

Before a connection can be accepted, we must give some consideration to device security. From the earliest Bluetooth specification, device security has always featured heavily. Typically connections cannot be made between devices without the consent of the user. For this reason, we will need to create a session with the Bluetooth security manager (RBtMan) to modify security settings for incoming GDP connections, to remove the requirement for user interaction, as well as for disabling additional security features including data encryption.

As we can see in CGdpBTReceiver::OpenL(), we follow a standard approach in setting up our socket to listen for a connection:

```
void CGdpBTReceiver::OpenL(MGdpPacketHandler& aHandler)
    {
    DEBUG_PRINT(_L("Entering CGdpBTReceiver::OpenL"));

    // Make sure the current iHandler is NULL - if it isn't then
    // we've already got a handler so Panic
    __ASSERT_DEBUG(iHandler == NULL, GdpUtil::Panic
        (GdpUtil::EReceiverInUse));

    iHandler = &aHandler;

    // Get the descriptor for protocol we are using. Better be
        Bluetooth...
    TProtocolDesc& desc = iResMan.ProtocolDesc();

    // Open the Listen Socket
    User::LeaveIfError(iListenSocket.Open(iResMan.SocketServer(),
                        desc.iAddrFamily,
                            desc.iSockType,
                            desc.iProtocol));

    // Set up the port to listen on
    TInt port = 0x0b; // Port 11 for GDP

    iAddr.SetPort(port);

    // Now set the socket to listen on the right port
    TInt ret = iListenSocket.Bind(iAddr);

    // If the Bind() didn't succeed we're outta here!

    if (ret != KErrNone)
        User::Leave(KErrInUse);              // Port is in use

    // Listen with a queue size of 4 (random) to allow more than
    // one connection
    err = iListenSocket.Listen(4);

    User::LeaveIfError(err);
    DoSecurityParams();

    // Now start the state machine off to wait for an incoming connection
    ChangeState(&iAcceptState);
    }
```

The main additional step was to make a call to DoSecurity-Params(), which will create our session with the security manager to make our modifications:

```
void CGdpBTReceiver::DoSecurityParams()
    {
    RBTSecuritySettings secset;
    RBTMan btman;
    TBTServiceSecurity service;
```

```
TUid uid;
uid.iUid = 0x1234;
service.SetUid(uid);
service.SetChannelID(11);
service.SetProtocolID(KSolBtL2CAP);

service.SetAuthentication(EFalse);
service.SetAuthorisation(EFalse);
service.SetEncryption(EFalse);
User::LeaveIfError(btman.Connect());
User::LeaveIfError(secset.Open(btman));
TRequestStatus requestStatus;
secset.RegisterService(service, requestStatus);
User::WaitForRequest(requestStatus);
User::LeaveIfError(requestStatus.Int());
}
```

We have informed the security manager that L2CAP connections on port 11 will not require user authorization, that the device will not need to be authenticated, and that communication will not be encrypted.

```
void CGdpBTReceiver::TAcceptState::EnterL()
    {
    RSocket& readSock = iReceiver.iReadSocket;

    readSock.Close();
    User::LeaveIfError(
        readSock.Open(iReceiver.iResMan.SocketServer()));

    // Tell socket we'll accept an incoming connection...
    iReceiver.iListenSocket.Accept(readSock, iReceiver.iStatus);
    // ...and wait
    iReceiver.SetActive();
    }
```

In the sockets API, Listen() simply queues incoming connections. In TAcceptState::EnterL(), we create a new socket that will accept one of the queued incoming connections. When this happens, we are ready to read our GDP data packet.

```
CGdpStateMachine::TState* CGdpBTReceiver::TAcceptState::CompleteL()
    {
    // Got an incoming connection!

    // Where'd it come from?
    TL2CAPSockAddr sockAddr;
    iReceiver.iReadSocket.RemoteName(sockAddr);

    // Convert and store as hex in the remote address descriptor
    TBTDevAddr remAddr = sockAddr.BTAddr();
    iReceiver.iRemAddr.Zero(); // Set length to zero
```

```
for (TInt i=0; i<KBTMaxDevAddrSize; ++i)
    {
    // 00 d0 b7 03 0e c5
    TUint16 num = remAddr[i];
    // Hack to pre-pend with a 0 if single digit hex
    if(num < 0x10)
            iReceiver.iRemAddr.AppendNum(0, EHex);

    iReceiver.iRemAddr.AppendNum(remAddr[i], EHex);
    }

// Now read some data
return &(iReceiver.iReadState);
}
```

Before reading the data we have the perfect opportunity to extract the address of the BT Device that sent the GDP packet. This address will be passed back to the handler.

```
void CGdpBTReceiver::TReadState::EnterL()
    {
    iReceiver.iReadSocket.Read(iReceiver.iPacket, iReceiver.iStatus);
    iReceiver.SetActive();
    }
```

We read the GDP packet with a standard socket Read().

```
CGdpStateMachine::TState* CGdpBTReceiver::TReadState::CompleteL()
    {
    // _LIT(KNullAddress,");
    // Complete read by telling handler where the packet came from
    // and what was in it
    iReceiver.iHandler->GdpHandleL(iReceiver.iRemAddr,
        iReceiver.iPacket);
    return this;        // Queue another read, while the channel is open
    }
```

We finally pass the received address and data to the handler.

Resource manager

As with GDP-SMS, there is a resource manager in GDP-BT that holds the shared resources used by both the sender and the receiver.

```
class CGdpBTResourceManager : public CBase
    {
public:
    CGdpBTResourceManager();
    ~CGdpBTResourceManager();
```

```
    void OpenL();
    void Close();
    // Resource access
    inline RSocketServ& SocketServer();
    inline TProtocolDesc& ProtocolDesc();

private:
    RSocket iSocket;
    // Shared resources
    RSocketServ     iSocketServer;
    TProtocolDesc   iProtocolDesc;
    };
```

Inline functions are provided to access these resources:

```
RSocketServ& CGdpBTResourceManager::SocketServer()
    { return iSocketServer; }
TProtocolDesc& CGdpBTResourceManager::ProtocolDesc()
    { return iProtocolDesc; }
```

The resource manager is responsible for finding the appropriate protocol we will use to connect our sender and receiver. As I mentioned before, we will be using the L2CAP protocol.

```
void CGdpBTResourceManager::OpenL()
//
// Allocate shared resources
//
    {
    // Connect to the Socket Server
    User::LeaveIfError(iSocketServer.Connect());
    // Find something in there that will give us a protocol called
    // "L2CAP" (hopefully, Bluetooth :) )
    // This fills in the Protocol Descriptor, iProtocolDesc
    User::LeaveIfError(iSocketServer.FindProtocol(_L("L2CAP"),
        iProtocolDesc));
    }
void CGdpBTResourceManager::Close()
//
// Release shared resources
//
    {
    iSocketServer.Close();
    }
```

20.4   Summary

In this chapter we have presented two GDP Protocol implementations – GDP SMS, and GDP BT. We have seen how both protocol implementations while based upon very different underlying technologies, can be accessed in the same way via the generic ESOCK plug-in interface.

In the SMS section, we have seen how the Symbian OS ESOCK sockets API is used to send and receive SMS messages via a GSM network. We have seen some utility classes including `CSmsBuffer` and `CSmsMessage`, which encapsulate and hide unnecessary network implementation details.

In the Bluetooth implementation, we have seen how it is possible to connect a sender and receiver and send a packet over Bluetooth. We have seen how we use the socket API's to connect devices and transfer data, and how we can change Bluetooth security settings using the Bluetooth Security Manager (`RBtMan`).

Appendix 1

Example Projects

This appendix details the projects described throughout the book.

Source code is available from `http://www.symbian.com/books/scmp/support/scmpdownloads.html`. You can download the examples to any location you choose, although the text of this book assumes that they are located in subdirectories of a `\scmp` top-level directory of the drive into which the UIQ SDK is installed.

The example subdirectories usually contain one project each. About half are independent example projects, which are covered in association with various topics in the book. The remaining projects all build up to Battleships.

If there are any specific instructions, or additional information about a project, they will be found in a `readme.html` file in that project's top-level directory.

Programs have been tested under the emulator on Windows NT and on a Sony Ericsson P800 phone.

Symbian OS C++ for Mobile Phones.   Edited by Richard Harrison
© 2003 John Wiley & Sons, Ltd   ISBN: 0-470-85611-4

The Independent Projects

Example	Purpose	Chapter
`active`	Basic active objects example.	17
`Bossfile`	Contains `dfbosswrite` subproject. For completeness, includes other subprojects that contain example code related to the Boss Puzzle.	13
`buffers`	Dynamic buffers.	8
`drawing`	Device-independent drawing, with a reusable view and support for zooming.	15
`exeloader`	An application that loads and runs text-based applications, such as `hellotext`.	1
`hellogui`	Hello World, GUI version.	4
`helloguifull`	Hello World, GUI version, with finishing touches.	14
`hellotext`	Hello World, text version. The code framework is used as a basis for the `buffers` and `string` examples.	1
`memorymagic`	How to allocate memory and clean it up again – and how not to.	6
`Streams`	Using files and stream APIs to save and load data.	13
`Strings`	Symbian OS strings, using descriptors.	5

The Battleships Projects

Example	Purpose	Chapter
`battleships`	The full, communicating, two-player Battleships game.	16
`soloships`	Solo Ships, a single-player game.	9
`tp-ships`	A simple two-player version.	9
`tp-viewtest`	A program to test the Battleships application's views.	9

The TOGS Projects

These projects are related to the TOGS reference material in Appendix 3. They are all needed for `battleships`, but may also be used for other purposes:

Example	Purpose
`gdp`	Game Datagram Protocol (GDP) – the basic comms interface, plus three implementations (loopback, Bluetooth and SMS).
`gdpchat`	Test chat program using GDP.
`gsdp`	Game Session Datagram Protocol (GSDP) – links packets into sessions, and distinguishes session types so that different games can be played.
`gsdpchat`	Test chat program using GSDP.
`rgcp`	Reliable Game Conversation Protocol (RGCP) – adds acknowledgements, resending, and piggy-backing to the GSDP session, so that an RGCP client can rely on the packet data it handles.
`rgcpchat`	Test Converse program using RGCP.

Appendix 2

Developer resources

Symbian Developer Network

Symbian Developer Network is at the hub of the Symbian OS development. With partners that include network operators, tool providers and mobile phone manufacturers, you can find the resources you need in just one place. This appendix is only a snapshot of the resources available to you and more are being released all the time. Check the Symbian DevNet website regularly for announcements and updates, and subscribe to the Symbian Community Newsletter for updates.

http://www.symbian.com/developer

Symbian OS developer tools

For the latest see *http://www.symbian.com/developer/tools.html*
Symbian DevNet and its partners offer various tools:

AppForge

AppForge development software integrates directly into Microsoft Visual Basic, enabling you to immediately begin writing multiplatform applications using the Visual Basic development language, debugging tools and interface.

http://www.appforge.com

Symbian OS C++ for Mobile Phones. Edited by Richard Harrison
© 2003 John Wiley & Sons, Ltd ISBN: 0-470-85611-4

Borland

Borland offers a range of tools for all developers including JBuilder Mobile-Set.
http://www.borland.com

Forum Nokia

In addition to a wide range of SDKs, Forum Nokia offers various development tools for download, including the Nokia Developer Suite for J2ME, which plugs in to Borland's JBuilder MobileSet or Sun's Sun One Studio integrated development environment.
http://www.forum.nokia.com

Metrowerks

Metrowerks offer the following products supporting Symbian OS development:

- CodeWarrior Wireless Studio, Nokia 9200 Communicator Series Edition
- CodeWarrior Development Tools for Symbian OS Professional Edition
- CodeWarrior Development Tools for Symbian OS with Personal-Java Technology

http://www.metrowerks.com

Sun Microsystems

Sun provides a range of tools for developing Java 2 Micro Edition applications including the J2ME Wireless Toolkit and Sun One Studio Mobile Edition.
http://java.sun.com

Texas Instruments

Easy-to-use software development environments are available for OMAP application developers, OMAP media engine developers and manufacturers. Tool suites including familiar third party tools and TI's own industry leading eXpressDSP tools are available, allowing developers to easily develop software across the entire family of OMAP processors.
http://www.ti.com

Tools archive

Symbian DevNet offers the following tools as an unsupported resource to all developers:

- Symbian OS SDK add-ons
  http://www.symbian.com/developer/downloads/v6sdks.html

- Symbian OS v5 SDK patches and tools archive
  http://www.symbian.com/developer/downloads/archive.html

Symbian OS SDKs

For the updates and links see *http://www.symbian.com/developer/SDKs.html*

The starting point for developing applications for Symbian OS phones is to obtain a software development kit (SDK). Symbian OS SDKs support development in both Java and C++. They provide binaries and tools to facilitate building and deployment of Symbian OS applications, full system documentation for APIs and tools PC-based emulation of Symbian OS phones. Example applications with supporting documentation SDKs for the following versions of Symbian OS are currently available.

Development languages other than Java and C++ are supported through other SDKs and SDK extensions.

Symbian OS v7.0

The UIQ SDK for Symbian OS v7.0 is available for download from Ericsson Mobility World.

This facilitates development in Symbian OS C++ or Java for the Sony Ericsson P800 and P802 smartphones. Symbian OS C++ developers need to obtain CodeWarrior Development Studio for Symbian OS from Metrowerks. This is available in Professional and Personal editions.

Java developers developing PersonalJava applications (optionally taking JavaPhone APIs) will need JDK 1.1.8, which is available for free download from Sun. Java developers developing MIDlets will need Sun's J2SE SDK, version 1.3 or higher, and Wireless Developer Toolkit, both of which are available for free download. A selection of IDEs is also available for use in conjunction with the Wireless Developer Toolkit.

Symbian OS v6.1

The Series 60 SDK for Symbian OS is available from Forum Nokia. This enables development in Symbian OS C++ for the Nokia 7650, Nokia 3650 imaging phones and the N-Gage mobile game deck.

C++ developers need to obtain Microsoft Visual C++ 6.0. Alternatively, Forum Nokia is making available the Nokia Series 60 C++ Toolkit 1.0, which bundles the new Borland C++ Mobile Edition, an extension to their popular C++ Builder, with a Borland-compatible build of the Series 60 emulator and associated binaries.

For Java MIDP development, the Nokia Series 60 MIDP SDK Beta 0.1 for Symbian OS is available from Forum Nokia. Java developers will also need Sun's J2SE SDK, version 1.3 or higher, and Wireless

Developer Toolkit, both of which are available for free download. A selection of IDEs is also available for use in conjunction with the Wireless Developer Toolkit.

Symbian OS v6.0

The Nokia 9200 Communicator Series SDK for Symbian OS v6.0 is available from Forum Nokia. This enables development in Symbian OS C++ or Java for the Nokia 9210, 9210c, 9210i and 9290 communicators. Localized Chinese versions of the SDK are available too.

C++ developers will need to obtain Microsoft Visual C++ 6.0.

Java developers developing PersonalJava applications (optionally taking advantage of JavaPhone APIs) will need JDK 1.1.8, which is available for free download from Sun. MIDP development is not currently supported for the 9200 Series Communicator.

Symbian OS v5

Psion has a long history of using Symbian OS, and several leading PDAs still use Symbian OS v5. Current products include the netBook and the netPad. Developers can make use of Symbian OS v5 SDKs and SDK extensions to target Psion PDAs in C++, Java or OPL.

Other SDKs and SDK extensions

OPL

OPL is a BASIC-like language designed to allow rapid application development, with provision for on-target application creation. OPL was included in all open Symbian OS products up to and including Symbian OS v5 and is now open source for the Nokia 9200.

OPL development for Symbian OS v5 requires the v5 OPL SDK. The SDK includes documentation, tools, example code and a Windows emulator for testing. No further tools are required. If you are already familiar with OPL, you can begin developing on Symbian OS v5 directly using the supplied program application.

For Symbian OS v6.0 phones (i.e. the Nokia 9200 Series Communicator), OPL is supplied as an unsupported add-on by Symbian. End users can download the runtime, which allows them to run OPL applications. OPL developers can download a development kit and the necessary tools and example code required to create OPL applications. A program editor for the Nokia 9200 Series is also supplied.

Visual Basic

A Visual Basic development environment is available from AppForge. You will need the MobileVB Mobile Application Development Software, which integrates with Visual Basic 6.0.

To deploy and test your application you will need to download AppForge's free Booster software. The Symbian OS download provides versions that can be installed onto the PC-based phone emulator provided with Symbian OS SDKs and onto the Nokia 9200 Series.

Intel PCA Development Kit

The Intel DBPXA250 Development Platform is a tool aimed at software developers, system integrators and OEMs targeting Intel PCA processors. Board support packages are available for Symbian OS.

Texas Instruments Innovator Development Kit

Texas Instruments' Innovator Development Kit for the OMAP platform is a tool aimed at software developers, system integrators and OEMs targeting TI's OMAP processors. Support is available for Symbian OS v6.x and v7.0.

Developer support

For updates and links see ***http://www.symbian.com/developer/support.html***

Symbian DevNet offers two types of support forum:

- support newsgroups
- support forum archive

Symbian DevNet's partners also offer support:

Ericsson Mobility World

As well as tools and SDKs, Ericsson Mobility World provides a range of services including newsletters and support packages for developers working with the latest Sony Ericsson products such as the Symbian OS powered P800.
> ***http://www.ericsson.com/mobilityworld***

Forum Nokia

As well as tools and SDKs, Forum Nokia provides newsletters, the Knowledge Network, fee-based case-solving, a knowledge base of resolved support cases, discussion archives and a wide range of C++ and Java-based technical papers of relevance to developers targeting Symbian OS.
> ***http://forum.nokia.com***

Appendix 3

TOGS Guide and Reference

In this appendix, I describe the main constituents of TOGS – Transaction Oriented Games Stack:

- GDP, providing unreliable datagram service
- GSDP, adding sessions
- RGCP, adding reliability and request/response conversation support
- BSP, implementing the processing for Battleships.

I'll describe each of these protocols in the following way:

- Introduce it and say why it's there
- Describe the protocol in the abstract
- Outline the Symbian OS implementation
- Point out what future development is needed on this protocol.

GDP

GDP is the **Game Datagram Protocol**. Its purpose is to provide the simplest possible interface for sending and receiving packets of data. As a client, you call a `SendL()` function, specifying a to-address and some data – a datagram (see Figure A3.1). A GDP implementation transfers this packet to the target address, where software executes a `GdpHandleL()` function whose parameters include the from-address and the data.

The address received by `GdpHandleL()` should be such that it can be used to generate a reply to the sender.

The address format is defined by the GDP implementation. A **networked** GDP implementation requires addresses. A **point-to-point** GDP

Symbian OS C++ for Mobile Phones.   Edited by Richard Harrison
© 2003 John Wiley & Sons, Ltd   ISBN: 0-470-85611-4

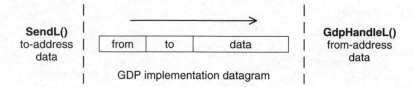

Figure A3.1

implementation doesn't require addresses: it relies on physical connectivity to get a datagram to the other endpoint. Both the SMS and
Bluetooth protocols are address-based. Clearly, loopback is a point-to-
point protocol.

GDP is not limited in its application to Symbian OS machines – a
GDP implementation may communicate with machines running other
operating systems as well. For this reason, a concrete GDP implementation should specify its physical data formats with sufficient precision for
another system to be able to implement a corresponding GDP stack that
connects to it.

GDP is unreliable. That means a request through SendL() is sent on a
best-efforts basis, but it's not guaranteed to arrive precisely once – it may
never arrive, or it may (in rare circumstances) arrive more than once. GDP
is not responsible for taking recovery action or for returning error codes
for these events to its clients. However, GDP *should* make reasonable
efforts and it *should* be possible for the end user to understand its reason
for failing. (The destination machine is turned off; a cable isn't connected;
the Bluetooth link goes out of range; etc.)

A GDP implementation should not fail because of timeouts in lower-
level protocol stacks. GDP is designed for sending packets in games
with a high concentration-to-action ratio such as chess. The time
between sending datagrams may be anything from several seconds to
minutes, quarter-hours, days, or even weeks, and all the while both
ends have a GDP session active. If a lower-level stack does have a
timeout so it can't be kept open indefinitely, the GDP implementation should manage the lower-level stack in such a way as to hide
this problem from the GDP client code. It could achieve this, for
instance, by reopening the lower-level stack whenever a datagram
is sent.

For receiving packets, GDP supports both **push** and **pull** protocols. All the concrete examples presented elsewhere are push protocols – the client is automatically notified of an incoming message.
Pull protocols require some action on the part of the user to initiate message retrieval and are therefore less desirable from a usability
perspective.

Symbian OS Implementation

The Symbian OS implementation of GDP is in \scmp\gdp\. It generates gdp.dll and gdp.lib, and exports gdp.h. It uses the ECom framework to manage the plug-in protocol implementations.

The GDP API consists of two interfaces. A client will use CGdpSession to send packets via the implementation, and implement MGdpPacketHandler to handle packets received from the implementation. A GDP implementation derives from CGdpSession and uses MGdpPacketHandler.

Here is the C++:

```
class MGdpPacketHandler
    {
public:
    virtual void GdpHandleL(const TDesC8& aFromAddress, const TDesC8&
        aData) = 0;
    virtual void SendComplete(TInt aErr) = 0;
    };
class CGdpSession : public CBase
    {
public:
    IMPORT_C static CGdpSession* NewL(TUid aUid);
    virtual void OpenL(MGdpPacketHandler* aHandler) = 0;
    virtual void SendL(const TDesC8& aToAddress, const TDesC8&
                        aData) = 0;
    virtual TInt ReceiveAllL() = 0;
    virtual TInt GetMaxPacketLength() const = 0;
    virtual TBool IsNetworked() const = 0;
    inline TUid Uid() const {return iDtor_ID_Key;};
protected:
    TUid iDtor_ID_Key;
    };
```

CGdpSession contains the following functions:

Function	Description
NewL()	Acts as a factory function for concrete implementations.
OpenL()	Constructs everything needed for the GDP implementation, and specifies the handler for received packets.
SendL()	Makes best efforts to send a datagram. You need to specify a to-address (needed by networked implementations) and data. This function may leave if resources need to be reallocated (because, for

	instance, they have timed out since `OpenL()`). Returns synchronously, but may cause an asynchronous process to be initiated for sending the datagram: errors are not reported. An implementation must make a private copy of the `aData` that is to be sent. The caller may reuse the data buffer any time after calling `SendL()`.
`ReceiveAll()`	Initiates an asynchronous process to cause any outstanding datagrams to be received. May return an error if resources need to be reallocated (because, for instance, they have timed out since `OpenL()`).
`IsNetworked()`	Returns `ETrue` if the protocol is networked. In this case, a nonempty to-address is required for `SendL()` calls, and a nonempty from-address is passed to `GdpHandleL()`.
`GetMaxPacketLength()`	Returns the maximum length of a GDP datagram, excluding addresses, which can be transmitted by the protocol implementation.
`Uid()`	Returns the Uid of the derived, concrete, implementation.

`MGdpPacketHandler` contains the following functions:

Function	Description
`GdpHandleL()`	Called to handle a packet that has been received, specifying the data and the from-address. For networked protocols, this enables you to reply to the sender. A handler should make a copy of the `aData` passed to it: the buffer may be reused by a GDP implementation immediately after `GdpHandleL()` returns.

SendComplete()   Called when the send initiated in SendL has
                      been completed. Not currently implemented.

GDP Loopback Implementation

The loopback implementation of GDP is useful for testing and also for
local game play using the GSDP server. CGdpLoopback is declared
in gdploop.h:

```
class CGdpLoopback : public CGdpSession
    {
public:
    CGdpLoopback();
    ~CGdpLoopback();
    static CGdpSession* NewL();
    // from CGdpSession
    void OpenL(MGdpPacketHandler* aHandler);
    void SendL(const TDesC8& aToAddress, const TDesC8& aData);
    TInt ReceiveAll();
    TInt GetMaxPacketLength() const;
    TBool IsNetworked() const;
private:
    MGdpPacketHandler* iHandler;
    };
```

It's simply a class that implements the pure virtual functions from
CGdpSession interface together with the infrastructure we need for
ECom. It's implemented in gdploop.cpp. Here is the complete imple-
mentation:

```
CGdpLoopback::CGdpLoopback()
    {
    }

CGdpLoopback::~CGdpLoopback()
    {
    REComSession::DestroyedImplementation(iDtor_ID_Key);
    }

void CGdpLoopback::OpenL(MGdpPacketHandler* aHandler)
    {
    iHandler=aHandler;
    }

void CGdpLoopback::SendL(const TDesC8& /*aToAddress*/, const TDesC8&
    aData)
    {
    _LIT8(KNullAddress,"");
    iHandler->GdpHandleL(KNullAddress, aData);
    }
```

```
TInt CGdpLoopback::ReceiveAll()
    {
    return KErrNone;
    }

TInt CGdpLoopback::GetMaxPacketLength() const
    {
    return KMaxTInt;
    }

TInt CGdpLoopback::IsNetworked() const
    {
    return EFalse;
    }

CGdpSession* CGdpLoopback::NewL()
    {
    return new (ELeave) CGdpLoopback();
    }
```

The key function, SendL(), ignores the address given, and sends a packet that looks as though it has come from an empty origin address. Provided the receiver copies the data from the sender's buffer (which is required by the GdpHandleL() contract), SendL() fulfills its contract to not use the sender's data after SendL() has returned.

GDP Chat

gdpchat is a straightforward test program that enables you to send text and displays the last-sent and last-received message.

The version provided on the book's CD requires you to rebuild it to select a new GDP protocol, or to change the phone number when using a protocol requiring an address.

Taking GDP Forward

Send() should be asynchronous

To make GDP more robust, GDP should post a flag when it is capable of sending another datagram. With the present design, it isn't possible to implement a transmit queue in the GSDP server, which means that a datagram is bound to be lost if one is already in the process of being sent. While GDP is allowed to lose packets, this kind of unreliability is arbitrary, and can only be fixed by making Send() report when it has finished. The SendComplete callback in MGdpPacketHandler is a starting point for the implementation.

GSDP

GSDP adds session capability to GSDP datagrams. On a single Symbian OS machine, all GDP implementations are run in a server. A client uses the GSDP client interface, not the GDP interface, to send data.

To the GDP datagram payload, GSDP adds the following IDs:

- A from-port ID: this is the nonzero ID of a port used by the sending client.

- A to-port ID: during an established session, this is a nonzero ID that identifies the client on the target Symbian OS phone (or other entity specified by a GDP address), which will receive the datagram.

- A game protocol ID: this is used in-session setup. When a client connects to the GSDP server on its machine, it specifies the game protocol it will use. If the client listens with its zero port ID, then an incoming packet with a zero port ID will be matched with a client's game protocol ID. Thus, the game protocol ID ensures that the session is set up with a compatible partner.

These IDs are all 32-bit numbers. The from-port and to-port IDs are allocated by the GSDP server in ascending sequence (see Figure A3.2). The game protocol ID is a UID. Theoretically, neither the game ID nor the from-port ID is needed throughout the session but they provide a useful redundancy check.

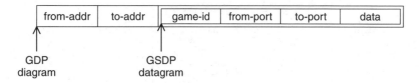

Figure A3.2

The GSDP datagram contents are passed as arguments to the GSDP send and handle functions. The session between the GSDP client and the GSDP server on the same machine carries the state required to set the nondata fields – from-port ID, to-address, to-port ID, and game protocol ID.

There is terminology confusion here: 'session' can be used to mean a GSDP session between two GSDP send-points, or a client-server session between an application and a server on a Symbian OS phone. The word 'session' is justified in both cases. I will try to be unambiguous.

By using a GDP loopback implementation inside the GSDP server, two GSDP clients on the same machine may communicate with each other.

Although GSDP is session-based, it is *not* reliable. A GSDP send is no more reliable than a GDP send: it may result in zero, one, or more receives at the destination.

The GSDP client API allows you to specify the GDP implementation to be used for a GSDP session. GDP implementations are managed by the GSDP server. The GSDP server uses GDP implementations, as specified above, without any change.

For a packet to reach a particular GSDP client successfully, its sender must specify the correct address, port, and game protocol ID. There are two interesting cases here:

- The to-port is nonzero: a GSDP client must be listening with the correct port ID. This is used for communication after a session has been set up.
- The to-port is zero: a GSDP client must be listening with a zero port ID and a matching game protocol ID. This is used for session setup.

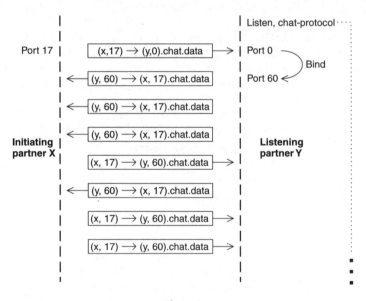

Figure 3

These two possibilities allow a session between two partners to be set up and then maintained. The session is set up by the **initiating** partner, which sends a packet with a nonzero from-port, a game protocol ID, and a zero to-port (see Figure A3.3). The session is accepted by a **listening** partner, which has a matching game protocol ID and a zero port ID. Once accepted, the listening partner allocates its own nonzero port ID, and sends back a packet to the initiating partner: this **binds** the session. Subsequent communication uses nonzero port IDs on both sides.

GSDP is intended to support two-player games with an arbitrarily long think time between moves. A GSDP session must persist even when the games are saved and closed at either or both ends. A client must save its GSDP state, including addresses and port IDs. The server provides functions that are able to support the client restoring its state. The server does not hold state information on clients' behalf beyond client-server session termination.

The server queues incoming datagrams and holds them so that they can be received by a dormant client, or by a new client launched in listening mode. The queue is managed as follows:

- If an incoming datagram has a nonzero to-port ID, which matches the port ID of an active listening client, then that request is satisfied by the datagram – that is, the datagram is received.

- If the datagram has a zero to-port ID, and a game protocol ID that matches an active client with a zero port ID, a matching game protocol ID and an outstanding receive request, then the request is satisfied and the datagram is received.

- If neither of these conditions is true, then the datagram is added to the queue.

- Whenever the game protocol ID or port ID of a client is changed, or a new receive request is issued, the queue is scanned to see if any datagrams in it match the rules above: if so, such datagrams are received.

- If a datagram is received by matching the above rules, but it doesn't match other sensible rules, then the datagram is dropped – that is, it's absorbed by the GSDP server and not sent to the client, and the client's receive request is not fulfilled. Examples of such 'sensible rules' include that the game protocol IDs must match when the port ID is nonzero, and the from-address should be as expected when the to-port ID is nonzero. These rules are based on redundant information in the GSDP packet, which allows a useful check to be performed.

- Packets on the queue may be expired according to rules at the discretion of the GSDP server implementer. If the queue is too large, then packets may not be accepted onto it when received by a GDP implementation. The present Symbian OS implementation expires packets only when the GSDP server is stopped – which happens when all its clients are stopped. The present Symbian OS implementation has a maximum queue length of 10: any additional packets are dropped.

- These awkward management issues notwithstanding, the queue is necessary because a client may not be started when an initiate request from a GSDP game on another Symbian OS machine arrives. Also, when a client's receive request is fulfilled, it doesn't issue a new one until it has handled the previous one, which causes a transient condition whereby the client is unable to receive.

GSDP specifies no formal mechanism for releasing a GSDP port ID, which means it's important that port IDs be allocated uniquely by a given GSDP server. The Symbian OS GSDP server maintains the last-allocated port ID in a file in `c:\System\Data\`: the port ID is incremented and restored every time a new session is initiated or accepted. This file persists across GSDP server invocations and guarantees unique port ID allocation provided that all port IDs are dropped by the time wraparound occurs. In practice, this is sufficient: if one port ID was allocated per second, it would take over 143 years to wraparound.

Symbian OS Implementation – Client-side

On Symbian OS, the GSDP client API is defined in `gsdp.h` and delivered in `gsdp.dll`. The client API consists of two classes: a concrete `RGsdpSession` class to control the session and send packets, and an abstract `MGsdpPacketHandler` class that you should implement for handling received packets (see Figure A3.4).

Additionally, a private, client-side active object, `CGsdpReceiveHandler`, turns the client listen function into a continuously renewed receive request. This removes any responsibility from the client to implement its own active object to handle or renew the request.

Here's `RGsdpSession`'s declaration in C++, showing additional housekeeping and getter functions:

```
class RGsdpSession : public RSessionBase
    {
public:
    // construct
    inline RGsdpSession() : iHandler(0) { };
    // open/close
    IMPORT_C void ConnectL(MGsdpPacketHandler& aHandler);
    IMPORT_C void Close();
    // Query supported protocols
    IMPORT_C TInt CountGdpProtocols() const;
    IMPORT_C TInt GetGdpProtocolInfo(TInt aProto,TGdpProtocolInfo&
            aInfo) const;

    // load and get GDP protocol
    IMPORT_C TInt SetGdpProtocol(TUid aProtocol);
    IMPORT_C TUid GetGdpProtocol() const;
    IMPORT_C TBool GdpIsNetworked() const;
```

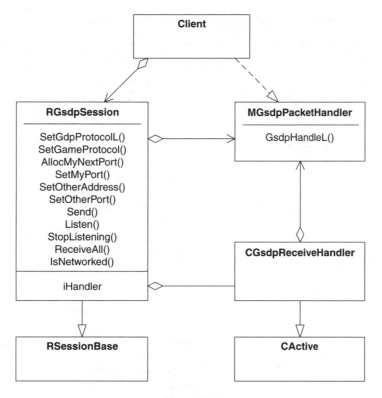

Figure A3.4

```
    // game protocol
    IMPORT_C void SetGameProtocol(TUint32 aProtocol);
    IMPORT_C TUint32 GetGameProtocol() const;
    // set and get my address and port
    IMPORT_C void SetMyPort(TUint32 aPort);
    IMPORT_C TUint32 GetMyPort() const;
    IMPORT_C TUint32 AllocMyNextPort();
    // set and get other address and port
    IMPORT_C void SetOtherAddress(const TDesC& aAddress);
    IMPORT_C void GetOtherAddress(TDes& aAddress) const;
    IMPORT_C void SetOtherPort(TUint32 aPort);
    IMPORT_C TUint32 GetOtherPort() const;
    // main protocol functions
    IMPORT_C void Listen();
    IMPORT_C void StopListening();
    IMPORT_C void Send(const TDesC8& aData);
    // initiate receive-all for "pull" protocols
    IMPORT_C void ReceiveAll() const;
private:
friend class CGsdpReceiveHandler;
    void Receive(TDes8& aBuffer, TRequestStatus& aStatus);
    void CancelReceive();
    CGsdpReceiveHandler* iHandler;
    };
```

The functions of `RGsdpSession` include the following:

Function	Description
ConnectL()	Connects to the GSDP server and specifies a GSDP packet handler. The server is launched if it is not already active.
Close()	Closes the session with the server. The server may choose to terminate if it has no more clients, but the client API does not mandate this.
GetGdpProtocolInfo()	Together with CountGdpProtocols(), can be used to enumerate the installed GDP protocols at run time.
SetGdpProtocol()	Sets the GDP implementation to be used by the session. Will return an error if the implementation is not available, or cannot be initialized.
GdpIsNetworked()	Indicates whether the GDP implementation is networked (requires valid addresses) or point-to-point (ignores addresses).
SetGameProtocol()	Sets the game protocol, specifying a 32-bit UID.
SetMyPort()	Sets my port ID, specifying the port ID as a 32-bit unsigned value. Use this function when restoring a session previously established.
AllocMyNextPort()	Allocates a unique port ID for me, and returns its value. Use this function when starting a new session.
SetOtherAddress()	Sets the address of the other partner in the communication, with a string. Used only by networked GDP implementations.
SetOtherPort()	Sets the port of the other partner in the communication, with a 32-bit

	unsigned value. Use this when setting up a session that has been previously established; set it to zero when starting a new session.
Listen()	Causes a Receive() request to be issued. When the request completes, the received datagram will be handled using MGsdpPack-etHandler::GsdpHandleL(). The receive request will then be renewed so that without client intervention any number of packets can be received.
StopListening()	Cancels any outstanding receive request started by Listen(). You may issue this function even if no call to Listen() has been issued or if it has already been canceled.
Send()	Sends a datagram using whatever current address, port, GDP implementation, and game protocol ID are specified.
ReceiveAll()	Causes the GDP implementation to do a ReceiveAll(). This may result in received packets that need to be handled for *any* GSDP session that uses the same GDP implementation, not only the GSDP session that issued the ReceiveAll().
GetXxx()	Getter functions for address, port, GDP implementation, and game protocol ID.

MGsdpPacketHandler is declared as,

```
class MGsdpPacketHandler
    {
public:
    virtual void GsdpHandleL(const TDesC8& aData) = 0;
    };
```

with the following function:

Function	Description
GsdpHandleL()	Handles a received packet. The server guarantees that the packet matches the game protocol and port numbers required by the protocol.

Server-side Symbian OS Implementation

The server is described elsewhere. It implements the requirements of the GSDP protocol's external behavior and that of the client interface.

GSDP Chat

GSDP chat is the test code for GSDP. You can find the C++ source in \scmp\gsdpchat\. In the GUI, the caption in English is Chat.

GSDP chat uses GSDP to set up and continue a conversation. All GSDP chat packets use the distinguishing game protocol ID 0x10005405.

Partners to a chat session may be in one of four states:

State	Meaning
Blank	Nothing happening, not even connected to GSDP.
Bound	A session is bound and active.
Initiating	The initiating partner has opened a session with the GSDP server, selected a GDP implementation, had a port ID allocated by the GSDP server, set the GSDP chat protocol ID to 0x10005405, and has sent a GSDP packet to a GDP address specifying a zero port ID. The initiating partner is now waiting for a reply from that GDP address: the reply will cause the session to bind.
Listening	A listening partner has opened a session with the GSDP server, selected a GDP implementation, is using a zero port ID, and has issued Listen(). The listening partner is expecting an incoming packet from any GDP address (using the same protocol), any GSDP port, and the GSDP chat protocol ID 0x10005405. When the packet comes in, the listening partner will accept it and bind the session.

A chat partner can only change to the bound state via listening or initiating. A bound session can only be established between two partners, one of whom initiates while the other listens (see Figure A3.5).

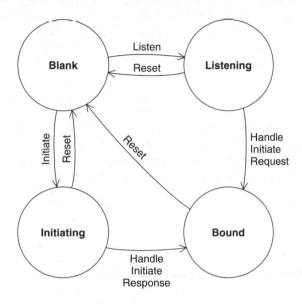

Figure A3.5

If two Chat users both try to initiate, or both just sit and listen, a session will not be set up. They must decide how to recover and start a session. A reset can be issued at any time, and will set the partner to the blank state.

Settings can be saved to file and later restored. When two chat users end their session, there is no formal way to tell GSDP of this fact. The partners simply cease to use their allocated port IDs.

The GSDP chat program provides the following user interface:

- Reset: reset everything and go into blank state

- Initiate: specify GDP implementation and my address (if appropriate), other partner's address and initial message to send

- Listen: specify GDP implementation and my address if appropriate

- Display settings: show GDP implementation, session state, my GDP address and GSDP port ID and the other GDP address and GSDP port ID

- Send: send a datagram

- Receive All: initiate receive all on the GDP protocol in use.

Taking GSDP Forward

The most urgent requirement in GSDP is a send queue. If a client tries to send a datagram when the GDP implementation for that client session is already busy sending a datagram, then the later datagram is lost. This should be managed by adding a transmit queue. There is a knock-on effect: GDP `Send()` must be able to tell the GSDP server when it can send another packet so that the GSDP transmit queue will know when to send the next packet.

The receive queue management could do with some improvement. Two clients listening with the same game protocol ID should perhaps be forbidden. Incoming packets should perhaps be dropped instead of being held in a queue until an application is ready to receive them.

The port ID allocator has a subtle bug that will show in about a century's time: it doesn't prevent zero being allocated on ID wraparound.

RGCP

TOGS is designed to support turn-based games for which a reliable conversation protocol is the most suitable means of presenting the communications API to any game application. The role of RGCP is to present a reliable conversation protocol to applications.

Protocol Overview

After session setup, conversations follow a strict request-response sequence: (see Figure A3.6).

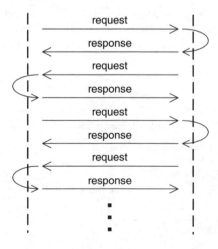

Figure A3.6

In order to save costs (a GSDP datagram using SMS has a financial cost to its sender), the response to the previous request by one partner shares a packet with the responding partner's next request, so that in terms of the lower-level GSDP communication, the picture is as seen in Figure A3.7.

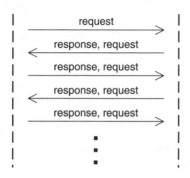

Figure A3.7

RGCP uses essentially the same methods as GSDP for session setup, but it adds reliability by allowing packets to be resent. RGCP also provides for normal session termination (rather than simply abandoning communications, which is the best you can do in GSDP).

Because RGCP is a conversation protocol, it is relatively easy to provide reliability. At most, one packet needs to be stored for potential resending (though in bad cases it may be resent many times). Contrast this with the kind of reliability mechanisms needed to implement FTP efficiently, in which **windowing** is used to send a potentially large number of outstanding packets before any acknowledgement is required – and all these outstanding packets may need to be resent. This means RGCP's job in adding reliability to GSDP is much easier than, say, transmission control protocol's (TCP) job in adding reliability to Internet protocol (IP).

The conversation protocol paradigm is clearly suitable for two-player, turn-based games. It can also be adapted for various situations that are not strictly conversational. For instance, a player may take multiple moves either by piggybacking them into a single RGCP request, or by sending them in individual RGCP requests with the other player acknowledging each one and sending a 'no-move' request in turn.

A single RGCP session may be used to play more than one game, if the request-response protocol for the game protocol ID supports it. The request-response protocol would then have to support, not only game moves but also game initialization, termination, and choosing who has the first move. Battleships protocol (BSP) does this.

Resend management

Many reliable protocols manage resends by waiting a fixed time for an acknowledgement and then resending if none is received. This approach is not open to RGCP because thinking time between moves may be long, and the response, which serves as an acknowledgement, is not sent until the next move request.

Therefore, RGCP users must agree when a packet is lost, and resend it manually. Since RGCP packets represent game moves – something quite tangible to each player – it is easy enough to determine when a packet has been lost and easy enough to resend. Thus, the manual intervention in a communications protocol has clear meaning for the user.

Although RGCP resends are manual, RGCP does provide reliability in the usual communications sense. RGCP will not allow either party in a conversation to send requests or responses *out of turn* and will prevent any packet from being delivered to the RGCP receiver more than once, even if it is resent at the GSDP level.

Packet structure

RGCP uses GSDP as a transport. An RGCP packet has the following structure:

- GDP header with from-address and to-address
- GSDP header with game protocol ID, from-port ID, and to-port ID
- RGCP sequence number (or zero for unsequenced packets)
- RGCP response packet
- RGCP request packet.

The request and response packets have identical formats:

- One-byte length in range 0–127, giving the total length of opcode plus data
- One-byte opcode specifying the operation to be performed (if the length is 1 or greater)
- Data bytes, 0–126 bytes long.

Certain opcodes are reserved for RGCP protocol functions. The meaning of all other opcodes depends on the particular type of conversation, as identified by the GSDP game protocol ID.

The minimum length of an RGCP datagram is 7 bytes, plus GSDP headers, plus GDP headers (see Figure A3.8). The 7 bytes comprise 4 for

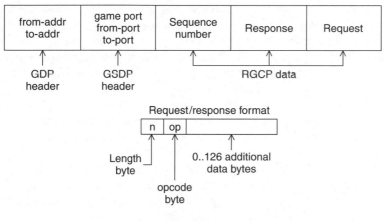

Figure A3.8

the sequence number, 1 for a zero-length response, and 2 for the request (a length of 1, and the request opcode).

Sequencing and resending

Once a GSDP session has been bound, only sequenced packets are accepted with sequence numbers starting at 1 after session establishment. The receiver maintains a record of the last-received sequence number and drops packets whose sequence number is less than or equal to this, eliminating duplicates.

The RGCP transmitter maintains the last-sent packet and resends it identically on request. The last-sent packet is deemed acknowledged when a corresponding response/request packet is received. The last-sent packet is deleted when it has been acknowledged and cannot then be resent.

Session setup

To set up a session, the initiating partner sends an 'initiate' request with a zero RGCP sequence number (and a zero GSDP to-port ID). This is received by a listening partner, who sends an 'initiate' response, any request, and sequence number one. The listening partner then considers itself bound in a session. When the initiating partner receives the 'initiate' response, it too considers itself bound in a session: the next response/request datagram it sends will be sequenced number one.

The listening partner sends the first sequenced packet. For the listening partner, receiving response n (from the initiating partner) acknowledges its request n. For the initiating partner, receiving response $n + 1$ (from the listening partner) acknowledges its request n. The initiate-request and initiate-response datagrams may be resent using the usual RGCP resend facility.

To terminate a session normally, the terminating partner sends a 'terminate' request. After sending a terminate request, the terminating partner is in the blank state: it does not expect to receive a response. The partner receiving a terminate request must therefore terminate without sending a response. A terminate request has a zero sequence number because it can be sent at any time, without waiting for the terminating partner's turn in the conversation. A terminate request cannot be resent.

Although a terminate request has a zero sequence number, it is uniquely identified with a particular RGCP session because it specifies GSDP port numbers that are unique to that session.

Abnormal termination may occur through either partner in a session – or during session setup – simply ceasing to communicate. The other partner must realize by other means that this has happened, abandon their current session, and decide what further action to take.

Standard opcodes

RGCP defines three request opcodes:

- $0 \times ff$: initiate – also used to indicate initiate-response
- $0 \times fe$: terminate
- 0×00: should not be used.

Other opcodes may be defined by specific RGCP implementations.

Responses are uniquely associated with their corresponding requests, so a response opcode system is not strictly necessary. RGCP protocol designers may, however, find it convenient and/or safer to use the same opcodes for responses as they use for requests.

States and transitions

As perceived by either endpoint, an RGCP session may have the following states:

State	Meaning
Blank	No GSDP connection.
Initiating	An initiate-request has been sent, but no response has been received.
Listening	Ready to receive an initiate-request, but none has yet been received.
Responding	The session has been bound, a request is being handled, and the partner must compose a response. This state lasts only for the duration of a single synchronous function call.

Requesting	The session has been bound, and it is this partner's turn to compose and send a request.
Waiting	The session has been bound, this party has sent a response/request, and is now waiting for its response and another request. Incoming responses are handled in waiting state.

Here are the allowable states, and all allowable transitions between them (except reset transitions, which have been omitted for clarity, but which are possible from any state to blank) (see Figure A3.9).

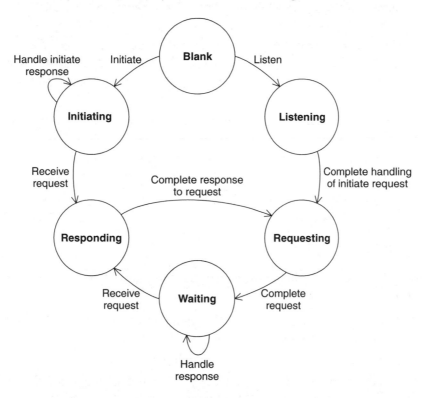

Figure A3.9

RGCP states correspond precisely to underlying GSDP states, with the distinction that GSDP's **bound** state is divided into two major states, **requesting** and **waiting**, that correspond to whether it's this partner's turn to 'talk' or not. That's natural: this is precisely what a conversation protocol is all about. The additional **responding** state is a brief transition state between **waiting** and **requesting**.

All these states may be of arbitrary duration, except **responding**, which must complete in the duration of a single synchronous function call. This implies that an RGCP response should be a response to the request that contains *information* about the result of the other partner's move – not a responding *move*, which would take arbitrary time.

The following state transitions are possible:

- Blank to **initiating** or **listening**: similar to GSDP

- **Listening** or **initiating** to **blank**, by a simple reset (omitted from the diagram for clarity)

- **Responding**, **requesting** or **waiting** to **blank**, by a terminate that's either called through this partner's API, or received from the other partner as a request (omitted from the diagram for clarity)

- **Listening** to **requesting**, by handling an initiate request (the responding state that you might expect is handled internally by the RGCP stack)

- **Initiating** to **responding**, by receiving an initiate response and handling the incoming request

- **Waiting** to **responding**, by receiving a response and handling the incoming request

- **Responding** to **requesting**, by completing the handling of an incoming request

- **Requesting** to **waiting**, by completing the formulation of a request and sending it.

Packet handling

RGCP's rules for handling incoming packets delivered to it by GSDP may be summarized as follows:

- If the port IDs and game protocol ID don't match those for the session, incoming packets won't even be received.

- If the packet's RGCP headers are malformed, it is dropped.

- If the sequence number is zero and it's a terminate request, then the RGCP session is immediately terminated: no response is sent.

- If the sequence number is zero and it's an initiate request, and the current state is **listening**, then an initiate response is formulated accepting the initiate, the GSDP session is bound and the RGCP state changes to **requesting**. The sending sequence number is incremented to 1 so that this becomes the first sequence number of an ordinary packet from the listening partner. In all other circumstances, a zero sequence number packet is dropped.

- If the sequence number is less than or equal to the last-received sequence number, then the packet is dropped.

- If the sequence number increments the last-received sequence number by more than 1, then the packet is dropped.

- We now assert that it is a normal in-session response/request datagram, so the state must be **waiting**.

- The last-received packet sequence number is incremented.

- The response is handled, and then we check that the RGCP stack hasn't been terminated by higher-level function calls.

- The state is changed to **responding**; the request is handled, and we check again that the stack hasn't been terminated. If no response has been written, we produce a null response.

- The response is not sent at this stage: rather, the partially written send buffer is maintained until an RGCP request is sent. At that point, the response and request are sent in a GSDP datagram.

- The state changes to **requesting**.

RGCP Symbian OS Implementation

The Symbian OS RGCP implementation is defined in `rgcp.h` and delivered in `rgcp.dll`.

The Symbian OS RGCP implementation operates entirely client-side, and builds on the GSDP client API. As usual, one concrete class (`CRgcpSession`) owns the session and handles sending, while another abstract class (`MRgcpHandler`) specifies virtual functions that should be implemented by the client (see Figure A3.10).

Here's `CRgcpSession`'s declaration in C++:

```
class CRgcpSession : public CBase, public MGsdpPacketHandler
    {
public:
    enum TState
        {
        EBlank, EListening, EInitiating, EResponding, ERequesting,
          EWaiting
        };
    enum TOpcode
        {
        EReserved=0,
        EInitiate=0xff,
        ETerminate=0xfe
        };
public:
    // Construct/destruct
    IMPORT_C CRgcpSession();
```

```
    IMPORT_C void ConstructL(TUint32 aGameProtocol);
    IMPORT_C ~CRgcpSession();
    IMPORT_C void SetHandler(MRgcpHandler* aHandler);

    // Initialization
    IMPORT_C void SetGdpProtocolL(TUid aGdpProtocol);

    // State
    inline TState State() const;
    inline TBool IsBlank() const;
    inline TBool IsInitiating() const;
    inline TBool IsListening() const;
    inline TBool IsResponding() const;
    inline TBool IsRequesting() const;
    inline TBool IsWaiting() const;
    inline TBool IsBound() const;

    // State transition functions
    IMPORT_C void Initiate(const TDesC& aOtherAddress);
    IMPORT_C void Listen();
    IMPORT_C void Terminate();
    IMPORT_C void SendResponse(TInt aOpcode, const TDesC8& aData);
    IMPORT_C void SendResponse(TInt aOpcode);
    IMPORT_C void SendResponse();
    IMPORT_C void SendRequest(TInt aOpcode, const TDesC8& aData);
    IMPORT_C void SendRequest(TInt aOpcode);
    IMPORT_C void Resend();

    // Persistence
    IMPORT_C void ExternalizeL(RWriteStream& aStream) const;
    IMPORT_C void InternalizeL(RReadStream& aStream);
    IMPORT_C TStreamId StoreL(CStreamStore& aStore) const;
    IMPORT_C void RestoreL(const CStreamStore& aStore, TStreamId
      aStreamId);


    // Access to GSDP stuff
    inline const RGsdpSession& Gsdp() const;
private:
    // Help with sending
    void DoSendRequest(TInt aOpcode, const TDesC8& aData);
    void DoSendResponse(TInt aOpcode, const TDesC8& aData);
    void DoTerminate(TBool aClientInitiated);


    // Handle incoming datagrams
    void GsdpHandleL(const TDesC8& aData);
    void CrackPacketL(const TDesC8& aData, TInt& aSeqNo,
    TInt& aResponseOpcode, TInt& aRequestOpcode);
    void HandlePacket(TInt aSeqNo,
    TInt aResponseOpcode, TInt aRequestOpcode);
    void HandleTerminateRequest();
    void HandleInitiateRequest();
private:
    TState iState;
    RGsdpSession iGsdp;
    MRgcpHandler* iHandler;
```

```
    // Send apparatus
    TInt iNextSendSequenceNo;
     RDesWriteStream iSendWriter;
    TBuf8< KMaxGsdpData>  iSendBuffer;


    // Receive apparatus
    TInt iLastReceivedSequenceNo;
    TBuf8< KMaxGsdpData>  iReceiveBuffer;
    TBuf8< 126>  iResponse; // Rather wasteful
    TBuf8< 126>  iRequest; // Rather wasteful
};
```

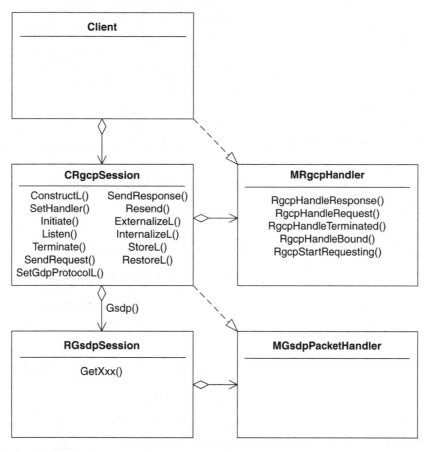

Figure A3.10

The main functions are:

Function	Description
ConstructL()	Construct the session class, specifying a game protocol. Internally, the function constructs an RGsdpSession object for communicating down the stack.
SetHandler()	Sets a handler for incoming packets. Prior to setting the handler, you can store and restore state, but cannot invoke any communication functions. Specify 0 to unset the handler: communications functions will be stopped (without issuing a Terminate()). Specify a nonzero value to set a handler: if the state demands it, a GSDP Listen() will be issued.
SetGdpProtocolL()	Set the GDP protocol to be used.
State()	Get current state. A group of IsXxx() functions also allows you to test whether the session is in a particular state. IsBound() is included, to mean any of the states, responding, requesting, or waiting.
Initiate()	Valid only from blank state. Specify the other address and the state changes to initiating. An initiate request with zero sequence number is sent to the other address in a GSDP datagram (together with a null response). The datagram is held in the resend buffer.
Listen()	Valid only from blank state. State changes to listening.
Terminate()	Valid from any state. From bound states, sends a terminate request to the other partner. State changes immediately to blank. The handler function RgcpHandleTerminated() is called.
SendResponse()	Valid only from responding state. State changes to requesting. Specify an

	opcode and data up to 126 bytes in length. If specified, the opcode must *not* be one of those reserved by RGCP. An overload with no data parameter is provided: the data defaults to 0 bytes. A further overload with no opcode parameter can be used to generate a minimal response with no opcode.
`SendRequest()`	Valid only from requesting state. State changes to waiting. Specify an opcode and data up to 126 bytes in length. The opcode must not be one of those reserved by RGCP. An overload with no data parameter is provided: the data defaults to 0 bytes. The previous response (and this request) is sent in a single GSDP datagram. The datagram is held in the resend buffer.
`Resend()`	Valid only from initiating and waiting states. Resends the last GSDP datagram from the resend buffer.
`ExternalizeL()`	Externalize state to stream, including all GSDP information and the resend buffer.
`InternalizeL()`	Internalize state from stream, including all GSDP information and the resend buffer.
`StoreL()`	Store state by creating a stream, externalizing, closing the stream, and returning its ID.
`RestoreL()`	Restore state by opening the specified stream, internalizing from it, and then closing the stream.
`Gsdp()`	Get a `const` version of the underlying GSDP session so that GSDP settings can be interrogated.

Here's `MRgcpHandler`'s declaration in C++:

```
class MRgcpHandler
    {
```

```
public:
    virtual void RgcpHandleResponse(TInt aOpcode, const TDesC8&
        aData) = 0;
    virtual void RgcpHandleRequest(TInt aOpcode, const TDesC8& aData) = 0;
    virtual void RgcpHandleTerminated(TBool aClientInitiated) = 0;
    virtual void RgcpHandleBound() = 0;
    virtual void RgcpStartRequesting() = 0;
    };
```

`MRgcpHandler` includes the following functions:

Function	Description
`RgcpHandleResponse()`	Handle response. Called in waiting state, this function takes opcode and data parameters. If the response was null, the opcode passed to this function is zero. After this function returns, state changes to responding (unless you called `Terminate()` to set the state to blank).
`RgcpHandleRequest()`	Handle request. Called in responding state, this function takes opcode and data parameters. You can call `SendResponse()` from within this function to send a response. If you do not call `SendResponse()`, then after this function returns, a default null response will be constructed. After this function returns, state changes to requesting (unless you called `Terminate()` to set the state to blank).
`RgcpHandleTerminated()`	Handle termination. Called in any state (except blank). Takes a parameter indicating whether the termination resulted from a client API call, or in response to a terminate request received from the other partner.

`RgcpHandleBound()`	Handle session binding. Called in initiating or listening states. If called in initiating state (because an initiate-response was received), then state afterwards changes to responding (unless you called `Terminate()` to set the state to blank). If called in listening state (because an initiate-request was received), then state afterwards changes to sending (unless you called `Terminate()` to set the state to blank).
`RgcpStartRequesting()`	Called when the state has changed to requesting. May be used to set an indicator to indicate the state transition. May be used to generate a synchronous request. You cannot generate such a request from `RgcpHandleBound()` or `RgcpHandleRequest()`.

RGCP Converse

The Converse application is in `\scmp\rgcpchat\`. It's not the most spectacular piece of test code and hasn't yet been updated to support SMS. For a better RGCP test application, see Battleships!

Taking RGCP Forward

RGCP's manual resend system and response piggybacking are tailored to the requirement of relatively expensive protocols, to avoid sending more datagrams than necessary.

For protocols such as Bluetooth, which are fast and free, this is too heavyweight. It produces the oddity that, when playing Battleships over Bluetooth, you don't see a response to your move until the other player has decided what move to take. That's OK over SMS, but over Bluetooth it feels odd.

It would be possible to add either another layer, or a fairly compatible modification, to RGCP:

- This could allow responses to be sent immediately, without waiting for a piggyback packet.

- Given the possibility of an immediate response, *requires* an immediate response in some situations and, if one does not arrive, resend automatically.

I have provisionally called this protocol QRGCP (Q = quick), but haven't implemented it.

CRgcpSession could use better construction encapsulation. Rather than providing a C++ constructor to be followed by either ConstructL() or RestoreL(), it would be better to provide two public NewL()s, one with construct semantics and the other with restore semantics.

The Battleships Protocol

BSP, the Battleships Protocol, builds on RGCP to allow players to play the two-player game. Within the span of a single conversation (or 'session'), multiple Battleships games may be played. Outside a session, you cannot play a game.

A session is set up by one partner initiating and the other listening. Partners specify whether they want to move first, or move second, or don't care. The decision is arbitrated in favor of the initiating partner (who will go first if they so requested, or if neither party cares, or if the other partner asked to go second).

Normal play consists of a sequence of move requests. The game may be finished by being won by one partner (and therefore lost by the other) or by being abandoned by the player whose turn it is to move. If the game is finished in either of these ways, the session remains active and another game can be started. This time, the player who lost or abandoned gets preference for the first move.

The RGCP session may be terminated, at any time, which of course terminates any game currently in progress.

Protocol Overview

Here's a brief look at the details of BSP.

States

The possible states of a Battleships application are

State	Meaning
Blank	No conversation is established. The game is in an arbitrary state (either a neutral state if the program has just started, or the state at the end of the last game if a conversation has been terminated).

Initiating	You have chosen whom to play against and initiated a conversation with them. You have also specified your first-move preferences. The game is initialized to a neutral position.
Listening	You have chosen to play and are waiting to accept a conversation initiated by someone else. You have also specified your first-move preferences. The game is initialized to a neutral position.
Starting	You are bound in a session and have sent a start request to the other player specifying your first-move preferences. The game is initialized in a neutral position.
Accepting	You are bound in a session and are expecting to receive a start request from the other player. You have specified your first-move preferences and the game is in a neutral position.
My-turn	A game is in progress, and it's your turn to move.
Opp-turn	A game is in progress, and it's your opponent's turn to move.
Finished	The game has been won, lost, or abandoned (which amounts to substates of finished).
Restarting	You are going to restart the game, but you don't yet have your first-move preferences.

State transitions

The transitions in Figure A3.11 are described in more detail below:

From	To	Occurs When
Blank	Initiating	Player specifies first-move preferences and the address of player to connect to. An RGCP initiate-request is sent.
Blank	Listening	Player specifies first-move preferences and starts RGCP listening.
Listening	Starting	RGCP initiate request is received and a response sent, so that the RGCP

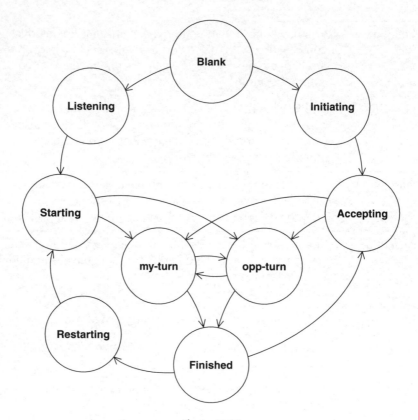

Figure A3.11

session is bound. You send a start request, specifying your first-move parameters.

Initiating	Accepting	RGCP initiate response is received, so that the RGCP session is bound.
Accepting	My-turn	You handle a start request from the other player. You arbitrate who has the first turn, and it's you. You send a start response indicating this.
Accepting	Opp-turn	You handle a start request from the other player. You arbitrate who has the first turn, and it's the other player. You send a start response indicating this, and then immediately send a no-operation request.

Starting	My-turn	You receive a start response from the other player that indicates you have first turn: the next request from the other player will be a no-operation request.
Starting	Opp-turn	You receive a start response from the other player that indicates the other player has first turn: the next request from the other player will be their move request.
My-turn	Opp-turn	Through the GUI, you specify a move: this is composed as a move request and sent to the other player.
Opp-turn	My-turn	You receive a move request from the other player, to which you respond.
My-turn	Finished	You abandon the game: you send an abandon request to the other player.
Opp-turn	Finished	You receive a move request that causes you to lose the game (in which case you immediately send a move-response and a no-operation request, so the other player knows they have won); or you receive a move-response indicating that you won the game; or you receive an abandon request.
Finished	Accepting	You are in BSP finished state and RGCP requesting state. This can happen because the other player abandoned, or you won. Through the UI, you specify your first-move preferences. You send a restart request with no parameters.
Finished	Restarting	You are in BSP finished state and RGCP waiting state. This can happen because you abandoned, or you lost. You receive a restart request from the other player, to which you respond.

Restarting	Starting	You have first-move parameters from the UI. You send a start request specifying your first-move parameters.
Any	Blank	You terminate the game, or you receive a terminate request from the other player to terminate it. If you terminated the game, you send an RGCP terminate request to the other player.

Some scenarios

It's worth seeing how this works in a few scenarios. Here's how the game starts up (see Figure A3.12), if the initiating player gets the first move:

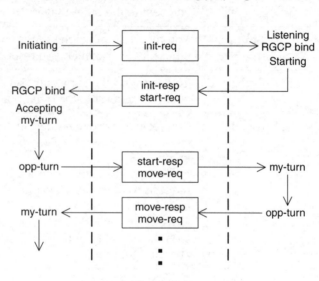

Figure A3.12

Here's how things work (see Figure A3.13) if the initiating player gets the second move:

And here's what happens when one player wins the game (see Figure A3.14):

What happens after this depends on the result of the accepting player's arbitration. I haven't illustrated what happens when one player abandons the game, but the specifications and diagrams above should be sufficient to work it out.

Requests and responses

The requests involved here are as follows:

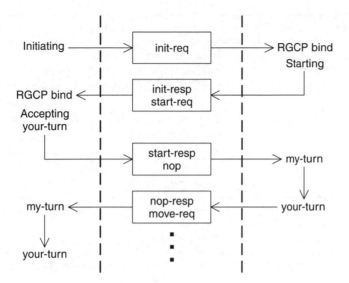

Figure A3.13

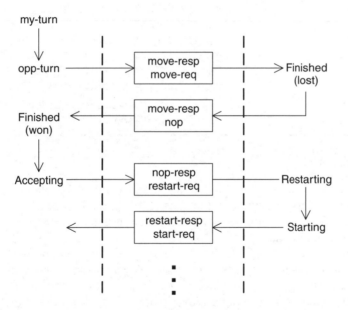

Figure A3.14

Opcode	Name	Details
1	Start	Indicates that you want to start a game. With request, pass your first-move preference. With response, indicates who actually got first move. Request has one byte: 0 for don't-care, 1 for want-first, and 2 for want-second. Response has 0 to indicate the responder has first move, 1 to indicate the original requester has first move.
2	Restart	Indicates that you want to restart the game.
3	Nop	No operation.
4	Abandon	Indicates that you want to abandon the current game.
10	Hit	Indicates that you want to hit a particular square. With request, passes the square to hit. With response, indicates what was there. Request has 2 bytes: row, then column. Either may be in range 0–7. Response has 3 bytes: row and column as above, while a third byte indicates 0 for sea, 1 for frigate, 2 for destroyer, 3 for cruiser, and 4 for battleship.

First-move arbitration

As we saw at the start of this section, the first move is arbitrated by the accepting player. At game setup, this is the initiating partner. At game restart, this is the partner who won the previous game, or the partner who did not abandon it.

Both players specify their first-move preference: want-first, don't-care, or want-second. The arbitrating player decides the first move when accepting the other player's start-request. In the event of a clash in preferences, the accepting player's preference overrides the starting player's preference. So the rules are as follows:

- If the accepting player wants first move, they get it.
- If the accepting player doesn't care, then they go first unless the starting player specified that they wanted to go first.
- If the accepting player wants second move, then they go second.

Game UI

The game UI needs to support the following functions, on top of the BSP protocol. The view displays the following:

- A status indicator
- My fleet
- My opponent's fleet.

The commands are as follows:

Command	Description
Start session	Starts game session. Valid in BSP blank state. Specify whether listening or initiating, and GDP protocol. If initiating and using networked GDP, specify to-address. Specify first-move preference.
Move	Does a move. Valid in BSP my-turn state. Command initiated by pointer.
Abandon game	Abandons current game. Valid in BSP my-turn state.
Start new game	Starts new game when session is already connected. Valid in BSP finished state. Specify first-move preference.
Terminate	Stops game session. Valid in any state.
Resend	Resend last RGCP response/request packet. Valid in RGCP initiating or waiting states.
Receive all	Initiates receive of any packets. Valid in any state except BSP blank.

Program Structure

In the Battleships program, the CGameController handles all the BSP protocol, including functions called by the UI and RGCP handler functions. The controller also implements all send-request and send-response functions implied by BSP and maintains state transitions in accord with the design above.

A key aspect of the controller design is that every public function and many private functions assert the validity of the requested operation, given current BSP and RGCP states. The app UI and views are responsible for prechecking user-initiated commands, so that no invalid command can be issued to the controller.

Taking BSP Forward

BSP combines a multigame session protocol with the specifics of the Battleships game. It would be possible to separate these aspects into two layers.

A truly general protocol may or may not be worthwhile: other games could reuse BSP's patterns without reusing its code. In any case, patterns differ between games. A game that can be tied, for instance (as opposed to only won or lost), may require a different approach.

It would be possible to improve on the first-move selection here, by changing only the UI, to select random first-move preferences. BSP itself would not have to be altered.

Summary

We've now seen TOGS components described in detail – with the exception of the Bluetooth and SMS GDP protocol implementations that are described in Chapter 20.

The content of this appendix is heavier and more precise than that of many of the chapters in this book. I wrote it *before* I wrote most of the code it describes, and have not changed it substantially since – except that I've maintained some parts as the code has evolved. That's a healthy (if unusual!) software engineering practice. In this case, doing the documentation before the code saved me a *lot* of time.

Appendix 4

Emulator Reference

As we've seen, the emulator is a vital tool in the Symbian OS software development process. For non-privileged code, the emulator is almost 100 percent source compatible with real Symbian OS phones. Many functions that used privileged code on other systems use servers on Symbian OS, facilitating source code compatibility through the provision of machine-independent APIs. This means that for the system as a whole, source compatibility is very high indeed.

The emulator uses Win32 APIs and services to emulate real-machine hardware and Symbian OS services, including

- a Windows window to emulate the Symbian OS screen and surrounding machine fascia,
- the Windows mouse to emulate the Symbian OS pointer,
- the Windows keyboard to emulate the Symbian OS keyboard, plus some control functions,
- directories in the Windows file system to emulate standard Symbian OS drives,
- PC sound card to emulate Symbian OS sound codec,
- PC communications ports to emulate Symbian OS communications ports,
- Win32 threads to emulate Symbian OS threads,
- a single Win32 process to emulate a single Symbian OS process containing *all* Symbian OS threads: this is an important difference between the emulator and a real Symbian OS phone, which I'll return to later in this appendix,
- a single `.exe` to include Symbian OS startup code: this is `epoc.exe` in the case of any GUI program or a custom `.exe` for any text console program,

Symbian OS C++ for Mobile Phones.   Edited by Richard Harrison
© 2003 John Wiley & Sons, Ltd   ISBN: 0-470-85611-4

- Win32 DLLs to emulate Symbian OS DLLs, with Win32 DLL search order used instead of Symbian OS search order,

- Win32 DLLs to emulate Symbian OS server `.exes`, with Win32 DLL search order used to find the Win32 DLL,

- the debug window, if active, for debug prints generated by `RDebug::Print()` (see `e32svr.h`).

For C++ development, the emulator's (Figure A4.1) debug build is typically used. Debug builds contain symbolic information and debugging statements, and the debug emulator contains useful debugging support such as many debugging keys.

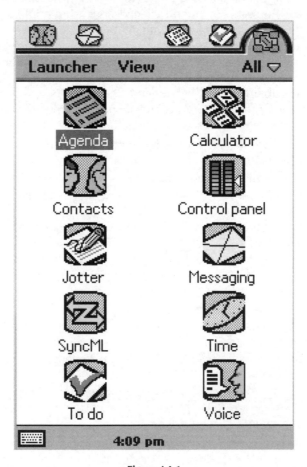

Figure A4.1

Inside the Emulator

As a C++ developer, you build C++ code for the emulator and debug it by running it natively. You then have to rebuild it for the intended target phone. This has a few implications that you need to understand at this point.

Source Compatibility

C++ is a platform-independent language that can be compiled to any instruction set. Symbian OS currently supports three ARM instruction sets: ARM4, THUMB, and ARMI. ARM4 is the StrongArm instruction set, THUMB is the 16-bit ARM instruction set, and ARMI is an interworking build (not strictly a separate instruction set) that allows programs to call DLLs that are built either in ARM or THUMB. The PCs used to develop Symbian OS use the x86 instruction set.

So, a C++ program for the *emulator* is compiled to native x86 machine code. When a program has been debugged under the emulator, the same source code is simply recompiled to use one of the ARM instruction sets, and transferred to a Symbian OS phone from which it can be run.

The following figure shows the steps involved in compiling for the two platforms, when using the CodeWarrior compiler (Figure A4.2):

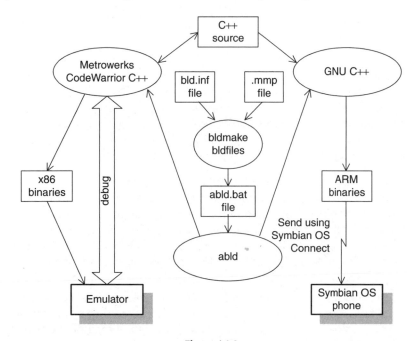

Figure A4.2

This **source compatibility** is achieved by delivering base Symbian OS services to all user programs through the same APIs (Figure A4.3). On a real Symbian OS phone, the base services are implemented by the kernel, file server, and device drivers using the real machine's hardware. On the emulator, the base services use Win32 APIs, PC hardware, and the PC's file system. Although there may be significant differences in the way these base services are implemented, the APIs offered are the same at source code level.

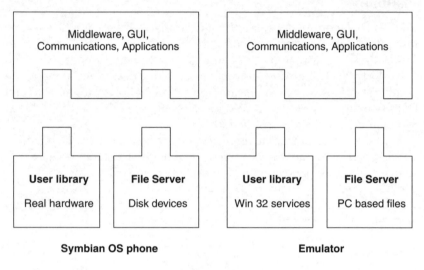

Figure A4.3

Source compatibility between the emulator and a real Symbian OS phone applies to the vast majority of Symbian OS. Only parts of the kernel, the file server, and device drivers need to be changed substantially. Applications, servers, and middleware libraries are completely source-compatible, and need only a rebuild to transfer them from emulator to target machine. Some things don't even need to be rebuilt:

- Data files should be implementation independent. You can easily ensure this by using Symbian OS stream and store APIs for saving data – don't just 'struct dump', because that puts internal formats into files and internal formats may differ between Symbian OS implementations. A `double`, for example, has different internal formats on x86 and ARM. See Chapter 13 for more on data management in Symbian OS.

- Interpreted programs such as those in pure Java are just data as far as the emulator is concerned. You can transfer them freely from emulator to target phone without conversion. Java native methods are really binary programs, and these clearly need to be rebuilt.

Drive Mapping

The emulator maps features of the target machine onto features of your PC environment. For software development, it's particularly important to know how the emulator maps drives and directories onto your PC's filing system.

On a real Symbian OS phone, there are two important drives (Figure A4.4):

- z: is the ROM, which contains a bootstrap loader and all the .exe, .dll, and other files required to boot and run Symbian OS and its applications. All files on z: are read-only; program files are executed directly from the ROM rather than first being loaded into RAM.

- c: is the read/write drive that's allocated from system RAM. c: contains application data, application and system .ini files, and user-installed applications.

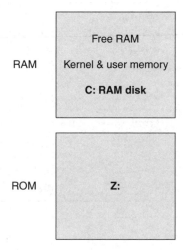

Figure A4.4

On the emulator, these drives are mapped onto subdirectories of the drive on which you installed the SDK (Figure A4.5). This is to separate different builds, and for safety, so that your Symbian OS programs can't write just anywhere on your Windows c: drive, and because most PC's don't have a z: drive anyway.

An emulator configuration directory and startup directory completes the list of directories required by the emulator. Here's an example of the directories involved when using CodeWarrior:

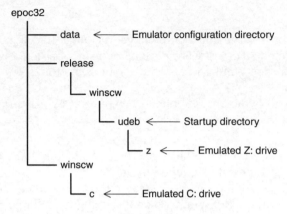

Figure A4.5

Directory	Description	Contents
\epoc32\data\	Emulator configuration directory	epoc.ini (the initialization parameters for the emulator), epoc.bmp (the bitmap used as the fascia surround for the screen), and variants for screens of different sizes.
\epoc32\release\ winscw\udeb\	Emulator startup directory	epoc.exe, the program you invoke from Windows to bring up the emulator, is in here. So are all the shared library DLLs.
\epoc32\release\ winscw\udeb\z\	Emulated z: drive	Everything that the EPOC z: drive should contain, except shared library DLLs that are in the parent directory.
\epoc32\winscw\c\	Emulated c: drive	Any data and files. No compiled C++ programs – those should all be on z:. In the emulator, all compiled applications become part of the pseudo-ROM that is the emulated z: drive.

Why those long names?

You might wonder why there are so many deeply nested directories. The directories categorize the Symbian OS SDK materials, as follows:

- `\epoc32\` sets apart all Symbian OS SDK runtime software from anything else on the same drive.
- `release\` sets apart released code from documentation, temporary build files, configuration files, and so on.
- `winstarget\` sets apart the emulator from target machine builds. Different targets use different directories; in the case of Metrowerks CodeWarrior it's `winscw`.
- `udeb\` sets apart the debug build from other builds.
- `z\` attempts to mirror the structure of `z:` on a real Symbian OS phone.

An executable C++ program built for the emulator debug build won't run in any other execution environment, so the debug build is kept distinct from any other build.

It would be nice if `\epoc32\release\winstarget\udeb\z\` could contain the entire emulated `z:` drive. Unfortunately, it can't: `.exes` and shared library DLLs can't use the file structure of Symbian OS, without impractical implications for the path environment in Win32. The only practical thing to do is to place all `.exes` and shared library DLLs in the startup directory – `\epoc32\release\winstarget\udeb\`.

Finally, data is independent of build, so `c:` is mapped to the `\epoc32\winstarget\c\` directory and shared between all builds.

The Directory Scheme

SDK tools require the default directory scheme, so in practice you have to use it for software development. However, you can choose to start the emulator from whichever directory you choose – say, `\mystuff\`. This can be useful for delivering demonstration software, for example. Once you've finished developing software, you can copy the emulator to a different location and run your emulator package independently of the development environment. The directories for emulated `c:` and `z:` drives, and emulator startup information, are subdirectories of the startup directory (Figure A4.6).

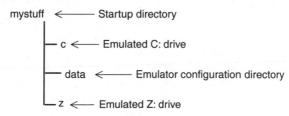

Figure A4.6

These are the rules that come into force if the startup directory is anything other than `\epoc32\release\winstarget\build\`:

Directory	Description	Contents
`mystuff\`	Startup directory	Startup `.exe` and all shared library DLLs
`mystuff\c\`	Emulated `c:` drive	Data disk
`mystuff\z\`	Emulated `z:` drive	Program and program data disk for all files except shared libraries
`mystuff\data\`	Initialization	Emulator `.ini` and fascia `.bmp` files

Emulator Startup

The emulator uses only a single Win32 process. That means it uses only a single Windows `.exe`, which you launch somehow – from a command line, for example, or using Windows Explorer.

The `.exe` that you specify is the startup `.exe`, and it resides in the **startup directory**; in the case of Metrowerks CodeWarrior, this is `\epoc32\release\winscw\udeb\`. You have two choices for startup `.exe`:

- build your own, like `hellotext.exe`, `string.exe`, or `buffers.exe`,
- use `epoc.exe`, which starts everything for you; then you can use the application launcher to launch your Symbian OS application.

Each Symbian OS `.exe` built for the emulator includes a stub file, `eexe.obj`. This file includes code to start up the emulated Symbian OS kernel and file server, and then call your `E32Main()` function. So, in fact Windows starts by calling `eexe.obj`, and your code is called by `eexe.obj`.

As part of kernel startup, the emulator reads an initialization file and a fascia bitmap from its configuration directory. By default, the configuration directory is `\epoc32\data\`, the initialization file is `epoc.ini`, and the fascia bitmap is `epoc.bmp`.

`epoc.exe` uses exactly the same `eexe.obj` as any `.exe` you build yourself. In fact, the source code for `epoc.exe` contains only a single line in its `E32Main()`, which starts the Symbian OS window server. The window server in turn starts the servers it needs (such as the font and bitmap server) and then reads its initialization data from

the *emulated* z:\system\data\wsini.ini. In the case of UIQ, this
.ini file contains instructions to the window server to start Qstart
(\system\programs\qstart.dll). Amongst the things that QStart
launches are the application picker (which shows the application icons
at the top of the screen) and the application launcher.

That startup sequence begs two questions:

- What are the contents of the emulator's .ini file?
- What command-line arguments can I specify to override this?

Emulator Startup Parameters

You can control the emulator startup by parameters in epoc.ini (in
\epoc32\data\). Here's an excerpt of the epoc.ini as supplied on
the UIQ C++ SDK:

```
ScreenWidth 208
ScreenHeight 320
PhysicalScreenWidth 3328
PhysicalScreenHeight 5120
ScreenOffsetX 38
ScreenOffsetY 92
LedOffsetX 0
LedOffsetY 0
LedSize 7
LedGap 5
LedArrangeHorizontally

# could be decreased to reflect the amount of memory available
# on actual hardware
MegabytesOfFreeMemory    16
```

This specifies the screen size, the offset of the screen area from the
top left of the fascia bitmap, and its position. It also specifies the position
and size of the two emulated LEDs. The amount of memory available on
the real phone is set in megabytes. The supplied epoc.ini then goes
on to define names and positions for the clickable virtual keys on the
fascia bitmap and keyboard mappings (not shown here). If you want to
emulate a different size of screen, you need to specify it in the .ini file.
An example is console.ini supplied in the UIQ C++ SDK.

Besides c: and z:, you can make additional emulated drives available
to Symbian OS programs under the emulator, by adding them to your
.ini file. For instance,

```
_EPOC_DRIVE_D a:\
```

allows me to code a PC diskette drive as my Symbian OS d:, and thereby test the effects of removable media. Be careful to specify the _EPOC_DRIVE_D in upper case: the specification will be silently ignored if you don't.

Emulator Command Line Syntax

The command line for the emulator syntax is

```
epoc [-M<machine name>] [-T] | [-C<emulated C drive>] --
```

Note that the command line must be terminated with two hyphens. These command line options are not available with the command line tools that get installed with the SDK. In order to use them, you need to invoke epoc.exe from the directory where it is located to prevent the developer kit tools being picked up.

If you want the emulator to use console.ini, start it up with this (and yes, you really do need all those – signs):

```
epoc - Mconsole --
```

You can override the emulated c: drive from the command line; use,

```
epoc -T --
```

to map the emulated c: drive to your PC's system temporary directory. This allows you to quickly boot up Symbian OS from a CD-ROM, which can be quite handy for demo purposes.

If you care about a specific emulated c:, use

```
epoc -C pcpath --
```

Key Mapping

The majority of the PC keyboard is mapped in a straightforward way to the Symbian OS keyboard. However, some special keys are available:

PC Key	Symbian OS Facility
F1	Menu key
Alt + F2	Help key (this doesn't work in the UIQ emulator as there isn't a system help file)

$Alt + F4$	Close the emulator window
$F9$	Power on
$F10$	Power off
$F11$	Case close toggle: when case is closed, window title changes to indicate it, and emulated keyboard/pointer becomes inactive

In different UIs, there can be additional keys that are part of the hardware. These are specified in the `.ini` file. For UIQ emulator the keys are

PC Key	UIQ Emulator Key
Numeric keypad −	Page up (TwoWayUp)
Numeric keypad +	Page down (TwoWayDown)
Numeric keypad 8	Cursor up (FourWayUp)
Numeric keypad 4	Cursor left (FourWayLeft)
Numeric keypad 6	Cursor right (FourWayRight)
Numeric keypad 2	Cursor down (FourWayDown)
Numeric keypad 5	Enter (FourWayConfirm)
Numeric keypad Enter	Enter (FourWayConfirm)

Communications

On a Symbian OS phone, there can be a variety of external communication methods such as Bluetooth or infrared. On the emulator, you use a PC's serial ports instead of real device hardware:

- you can test Symbian OS RS232 programs using PC RS232 ports,
- you can use a PC-style serial cable, often available for data-capable mobile phones, to connect the Symbian OS emulator to a mobile phone,
- you *can't* use your LAN for TCP/IP; instead, you have to use dial-up networking and a modem,
- you can use an infrared pod for beaming or communication with an infrared-enabled mobile phone,
- you can use a Bluetooth device provided you have the correct `hci.dll` for your Bluetooth hardware.

The communications settings are defined in a database called `commdb`, which can be set up using `\epoc32\release\winstarget\build\`

ced.exe or \epoc32\tools\setupcomms.bat. You can check all the current settings by running ceddump.exe that writes the contents of the database into a text file, cedout.cfg in the emulated c: drive. If you wish to make changes, you can edit the file using a text editor, and save it as ced.cfg. Running ced.exe reads ced.cfg from the emulated c: drive and uses it to create a new commdb, recording what happened in ced.log. In ced.cfg sections specify settings such as those for modems, locations, and ISPs. The communications database editor tools (ced*)need to be run from the directory they are in.

Some points to note when using the emulator for communications:

- There can be problems if the assigned Bluetooth port is already being used by another device such as a modem, causing the emulator to crash. This can be resolved by moving the modem to a different port or removing the bt.esk file from \epoc32\winstarget\c\system\data to disable support for Bluetooth and avoid any conflicts.

- Comms ports are indexed from 0, not 1 as is usual on a PC. This means that in commdb COMM::0 maps to COM:1 on the PC and COMM::1 maps to COM:2. Similarly, physical ports in bt.esk and irda.esk are numbered from 0, so irPhysicalComPort= 0 and port= 0 both map to the PC's COM:1 port.

- If you run out of ports, you may need to install extra ports on your PC.

Some kinds of TCP/IP testing are much more convenient – not to say cheaper – if you can access TCP/IP services on your LAN, rather than using dial-up. If you need to do this, the best solution is to use RAS on Windows NT. You need a PC with two communications ports, connected to each other. Use one port for the emulator serial 'out' and another for RAS 'in'. Check out Symbian Developer Network www.symbian.com/developer for instructions on how to set up RAS on Windows NT for this purpose.

How Good is the Emulator?

The answer is, pretty good.

- Because the emulator uses straightforward mappings of Symbian OS facilities to Windows, there is relatively little fancy Win32 code at the bottom of the emulator, and therefore relatively little that can go wrong.

- The emulator runs in its own process. It's a process just like any other Windows process, so you can have multiple emulators all running on one machine.

- The emulator uses its own window. It's not resizable, but it can be dragged, and moved to front and back, just like any other window.

- The emulator uses files that you simply copy around on your PC directories. There's no special file pool that requires tools to check files in and out. If you copy a file to a directory used by the emulator for an emulated drive, it's instantly available for use.

There are a few things to watch out for, though:

Files copied to the emulator are instantly ready for use. However, if you use the PC to copy a file, the emulator's F32 won't know that you've done this, and so no file-system notifications will trigger. So if the file is, say, a new application icon, the icon won't get refreshed automatically. In the UIQ emulator, you can force a refresh in the file manager by clicking on the location in the menu bar.

The emulator keeps application resource files (`.aif` and `.mbm` as well as `.rsc`) open. So, if you have the emulator running, you have used an application, and you want to update that application's resource file, you have to close the emulator – not just the application. This doesn't apply to any other type of file – you can update the `.app`, for instance, without restarting the emulator.

The emulator is too fast. You can be deceived into thinking you have an application that performs well enough, especially if you use a superfast PC for development work. And, of course, you *will* use a superfast PC if you can, because C++ compilers use all the PC power you can throw at them.

PC screens and pointers aren't the same as Symbian OS phone screens and pointers. If an application looks good on a PC screen and can be used effectively with a mouse, that doesn't mean it will also look good on a real device, with a real pointer (or your finger).

The emulator is only source compatible, not binary compatible. So you can't install compiled programs built for a real phone onto the emulator.

There are subtle incompatibilities in source code due to alignment restrictions on ARM and the single-process restriction on the emulator.

The C++ dialect and warnings differ between compilers such as Metrowerks and GCC. So, before you commit to a design that relies on too many C++ tricks, check out your ideas on all the compilers you're planning to use.

ARM data alignment rules are strict; x86 alignment rules are more relaxed. If you need alignment, copy potentially unaligned data byte by byte into an aligned area. For example,

```
TBuf8<200> buffer;
TInt index = 39; // not a multiple of 4!!
TInt* p = (TInt*)(buffer.Ptr() + index);
TInt i = *p;
```

The cast ought to be a warning of trouble; this code is platform dependent. It happens to work on x86, but not on ARM, where it generates an alignment fault. You need to use code like:

```
TBuf8<200> buffer;
TInt index = 39; // not a multiple of 4!!
TInt i;
Mem::Copy(&i, buffer.Ptr() + index, 4);
```

On a target phone, each application and server is a separate process, with its own `.exe`. On the emulator, there is only one process, so applications use `apprun.dll` instead of `apprun.exe` and most servers are delivered as DLLs. There are established patterns for working around these issues; they affect very few lines of Symbian OS code. For applications, the issue is addressed for you, so you don't need to worry. If you write a server, you'll need to copy one of the standard patterns, such as the one in Chapter 19.

On the emulator, all Symbian OS threads run in the same PC address space. Standard Symbian OS programming uses active objects and the client-server architecture, which means that *deliberate* use of shared memory is very rare. So very few programs are by nature difficult to debug because of differences between the emulator and target machines. Some very obscure and awkward bugs can result, however, whereby code with random reads or writes appears to work under the emulator, but crashes quickly on a target machine.

The emulator builds on top of Win32 APIs; real Symbian OS builds on a microkernel, hardware, and device drivers. If you're programming these, then the emulator won't help; you need the real hardware. Likewise, the emulator is of less help if you're working with communications in which there are always device-specific issues. Timing resolution is different on the real hardware and on a PC. This makes a big difference when implementing anything that is timer-driven, particularly any communications software. On the real hardware, the timing resolution is finer (typically 1/64 s compared with 1/10 s on the PC), so fast timers can be very slow on the PC. This makes it especially important to test communications software on the real hardware as early as possible.

On target phones, Symbian OS controls the executable image format and has incorporated UIDs into it. On the emulator, Symbian OS clearly doesn't control the executable image format, so we use the stub `.uid.cpp` file to generate UIDs into the `.E32_UID` data segment. Files on the emulator are recognized as such by the application architecture, but aren't checked by the Windows loader as they are on a target machine.

For some projects, you can build for the emulator only, until the day before you ship. A simple rebuild, and you're up and running on the target hardware and everything's fine. But it would be unwise to think all projects are like that, even all those that 'ought to be'. Any of the issues I mentioned above could be a factor in your project. The best way to tackle issues is to see them a long way in advance. Build periodically for the real hardware. Check for obvious bugs, UI considerations, and performance. Take action early on, while you can still make a difference.

Debug Keys

In debug builds, (udeb), the following keys used with $Ctrl + Alt + Shift$ can be useful:

Key	Description
A	Display allocation info-message, saying how many heap cells have been allocated by the current application's main thread.
B	Display file server resource info-message, saying how many file server resources are allocated on behalf of the current application's built-in RFs.
C	Display the window server resource info-message, saying how many window server resources are allocated on behalf of the current application's built-in window server session.
D	Stop (disable) window server logging.
E	Stop (enable) window server logging.
F	Enable window server autoflush for this application, so that each drawing command is immediately sent to the window server.
G	Disable window server autoflush for this application (the default state), so that drawing commands are batched together and only sent to the window server when either the buffer is full, the application requests an explicit flush, or the application waits for another event.
H	Dump the contents of the window server's heap.
K	Causes the window server to kill the current application. Also available in release builds.
M	Display **Move me!** Dialog, which you can drag around the screen, causing your application to redraw in its wake.

O	Cycles between the different rotations of the emulator for the current screen size mode (for example, the flip state – open or closed). This will vary for different UIs – with UIQ it rotates the emulator through 180°.
P	Display heap failure dialog, so you can specify systematic failures of heap allocations, file server allocations, and window server allocations.
Q	Cancel any heap failure mode settings, so that allocations only fail if the system is genuinely out of memory.
R	Display and immediately remove a blank window, so that your application has to redraw its entire view.
S	A window server hotkey that takes a screenshot.
T	Display system task list, a way to switch to or close down running applications. Once in the system task list, if you can't close a task, you can use $Ctrl + Shift + E$ to kill it in both debug and release versions. Within the system task list, $Ctrl + Alt + R$ shows information about the last reset of the machine.
U	Cycles between the screen size modes (for example, toggles the flip state between open and closed, if applicable to the UI).
V	Toggle display of verbose info-messages.
W	Dump full window tree.
X	Brings down the emulator.
Y	Mount simulated removable media device on x:
Z	Send keys ABCDEFGHIJ in rapid sequence to the application to ensure it can handle rapid key events.

Index

Symbian OS C++ for Mobile Phones. Edited by Richard Harrison
© 2003 John Wiley & Sons, Ltd   ISBN: 0-470-85611-4